Handbook of Chalcogen Chemistry

New Perspectives in Sulfur, Selenium and Tellurium

2nd Edition, Volume 2

Handbook of Chalcogen Chemistry
New Perspectives in Sulfur, Selenium and Tellurium
2nd Edition, Volume 2

Edited by

Francesco Antonio Devillanova
Complesso Universitario di Monserrato, Italy
Email: devilla@unica.it

Wolf-Walther du Mont
Technische Universität Braunschweig, Germany
Email: w.du-mont@tu-bs.de

RSC Publishing

ISBN: 978-1-84973-624-4

A catalogue record for this book is available from the British Library

Published by The Royal Society of Chemistry,
Thomas Graham House, Science Park, Milton Road,
Cambridge CB4 0WF, UK

Registered Charity Number 207890

Visit our website at www.rsc.org/books

Printed in the United Kingdom by CPI Group (UK) Ltd, Croydon, CR0 4YY

Preface

This second edition of the *Handbook of Chalcogen Chemistry* appears only six years after the first edition. The need to prepare a new edition so soon has two causes: first, some important topics of the chemistry of chalcogen elements were not included in the first edition for various reasons and secondly, chalcogen chemistry continues to grow rapidly. The graph referred to in Chapter 11.3, illustrating the remarkable increase of interest in dichalcogenolene chemistry, shows that the number of papers published in the last decade has been greater than that of all papers on the same topic published up to 2000. At the same time, the chemistry of selenium and tellurium is becoming increasingly important with respect to that of sulfur in some particular fields such as semiconductors or nanoparticles and quantum dots based on polychalcogenides. In the preface to the first edition, we indicated the main factors that had led to a marked increase of interest in the chemistry of selenium and tellurium. Here, we confirm that two main factors, closely connected with each other, have contributed further to this development: (1) the considerable progress in crystallographic techniques and (2) the great variety of technological applications that have been found for many compounds of these elements. As our knowledge of the structural features of particular aggregates and of their functional mechanisms increases, new chalcogen compounds are being designed and organized in the solid state to improve their performance. The continuous demand for new compounds has stimulated the search for new sophisticated synthetic methodologies, that, particularly for the selenium and tellurium compounds, utilize the elements in their elemental state and thus avoid the use of noxious reagents.

This second edition is organized in a similar way to the first. Maintaining the periodic table as a guideline, the chapters of the first part have been substantially updated to the end of 2011. The systematic overview of chalcogen compounds in combination with the main group elements of the periodic table, from boron to the halogens, including the chalcogens themselves, offers a complete account of

Handbook of Chalcogen Chemistry: New Perspectives in Sulfur, Selenium and Tellurium,
2nd Edition, Volume 2
Edited by Francesco Antonio Devillanova and Wolf-Walther du Mont

Published by the Royal Society of Chemistry, www.rsc.org

all the relevant chalcogen compounds. The overview is completed by the two chapters on metal chalcogenides and polychalcogenides. The second part has been extended with respect to the first edition by some important chapters in the fields of biological chemistry and material chemistry, and by a new chapter on the theoretical calculations and NMR spectroscopy. All these chapters have been included not only to improve the balance between the two parts of the handbook, but also to consider important aspects of the chemistry of the chalcogen elements that are increasing the scope for their potential applications. We can maintain that, while the first part can be considered to contain mainly aspects of fundamental research, the second part emphasizes the ambitious purpose of the handbook in pointing out the increasing role of chalcogen chemistry in satisfying the continuous need for new materials for specific applications in many fields. However, the apparent separation between the two parts is simply one of convenience since they are intimately connected; the chapters in the second part represent further developments of those in the first part. Furthermore, fundamental research often opens up new and unexpected developments, thus proving once more, if proof were necessary, that the separation between fundamental and applied research is arbitrary and irrelevant. The recent discovery that the ring-opening polymerization of S_2N_2 to form a conducting material has potential application in forensic science (rapid imaging of latent fingerprints) represents a surprising development in chalcogen/nitrogen chemistry and underlines yet again the importance of fundamental research. Instead of an artificial separation between fundamental and applied research, we hope the *Handbook* clearly demonstrates the necessity of a multidisciplinary approach to tackle any research problem. As a unique example, we point out that in the explosive development of nanoscience and nanotechnology over the last few years, metal chalcogenide quantum dots have proved to be interesting materials in terms of their peculiar optical, magnetic, electronic, and catalytic properties. Their characteristics as light-emitting diodes, or as solar cells, or in biomedicine, will stimulate synthetic chemists to develop new compounds with improved efficiency. Though not exhaustive, the book includes mainly the results from the last decade and illustrates the trends of the most striking research work. Particular attention has been paid by the authors to updating the literature as completely as possible to the end of 2011.

It is our hope that the book will serve as a reference work for years to come. The multidisciplinary approach, with chapters devoted to biological, materials, and supramolecular chemistry, renders the book an important source of information not only for chemists but also for physicists, biochemists, and other researchers who deal with chalcogen compounds. We hope the book may induce sufficient curiosity in readers to attract them towards this branch of chemistry.

We are greatly indebted to all the authors for their commitment.

Francesco Antonio Devillanova
Complesso Universitario di Monserrato, Italy

Wolf-Walther du Mont
Technische Universität Braunschweig, Germany

Contents

Handbook of Chalcogen Chemistry: New Perspectives in Sulfur, Selenium and Tellurium, 2nd Edition, Volume 2
Edited by Francesco Antonio Devillanova and Wolf-Walther du Mont

Published by the Royal Society of Chemistry, www.rsc.org

Compounds Containing the Chalcogen–Chalcogen E–E Bond (E = S, Se, Te)

Volume 2

Material Chemistry

Miscellaneous Aspects

Introduction

FRANCESCO ANTONIO DEVILLANOVA*[a] AND
WOLF-WALTHER DUMONT*[b]

[a] Dipartimento di Chimica Inorganica ed Analitica, Università di Cagliari, S.S. 554 Bivio Sestu, I-09042 Monserrato-Cagliari, Italy; [b] Institut für Anorganische und Analytische Chemie der Technischen Universität Braunschweig, D-38023 Braunschweig, Germany
*Email: devilla@unica.it; w.du-mont@tu-bs.de

The Group 16 elements are usually named *chalcogens* ('ore builders'), but *chalcophiles* in the geochemical sense are *soft* elements (in the terminology of the hard and soft acids and bases concept, HSAB), that preferably form minerals with sulfur and with the other heavier chalcogens selenium and tellurium. Oxygen, however, tends to bind preferably to *lithophiles*, which are *hard* elements such as the alkali metals, early transition metals, and silicon. The special position of oxygen among the chalcogen elements is associated with its very high electronegativity, which in turn correlates with its high ionization energy and the small atomic radius, and makes it difficult for oxygen to act as a central atom in electron-rich multicentre bonds.[1–3] In contrast to the 3p orbitals of sulfur, which are appreciably larger than the 3s orbital, the 2p and 2s orbitals of oxygen have similar radial extents, because 2p shells (similarly to 1s, 3d, and 4f shells), lacking a core shell of the same angular momentum, have no radial node and are exceptionally compact.[1,3,4]

Among the non-oxygen chalcogens covered in this *Handbook of Chalcogen Chemistry*, the atomic and covalent radii increase gradually from sulfur to tellurium, whereas the electronegativities of sulfur and selenium are very similar

Handbook of Chalcogen Chemistry: New Perspectives in Sulfur, Selenium and Tellurium, 2nd Edition, Volume 2
Edited by Francesco Antonio Devillanova and Wolf-Walther du Mont

Published by the Royal Society of Chemistry, www.rsc.org

($\chi = 2.4$–2.5, depending on the scale) and only tellurium is noticeably less electronegative ($\chi = 2.1$).[5] The similarity of ionization energies and electronegativities of sulfur and selenium is explained by the presence in the latter of the filled $3d^{10}$ subshell. Because the 10 electrons in the 3d shell shield the 4s and 4p valence shell imperfectly from the additional 10 positive nuclear charges, the 4s and 4p ionization energies of selenium are increased to values close to those of the smaller sulfur atom. The consequence is that there is a strong similarity between S and Se isologues, whereas these differ from the corresponding O/Te derivatives. By analogy with 'deuteration', the term 'selenation' was coined to draw attention to this similarity, particularly as regards the vibrational properties.[6] In fact, the S/Se substitution works like a quasi-isotopic substitution, the infrared spectra of S/Se isologues being almost completely superimposable with the exception of the bands involving the S/Se atoms, with those of the selenium compounds occurring at lower frequencies with respect to those of the sulfur compounds. A practical consequence in synthetic chemistry is that a simple overlap of the infrared spectra of two S/Se compounds can be used as a diagnostic tool to ascertain their nature as isologues. Moreover, the polarization of bonds from sulfur and selenium to other elements is very similar, *e.g.* their nearly non-polar bonds to carbon and iodine and the only slightly polar bonds to nitrogen. Because of the close similarity of sulfur and selenium derivatives, sulfur in minerals tends to be accompanied by trace amounts of selenium, and in organisms the trace selenium contents can be divided into a ubiquitous 'statistical' selenium incorporation at various sulfur sites, contrasted with the genetically programmed selenium content of essential selenoenzymes.

As an element with a stable polymeric semiconducting modification and several molecular polymorphs, selenium lies between sulfur (for which many non-metallic polymorphs are known under standard conditions) and tellurium, which exists as a semiconducting polymer with significant interchain contacts. The ability of the chalcogen atoms to establish secondary bonds, ranging from van der Waals-like interactions to hypervalent 3c–4e bonding systems, increases from sulfur to tellurium. When these secondary interactions involve only three chalcogen atoms (interaction of an E-donor with an E–E acceptor pair), a great variety of linear trichalcogen systems can be observed in the solid state. As an example, a continuum of distances from very asymmetric to perfectly symmetric has been observed for E–E–E linear three-body systems, with no indication of a critical distance at which the bond switches from a predominantly electrostatic to a substantially covalent nature.[7,8] The same observation has been made for selenium and iodine: a continuum of Se···I contacts from van der Waals-like interactions via 'hypervalent' three-centre-bond systems to slightly perturbed Se–I single bonds was established experimentally with help of model compounds.[9] The correlation of increasing homonuclear or heteronuclear secondary interactions with a weakening of the neighbouring covalent bonds can be easily explained by intermolecular overlap of p-orbitals with antibonding functions of single bonds ($n \rightarrow \sigma^*$ interactions).[10,11]

Tellurium, which offers among the chalcogens the weakest single bonds for $n \rightarrow \sigma^*$ attack, has the greatest tendency to form stable species with hypervalent/hypercoordinated structures. Based on the weaker bonds, smaller HOMO–LUMO gaps (such as n/σ^*) are associated with a greater frequency of coloured compounds for the heavier elements. The significant intermolecular interactions displayed by the chalcogen elements, particularly by the heavier Se and Te, are very useful in crystal engineering for creating novel materials with extended structures, modifying the solid state electronic properties, or building functional solid state networks. In the latter respect, the synthetic chemistry developed at the tetrathiafulvalene core illustrates well how the chalcogen elements can be used in the design of electroactive materials.[12] The remarkable variety of structural motifs found in the polymeric networks of polyselenides and polytellurides, demonstrating the much richer homopolyatomic anion chemistry for selenium and tellurium than for sulfur, can promote the development of new features such as charge density waves, low-dimensional metallic properties, and even superconductivities in these materials.[13] An important and rapidly growing field within the chemistry of the chalcogen elements involves chalcogen radicals, which can be isolated as stable compounds in the solid state.[14,15] Their properties are very attractive since they can be used as building blocks for the design of new magnetic materials. In recent years, the advances of nanoscience and nanotechnology have opened unexpected opportunities for chalcogenide nanomaterials, such as nanowires and quantum dots, to provide important functional nanomaterials with potential applications in optoelectronics and life sciences.[16,17]

In biological systems, detailed structures of a number of enzymes have been recently clarified thanks to the remarkable progress in the techniques of single-crystal X-ray analysis, and great attention has been paid to the preparation of organo-sulfur[18] and organo-selenium compounds[19,20] as enzyme mimics. The weakness of bonds to selenium compared with those to sulfur can be an advantage when bond-making/bond-breaking processes are exploited enzymatically, as in antioxidant enzymes. Under physiological conditions the weaker Se–H bonds of selenols lead to significantly larger extents of protolytic dissociation than with thiols, and additionally the resulting selenolate anions RSe^- are *softer* nucleophiles than thiolates in the HSAB sense. This is also relevant to the particular role of selenium in the thyroid, catalysing a hormone-activing deiodination step.[21]

References

1. M. Kaupp, *J. Comput. Chem.*, 2006, **28**, 320–325.
2. P. Pyykkö, *Phys. Scripta*, 1979, **20**, 647.
3. P. Pyykkö, *J. Chem. Res. Synop.*, 1979, 380.
4. W. Kutzelnigg, *Angew. Chem.*, 1984, **96**, 262; *Angew. Chem. Int. Ed. Engl.* 1984, **23**, 272.
5. L. Pauling, *The Nature of the Chemical Bond*, 3rd Ed.; Cornell University Press, Ithaca, 1960.

6. K. A. Jensen, *Ann. N.Y. Acad. Sci.*, 1972, **192**, 115.
7. M. C. Aragoni, M. Arca, A. F. Devillanova, F. Isaia and V. Lippolis, *Phosph. Sulf. and Silic.*, 2008, **183**, 1036–1045.
8. M. C. Aragoni, M. Arca, A. F. Devillanova, F. Isaia and V. Lippolis, *Cryst. Growth Des.*, 2012, **12**, 2769–2779.
9. W.-W. du Mont, A. Martens-von Salzen, F. Ruthe, E. Seppälä, G. Mugesh, F. A. Devillanova, V. Lippolis and N. Kuhn, *J. Organomet. Chemistry*, 2001, **623**, 14–28.
10. R. E. Rosenfield, R. Parthasarathy and J. D. Dunitz, *J. Am. Chem. Soc.*, 1977, **99**, 4860–4862.
11. D. B. Werz, B. J. Rausch and R. Gleiter, *Tetrahedr. Letters*, 2002, **43**, 5767–5769.
12. T. D. Westgate, D. Cortizo Lacalle and P. J. Skabara, Chapter 11.2 of this Handbook.
13. W. S. Sheldrick, Chapter **9.2** of this Handbook.
14. T. Chivers and R. S. Laitinen, Chapter 4 of this Handbook.
15. J. M. Rawson and J. J. Hayward, Chapter 11.1 of this Handbook.
16. B. He and W.-J. Zhang, Chapter 11.4 of this Handbook.
17. L. Yuwen and L. Wang, Chapter 11.5 of this Handbook.
18. Y. Mizobe and H. Seino, Chapter 10.1 of this Handbook.
19. B. J. Bhuyan, D. S. Lamani, G. Mugesh and T. Wirth, Chapter 10.2 of this Handbook.
20. S. Hayashi and W. Nakanishi, Chapter 12.3 of this Handbook.
21. G. Mugesh, Chapter 10.3 of this Handbook.

Biological Chemistry

CHAPTER 10.1

Metal–Sulfur Clusters as the Model for the Active Sites of Metalloenzymes

YASUSHI MIZOBE AND HIDETAKE SEINO*

Institute of Industrial Science, The University of Tokyo, Komaba, Meguro-ku, Tokyo 153-8505, Japan
*Email: seino@iis.u-tokyo.ac.jp

10.1.1 Introduction

Nitrogen is one of the essential elements for all living things and a significant part of the nitrogen needed on Earth is supplied in the form of ammonia produced from atmospheric nitrogen by biological nitrogen fixation. This highly important reaction, converting a quite inert N_2 molecule into ammonia through coupled protonation and electronation, is known to proceed under ambient conditions by the catalysis of the metalloenzyme nitrogenase.[1] This presents a sharp contrast to industrial nitrogen fixation,[2] *i.e.* the Haber–Bosch process, which requires an extremely drastic condition to produce ammonia from gaseous nitrogen and hydrogen in the presence of an Fe-based heterogeneous catalyst. From the 1970s, certain Mo–Fe–S aggregates had been proposed as the active site structure of nitrogenase mainly on the basis of EXAFS data.[3] It was in 1992 that the first report appeared about the results of the single-crystal X-ray diffraction study (2.7 Å resolution) for the most common nitrogenase, *viz.* Mo nitrogenase, which disclosed the surprising $MoFe_7S_9$ mixed-metal sulfido core present at the active site.[4] From higher-resolution (1.16 Å) crystallographic results of in 2002, one light

Handbook of Chalcogen Chemistry: New Perspectives in Sulfur, Selenium and Tellurium, 2nd Edition, Volume 2
Edited by Francesco Antonio Devillanova and Wolf-Walther du Mont

Published by the Royal Society of Chemistry, www.rsc.org

atom X (C, N, or O) was found to be at the centre of this core, as depicted in Figure 10.1.1.[5] Although nitrogen was proposed to be most probable for the interstitial atom X at that time, it was characterized as carbon in more recent studies (2011) by using crystal structure at the stage of more accurate resolution (1.0 Å) and X-ray emission spectroscopy.[6]

Owing to the remarkable progress in single-crystal X-ray analysis techniques, detailed structures have recently been clarified for a number of enzymes, which include several metalloenzymes containing the cluster cores with sulfur-bridged multimetallic centres at their active sites such as hydrogenase,[7] sulfite reductase,[8] and carbon monoxide hydrogenase/acetyl-CoA synthase[9] (Figure 10.1.2), as well as nitrogenase. It is likely that the high catalytic activities of these enzymes result from the cooperation of two or more metal centres in proximity, making sulfur

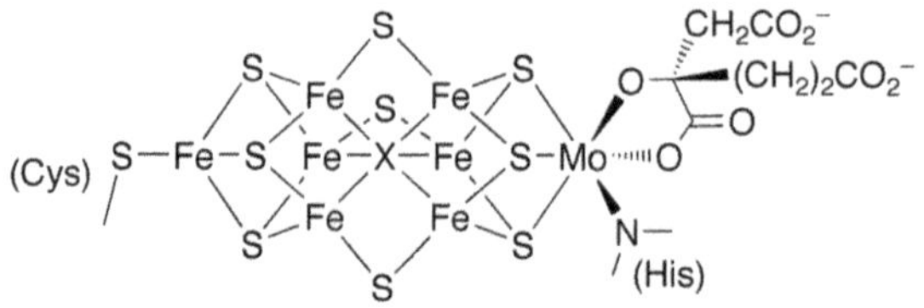

Figure 10.1.1 Structure of the Fe–Mo cofactor in Mo nitrogenase.

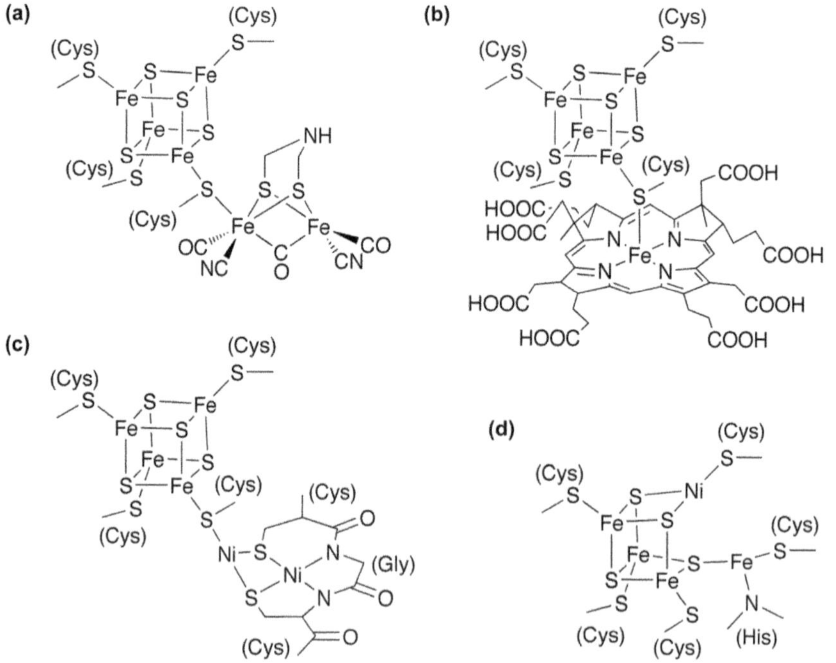

Figure 10.1.2 Structures of the active sites of metalloenzymes containing metal–sulfur cluster units: (a) FeFe hydrogenase, H-cluster (H^{ox} form); (b) Sulfite reductase; (c) A-cluster in acetyl-CoA synthase; (d) C-cluster in carbon monoxide dehydrogenase.

ligands such as sulfides and thiolates the choice as the bridges to maintain these multimetallic cores intact during catalysis. This is presumably because of the characteristics of the S atom, *i.e.* its strong affinity with transition metals and high bridging ability. However, although X-ray crystallographic analyses have successfully disclosed the active site structures of these enzymes in the resting state, the structures during catalytic turnover may possibly be different. Furthermore, their function mechanisms are essentially unknown. It is difficult to observe directly what is occurring at active sites embedded within huge proteins, so studies to synthesize model compounds and clarify their reactivity are of much importance. In this chapter, recent advances in the chemistry of metal–sulfur clusters as synthetic analogues to natural enzymes are briefly summarized.

10.1.2 Metal–Sulfur Clusters in Metalloenzymes and Syntheses of Their Structural Models

Iron–sulfur proteins are ubiquitous in all life forms, and at their active sites they most commonly contain the Fe_2S_2, Fe_3S_4, and Fe_4S_4 cluster cores shown in Figure 10.1.3 to mediate the electron transfer as their predominant role. In the early 1970s a synthetic approach to these iron–sulfur cluster sites was initiated, and analogues of these rhombic, incomplete cubane-type, and cubane-type Fe clusters having certain thiolate groups as ancillary ligands in place of the cysteinyl residues have already been successfully prepared.[10]

On the other hand, such approaches to the metalloenzymes shown in Figures 10.1.1 and 10.1.2 are still under way.[11] Thus, the model clusters reproducing precisely their complex metal–sulfur assemblies in the native form have not yet been isolated. In this section, studies aiming at the syntheses of the model compounds of two clusters in nitrogenase, FeMo cofactor (FeMo-co) and P cluster, are surveyed. The choice of these clusters as the representatives of the metal–sulfur clusters in metalloenzymes arises from the fact that these are the largest and most complicated metal–sulfur clusters known at present among those observed in natural enzymes.

10.1.2.1 Preparation and Reactions of the FeMo Cofactor Model Clusters

The most common nitrogenase is composed of two proteins, Mo–Fe protein and Fe protein, both of which contain metal–sulfur clusters. The former

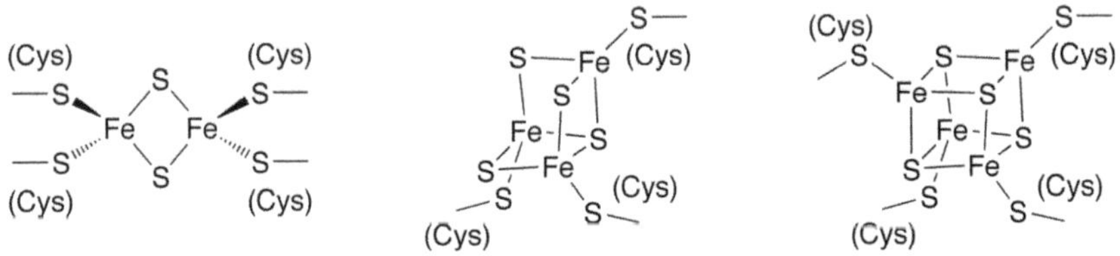

Figure 10.1.3 Three representative Fe–S multimetallic cores observed in iron–sulfur proteins.

protein contains two kinds of clusters: FeMo-co, a Mo–Fe–S cluster which is believed to be the site for N_2 activation and reduction, and P cluster, a Fe–S cluster mediating the transfer of electrons for N_2 reduction from Fe protein to FeMo-co, while the latter protein has the Fe_4S_4 cluster. Before its elucidation by crystallography, the structure of FeMo-co had been proposed mostly based on the EXAFS results, and a cubane-type $MoFe_3S_4$ core was one of those suggested as its partial structure using the data around Mo in 1978.[3] Synthesis of the $MoFe_3S_4$ clusters was reported in the same year,[12,13] and their reactivity toward nitrogenase substrates has been investigated since then. Importantly, this led not only to the isolation of clusters with these substrate molecules bonded to the Mo site but also the demonstration of the reaction systems containing the $MoFe_3S_4$ clusters that can reduce nitrogenase substrates. Although it is now known that, as shown in Figure 10.1.1, the Mo–Fe–S aggregate present in FeMo-co does not have the cubane-type $MoFe_3S_4$ fragment but an incomplete cubane-type $MoFe_3(\mu_3\text{-}S)_3$ chromophore that is connected by one μ_6-C and three μ_2-S ligands to the $Fe_4(\mu_3\text{-}S)_3$ unit, these studies nevertheless represent one of the most successful examples of the synthetic approach mimicking the metalloenzyme system.

The cubane-type $MoFe_3S_4$ cluster was first prepared in the form of the double cubane clusters $[Mo_2Fe_6S_8(SR)_9]^{3-}$,[12,13] and $[Mo_2Fe_6S_9(SEt)_8]^{3-}$,[13] which contain two cubanes connected at each Mo site by three μ_2-SR bridges or one μ_2-S and two μ_2-SEt bridges (Figure 10.1.4, left). These clusters were obtained by self-assembly reactions using tetrathiomolybdate, $FeCl_3$, and thiols. The $[Mo_2Fe_6S_8(SR)_9]^{3-}$ cluster undergoes chemical and electrolytic, one- and two-electron reductions to give the clusters $[Mo_2Fe_6S_8(SR)_9]^{n-}$ ($n=4$ and 5). Hydrogen evolution by treatment of these reduced clusters with PhSH or Et_3NH^+ in solution was demonstrated as the model of hydrogenase function observed for nitrogenase.[14] Reduction of acetylene to ethylene[15] and of hydrazine[16] or N_2[17] to ammonia was also attained under electrolytic conditions, *e.g.* in THF–MeOH by using $[Mo_2Fe_6S_8(SPh)_9]^{3-}$ as catalyst precursor. However, the efficiencies were still poor and the mechanism operating in these substrate reductions is uncertain. It is noteworthy that the cluster contains only coordinately saturated Mo and Fe centres, even in the reduced form $[Mo_2Fe_6S_8(SPh)_9]^{3-}$.

The synthesis of $MoFe_3S_4$ single cubane clusters having the potential substrate-binding site was also attempted, which provided more sophisticated models for FeMo-co than the above-mentioned double cubane with two Mo

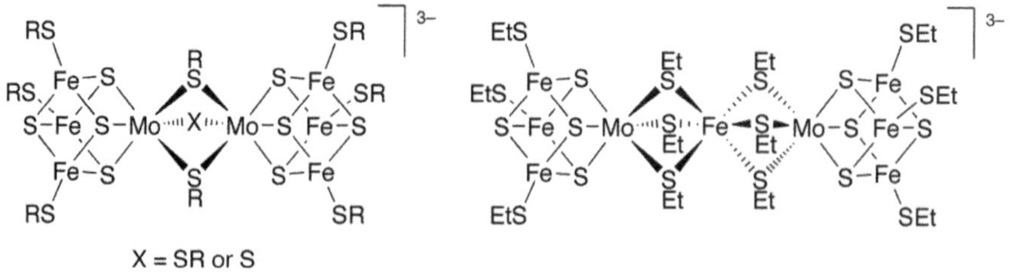

Figure 10.1.4 Mo–Fe–S clusters containing two cubane-type $MoFe_3S_4$ cores, which are the extensively studied synthetic models of FeMo-co.

atoms each surrounded by six sulfur ligands. By starting from the $Fe(SEt)_6$-bridged double cubane cluster $[Mo_2Fe_7S_8(SEt)_{12}]^{3-}$ (Figure 10.1.4, right), the desired single cubane clusters $[MoFe_3S_4(SR)_3(cat)(L)]^{n-}$ (cat = substituted catecholate; $n = 2, 3$) were able to be derivatized, in which not only the RS^- and RO^- anions but also nitrogenase substrates such as MeCN, N_2H_4, N_3^-, and CN^- can bind to the Mo site as L (Figure 10.1.5).[18] Hydrazine-bridged double cubane clusters $[\{MoFe_3S_4Cl_2(Cl_4\text{-cat})\}_2(\mu\text{-}N_2H_4)(\mu\text{-}S)]^{4-}$ (Cl_4-cat = tetrachlorocatecholate) and $[\{MoFe_3S_4Cl_3(Cl_4\text{-cat})\}_2(\mu\text{-}N_2H_4)]^{4-}$ were also isolated.[19] By the use of $[MoFe_3S_4Cl_3(Cl_4\text{-cat})(MeCN)]^{2-}$ as catalyst, reduction of acetylene to ethylene,[20] *cis*-dimethyldiazene to methylamine,[21] and hydrazine to ammonia[22] were attained at room temperature in the system using 2,6-lutidinium chloride ([LutH]Cl) and cobaltocene as external proton and electron sources, respectively.

Upon elucidation of the presence of a homocitrato ligand at the Mo atom in FeMo-co, synthesis of the $MoFe_3S_4$ cubane clusters with a polycarboxylato ligand was subsequently attempted, which led to the isolation of the clusters $[MoFe_3S_4Cl_3(L^1)(L^2)]^{n-}$ (L^1 = oxalate and L^2 = Cl, CN; L^1/L^2 = methyliminodiacetate, thiodiglycolate).[23] Interestingly, a series of the clusters of this type consisting of the tridentate ligands at the Mo site can catalyse the reduction of hydrazine into ammonia under conditions analogous to those described above, for which reduction of hydrazine is presumed to proceed on the single Mo site. Other routes to the $MoFe_3S_4$ single cubanes have been developed recently, which use $[TpMoS(S_4)]^-$ (Tp = hydrotris(pyrazol-1-yl)borate)[24] or $[Cp^*Mo(S^tBu)_3]$ ($Cp^* = \eta^5\text{-}C_5Me_5$)[25] as synthetic precursors. The Mo atom of the resulting clusters, $[TpMoFe_3S_4Cl_3]^-$ or $[Cp^*MoFe_3S_4(S^tBu)_3]^-$, is tightly capped by Tp or Cp*, and only the Fe sites undergo ligand substitution and substrate coordination.

The Mo–Fe–S clusters containing the cores other than the $MoFe_3S_4$ cubane are also known. These include, for example, $[Mo_2Fe_6S_6L_6(CO)_6]^{3-}$ (L = Cl, Br, OR) with the bicapped prismatic core[26] and $[MoFe_5S_6(CO)_6L_3]^{n-}$ ($L = PEt_3$, $n = 0$; L = I, $n = 2$), which consists of a $MoFe_3S_4$ cubane and a Fe_2S_2 unit (Figure 10.1.6).[27] The cuboidal (defect cubane) $MoFe_3S_3$ clusters such as $[MoFe_3S_3(Cl_4\text{-cat})(CO)_6(PEt_3)_2]$ and $[MoFe_3S_3(Cl_4\text{-cat})(CO)_4(PPr_3)_3(pyridine)]$

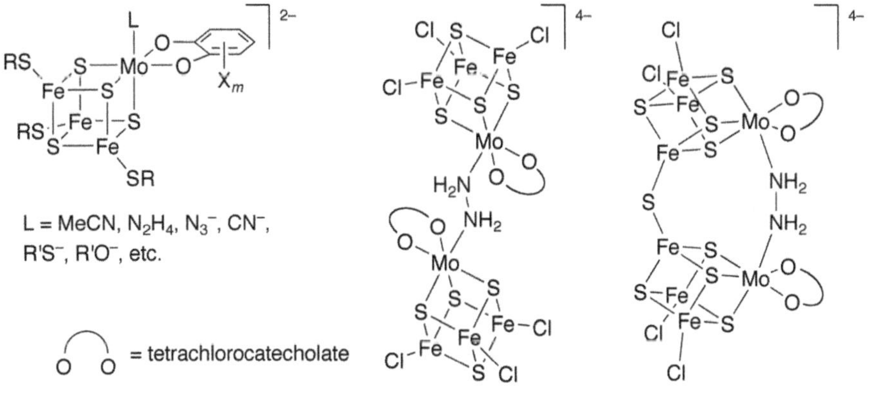

Figure 10.1.5 $MoFe_3S_4$ clusters with the nitrogenase substrates bonded to their Mo site.

are of much interest as the model of the $MoFe_3S_3$ fragment in FeMo-co.[28] In $[MoFe_4S_6(PEt_3)_4Cl]$, a cuboidal Fe_4S_3 fragment is connected to a Mo atom by three μ_2-S atoms.[29] Relevantly, the cubane-type Fe_4S_3N and $Fe_4S_2N_2$ clusters were synthesized recently as a partial representation of FeMo-co, in which the interstitial μ_6-X atom is postulated to be N.[30]

By treatment of the single cubane cluster $[MoFe_3S_4Cl_3(Cl_4\text{-cat})(MeCN)]^{2-}$ with PEt_3, the edge-fused double cubane cluster $[\{MoFe_3S_4(PEt_3)_3(Cl_4\text{-cat})\}_2]$ was obtained,[31] which was converted upon reaction with $[Et_4N]SH$ to the higher-nuclearity cluster $[Mo_6Fe_{20}S_{30}(PEt_3)_6(Cl_4\text{-cat})_6]^{8-}$ containing the $Mo_2Fe_6S_9$ fragments relating topologically to the P^N cluster core in nitrogenase described below.[32] Similarly, a $Mo_2Fe_6S_9$ core was also found in the relating cluster $[(TpMo)_2Fe_6S_9(SH)_2]^{3-}$ that is derivatized from $[TpMoFe_3S_4Cl_3]^{-}$ by an analogous route, as depicted in Figure 10.1.7.[33]

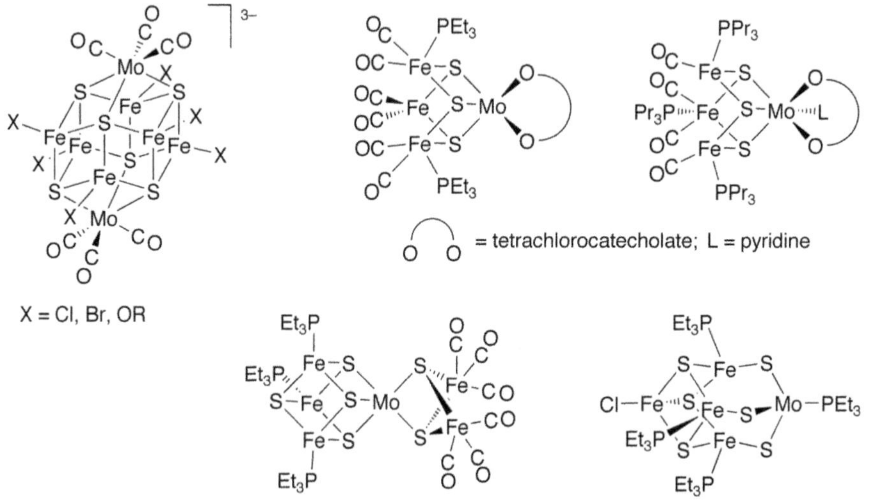

Figure 10.1.6 Mo–Fe–S clusters prepared as models of FeMo-co, representing the partial structures.

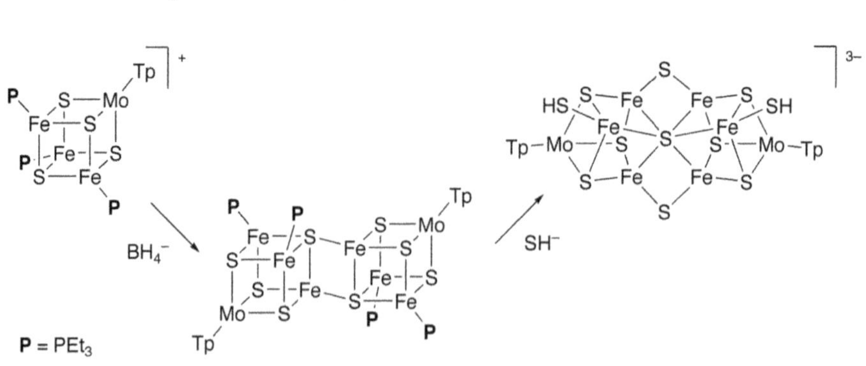

Figure 10.1.7 Mo–Fe–S–clusters consisting of the metal–sulfur connecting scheme akin to that in the P^N cluster.

10.1.2.2 Preparation of P^N Cluster Models

X-ray analysis disclosed the structures of the P cluster in the dithionite-reduced state (P^N cluster) and the oxidized state (P^{OX} cluster) (Figure 10.1.8).[34] The P^N cluster core has a corner-shared double cubane structure connected further by two cysteinyl μ-S atoms. As described above, the $Mo_2Fe_6S_9$ fragments observed in $[Mo_6Fe_{20}S_{30}(PEt_3)_6(Cl_4\text{-cat})_6]^{8-}$ and $[Mo_2Fe_6S_9(Tp)_2(SH)_2]^{3-}$ closely resembled the metal–sulfur connecting scheme of the P^N cluster, and analogous clusters having $W_2Fe_6S_9$, $V_2Fe_6S_9$, and $Mo_2Fe_4Cu_2S_9$ cores of the same topology are known.[35] Apparently, the core structures of FeMo-co and the P cluster are mutually intimately related: two incomplete cubane-type M_4S_3 cores are connected by either μ_6-X or μ_6-S, respectively.

The all-iron corner-shared Fe_8S_7 core itself has been constructed by a self-assembly reaction using $[Fe\{N(SiMe_3)_2\}_2]$, tetramethylthiourea, 2,4,6-triisopropylbenzenethiol, and elemental sulfur. The product $[Fe_8S_7\{SC(NMe_2)_2\}_2\{N(SiMe_3)_2\}_4]$ contains two cuboidal Fe_4S_3 cores bridged by one μ_6-sulfide and two μ_2-bis(trimethylsilyl)amides (Figure 10.1.9, left).[36] The thiourea and

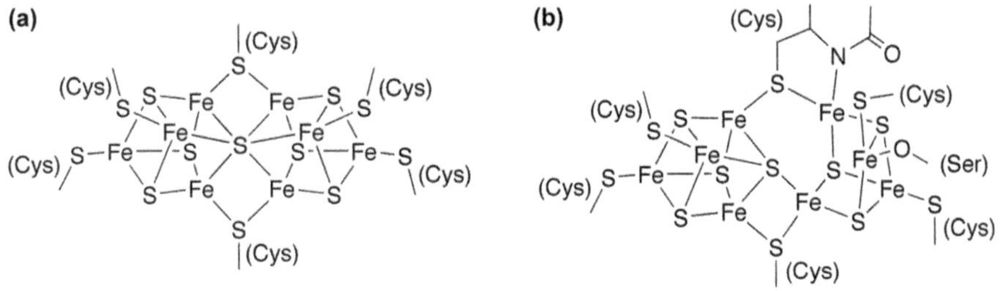

Figure 10.1.8 Structure of the P-cluster: (a) reduced form (P^N); (b) oxidized form (P^{OX}).

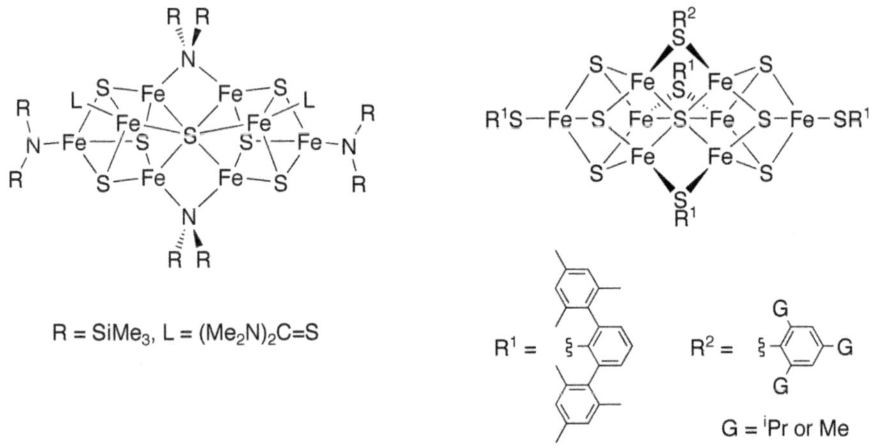

Figure 10.1.9 Fe–S clusters containing a Fe_8S_7 core analogous to that in the P^N cluster.

terminal amide ligands in this cluster are replaceable by thiolate ligands. The amide-free clusters $[\{Fe_4S_3(SR^1)\}_2(\mu_6\text{-}S)(\mu_2\text{-}SR^1)_2(\mu_2\text{-}SR^2)]$ (R^1 = 2,6-(2,4,6-$Me_3C_6H_2)_2C_6H_3$, R^2 = 2,4,6-$^iPr_3C_6H_2$ or 2,4,6-$Me_3C_6H_2$) have been isolated under modified reaction conditions.[37] The topology of these clusters (Figure 10.1.9, right), with three μ_2-S atoms connecting two cuboidal Fe_4S_3 units, closely resembles that of FeMo-co.

For further details of progress in the syntheses of the structural models of the clusters in FeMo protein, see other reviews,[38] together with those dealing with the reactions of metal–sulfur clusters relating to nitrogen fixation chemistry.[39]

10.1.3 Metal–Sulfur Clusters as the Functional Models for Metalloenzymes

Approaches to the chemistry of metal–sulfur clusters relevant to the active sites of metalloenzymes include not only attempting to synthesize precise structural models but also preparing functional models. The latter approach stems from the view that although the enzymes presumably possess the optimum catalytic systems that they have finally acquired after a long evolutionary process, these have been attained under conditions available only in the natural environment. Thus, the metals that can be selected in nature are limited to those with high natural abundance, and the construction of the active sites must be carried out under ambient conditions. This means that it may be possible for us, as synthetic chemists, to create synthetic catalysts that are better than natural enzymes. Indeed, in many organic reactions higher activities are found for catalysts consisting of noble metals than for those containing the 3d transition metals in the same triad, although noble metals are not the chosen key metals for metalloenzymes. It is therefore of great interest to synthesize metal–sulfur clusters that contain a wide range of transition metals and survey their reactivity as the functional models of the enzyme reactions. It is to be noted that the metal–sulfur clusters have also been investigated extensively in terms of the active site models of industrial hydrodesulfurization catalysts.[40]

A large number of transition metal–sulfur clusters have been synthesized, most of which have been obtained using self-assembly methods. On the other hand, to construct cluster cores with the desired metal–sulfur compositions and connecting schemes, rational pathways leading to the high-yield syntheses of tailored metal–sulfur clusters have recently been explored. Fragment condensations have been demonstrated to be powerful methods of obtaining many such clusters,[41] some examples of which are shown in this section. This methodology is intriguing in connection with recent findings in biological syntheses of iron–sulfur enzymes, showing that highly complex clusters such as FeMo-co and P cluster are elaborately assembled from small modules.[42]

10.1.3.1 Preparation of Metal–Sulfur Clusters from Dinuclear Precursors

Among the dinuclear metal–sulfur complexes that serve as the excellent precursors to tailored metal–sulfur clusters, those with M_2S_4 cores are known to be useful for syntheses of cubane-type and relating sulfido clusters. These include reactions that start from the complexes having Mo(μ-SH)$_2$(μ-S)$_2$Mo,[43] M(μ-S$_2$)(μ-S)$_2$M (M = V,[44] Cr, Mo[45]), Ru(μ-S$_2$)$_2$Ru,[46] and MS(μ-S)$_2$MS (M = Ti,[47] Mo, W,[48] Re[49]) cores, which are summarized in Figure 10.1.10.

Versatile dinuclear precursors of the other type are those containing M_2S_2 cores. Cubane-type clusters [(Cp*M)$_4$(μ$_3$-S)$_4$] are available through dimerization of the M(μ-S)$_2$M complexes [Cp*M(PMe$_3$)(μ-S)$_2$IrCp*] (M = Rh, Ir)[50] and [{Cp*Ir(CO)$_2$}$_2$(μ-S)$_2$].[51] Although bridging hydrosulfido complexes are less common than bridging sulfido complexes, a series of M(μ-SH)$_2$M complexes [(LMCl)$_2$(μ-SH)$_2$] (L = Cp*, M = Ru, Rh, Ir; LM = (η^6-arene)Ru) has been isolated and found to serve as versatile precursors of not only the cubane-type clusters via dimerization in the presence of NEt$_3$ for dehydrochlorination[52] but

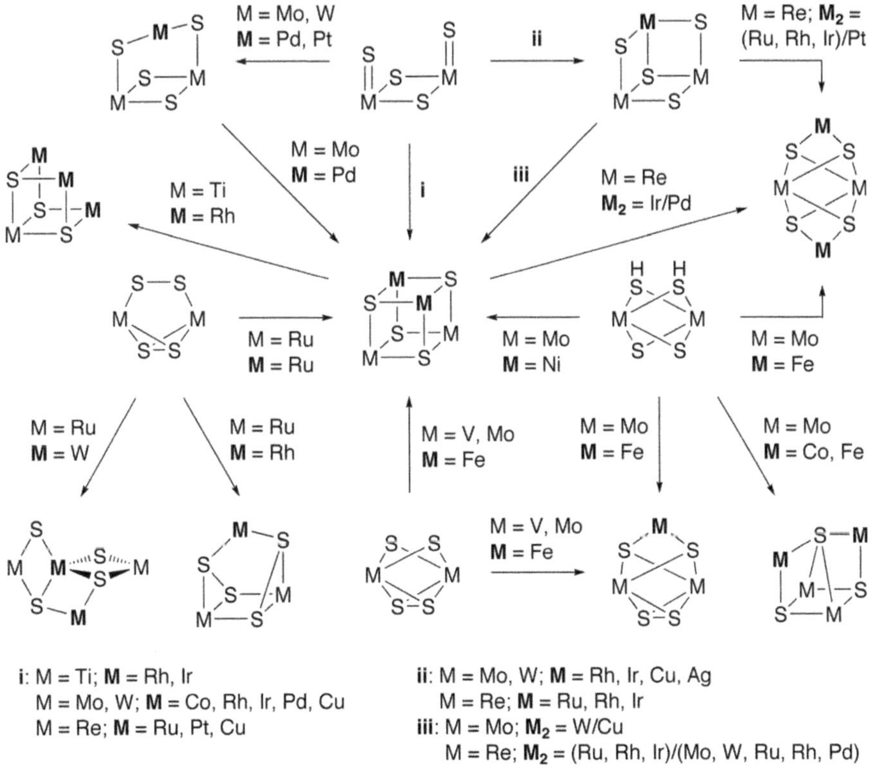

Figure 10.1.10 Schematic representation of the syntheses of cubane-type and relating sulfido clusters using M_2S_4 complexes as precursors.

also the numerous homo- and heterometallic trinuclear $M_3(\mu_3\text{-}S)_2$, tetranuclear $M_4(\mu_3\text{-}S)_2$, and pentanuclear $M_5(\mu_3\text{-}S)_4$ clusters by treatment with certain metal (Figure 10.1.11) species.[53] The trinuclear cluster $[(Cp^*Ir)_2(PdCl_2)(\mu_3\text{-}S)_2]$ catalyses addition of alcohol to internal 1-aryl-1-alkynes to give 2,2-dialkoxy-1-arylalkane with high regioselectivity (up to 99%).[54]

The complexes $[(Cp^*MCl)_2(\mu\text{-}SH)_2]$ undergo cross-condensation with $[\{MoO(dmf)_3\}_2(\mu\text{-}S)_2]I_2$ (dmf = $HCONMe_2$) in the presence of $[NMe_4]Cl$, yielding the mixed-metal cubane-type clusters $[(Cp^*M)_2\{MoCl_2(dmf)\}(MoOCl_2)(\mu_3\text{-}S)_4]$ (M = Rh, Ir) and $[(Cp^*Ru)_2\{MoCl_2(dmf)\}_2(\mu_3\text{-}S)_4]$.[55] Interestingly, the former oxocubanes can catalyse the nitrogenase model reaction. Thus, reduction of the hydrazine $MePhNNH_2$ proceeds smoothly at room temperature in the presence of [LutH]Cl and cobaltocene to give MePhNH and NH_3. By treatment of this cluster (M = Ir) with $MePhNNH_2$, the cubane cluster with the hydrazido(2–) ligand $[(Cp^*Ir)_2\{MoCl_2(dmf)\}\{Mo(NNMePh)Cl_2\}(\mu_3\text{-}S)_4]$ was isolated as the possible intermediate of the reductive cleavage of hydrazine (Figure 10.1.12).

Dinuclear Pt complexes of the type $[\{Pt(PR_3)\}_2(\mu\text{-}S)_2]$ are also known to serve as good precursors to the trinuclear clusters. A series of triangular cluster cores $M'Pt_2(\mu_3\text{-}S)_2$ has been constructed from them (M' = d- and p-block metals).[56]

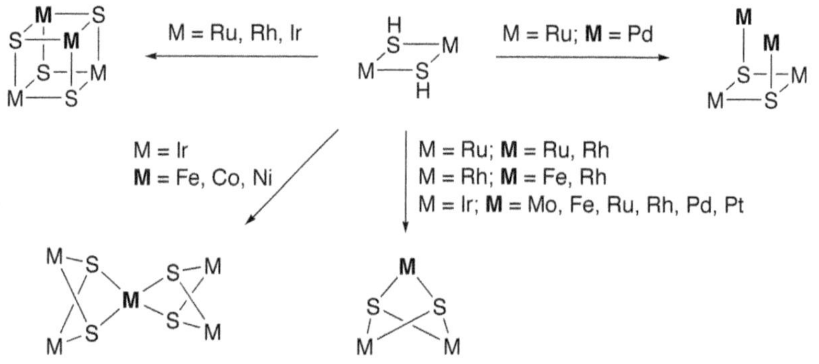

Figure 10.1.11 Construction of metal–sulfide cores starting from hydrosulfido-bridged dinuclear complexes.

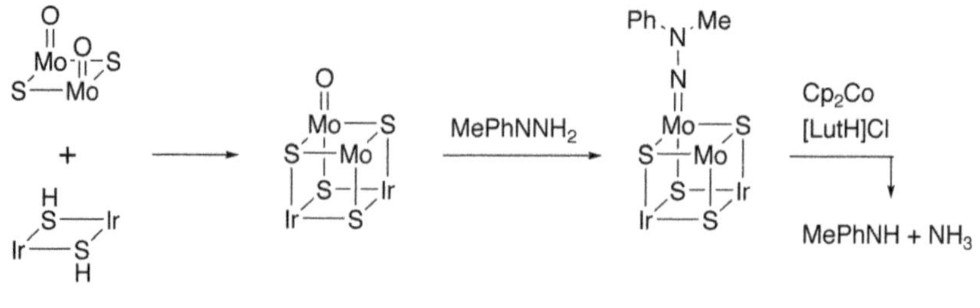

Figure 10.1.12 Reduction of $MePhNNH_2$ to MePhNH and NH_3 by cobaltocene and lutidinium chloride using the Mo site of cubane-type $Mo_2Ir_2S_4$ cluster via the hydrazido(2–) intermediate.

10.1.3.2 Preparation of Metal–Sulfur Clusters from Trinuclear Precursors

Construction of cubane-type clusters by incorporating the metal fragment into the void corner of the incomplete cubane-type M_3S_4 clusters (M = Mo, W) has been studied extensively. Thus, $[M_3(\mu_3\text{-}S)(\mu_2\text{-}S)_3(H_2O)_9]^{4+}$ was treated with elemental metals M′ in acidic aqueous solutions to give single cubane-type $M_3M'S_4$, corner-shared double cubane-type $M_6M'S_8$, and edge-fused double cubane-type $M_6M'_2S_8$ clusters, depending on the nature of the incorporated metals (Figure 10.1.13).[57] It has also been found that analogous clusters are available by using metal salt/$NaBH_4$ systems in place of elemental M′ in acidic media.[58]

Among the clusters derivatized by these methods, $[\{Mo(tacn)\}_3(PdCl)(\mu_3\text{-}S)_4]^{3+}$ (tacn = 1,4,7-triazacyclononane) is noteworthy, since this cluster can bind substrate molecules such as alkene and CO at its tetrahedral Pd site generated by dissociation of the Cl anion.[59] It is more interesting that addition of ROH and RCOOH to various alkynes proceeds at this Pd site to give vinyl ethers and vinyl carboxylates catalytically.[60] These are still rare examples of reactions that are catalysed by well-defined cubane-type clusters with the core structures retained.[61]

Organometallic precursors containing incomplete cubane-type cores, $[(Cp'M)_3(\mu_3\text{-}S)(\mu_2\text{-}S)_3]^+$ (Cp′ = $\eta^5\text{-}C_5H_5$, $\eta^5\text{-}C_5H_4Me$, M = Mo, W;[62] Cp′M = Cp*Mo[63]), have been prepared more recently, from which a new series of mixed-metal cubane-type clusters with the $M_3M'S_4$ cores (M′ = Group 6, 8—11 metals) have been derived.[64] With respect to these organometallic cubanes, reactions with a range of small molecules have been investigated in organic media, leading to the isolation of the clusters binding CO,[63,65] H_2,[65] NO,[66] and alkyne,[67] as well as substrates of hydrodesulfurization and hydrodenitrogenation.[68] Catalytic activity of the $(Cp^*Mo)_3RuS_4$ cluster for disproportionation of hydrazine into ammonia and N_2 has been shown, and the ammine cluster $[(Cp^*Mo)_3\{Ru(NH_3)(PPh_3)\}(\mu_3\text{-}S)_4]^+$ and the μ-amido-μ-hydrazido(1−) cluster $[\{(Cp^*Mo)_3(\mu_3\text{-}S)_4Ru\}_2(\mu\text{-}NH_2)(\mu\text{-}NHNH_2)]^{2+}$ have been obtained from the resulting reaction mixture (Figure 10.1.14).[69] The conversion of organic substrates using the Pd- or Ni-containing clusters as catalyst has also been reported.[61]

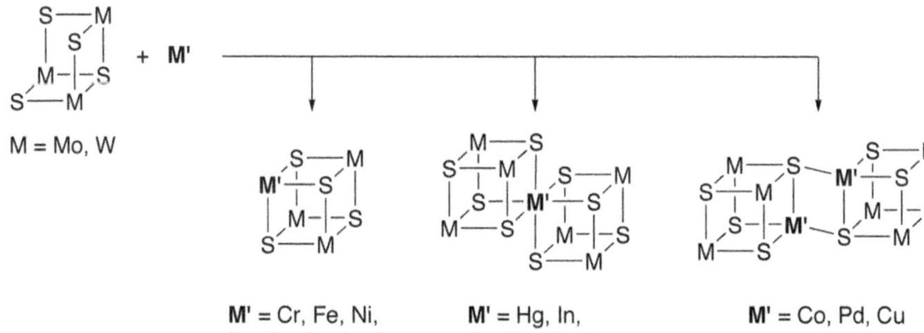

Figure 10.1.13 Syntheses of cubane-type clusters starting from incomplete cubane-type Mo_3S_4 and W_3S_4 aqua clusters.

Rational syntheses of mixed-metal cubane-type Fe_3MS_4 clusters starting from Fe_3S_4 clusters have also been attained with respect to the two precursors with different core structures; one is a linear $[Fe_3(\mu_2\text{-}S)_4(SR)_4]^{3-}$ cluster and the other is an incomplete cubane-type $[Fe_3(\mu_3\text{-}S)(\mu_2\text{-}S)_3(L)]^{3-}$ cluster where L = tris(thiolato) ligand. The former reacts with certain metal salts or complexes to give Fe_3MS_4 cubanes accompanied by core rearrangement, while incorporation of M into the missing corner of the latter precursor gives another series of Fe_3MS_4 cubanes (Figure 10.1.15).[10,11c,70]

Incomplete cubane-type μ-sulfido-μ-hydrosulfido cluster $[(Cp^*Ir)_3(\mu_3\text{-}S)(\mu\text{-}SH)_3]^+$ can also be used as the precursor to the mixed-metal cubane-type clusters,[71] those synthesized from which are single cubanes containing Mo,[72] Fe,[73] Ru,[74] or Sb, along with a corner-shared double cubane $[\{(Cp^*Ir)_3(\mu_3\text{-}S)_4\}_2Bi]^+$ and a cubanoid $[(Cp^*Ir)_3\{PdCl(PPh_3)\}(\mu_3\text{-}S)_3(\mu\text{-}SH)]$[75] (Figure 10.1.16). It is remarkable that the $RuIr_3S_4$ cluster binding a molecular N_2 at the Ru site has been derived, which is at present the only example of metal–sulfido cluster with an N_2 ligand. A DFT study on this N_2 cluster has confirmed that N_2 is reductively activated by back-donation from the d-orbital of Ru to the N–N π* orbital.[76]

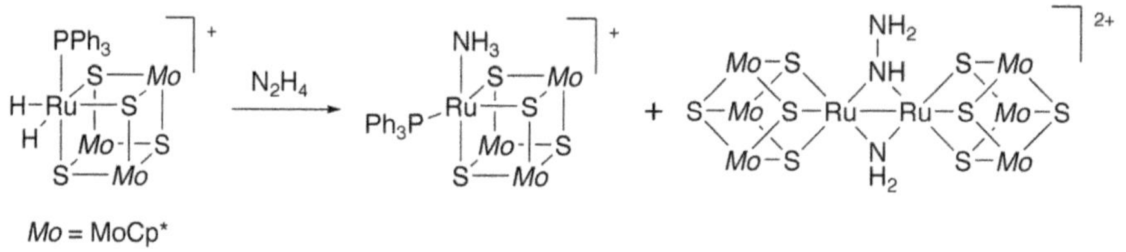

Figure 10.1.14 Reaction of the cubane-type Mo_3RuS_4 cluster with hydrazine affording ammine cluster and μ-amido-μ-hydrazido(1–) cluster.

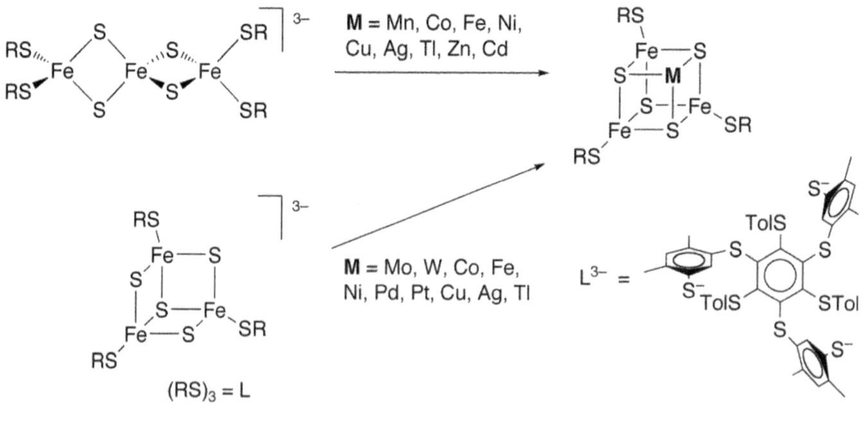

Figure 10.1.15 Syntheses of cubane-type Fe_3MS_4 clusters using linear or incomplete cubane-type Fe_3S_4 clusters.

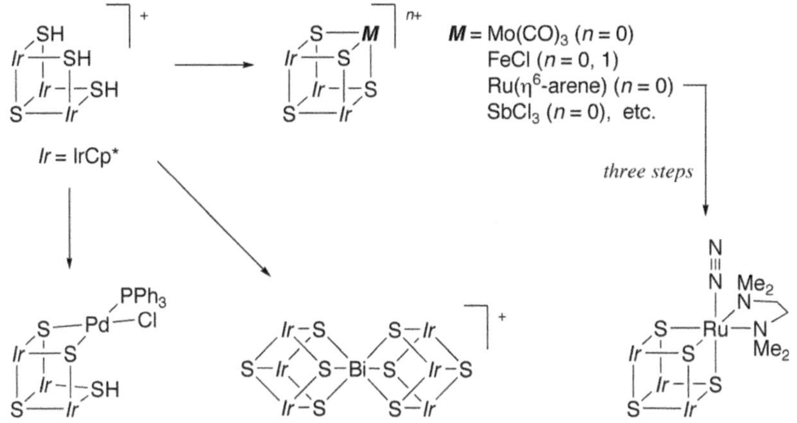

Figure 10.1.16 Clusters of cubane-type and the relating structures derived from the incomplete cubane-type sulfido–hydrosulfido cluster.

10.1.4 Conclusion

Metal–sulfur aggregates are known to occur in biological systems as the active sites of a significant number of metalloenzymes, and clarification of their detailed structures is currently progressing rapidly. In the light of these results, many researchers are attempting to synthesize direct structural models as well as functional models of the active sites of natural enzymes that promote various important reactions in the biological systems under ambient conditions. These studies will presumably lead in the near future to the development of novel synthetic catalysts that exhibit activities exceeding those of natural enzymes.

References

1. (a) B. K. Burgess and D. J. Lowe, *Chem. Rev.*, 1996, **96**, 2983; (b) B. E. Smith, *Adv. Inorg. Chem.*, 1999, **47**, 159; (c) D. C. Rees and J. B. Howard, *Curr. Opin. Chem. Res*, 2000, **4**, 559; (d) P. M. C. Benton, S. Sen and J. W. Peters in *Nitrogen Fixation at the Millennium*, G. J. Leigh (ed), Elsevier Science B. V., Amsterdam, 2002, 35; (e) L. C. Seefeldt, B. M. Hoffman and D. R. Dean, *Annu. Rev. Biochem.*, 2009, **78**, 701.
2. (a) M. Appl, *Ammonia*, Wiley-VCH, Weinheim, 1999; (b) R. Schlögl, *Angew. Chem. Int. Ed.*, 2003, **42**, 2004; (c) G. J. Leigh in *Catalysts for Nitrogen Fixation*, B. E. Smith, R. L. Richards and W. E. Newton (eds), Kluwer Academic Publishers, Dordrecht, 2004, 33.
3. S. P. Cramer, K. O. Hodgson, W. O. Gillum and L. E. Mortenson, *J. Am. Chem. Soc.*, 1978, **100**, 3398.
4. (a) J. Kim and D. C. Rees, *Science*, 1992, **257**, 1677; (b) J. Kim and D. C. Rees, *Nature*, 1992, **360**, 553.

5. O. Einsle, F. A. Tezcan, S. L. A. Andrade, B. Schmid, M. Yoshida, J. B. Howard and D. C. Rees, *Science*, 2002, **297**, 1696.
6. (a) T. Spatzal, M. Aksoyoglu, L. Zhang, S. L. A. Andrade, E. Schleicher, S. Weber, D. C. Rees and O. Einsle, *Science*, 2011, **334**, 940; (b) K. M. Lancaster, M. Roemelt, P. Ettenhuber, Y. Hu, M. W. Ribbe, F. Neese, U. Bergmann and S. DeBeer, *Science*, 2011, **334**, 974.
7. Y. Nicolet, J. B. Lemon, J. C. Fontecilla-Camps and J. W. Peters, *Trends Biochem. Sci.*, 2000, **25**, 138.
8. B. R. Crane, L. M. Stegel and E. D. Getzoff, *Science*, 1995, **270**, 59.
9. (a) H. Dobbek, V. Svetlitchnyi, L. Gremer, R. Huber and O. Meyer, *Science*, 2001, **293**, 1281; (b) C. Darnault, A. Volbeda, E. J. Kim, P. Legrand, X. Vernède, P. A. Lindahl and J. C. Fontecilla-Camps, *Nature Struct. Biol.*, 2003, **10**, 271; (c) V. Svetlitchnyi, H. Dobbek, W. Meyer-Klaucke, T. Meins, B. Thiele, P. Römer, R. Huber and O. Meyer, *Proc. Natl. Acad. Sci*, 2004, **101**, 446; (d) A. Volbeda and J. C. Fontecilla-Camps, *Dalton Trans.*, 2005, 3443.
10. (a) H. Beinert, R. H. Holm and E. Münck, *Science*, 1997, **277**, 653; (b) P. V. Rao and R. H. Holm, *Chem. Rev.*, 2004, **104**, 527.
11. (a) C. Tard, X. Liu, S. K. Ibrahim, M. Bruschi, L. De Gioia, S. C. Davies, X. Yang, L.-S. Wang, G. Sawers and C. J. Pickett, *Nature*, 2005, **433**, 610; (b) L. Cai and R. H. Holm, *J. Am. Chem. Soc.*, 1994, **116**, 7177; (c) R. Panda, Y. Zhang, C. C. McLauchlan, P. V. Rao, F. A. T. de Oliveira, E. Münck and R. H. Holm, *J. Am. Chem. Soc.*, 2004, **126**, 6448; (d) P. V. Rao, S. Bhaduri, J. Jiang, D. Hong and R. H. Holm, *J. Am. Chem. Soc.*, 2005, **127**, 1933.
12. (a) G. Christou, C. D. Garner and F. E. Mabbs, *Inorg. Chim. Acta*, 1978, **28**, L189; (b) G. Christou, C. D. Garner, F. E. Mabbs and T. J. King, *J. Chem. Soc., Chem. Commun.*, 1978, 740.
13. (a) T. E. Wolff, J. M. Berg, C. Warrick, K. O. Hodgson, R. H. Holm and K. B. Frankel, *J. Am. Chem. Soc.*, 1978, **100**, 4630; (b) T. E. Wolff, J. M. Berg, K. O. Hodgson, R. B. Frankel and R. H. Holm, *J. Am. Chem. Soc.*, 1979, **101**, 4140.
14. (a) T. Yamamura, G. Christou and R. H. Holm, *Inorg. Chem.*, 1983, **22**, 939; (b) Y. Mizobe, P. K. Mascharak, R. E. Palermo and R. H. Holm, *Inorg. Chim. Acta*, 1983, **80**, L65.
15. K. Tanaka, M. Tanaka and T. Tanaka, *Chem. Lett.*, 1981, 895.
16. Y. Hozumi, Y. Imasaka, K. Tanaka and T. Tanaka, *Chem. Lett.*, 1983, 897.
17. K. Tanaka, Y. Hozumi and T. Tanaka, *Chem. Lett.*, 1982, 1203.
18. (a) P. K. Mascharak, W. H. Armstrong, Y. Mizobe and R. H. Holm, *J. Am. Chem. Soc.*, 1983, **105**, 475; (b) R. E. Palermo, R. Singh, J. K. Bashkin and R. H. Holm, *J. Am. Chem. Soc.*, 1984, **106**, 2600–2612.
19. (a) P. R. Challen, S.-M. Koo, C. G. Kim, W. R. Dunham and D. Coucouvanis, *J. Am. Chem. Soc.*, 1990, **112**, 8606; (b) P. E. Mosier, C. G. Kim and D. Coucouvanis, *Inorg. Chem.*, 1993, **32**, 2620.
20. L. J. Laughlin and D. Coucouvanis, *J. Am. Chem. Soc.*, 1995, **117**, 3118.

21. S. M. Malinak, A. M. Simeonov, P. E. Mosier, C. E. McKenna and D. Coucouvanis, *J. Am. Chem. Soc.*, 1997, **119**, 1662.
22. D. Coucouvanis, K. D. Demadis, S. M. Malinak, P. E. Mosier, M. A. Tyson and L. J. Laughlin, *J. Mol. Catal. A: Chem*, 1996, **107**, 123.
23. K. D. Demadis, S. M. Malinak and D. Coucouvanis, *Inorg. Chem.*, 1996, **35**, 4038.
24. D. V. Fomitchev, C. C. McLauchlan and R. H. Holm, *Inorg. Chem.*, 2002, **41**, 958.
25. T. Komuro, H. Kawaguchi, J. Lang, T. Nagasawa and K. Tatsumi, *J. Organomet. Chem.*, 2007, **692**, 1.
26. (a) M. G. Kanatzidis and D. Coucouvanis, *J. Am. Chem. Soc.*, 1986, **108**, 337; (b) S. A. Al-Ahmad, A. Salifoglou, M. G. Kanatzidis, W. R. Dunham and D. Coucouvanis, *Inorg. Chem.*, 1990, **29**, 927.
27. K. S. Bose, S. A. Chmielewski, P. A. Eldredge, E. Sinn and B. A. Averill, *J. Am. Chem. Soc.*, 1989, **111**, 8953.
28. D. Coucouvanis, J. Han and N. Moon, *J. Am. Chem. Soc.*, 2002, **124**, 216.
29. E. Nordlander, S. C. Lee, W. Cen, Z. Y. Wu, C. R. Natoli, A. Di Cicco, A. Filipponi, B. Hedman, K. O. Hodgson and R. H. Holm, *J. Am. Chem. Soc.*, 1993, **115**, 5549.
30. X.-D. Chen, J. S. Duncan, A. K. Verma and S. C. Lee, *J. Am. Chem. Soc.*, 2010, **132**, 15884.
31. K. D. Demadis, C. F. Campana and D. Coucouvanis, *J. Am. Chem. Soc.*, 1995, **117**, 7832.
32. F. Osterloh, Y. Sanakis, R. J. Staples, E. Münck and R. H. Holm, *Angew. Chem. Int. Ed.*, 1999, **38**, 2066.
33. (a) Y. Zhang, J.-L. Zuo, H.-C. Zhou and R. H. Holm, *J. Am. Chem. Soc.*, 2002, **124**, 14292; (b) Y. Zhang and R. H. Holm, *J. Am. Chem. Soc.*, 2003, **125**, 3910.
34. J. W. Peters, M. H. B. Stowell, S. M. Soltis, M. G. Finnegan, M. K. Johnson and D. C. Rees, *Biochemistry*, 1997, **36**, 1181.
35. (a) J.-L. Zuo, H.-C. Zhou and R. H. Holm, *Inorg. Chem.*, 2003, **42**, 4624; (b) D. Hong, Y. Zhang and R. H. Holm, *Inorg. Chim. Acta*, 2005, **358**, 2303; (c) M. Koutmos and D. Coucouvanis, *Inorg. Chem.*, 2006, **45**, 1421.
36. (a) Y. Ohki, Y. Sunada, M. Honda, M. Katada and K. Tatsumi, *J. Am. Chem. Soc.*, 2003, **125**, 4052; (b) Y. Ohki, M. Imada, A. Murata, Y. Sunada, S. Ohta, M. Honda, T. Sasamori, N. Tokitoh, M. Katada and K. Tatsumi, *J. Am. Chem. Soc.*, 2009, **131**, 13168.
37. (a) Y. Ohki, Y. Ikagawa and K. Tatsumi, *J. Am. Chem. Soc.*, 2007, **129**, 10457; (b) T. Hashimoto, Y. Ohki and K. Tatsumi, *Inorg. Chem.*, 2010, **49**, 6102.
38. (a) S. C. Lee and R. H. Holm, *Chem. Rev.*, 2004, **104**, 1135; (b) D. J. Evans in *Catalysts for Nitrogen Fixation*, B. E. Smith, R. L. Richards and W. E. Newton (eds), Kluwer Academic Publishers, Dordrecht, 2004, 201; (c) S. Groysman and R. H. Holm, *Biochemistry*, 2009, **48**, 2310.

39. (a) R. A. Henderson, *Chem. Rev.*, 2005, **105**, 2365; (b) F. Barriere, *Coord. Chem. Rev.*, 2003, **236**, 71.
40. (a) M. D. Curtis, in *Transition Metal Sulfur Chemistry: Biological and Industrial Significance*, ed. E. I. Stiefel and K. Matsumoto, American Chemical Society, Washington DC, 1996, 154; (b) M. Brorson, J. D. King, K. Kiriakidou, F. Prestopino and E. Nordlander, in *Metal Clusters in Chemisry*, P. Braunstein, L. A. Oro and P. R. Raithby ed., Wiley-VCH, Weinheim, 1999, **vol. 2** , p. 741; (c) R. A. Sánchez-Delgado, *Organometallic Modeling of the Hydrodesulfurization and Hydrodenitrogenation Reactions*, Kluwer Academic Publishers, Dordrecht, 2002.
41. (a) X. Wu, P. Chen, S. Du, N. Zhu and J. Lu, *J. Clust. Sci.*, 1994, **5**, 265; (b) M. Hidai, S. Kuwata and Y. Mizobe, *Acc. Chem. Res.*, 2000, **33**, 46; (c) R. Llusar and S. Uriel, *Eur. J. Inorg. Chem.*, 2003, 1271; (d) M. Hidai, in *Perspectives in Organometallic Chemistry*, ed. C. G. Screttas and B. R. Steele, Royal Society of Chemistry, Cambridge, 2003, 62.
42. (a) Y. Hu, A. W. Fay, C. C. Lee, J. Yoshizawa and M. W. Ribbe, *Biochemistry*, 2008, **47**, 3973; (b) L. M. Rubio and P. W. Ludden, *Annu. Rev. Microbiol.*, 2008, **62**, 93; (c) R. D. Bethel, M. L. Singleton and M. Y. Darensbourg, *Angew. Chem. Int. Ed.*, 2010, **49**, 8567; (d) E. M. Shepard, E. S. Boyd, J. B. Broderick and J. W. Peters, *Curr. Opin. Chem. Biol.*, 2011, **15**, 319.
43. (a) B. A. Cowans, R. C. Haltiwanger and M. Rakowski DuBois, *Organometallics*, 1987, **6**, 995; (b) M. D. Curtis, P. D. Williams and W. M. Butler, *Inorg. Chem.*, 1988, **27**, 2853.
44. (a) T. B. Rauchfuss, T. D. Weatherill, S. R. Wilson and J. P. Zebrowski, *J. Am. Chem. Soc.*, 1983, **105**, 6508; (b) C. M. Bolinger, T. D. Weatherill, T. B. Rauchfuss, A. L. Rheingold, C. S. Day and S. R. Wilson, *Inorg. Chem.*, 1986, **25**, 634.
45. (a) H. Brunner, H. Kauermann and J. Wachter, *J. Organomet. Chem.*, 1984, **265**, 189; (b) H. Brunner, N. Janietz, J. Wachter, T. Zahn and M. L. Ziegler, *Angew. Chem. Int. Ed. Engl.*, 1985, **24**, 133.
46. (a) Q. Feng, T. B. Rauchfuss and S. R. Wilson, *J. Am. Chem. Soc.*, 1995, **117**, 4702; (b) A. Venturelli, T. B. Rauchfuss and A. K. Verma, *Inorg. Chem.*, 1997, **36**, 1360; (c) M. Yuki, M. Okazaki, S. Inomata and H. Ogino, *Organometallics*, 1999, **18**, 3728.
47. (a) T. Amemiya, S. Kuwata and M. Hidai, *Chem. Commun.*, 1999, 711; (b) M. A. Casado, M. A. Ciriano, A. J. Edwards, F. J. Lahoz, L. A. Oro and J. J. Pérez-Torrente, *Organometallics*, 1999, **18**, 3025.
48. (a) T. R. Halbert, S. A. Cohen and E. I. Stiefel, *Organometallics*, 1985, **4**, 1689; (b) N. Zhu, X. Wu and J. Lu, *J. Chem. Soc., Chem. Commun.*, 1991, 235; (c) H. Diller, H. Keck, H. Wunderlich and W. Kuchen, *J. Organomet. Chem.*, 1995, **489**, 123; (d) T. Ikada, S. Kuwata, Y. Mizobe and M. Hidai, *Inorg. Chem.*, 1998, **37**, 5793; (e) T. Ikada, S. Kuwata, Y. Mizobe and M. Hidai, *Inorg. Chem.*, 1999, **38**, 64; (f) D. Rehder, W. Zhang, X. Wu, M. Ebel and C. Bader, *Inorg. Chim. Acta*, 2005, **358**, 1981.

49. (a) H. Seino, T. Kaneko, S. Fujii, M. Hidai and Y. Mizobe, *Inorg. Chem.*, 2003, **42**, 4585; (b) R. Shibata, H. Seino, S. Fujii and Y. Mizobe, *Inorg. Chem.*, 2010, **49**, 6889.
50. D. A. Dobbs and R. G. Bergman, *Inorg. Chem.*, 1994, **33**, 5329.
51. M. Herberhold, G.-X. Jin and W. Milius, *Chem. Ber*, 1995, **128**, 557.
52. (a) Z. Tang, Y. Nomura, Y. Ishii, Y. Mizobe and M. Hidai, *Inorg. Chim. Acta*, 1998, **267**, 73; (b) S. Kuwata, M. Andou, K. Hashizume, Y. Mizobe and M. Hidai, *Organometallics*, 1998, **17**, 3429; (c) H. Seino, Y. Mizobe and M. Hidai, *New J. Chem.*, 2000, **24**, 907.
53. (a) K. Hashizume, Y. Mizobe and M. Hidai, *Organometallics*, 1996, **15**, 3303; (b) Z. Tang, Y. Nomura, S. Kuwata, Y. Ishii, Y. Mizobe and M. Hidai, *Inorg. Chem.*, 1998, **37**, 4909; (c) S. Kuwata, K. Hashizume, Y. Mizobe and M. Hidai, *Organometallics*, 2002, **21**, 5401; (d) H. Kajitani, H. Seino and Y. Mizobe, *Organometallics*, 2007, **26**, 3499 and references therein.
54. D. Masui, Y. Ishii and M. Hidai, *Chem. Lett.*, 1998, **27**, 717.
55. H. Seino, T. Masumori, M. Hidai and Y. Mizobe, *Organometallics*, 2003, **22**, 3424.
56. (a) S.-W. A. Fong and T. S. A. Hor, *J. Chem. Soc., Dalton Trans.*, 1999, 639; (b) Z. Li, S.-W. A. Fong, J. S. L. Yeo, W. Henderson, K. F. Mok and T. S. A. Hor, in *Modern Coordination Chemistry: the Legacy of Joseph Chatt*, ed. G. J. Leigh and N. Winterton, Royal Society of Chemistry, Cambridge, 2002, 355.
57. T. Shibahara, *Coord. Chem. Rev.*, 1993, **123**, 73.
58. R. Hernandez-Molina, M. N. Sokolov and A. G. Sykes, *Acc. Chem. Res.*, 2001, **34**, 223.
59. T. Murata, Y. Mizobe, H. Gao, Y. Ishii, T. Wakabayashi, F. Nakano, T. Tanase, S. Yano, M. Hidai, I. Echizen, H. Nanikawa and S. Motomura, *J. Am. Chem. Soc.*, 1994, **116**, 3389.
60. T. Wakabayashi, Y. Ishii, K. Ishikawa and M. Hidai, *Angew. Chem., Int. Ed. Engl.*, 1996, **35**, 2123.
61. H. Seino and M. Hidai, *Chem. Sci*, 2011, **2**, 847.
62. B. Rink, M. Brorson and I. J. Scowen, *Organometallics*, 1998, **18**, 2309.
63. I. Takei, K. Suzuki, Y. Enta, K. Dohki, T. Suzuki, Y. Mizobe and M. Hidai, *Organometallics*, 2003, **22**, 1790.
64. (a) K. Herbst, M. Monari and M. Brorson, *Inorg. Chem.*, 2001, **40**, 2979; (b) K. Herbst, P. Zanello, M. Corsini, N. D'Amelio, L. Dahlenburg and M. Brorson, *Inorg. Chem.*, 2003, **42**, 974; (c) I. Takei, K. Kobayashi, K. Dohki and M. Hidai, *Inorg. Chem.*, 2007, **46**, 1045; (d) P. Chen, Y. Chen, Y. Zhou, Y. Peng, J. Qu and M. Hidai, *Dalton Trans*, 2010, **39**, 5658 and references therein.
65. I. Takei, K. Kobayashi, K. Dohki, S. Nagao, Y. Mizobe and M. Hidai, *Chem. Lett.*, 2007, **36**, 546.
66. K. Herbst, E. Söderhjelm, E. Nordlander, L. Dahlenburg and M. Brorson, *Inorg. Chim. Acta*, 2007, **360**, 2697.

67. I. Takei, Y. Wakebe, K. Suzuki, Y. Enta, T. Suzuki, Y. Mizobe and M. Hidai, *Organometallics*, 2003, **22**, 4639.
68. K. Herbst, M. Monari and M. Brorson, *Inorg. Chem.*, 2002, **41**, 1336.
69. I. Takei, K. Dohki, K. Kobayashi, T. Suzuki and M. Hidai, *Inorg. Chem.*, 2005, **44**, 3768.
70. R. H. Holm, *Pure Appl. Chem.*, 1998, **70**, 931.
71. F. Takagi, H. Seino, Y. Mizobe and M. Hidai, *Organometallics*, 2002, **21**, 694.
72. Y. Manaka, H. Mori, H. Seino and Y. Mizobe, unpublished result.
73. (a) H. Seino, H. Mori, A. Shinozaki and Y. Mizobe, *J. Organomet. Chem.*, 2010, **695**, 1878; (b) P. M. Kozlowski, Y. Shiota, S. Gomita, H. Seino, Y. Mizobe and K. Yoshizawa, *Bull. Chem. Soc. Jpn.*, 2007, **80**, 2323.
74. H. Mori, H. Seino, M. Hidai and Y. Mizobe, *Angew. Chem. Int. Ed.*, 2007, **46**, 5431.
75. A. Shinozaki, H. Seino, M. Hidai and Y. Mizobe, *Organometallics*, 2003, **22**, 4636.
76. H. Tanaka, H. Mori, H. Seino, M. Hidai, Y. Mizobe and K. Yoshizawa, *J. Am. Chem. Soc.*, 2008, **130**, 9037.

CHAPTER 10.2

Current Research on Mimics and Models of Selenium-Containing Antioxidants

BHASKAR J. BHUYAN,[a] DEVAPPA S. LAMANI,[a] GOVINDASAMY MUGESH*[a] AND THOMAS WIRTH[b]

[a] Department of Inorganic and Physical Chemistry, Indian Institute of Science, Bangalore 560 012, India; [b] School of Chemistry Cardiff University, Main Building, Park Place, Cardiff CF10 3AT, UK, Email: wirth@cf.ac.uk
*Email: mugesh@ipc.iisc.ernet.in

10.2.1 Introduction

The chemistry of selenium, the next element to sulfur in the chalcogen group, is much less well explored than the chemistry of sulfur. Selenium was discovered by the Swedish scientist Jöns Jakob Berzelius in 1818.[1] The first synthetic selenium compound, diethyl selenide, was synthesized by Löwig in 1836 and was obtained in pure form in 1869.[2,3] Initially, selenium chemistry focused mainly on the synthesis of simple diselenides (RSeSeR), selenols (RSeH), *etc.* However, the unpleasant odour of selenols and aliphatic selenides was a serious drawback in selenium chemistry. Furthermore, selenium was associated with diseases such as livestock disease[4] and intoxication in experimental animals,[5–9] and it was therefore considered to be a toxic element. The beneficial effect of selenium for living organisms was established in 1954 by Pinsent with the discovery that certain bacteria grew faster in selenium-fortified medium.[10] However, the exact role of selenium in the growth of bacteria was not clear. Almost after 20 years of this

Handbook of Chalcogen Chemistry: New Perspectives in Sulfur, Selenium and Tellurium, 2nd Edition, Volume 2
Edited by Francesco Antonio Devillanova and Wolf-Walther du Mont

Published by the Royal Society of Chemistry, www.rsc.org

discovery, in 1973, it was found that two bacterial enzymes, formate dehydrogenase and glycine reductasen contain selenium in their active sites.[11,12] The importance of selenium to mammals was discovered almost at the same time by Flohé and co-workers.[13] They found that the mammalian enzyme glutathione peroxidase (GPx) contains a selenocysteine residue in its active site. The major selenoenzymes discovered to date include formate dehydrogenases,[14] hydrogenases,[15–17] glycine reductase,[18] iodothyronine deiodinases (ID),[19–22] thioredoxin reductases (TrxR),[23–26] selenophosphate synthetase,[27] and selenoprotein P,[28–29] and glutathione peroxidase (GPx)[13,30–33] Although synthetic mimics/models are not available for all of these enzymes, the antioxidant mechanism of GPx has been investigated thoroughly with the help of synthetic selenium compounds. An interesting feature of GPx is its remarkable redox properties, which are due to the presence of selenocysteine, the 21st amino acid. Selenocysteine (Sec) is structurally similar to cysteine (Cys),[34] but the redox property of Sec is remarkably different from that of Cys: the pK_a value of Sec (5.2) is significantly lower than that of Cys (8.5). Therefore, Sec residues in proteins exists predominantly in their selenolate form at physiological pH.[35]

10.2.2 Cellular Antioxidant Defence Mechanism

It is known that reactive oxygen species (ROS) such as superoxide radical anion ($O_2^{\bullet -}$), hydroxyl radical ($^{\bullet}OH$) and hydrogen peroxide (H_2O_2), which are generated during the metabolism of oxygen, can cause damage to cellular components.[36] In addition to these species, hypochlorous acid (HOCl), which is generated from H_2O_2 by myeloperoxidase (MPO) in neutrophils, and peroxynitrite ($ONOO^-$), which is generated from the diffusion-controlled reaction of superoxide ($O_2^{\bullet -}$) and nitric oxide ($^{\bullet}NO$), can also be considered as strong biological oxidants.[37–39] Antioxidants are species that control and can decompose cellular oxidants. Therefore, a balance in the concentration of oxidants and antioxidants is important for the normal biological function of an organism. There are several natural antioxidant defence mechanisms, which regulate the balance between oxidants and antioxidants (Scheme 10.2.1).[40–42] Although the conversion of oxygen to the superoxide radical anions by oxidases is essential for the catabolic mechanism, an excess amount of this species is extremely harmful as it can cause lipid peroxidation or oxidative damage to the living cell. Furthermore, the reaction of superoxide with nitric oxide produces peroxynitrite, which inactivates several enzymes by tyrosine nitration. Superoxide dismutase (SOD) converts superoxide radical anion to hydrogen peroxide, which is then reduced to water by two enzymes, catalase and GPx.

10.2.3 Glutathione Peroxidase

For the last four decades, extensive research has been carried out on the mammalian antioxidant enzymes GPx.[43–49] These enzyme catalytically reduce cellular peroxides to water or alcohols by using glutathione or other related thiols

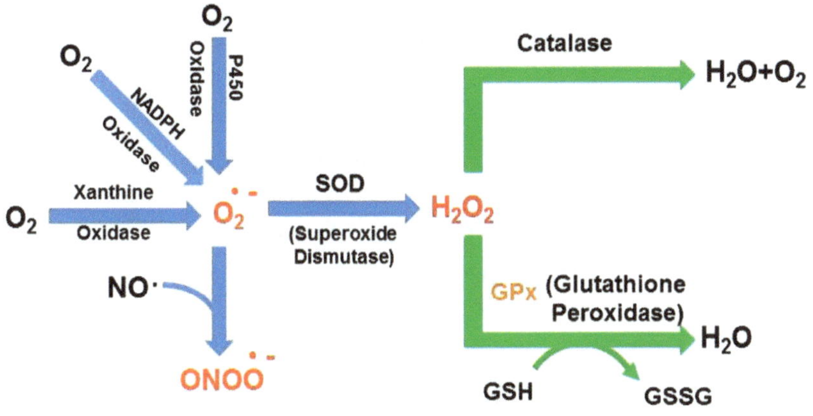

Scheme 10.2.1 Biological antioxidant defence cycle.

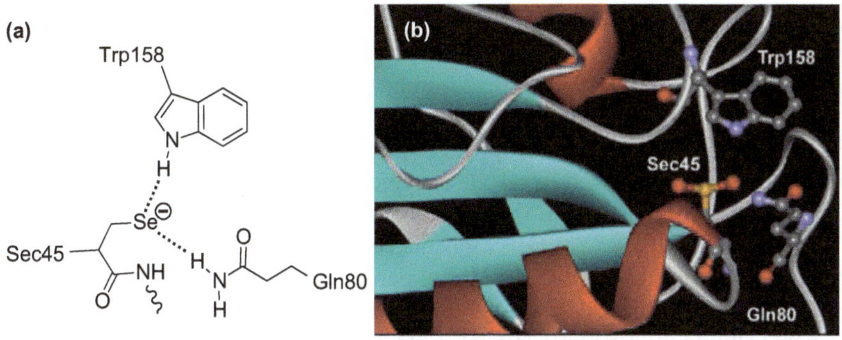

Figure 10.2.1 (a) Catalytic triad at the active site of GPx. (b) Active site of glutathione peroxidase (PDB Code 1GP1) determined by X-ray crystallography.

as cofactors. The GPx superfamily contains four enzymes: cytosolic GPx (cGPx), phospholipid hydroperoxide GPx (PHGPx), plasma GPx (pGPx), and gastrointestinal GPx (giGPx). All these enzymes have a conserved Sec residue at the active site, which is essential for their catalytic activity. The activity of these enzymes depends upon the nature of the hydroperoxide and the thiol cofactor. The cGPx utilizes GSH as reducing substrate exclusively for the reduction of H_2O_2 and organic hydroperoxides such as *tert*-butyl hydroperoxide (t-BuOOH) and cumene hydroperoxide (Cum-OOH). While PHGPx also uses GSH as cofactor, it has broad hydroperoxide substrate specificity. This enzyme exhibits good activity with all phospholipid hydroperoxides, fatty acid hydroperoxides, t-BuOOH, Cum-OOH, cholesterol hydroperoxides, and H_2O_2. In contrast, pGPx has a very narrow hydroperoxide substrate specificity. Although pGPx can reduce H_2O_2 and organic hydroperoxides, it is approximately one order of magnitude less active than the cGPx. The crystal structure of GPx indicates that the Sec residue (Sec45) forms a 'catalytic triad' with other two amino acids, glutamine (Gln80) and tryptophan (Trp158) (Figure 10.2.1a).[50] The crystal

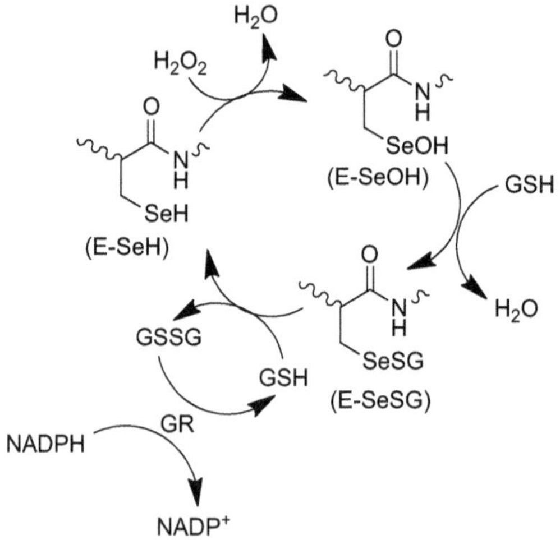

Scheme 10.2.2 Proposed mechanism for the GPx-catalysed reduction of H_2O_2.

structure of the seleninic acid form of human pGPx also indicates that Gln79 and Trp153 are located within hydrogen bonding distance of the selenium atom (Figure 10.2.1b). These residues appear to play functional roles in catalysis. These two residues are conserved in all the enzymes in the GPx superfamily and may account for the similarities in their catalytic mechanisms.

A typical catalytic cycle of GPx (Scheme 10.2.2) starts with the oxidation of the selenol (ESeH) moiety of Sec residue by peroxide to generate the selenenic acid (ESeOH),[50–52] which reacts with cellular thiol (glutathione, GSH) to generate a selenenyl sulfide intermediate (ESeSG). Another equivalent of GSH cleaves the –Se–S– bond in the selenenyl sulfide intermediate to regenerate the selenol with elimination of glutathione disulfide (GSSG). Concentration of GSH at the cellular level is maintained by glutathione reductase (GR), which reduces GSSG to GSH by using NADPH as cofactor. In the overall catalytic mechanism, two equivalents of NADPH are consumed to reduce one equivalent of peroxide. At very high hydroperoxide concentrations, the selenium centre in GPx may be overoxidized to produce seleninic acid ($ESeO_2H$) and selenonic acid ($ESeO_3H$). While the oxidation of selenenic acid to seleninic acid is reversible in the presence of GSH, further oxidation to selenonic acid may inactivate the enzyme.

10.2.4 Mimics and Models of Glutathione Peroxidase

10.2.4.1 Ebselen and its Analogues as GPx Mimics/Models

Synthetic selenium compounds with significant GPx activity have potential therapeutic applications. Furthermore, such compounds may help in understanding the chemistry at the active site of GPx. The first synthetic compound that was shown to mimic the GPx activity was ebselen [2-phenyl-1,2-benzisoselenazole-3-(2*H*)-one] (**1**).[53,54] Ebselen reduces hydrogen peroxide

(H_2O_2) and lipid peroxides, effectively scavenges the highly reactive peroxynitrite, and inhibits a variety of free radical generating enzymes such as NADPH oxidase,[55] lipoxygenase, and cyclooxygenase.[56,57] The initial success of ebselen was mainly due to its very low toxicity. The high stability of the selenazole moiety does not allow the elimination of selenium during the biotransformations. Therefore, this compound does not interfere with the selenium metabolism of the organism. Ebselen exhibited beneficial effects in clinical trials for the treatment of patients suffering from active ischaemic stroke.[58] Several animal model studies have demonstrated that ebselen reduces oxidative stress in ischaemia-reperfusion in heart[59] and that it exhibits promising neuroprotective effect in brain.[60,61] However, recent evidence suggests that ebselen can be toxic to cells: it has been shown to inhibit some cell growth and induce apoptosis.[62] While the mechanism underlying the toxicity of ebselen is not known, the cellular glutathione (GSH) level appears to be depleted by ebselen.[62] As reduced GSH is important for cell survival, GSH depletion increases the susceptibility of cells to oxidant injury.

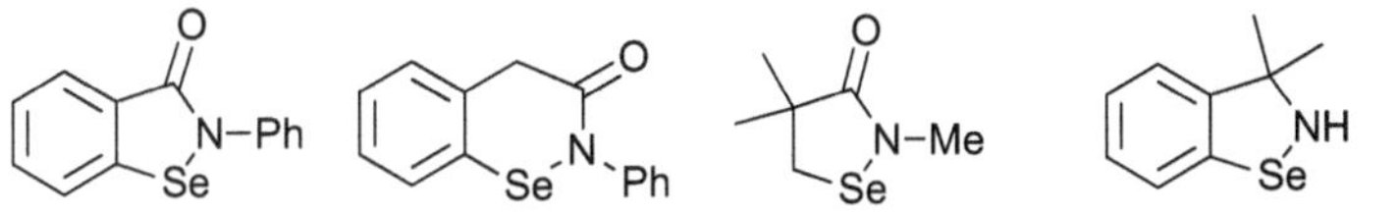

1, Ebselen **2** **3** **4**

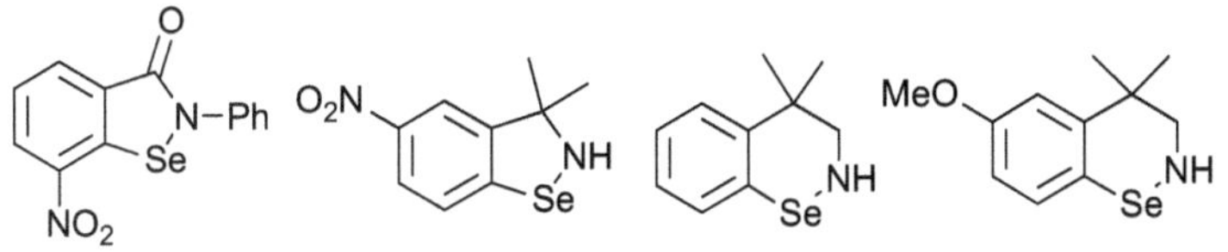

5 **6** **7** **8**

Since the discovery that ebselen exhibits significant antioxidant activity by mimicking the active site of GPx, much attention has been devoted to the design and synthesis of novel analogues of ebselen.[63,64] The ebselen homologue **2**, in which a supplementary tetrahedral carbon is incorporated into the heterocycle, retains the Se–N bond essential for GPx activity.[65] The selenazole model system **3** has been used extensively to understand the antioxidant redox chemistry of selenocysteine at the active site of GPx.[66] Compound **4**, having a covalent Se–N bond but without a C=O group, has been synthesized and evaluated for its GPx activity.[67] This compound has also been shown to protect endothelial cells from the toxic effect of hydroperoxides. The GPx activity of ebselen and its analogues can be altered by introducing substituents on the phenyl ring. The introduction of a nitro group in the *ortho*-position to selenium (compound **5**) has been shown to increase the GPx activity of ebselen.[68] The enhancement in the activity has been attributed to the electronic effect of the nitro group. In contrast, the introduction of a *para*-nitro substituent to compound **4** significantly decreases the activity.[67] The GPx activity of compound **6** was found to be almost 2.5 times lower than that of **4**. The ring size also appears to have an effect on the GPx activity. Compound **7**, with a six-membered ring, is much more active than the parent compound **4**. The GPx activity of **8**, containing a *para*-methoxy group, is almost identical to that of **7**, indicating that the introduction of an electron-donating substituent does not influence the activity.[67]

A structure–activity correlation study using a number of ebselen analogues revealed that the presence of a phenyl substituent on the nitrogen atom is important for the GPx activity of ebselen and related compounds.[69] Compounds **9–13** exhibited different catalytic behaviour with different thiols. When an aromatic thiol such as PhSH was used as the thiol co-substrate, very poor activity was observed. Compound **14**, having a six-membered heterocycle, also exhibited activity similar to that of ebselen. The GPx activity of **14** in the presence of PhSH was only 1.2 times higher than that of ebselen, indicating that the modification in the ring size does not improve the catalytic activity of ebselen analogues.[69]

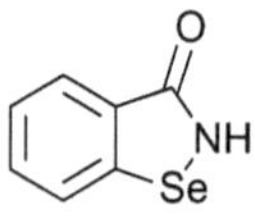

9

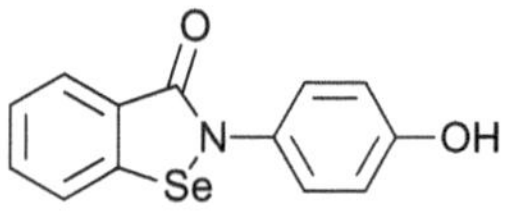

10

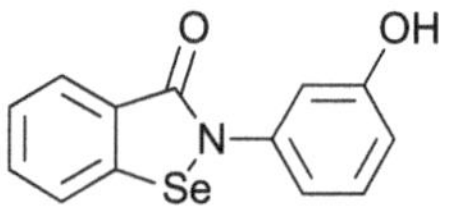

11

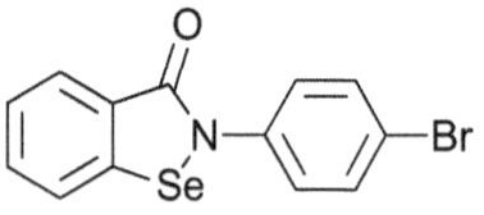

12

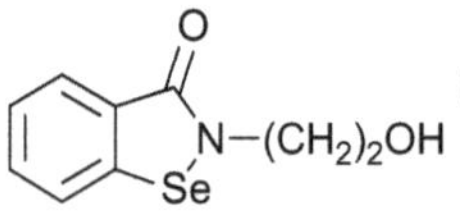

13

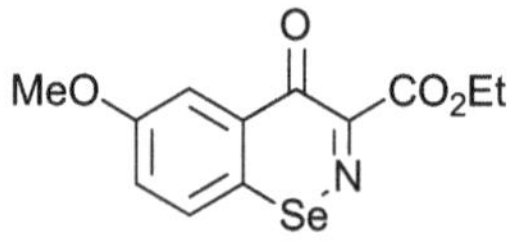

14

The relatively poor GPx activity of these compounds in the presence of aromatic thiols has been ascribed to the presence of strong Se···O non-covalent interactions in the selenenyl sulfide intermediates.[69] For examples, the reactions of compounds **10–13** with PhSH afford the corresponding selenenyl sulfides **15–18**. As shown in Figure 10.2.2, the carbonyl oxygen interacts non-covalently with the selenium atom. These interactions were confirmed by experimental and theoretical investigations. While the non-covalent interactions increase the nucleophilic attack of thiol at the selenium centres, the strength of Se···O interactions depends upon the substituents attached to the phenyl ring of ebselen. The presence of strong Se···O interactions in the selenenyl sulfide intermediates prevents the formation of catalytically active selenol. On the other hand, such interactions favour the exchange of the thiolate moiety in the selenenyl sulfide intermediates. These reactions are generally referred to as thiol exchange reactions.[70,71]

In contrast to the poor catalytic activity of ebselen and related compounds in the presence of aromatic thiols, these compounds exhibit good GPx activity when GSH is used as the thiol co-substrate.[69] The GPx activity of compounds **10–13** in the presence of GSH was found to be much higher that observed in the presence of PhSH. Although the nature of the thiol has an effect on the GPx activity, ebselen

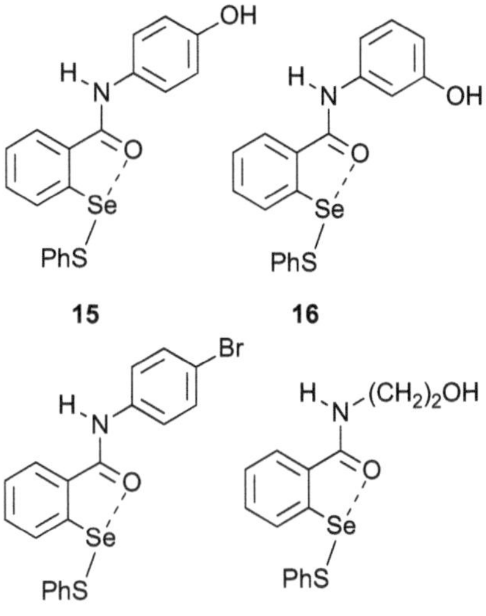

Figure 10.2.2 Structure of selenenyl sulfides showing non-covalent Se···O interactions.

and compounds **10–13** exhibited similar activities in the presence of different peroxides such as H_2O_2, t-BuOOH, and Cum-OOH, indicating that the peroxide substrate does not have a significant effect on GPx activity. It is known that the four different types of GPx—cytosolic GPx (cGPx), phospholipid hydroperoxids GPx (PHGPx), plasma GPx (pGPx) and gastrointestinal GPx(giGPx)—are highly specific to thiol co-substrates.[72–76] On the other hand, hydroperoxide substrate specificity is very broad for these enzymes as they accept H_2O_2 and a number of organic hydroperoxides such as t-BuOOH and Cum-OOH. Recent studies on the GPx activity of synthetic compounds suggest that the identification of thiol cofactor systems with enhanced reducing ability is important for the development of selenium-based antioxidants.[70,71,77]

Back and co-workers reported a different cyclic selenenamide as an efficient GPx mimic.[78] They showed that the camphor-derived selenenamide **19** promotes the oxidation of phenylmethane thiol (BnSH) to the corresponding disulfide with t-BuOOH. An interesting observation in this study was that the selenenamide **19** is merely a pro-catalyst, which undergoes a rapid reaction with BnSH in the first step to produce the selenenyl sulfide **20**. In this process, compound **20** acts as the true catalyst. The reaction of **20** with BnSH generates the selenol **21** with elimination of BnSSBn (Scheme 10.2.3).[78] The rapid oxidation of **21** with t-BuOOH produces the corresponding selenenic acid **22**. In the last step, the regeneration of **20** is achieved by the reaction of compound **22** with BnSH. While the mechanism for the reduction of hydroperoxide by **19** is different from that of ebselen, the catalytic cycle shown in Scheme 10.2.3 resembles the one proposed for the GPx enzyme. In the entire catalytic cycle, the conversion of the selenenyl sulfide intermediate to the selenol appears to be the rate-determining step.[78]

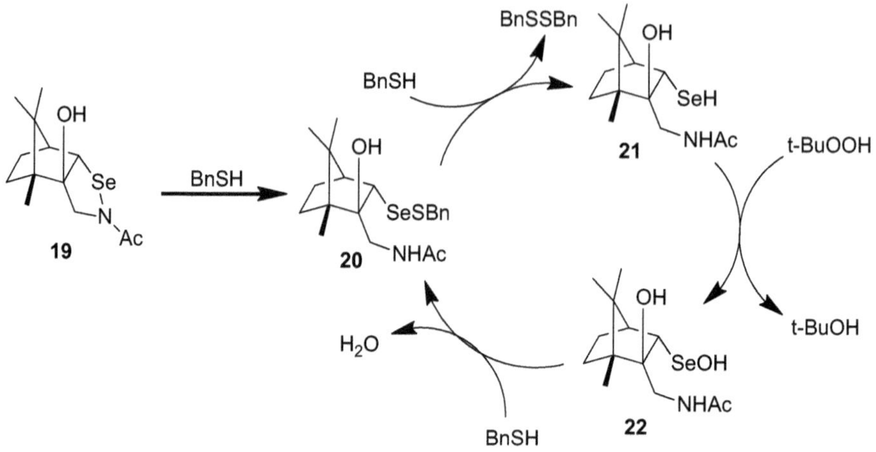

Scheme 10.2.3 Proposed mechanism for the GPx activity of the camphor–derived selenenamide **19**.[78]

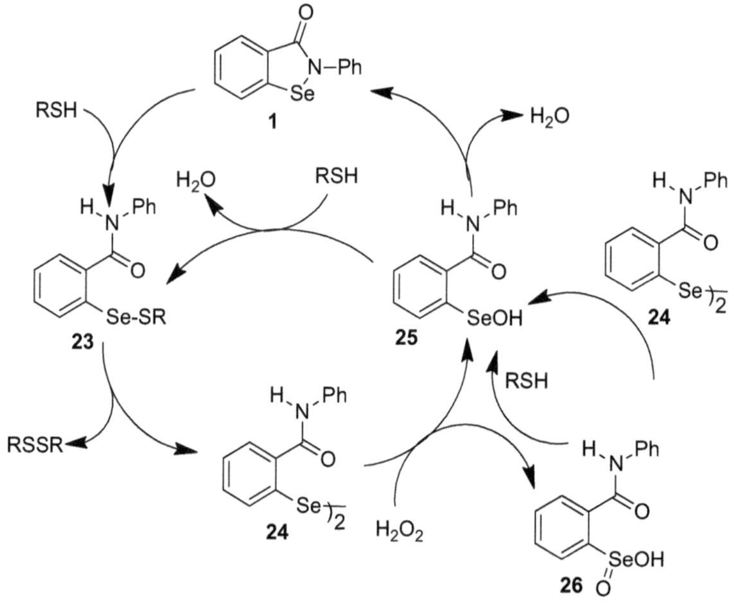

Scheme 10.2.4 Revised catalytic cycle of ebselen and related compounds.[79]

According to the initially proposed mechanism, ebselen reacts with thiol to produce the selenenyl sulfide, which upon reaction with a second equivalent of thiol produces the corresponding selenol. Mugesh and co-workers[79] revisited the mechanism for the reduction of hydroperoxide by ebselen. On the basis of new experimental findings, a revised mechanism has been proposed. According to this mechanism, ebselen (**1**) reacts with a thiol (RSH) to produce the corresponding selenenyl sulfide **23**. This compound is found to be unstable in the assay system, and therefore, undergoes a disproportionation reaction to generate the diselenide **24**. Subsequent reaction of **24** with peroxide produces the selenenic acid **25** and seleninic acid **26**. In the presence of an excess amount of thiol in the assay mixture, compounds **25** and **26** react with the thiol to produce the selenenyl sulfide **23**. When RSH is depleted in the reaction mixture, the seleninic acid **26** reacts further with the diselenide **24** to generate the selenenic acid **25**. Interestingly, the selenenic acid **25** having a free N-H moiety undergoes cyclization to regenerate ebselen. It is observed that the disproportionation of selenenyl sulfide **23** to diselenide **24** is the rate-limiting step in this catalytic cycle (Scheme 10.2.4).[79]

10.2.4.2 Diaryl Diselenides as GPx Mimics/Models

Spector and co-workers[80] reported that diaryl diselenides would function as effectively as the cyclic compounds in the production of the catalytically active species. They also proposed that the introduction of a strongly basic group in the proximity of selenium is important as this substituent may catalyse the reaction of thiols with diselenides and selenenyl sulfide. The basic function may provide a source of nucleophilic thiolate anion for the catalytic activity. On the basis of these assumptions, Spector and co-workers synthesized the diaryl diselenides **27** and **28** having *tert*-amino substituents and studied their GPx activity.[80] When GSH was used as co-substrate, compounds **27** and **28** were 10-fold more active than ebselen, indicating that it is possible to produce relatively simple compounds with significant GPx activity. This study also revealed that the amine may deprotonate the thiol group to provide a high local concentration of nucleophilic thiolate anion. Alternatively, the conjugate acid of the amine, the ammonium ion, may serve as a proton source, and therefore facilitate the reduction of H_2O_2 to water.

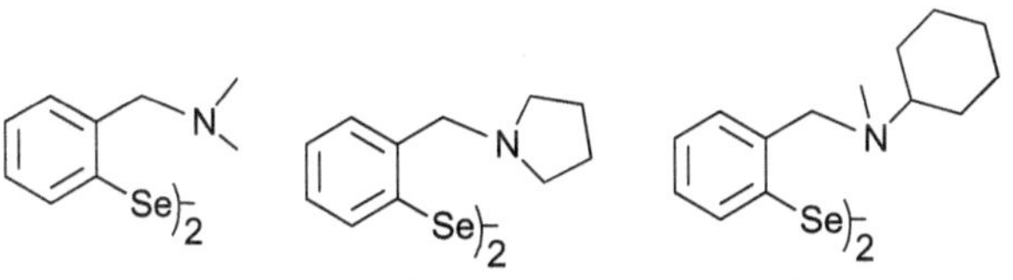

27 **28** **29**

Iwaoka and Tomoda[81] used ^{77}Se NMR spectroscopy to provide direct evidence for the involvement of basic amino groups in the catalysis by using the diaryl diselenide **29**. The authors identified all the crucial intermediates involved in the GPx activity of **29** and proposed that the non-covalent Se···N interactions play an important role in the catalytic activity. On the basis of experimental and theoretical investigations, Iwaoka and Tomoda proposed a mechanism for the reduction of H_2O_2 with PhSH in the presence of **29** as shown in Scheme 10.2.5.[81] The initial step is the reaction of **29** with PhSH to produce the selenenyl sulfide **30** and selenol **31**. The abstraction of proton by the amino

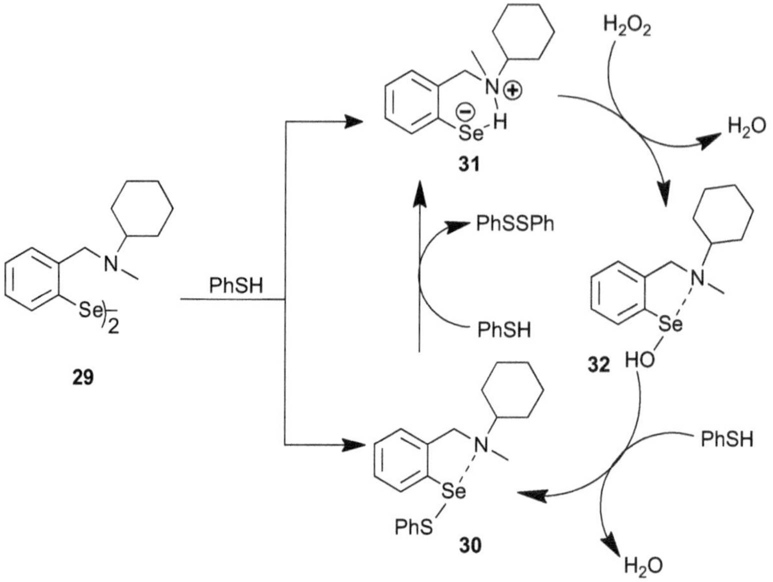

Scheme 10.2.5 The role of a basic amino substituent in the catalytic cycle of compound **29**.

group converts the selenol to a more reactive selenolate. Similar to the selenenamide **21**, the diselenide **29** acts as a pro-catalyst. While the reaction of the selenenyl sulfide **30** with PhSH produces the selenol **31**, the oxidation of the selenol with H_2O_2 produces the selenenic acid **32**. The rapid bimolecular displacement at the selenium centre in compound **32** with PhSH regenerates the selenenyl sulfide (Scheme 10.2.5).[81]

On the basis of a detailed mechanism investigation, Iwaoka and Tomoda proposed several possible roles for the amino group present at the active site of GPx:[81] (1) The proximal nitrogen base can activate the selenol into a more reactive and more nucleophilic selenolate anion. (2) The non-covalent Se···N interactions in the selenenic acid intermediate prevents the further oxidation of this species to other oxidized selenium species. (3) The non-covalent Se···N interactions may prevent the nucleophilic attack of thiol at the selenium in the selenenyl sulfide intermediate, which allows an effective regeneration of the selenol. The proposed roles for the amino group were in agreement with the report of Reich *et al.*[66] that the selenenyl sulfide and diselenide derived from the selenenamide **3** reacted with the thiol only in the presence of a strong base. These results helped in the design and synthesis of organoselenium compounds with high GPx activity.

Wendel and co-workers[50] reported that the selenocysteine residue in GPx is located at the N-terminal end of helix α_1. The amino acid residues near the selenium atom may interact with the selenium, stabilizing the selenolate and enhancing the nucleophilicity. Hilvert and co-workers[82] reported that selenosubtilisin, a selenoprotein prepared semi-synthetically by chemical conversion of a serine residue, mimics GPx. Although this study was aimed at demonstrating how a single atom change from oxygen to selenium at the active site of subtilisin can alter the enzyme activity, it highlighted the importance of a basic histidine residue for high GPx activity. Because the enzyme-bound selenol is deprotonated by His64 to from an imidazolium-selenolate ion pair, the selenocysteine residue is expected to be highly reactive. Another amino acid residue, Asn155, may also play a key role in stabilizing the selenolate form (Figure 10.2.3).[83] In the selenium analogues of GAPDH, the selenocysteine residue, Sec149, has been shown to interact with a histidine residue (His176) to form an imidazolium-selenolate ion pair.[84] Computer-assisted molecular modelling by Flohé and co-workers also supported the role of non-covalent interactions.[85] All kinetically defined intermediates and enzyme–substrate complexes were subjected to force field and molecular dynamics calculations. The results from this study indicate that the selenium atom in the active site interacts weakly with the imino group of Trp165 and the amido group of Gln87 (Se···N) as proven by the distances of 3.31 and 3.37 Å, respectively. Similar interactions were previously observed in the crystal structure of GPx from human plasma.[86]

Singh and co-workers[87,88] reported that the redox-active diferrocenyl diselenides **33** and **34** containing *tert*-amino substituents exhibit excellent GPx activity in the presence of PhSH. In contrast to the phenyl-based diselenide **29** that exhibits strong Se···N interactions, the diferrocenyl derivatives **33** and **34** do not have any non-covalent Se···N interactions. The Se···N distances observed for these compounds (3.697 and 4.296 Å for **33** and 3.98 and 4.12 Å for **34**) are greater than the sum of their van der Waals radii (3.54 Å). As the GPx activity of **33** and **34** was remarkably higher than that of **27**, it has been suggested that strong Se···N interactions may decrease the GPx activity. Mechanistic investigations reveal that the amine nitrogen interacts with the selenium only in the selenenic acid intermediate.[88] This interaction is expected

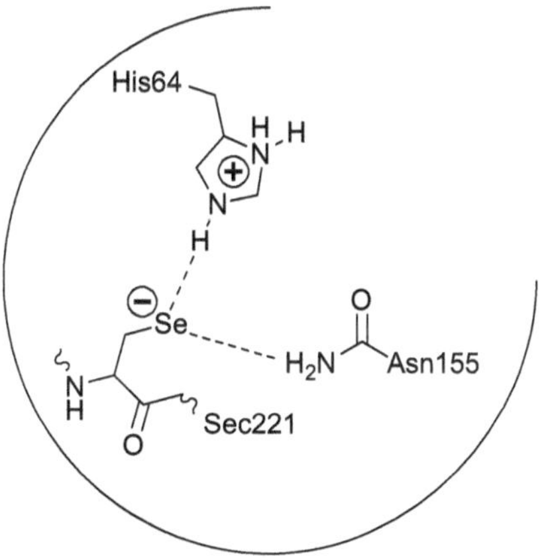

Figure 10.2.3 Non-covalent interactions at the active site of selenosubtilisin, involving Sec, His, and Asn residues.[83]

to favour the nucleophilic attack of thiol at selenium. On the basis of kinetic and NMR experiments, a catalytic cycle for compounds **33** and **34** has been proposed as shown in Scheme 10.2.6. According to this mechanism, the reaction of diselenide **33** with PhSH generates the selenol **35** and selenenyl sulfide **36**. Subsequently reaction of **35** with H_2O_2 produces the selenenic acid **37** which upon reaction with PhSH produces the selenenyl sulfide **36**. The reaction of **36** with PhSH regenerates the selenol with elimination of PhSSPh. While the catalytic cycle shown in Scheme 10.2.6 is identical to that of **29** (Scheme 10.2.5), the major difference is that the amine nitrogen does not interact with selenium in the selenenyl sulfide intermediate when the phenyl group is replaced by a ferrocenyl moiety.

Considering the GPx cycle, the reaction of enzyme-bound selenenic acid (E-SeOH) with GSH is expected to be facile, due to the high electrophilic reactivity of selenium in this intermediate. In contrast, the reaction of the enzyme-bound selenenyl sulfide (E-SeSG) with GSH to produce the selenol (ESeH) is not a favoured process as the selenium centre in the selenenyl sulfide intermediate is also electrophilic in nature. Therefore, a thiol exchange at the selenium centre is expected instead of a nucleophilic attack of thiol (or thiolate) at the sulfur centre. As the conversion of selenenyl sulfides to the corresponding selenols needs to overcome large energy barriers, the nucleophilic attack of thiol at the sulfur centres of the selenenyl sulfide is less favourable than at selenium.[89] It has been observed that the non-covalent Se···N or Se···O interactions in the selenenyl sulfide intermediates generally favour the undesired thiol exchange reaction, leading to a decrease in the GPx activity.[69–71] The thiol exchange reactions, which hamper the selenol formation, may be responsible for the poor GPx activity of several selenium compounds in the presence of aromatic thiols. For examples, the

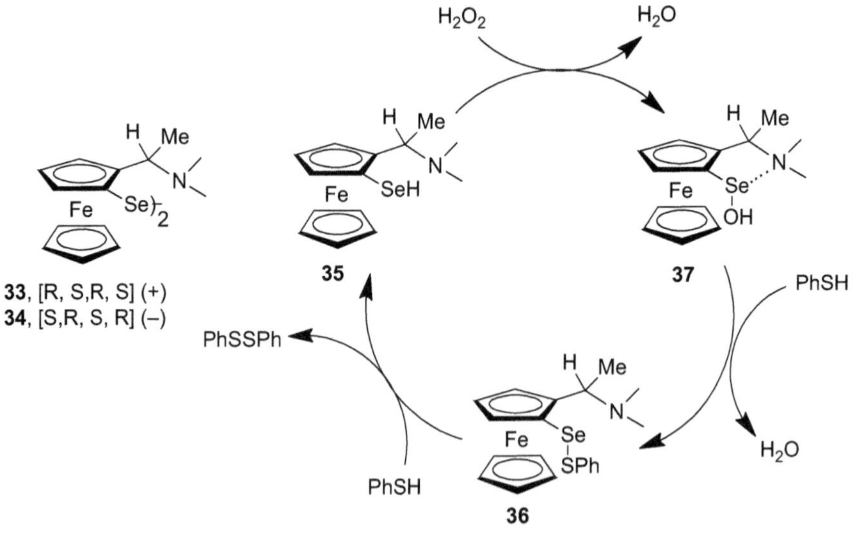

Scheme 10.2.6 Proposed mechanism for the GPx activity of compounds **33** and **34**.[88]

poor GPx activity of ebselen in the presence of aryl and benzyl thiols (such as PhSH and BnSH) can be ascribed to the thiol exchange reactions that take place at the selenenyl sulfide intermediates (Scheme 10.2.7).

The selenenyl sulfides derived from amine-substituted diselenides also undergo thiol exchange reactions. Therefore, the prevention of strong non-covalent Se···N and Se···O interactions in ebselen analogues or amine-based diselenides may enhance the GPx activity. However, it is very difficult to develop diaryl diselenides that have amino groups in close proximity to selenium, but do not exhibit any strong non-covalent interactions in the selenenyl sulfide intermediates. Bhabak and Mugesh[90] reported that the replacement of aryl proton in compounds **38–40** by a methoxy substituent effectively prevents the Se···N interaction in the selenenyl sulfide intermediates and enhances the GPx activity. They have shown that the methoxy-substituted compounds **41–43** are much better catalysts for the reduction of H_2O_2 than the parent compounds **38–40**.

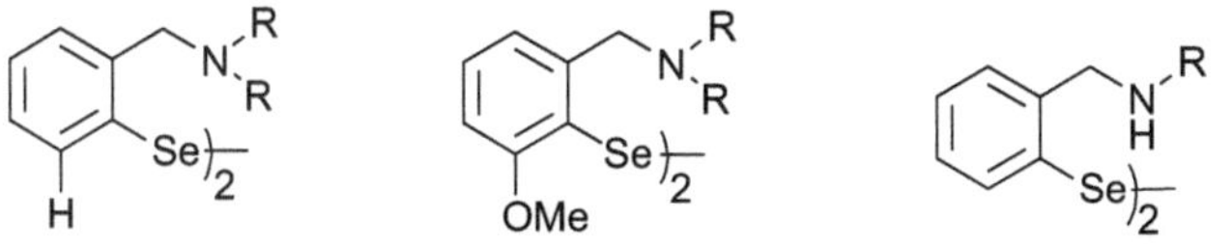

38, R = Me; **39**, R = Et
40, R = *n*Pr

41, R = Me; **42**, R = Et
43, R = *n*Pr

44, R = Me; **45**, R = Et
46, R = *n*Pr; **47**, R = *i*Pr

The simple replacement of hydrogen atoms in compounds **38–40** by methoxy substituents alters the reaction at the selenium centre. In compounds **41–43**, the methoxy group enhances the GPx activity by altering the electronic and steric environments around the sulfur and selenium atoms in some of the catalytic intermediates.[90] These include the protection of the selenenic acid from over-oxidation to seleninic and selenonic acids by peroxides and the prevention of the thiol exchange reaction at the selenium atoms in the selenenyl sulfide

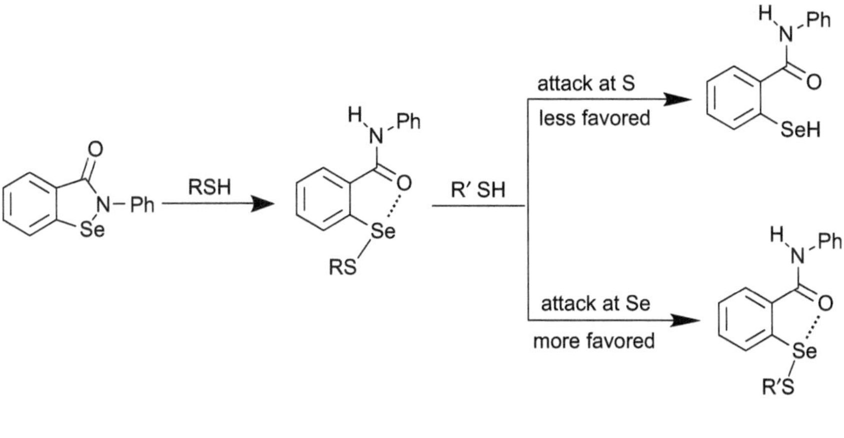

Scheme 10.2.7 Thiol exchange reactions in the selenenyl sulfide intermediates derived from ebselen.

intermediates. On the basis of experimental and theoretical investigations, Bhabak and Mugesh proposed the following revised roles for the basic amino groups in GPx mimics. The amino group should (1) not interact with selenium in the selenols, but should be basic enough for the deprotonation of the selenols to generate more reactive selenolate anions; (2) not interact with selenium in the selenenyl sulfide intermediates; (3) exhibit an interaction with selenium in the selenic acid intermediates to increase the electrophilic reactivity of selenium toward thiols.

Bhabak and Mugesh[91] also reported that the *sec*-amino substituted diselenides **44–47** exhibit much better GPx activity than the corresponding *tert*-amine-based diselenides. Compounds **44–47** readily react with thiols to produce catalytically active selenols. This study revealed that the replacement of the *tert*-amino groups by *sec*-amino alters the reactivity of key intermediates and the effect is somewhat similar to that of the methoxy substitution discussed earlier. Although the mechanism for the GPx activity of *sec*-amino substituted diselenides is similar that of *tert*-amine-based diselenides, the *sec*-amino moiety appears to be better than the *tert*-amino groups in generating the selenols.[91] In other words, the thiol exchange reaction that takes place in the selenium centres of selenenyl sulfide intermediates is much less for the *sec*-amino-substituted compounds as compared to that of *tert*-amine-based diselenides. A comparison of the GPx activity of amine-based compounds with that of ebselen analogues indicates that the amine-based compounds are more sensitive to the nature of peroxides than the ebselen analogues.[91]

Wirth and co-workers[92,93] reported another class of GPx mimics (**48–52**) having oxygen atoms in close proximity to selenium. In these compounds, the non-covalent Se···N interactions were shown to enhance the GPx activity. They developed a cellular model for Friedreich's ataxia and demonstrated that small-molecule GPx mimics such as **48–52** have potential in the treatment for this disease. It has been suggested that the GPx-mimic-based treatment strategy may be useful in the treatment of neurodegenerative diseases with mitochondrial impairment.[93]

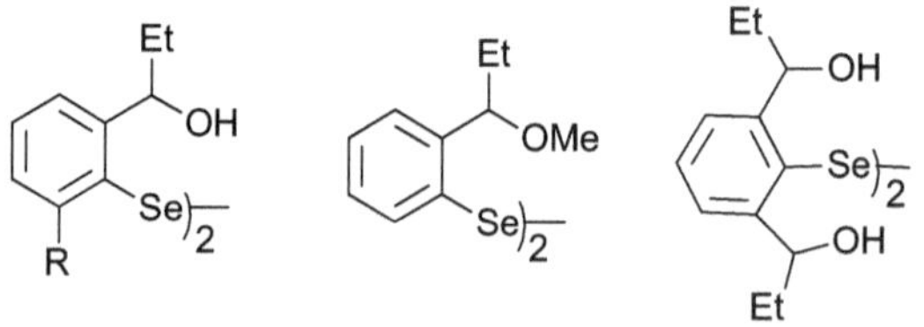

48, R = H; **49**, R = Me **51** **52**
50, R = CF_3

10.2.4.3 Allyl Selenides and Selenenate Esters as GPx Mimics/Models

Back and Moussa[94,95] reported that certain allyl selenides can be used to generate catalytically active species via [2,3]-sigmatropic rearrangements of the corresponding allyl selenoxides. The selenoxides could be generated by treating the selenides with peroxides. They reported that allyl 3-hydroxypropyl selenide **53** exhibits remarkable antioxidant activity. The catalytically active species is supposed to be the monocyclic seleninate ester **54**, which reacts with BnSH to produce thioseleninate **55**. The reaction of compound **55** with BnSH produces the selenenic acid **56** and BnSSBn. Oxidation/cyclization of **57** regenerates the cyclic seleninate **54** (Scheme 10.2.8). The allyl selenide **53**, therefore, acts as a pro-catalyst, which generates the seleninate ester **54** upon oxidation. Compound **54** was found to be significantly more active in catalysing the reduction of t-BuOOH than ebselen **1** and the camphor-based selenenyl amide **19**. These observations suggested that Se-O compounds can be more effective than the commonly studied Se–N compounds.[95]

Back and co-workers[96] also reported that the di(3-hydroxy-propyl) selenide **58** acts an efficient catalyst for the reduction of t-BuOOH in the presence of BnSH. The catalytic cycle of **58** involves the formation of an unusual spiro-dioxyselenurane **60**. The oxidation of compound **58** with t-BuOOH produces the transient selenoxide **59**, which undergoes a spontaneous cyclization to produce the dioxyselenurane **60**. The structure of **60** was confirmed by spectroscopic methods and single crystal X-ray crystallography. The reaction of **60** with BnSH produces an intermediate **61**, which upon reaction with second equivalent of BnSH regenerates the selenide **58** with elimination of BnSSBn (Scheme 10.2.9).[96] When t-BuOOH is present in the reaction mixture, compound **59** is recyclized to compound **60.** Although the reactivity of compound **58** was only about 15 times higher than that of ebselen under identical conditions, the catalytic mechanism involves the formation of an unusual spiro compound.

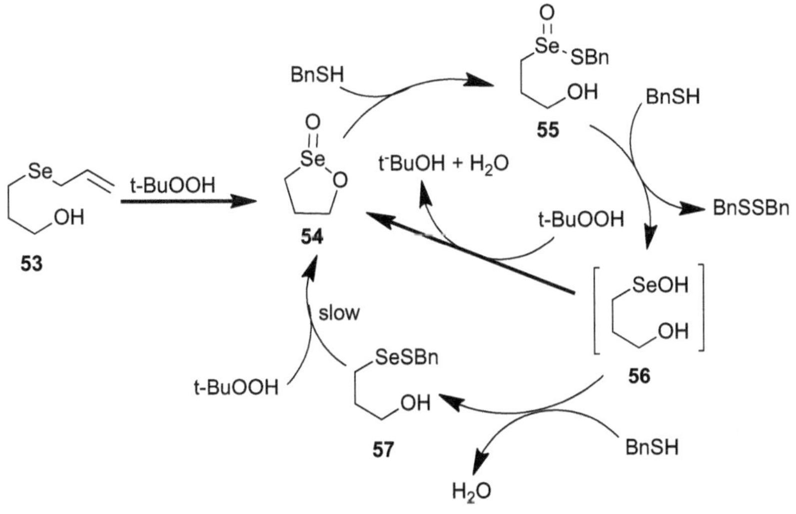

Scheme 10.2.8 Proposed mechanism for the GPx activity of the allyl selenide **53**.[94,95]

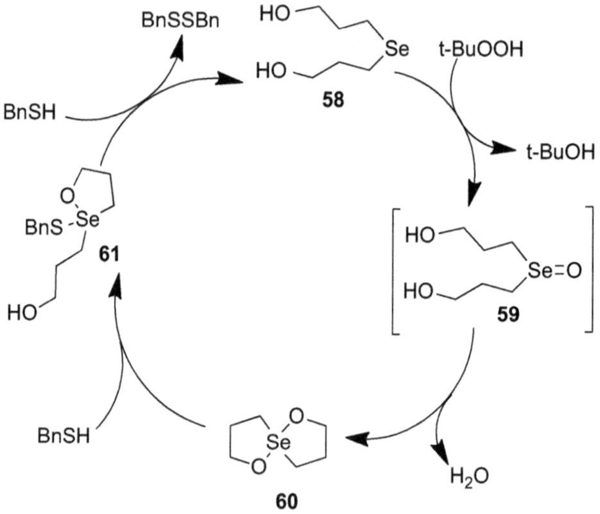

Scheme 10.2.9 Catalytic cycle of selenide **58**.[96]

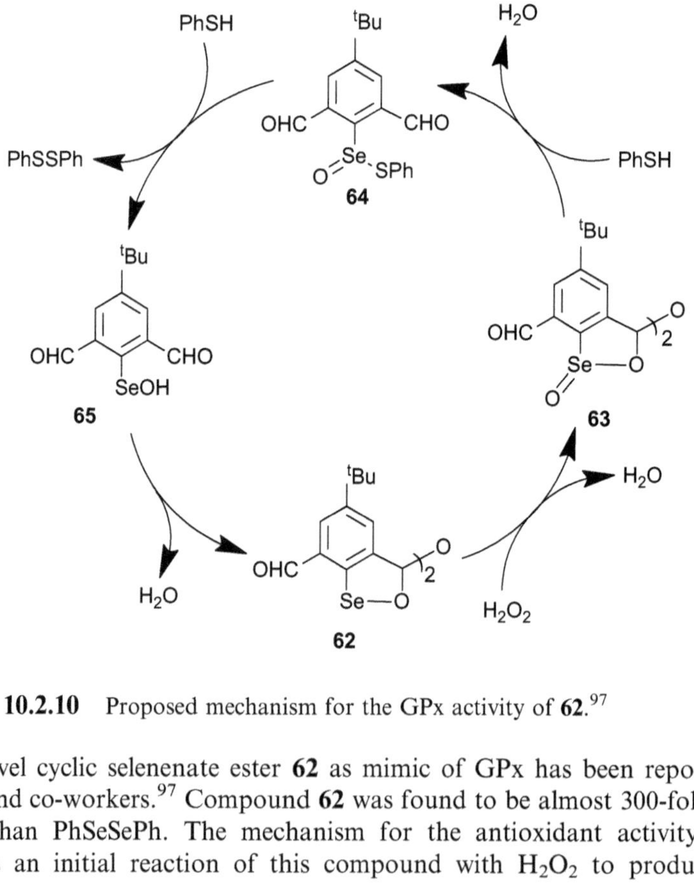

Scheme 10.2.10 Proposed mechanism for the GPx activity of **62**.[97]

A novel cyclic selenenate ester **62** as mimic of GPx has been reported by Singh and co-workers.[97] Compound **62** was found to be almost 300-fold more active than PhSeSePh. The mechanism for the antioxidant activity of **62** involves an initial reaction of this compound with H_2O_2 to produces the seleninate **63**, which upon reaction with PhSH produces the intermediate **64** (Scheme 10.2.10).[97] Subsequent reaction with PhSH produces the corresponding selenenic acids **65**. As shown previously for other selenenic acids, the cyclization of **65** regenerates the selenenate ester **62** with elimination of water. It is known that the selenium centre in selenenic acid undergoes further oxidation to produce the corresponding seleninic and selenonic acids. However, no such oxidation was observed for compound **65**, indicating that the cyclization protects the selenium moiety from overoxidation.[97] Such a cyclization was observed for the selenenic acid derived from ebselen. The cyclization of open-chain intermediates to ebselen has been proposed to be important for the protection of the selenium moiety from irreversible inactivation. Therefore, the selenium atom is not cleaved off from the aromatic moiety during the biotransformations.

10.2.5 Organoselenium Compounds as Peroxynitrite Scavengers

Nitrotyrosine is generated by enzymatic or non-enzymatic nitration of tyrosyl residues in proteins.[98–100] As nitrotyrosine is associated with a number of diseases such as Alzheimer's disease, or with modifications of biomolecules such as lipid peroxidation, it is treated as a biomarker for oxidative and nitrative stress in inflammatory, allergic, and other diseases.[98–100] Enzymatic tyrosine nitration is catalysed by a number of haem proteins, such as peroxidases (myeloperoxidase, eosinophil peroxidase, horseradish peroxidase, lactoperoxidase), haemoglobin, myoglobin, and cytochrome *c*. Non-enzymatic or chemical nitration of tyrosine is caused extensively by peroxynitrite (PN).[98–100] In a biological medium, PN is produced by the diffusion-controlled reaction of a superoxide anion radical and

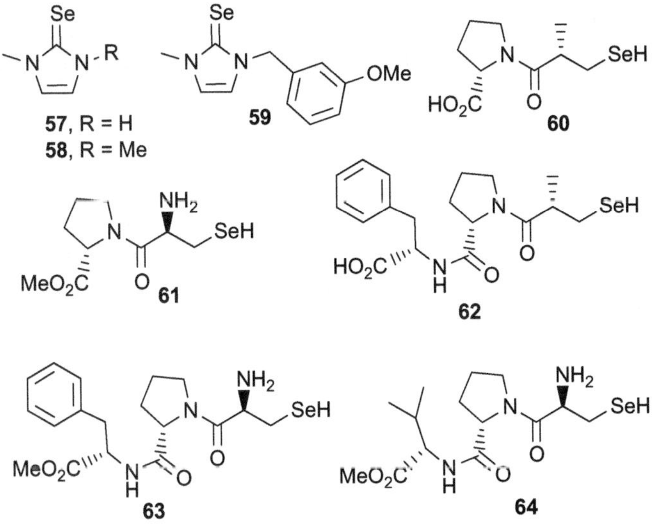

Figure 10.2.4 Organoselenium compounds used as PN scavengers.[107–111]

nitric oxide (NO) which is an endothelial-derived relaxing factor. Small biological molecules such as carbon dioxide, ascorbate, methionine, cysteine are shown to exhibit good PN scavenging activities.[98–100] Mammalian selenoenzymes GPx, selenoproteine P, and thioredoxin reductase (TrxR) are also known to exhibit PN scavenging behaviour.[101–106] Furthermore, it has been shown that organoselenium compounds such as ebselen act as efficient scavengers of PN, and therefore inhibit the PN-mediated damage.[104–106] Apart from ebselen and its analogues, a few other organoselenium compounds used as PN scavengers are shown in Figure 10.2.4.[107–111]

10.2.6 Conclusions

In this chapter, several mimics and models of selenium-based antioxidants are described. Particular emphasis has been given to compounds that mimic the activity of the selenoenzyme GPx. Although the antibacterial, antiviral, antifungal, and antiparastic properties of selenium compounds have been known for long time, the application of organoselenium compounds as enzyme mimics has only recently attracted much attention. After the finding that ebselen mimics the biological activity of GPx, there has been a major development in the area of GPx mimics/models. The synthetic selenium-based GPx mimics reported in the literature can be classified into three major categories: (1) cyclic selenenyl amides having a Se–N bond, (2) diaryl diselenides, and (3) aliphatic/aromatic monoselenides. Recent experimental evidences show that ebselen exhibits poor GPx activity in the presence of aryl or benzylic thiol cofactors such as PhSH or BnSH. The difference in the thiol cofactors causes the discrepancies observed in different studies as the catalytic activity of each GPx mimic is influenced by the nature of thiol. It is also now clear that the existence of strong Se···O/N interactions in the selenenyl sulfide intermediates in the catalytic cycle decreases the GPx activity, as such interactions facilitate an attack of thiol at selenium rather than at sulfur. A series of studies have shown that any substituent capable of facilitating the nucleophilic attack of thiol at sulfur in the selenenyl sulfide intermediate would enhance the GPx activity of selenium compounds.

Acknowledgements

The work in our laboratories and research on the GPx activity of organoselenium compounds was supported by the UK-India Education and Research Initiative (UKIERI) of the British Council and the Department of Science and Technology (DST), New Delhi.

References

1. J. J. Berzelius, *Afhandl. Fys. Kemi Mineralog.*, 1818, **6**, 42.
2. C. J. Löwig, *Poggendorff's Ann. Phys.*, 1836, **37**, 552.
3. B. Rathke, *Ann. Chem.*, 1869, **152**, 211.
4. J. G. Hutton, *Agron. J.*, 1931, **23**, 1076.

5. K. W. Franke, *J. Nutr.*, 1934, **8**, 597.
6. K. W. Franke, *J. Nutr.*, 1934, **8**, 609.
7. K. W. Franke and V. R. Potter, *J. Nutr.*, 1934, **8**, 615.
8. K. W. Franke, *J. Nutr.*, 1935, **10**, 223.
9. K. W. Franke and E. P. Painter, *J. Nutr.*, 1935, **10**, 599.
10. J. Pinsent, *Biochem. J.*, 1954, **57**, 10.
11. R. Andreesen and L. Ljungdahl, *J. Bacteriol.*, 1973, **116**, 867.
12. D. C. Turner and T. C. Stadtman, *Arch. Biochem. Biophys.*, 1973, **154**, 366.
13. L. Flohé, E. A. Günzler and H. H. Schock, *FEBS Lett.*, 1973, **32**, 132.
14. J. C. Boyington, V. N. Gladyshev, S. V. Khangulov, T. C. Stadtman and P. D. Sun, *Science*, 1997, **275**, 1305.
15. R. Wilting, S. Schorling, B. C. Persson and A. Böck, *J. Mol. Biol.*, 1977, **266**, 637.
16. E. Garcin, X. Vernede, E. C. Hatchikian, A. Volbeda, M. Frey and J. C. Fontecilla-Camps, *Structure*, 1999, **7**, 557.
17. M. Pfeiffer, R. Bingemann and A. Klein, *Eur. J. Biochem.*, 1998, **256**, 447.
18. M. Wagner, D. Sonntag, R. Grimm, A. Pich, C. Eckerskorn, B. Söhling and J. R. Andreesen, *Eur. J. Biochem.*, 1999, **260**, 38.
19. D. Behne, A. Kyriakopoulos, H. Meinhold and J. Köhrle, *Biochem. Biophys. Res. Commun.*, 1990, **173**, 1143.
20. J. R. Arthur, F. Nicol and G. J. Beckett, *Biochem. J.*, 1990, **272**, 537.
21. J. C. Davey, K. B. Becker, M. J. Schneider, D. L. Germain and V. A. Galton, *J. Biol. Chem.*, 1995, **270**, 26786.
22. W. Croteau, S. L. Whittemore, M. J. Schneider and D. L. Germain, *J. Biol. Chem.*, 1995, **270**, 16569.
23. A. Lescure, D. Gautheret, P. Carbon and A. Krol, *J. Biol. Chem.*, 1999, **274**, 38147.
24. T. Tamura and T. C. Stadtman, *Proc. Natl. Acad. Sci. USA*, 1996, **93**, 1006.
25. S. R. Lee, J. R. Kim, K. S. Kwon, H. W. Yoon, R. L. Leveine, A. Ginsburg and S. G. Rhee, *J. Biol. Chem.*, 1999, **274**, 4722.
26. D. Mustacich and G. Powis, *Biochem. J.*, 2000, **346**, 1.
27. S. Watabe, Y. Makino, K. Ogawa, T. Hiroi, Y. Yamamoto and S. Y. Takahashi, *Eur. J. Biochem.*, 1999, **264**, 74.
28. G. C. Mills, *J. Biol. Chem.*, 1957, **229**, 189.
29. M. A. Motsenbocker and A. L. Tappel, *J. Nutr.*, 1984, **114**, 279.
30. F. Ursini, M. Maiorino, M. Valente, L. Ferri and C. Gregolin, *Biochim. Biophys. Acta*, 1982, **710**, 197.
31. K. Takahashi, N. Avissar, J. Whittin and H. Cohen, *Arch. Biochem. Biophys.*, 1987, **256**, 677.
32. F. F. Chu, J. H. Doroshow and R. S. Esworthy, *J. Biol. Chem.*, 1993, **268**, 2571.
33. S. Gromer, J. K. Eubel, B. L. Lee and J. Jacob, *Cell. Mol. Life Sci.*, 2005, **62**, 2414.
34. C. Jacob, G. I. Giles, N. M. Giles and H. Sies, *Angew. Chem. Int. Ed.*, 2003, **42**, 4742.

35. S. D. Aust and B. A. Svingen, in *Free Radicals in Biology*, ed. W. A. Pryor, Academic Press: New York, 1982, Vol. 5, P1.
36. R. E. Huie and S. Padmaja, *Free Radic. Res. Commun.*, 1993, **18**, 195.
37. S. Goldstein and G. Czapski, *Free Radic. Biol. Med.*, 1995, **19**, 505.
38. R. Kissner, T. Nauser, P. Bugnon, P. G. Lye and W. H. Koppenol, *Chem. Res. Toxicol.*, 1997, **10**, 1285.
39. H. Sies, Oxidative stress: introductory remarks, in *Oxidative Stress*, ed. H. Sies, Academic Press: London, 1985, P1.
40. A. T. Diplock, Antioxidants and free radical scavengers, in *Free Radical Damages and Its Control*, ed. C. A. Rice-Evans and R. H. Burdon, Elsevier: Amsterdam, 1994, p113 and references therein.
41. R. J. Hondal, B. L. Nilsson and R. T. Raines, *J. Am. Chem. Soc.*, 2001, **123**, 5140.
42. R. Quaderer, A. Sewing and D. Hilvert, *Helv. Chim. Acta*, 2001, **84**, 1197.
43. M. Maiorino, K-D. Aumann, R. Brigelius-Flohé, R. D. Doria, L. van den Heuvel, J. McCarthy, A. Rovery, F. Ursini and L. Flohé, *Biol. Chem. Hoppe-Seyler*, 1995, **376**, 651.
44. H. Sies, *Angew. Chem. Int. Ed.*, 1986, **25**, 1058.
45. C. Rocher, J.-L. Lalanne and J. Chaudiére, *Eur. J. Biochem.*, 1992, **205**, 955.
46. K. R. Maddipati and L. J. Marnett, *J. Biol. Chem.*, 1987, **262**, 17398.
47. F. Ursini, M. Maiorino, R. Brigelius-Flohé, K.-D. Aumann, A. Roveri, D. Schomburg and L. Flohé, *Methods Enzymol.*, 1995, **252**, 38.
48. L. Flohé, in *Free Radicals in Biology, Vol. 5*, ed. W. A. Pryor. Academic Press: New York, 1982, p. 223.
49. M. Maiorino, A. Roveri, M. Coassin and F. Ursini, *Biochem. Pharmacol.*, 1988, **37**, 2267.
50. O. Epp, R. Ladenstein and A. Wendel, *Eur. J. Biochem.*, 1983, **133**, 51.
51. L. Flohé, in *Glutathione*, ed. D. Dolphin, O. Avramovic and R. Poulson, John Wiley and Sons, New York, 1989, p. 644.
52. L. Flohé, G. Loschen, W. A. Gunzler and E. Eichele, *Hoppe-Seyler's Z. Physiol. Chem.*, 1972, **353**, 987.
53. A. Müller, E. Cadenas, P. Graf and H. Sies, *Biochem. Pharmacol.*, 1984, **33**, 3235.
54. A. Wendel, M. Fausel, H. Safayhi, G. Tiegs and R. Otter, *Biochem. Pharmacol.*, 1984, **33**, 3241.
55. I. A. Cotgreave, S. K. Duddy, G. E. N. Kass, D. Thompson and P. Moldus, *Biochem. Pharmacol.*, 1989, **38**, 649.
56. C. Schewe, T. Schewe and A. Wendel, *Biochem. Pharmacol.*, 1994, **48**, 65.
57. V. Galet, J.-L. Bernier, J.-P. Hénichart, D. Lesieur, C. Abadie, L. Rochette, A. Lindenbaum, J. Chalas, J.-F. Renaud de la Faverie, B. Pfeiffer and P. Renard, *J. Med. Chem.*, 1994, **37**, 2903.
58. M. Parnham and H. Sies, *Exp. Opin. Invest. Drugs*, 2000, **9**, 607.
59. N. Maulik, T. Yoshida and D. K. Das, *Free Radical Biol. Med.*, 1998, **24**, 869.
60. D. A. Dawson, H. Masayasu, D. I. Graham and I. M. Macrac, *Neurosci. Lett.*, 1995, **185**, 65.

61. S. Namura, I. Nagata, S. Takami, H. Masayasu and H. Kikuchi, *Stroke*, 2001, **32**, 1906.
62. C. F. Yang, H. M. Shen and C. N. Ong, *Arch Biochem Biophys.*, 2000, **374**, 142.
63. G. Mugesh and H. B. Singh, *Chem. Soc. Rev.*, 2000, **29**, 347.
64. G. Mugesh, W.-W. du Mont and H. Sies, *Chem. Rev.*, 2001, **101**, 2125.
65. M. C. Fong and C. H. Schiesser, *Tetrahedron Lett.*, 1995, **36**, 7329.
66. H. J. Reich and C. P. Jasperse, *J. Am. Chem. Soc.*, 1987, **109**, 5549.
67. J. Chaudiere, J.-C. Yadan, I. Erdelmeier, C. Tailhan-Lomont and M. Moutet, in *Oxidative Processes and Antioxidants*, ed. R. Paoletti, Raven Press, New York, 1994, p. 165.
68. M. J. Parnham, J. Biederman, C. Bittner, N. Dereu, S. Leyck and H. Wetzig, *Agents Actions*, 1989, **27**, 306.
69. K. P. Bhabak and G. Mugesh, *Chem. Eur. J.*, 2007, **13**, 4594.
70. B. K. Sarma and G. Mugesh, *J. Am. Chem. Soc.*, 2005, **127**, 11477.
71. B. K. Sarma and G. Mugesh, *Org. Biomol. Chem.*, 2008, **6**, 965.
72. M. Maiorino, K.-D. Aumann, R. Brigelius-Flohé, D. Doria, J. van den Heuvel, J. McCarthy, A. Rovery, F. Ursini and L. Flohé, *Biol. Chem. Hoppe-Seyler*, 1995, **376**, 651.
73. C. Rocher, J.-L. Lalanne and J. Chaudiére, *Eur. J. Biochem.*, 1992, **205**, 955.
74. K. R. Maddipati and L. J. Marnett, *J. Biol. Chem.*, 1987, **262**, 17398.
75. F.-F. Chu, J. H. Doroshow and R. S. Esworthy, *J. Biol. Chem.*, 1993, **268**, 2571.
76. R. Brigelius-Flohé, *Free Radical Biol. Med.*, 1999, **27**, 951.
77. K. P. Bhabak and G. Mugesh, *Acc. Chem. Res.*, 2010, **43**, 1408.
78. T. G. Back and B. P. Dyck, *J. Am. Chem. Soc.*, 1997, **119**, 2079.
79. B. K. Sarma and G. Mugesh, *Chem. Eur. J.*, 2008, **14**, 10603.
80. S. R. Wilson, P. A. Zucker, R. Ruey, C. Huang and A. Spector, *J. Am. Chem. Soc.*, 1989, **111**, 5936.
81. M. Iwaoka and S. Tomoda, *J. Am. Chem. Soc.*, 1994, **116**, 2557.
82. K. L. House, R. B. Dunlap, J. D. Odom, Z. P. Wu and D. Hilvert, *J. Am. Chem. Soc.*, 1992, **114**, 8573.
83. I. M. Bell and D. Hilvert, *Biochemistry*, 1993, **32**, 13969.
84. S. Boschi-Muller, S. Müller, A. Van Dorsselaer, A. Böck and G. Branlant, *FEBS Lett.*, 1998, **439**, 241.
85. K.-D. Aumann, N. Bedorf, R. Brigelius-Flohé, D. Schomburg and L. Flohé, *L. Biomed. Environ. Sci.*, 1997, **10**, 136.
86. B. Ren, W. Huang, B. Åkesson and R. Ladenstein, *J. Mol. Biol.*, 1997, **268**, 869.
87. G. Mugesh, A. Panda, H. B. Singh, N. S. Punekar and R. J. Butcher, *Chem. Commun.*, 1998, 2227.
88. G. Mugesh, A. Panda, H. B. Singh, N. S. Punekar and R. J. Butcher, *J. Am. Chem. Soc.*, 2001, **123**, 839.
89. S. Bachrach, D. W. Demoin, M. Luk and J. V. Miller, *J. Phys. Chem.*, 2004, **108**, 4040.

90. K. P. Bhabak and G. Mugesh, *Chem. Eur. J.*, 2008, **14**, 8640.
91. K. P. Bhabak and G. Mugesh, *Chem. Eur. J.*, 2009, **15**, 9846.
92. T. Wirth, *Molecules*, 1998, **3**, 164.
93. M. L. Jauslin, T. Wirth, T. Meier and F. Schoumacher, *Hum. Mol. Genet.*, 2002, **11**, 3055.
94. T. G. Back and Z. Moussa, *J. Am. Chem. Soc.*, 2002, **124**, 12104.
95. T. G. Back and Z. Moussa, *J. Am. Chem. Soc.*, 2003, **125**, 13455.
96. T. G. Back, Z. Moussa and M. Parvez, *Angew. Chem., Int. Ed.*, 2004, **43**, 1268.
97. S. S. Zade, H. B. Singh and R. J. Butcher, *Angew. Chem. Int. Ed.*, 2004, **43**, 4513.
98. R. E. Huie and S. Padmaja, *Free Radic. Res. Commun.*, 1993, **18**, 195.
99. J. S. Beckman, The physiological and pathophysiological chemistry of nitric oxide, in *Nitric Oxide: Principles and Actions*, ed. J. Lancaster, Academic Press, San Diego, CA, 1996, p. 1.
100. H. Traulsen, H. Steinbrenner, D. P. Buchczyk, L.-O. Klotz and H. Sies, *Free Radic. Res.*, 2004, **38**, 123.
101. G. E. Arteel, K. Briviba and H. Sies, *Chem. Res. Toxicol.*, 1999, **12**, 264.
102. Y. S. Park, N. Fujiwara, Y. H. Koh, Y. Miyamoto, K. Suzuki, K. Honke and N. Taniguchi, *Biol. Chem.*, 2002, **383**, 683.
103. M. Whiteman and B. Halliwell, *Free Radic. Biol. Med.*, 1997, **22**, 1309.
104. K. Briviba, R. Tamler, L.-O. Klotz, L. Engman, I. A. Cotgreave and H. Sies, *Biochem. Pharmacol.*, 1998, **55**, 817.
105. G. E. Arteel, K. Briviba and H. Sies, *FEBS Lett.*, 1999, **445**, 226.
106. R. Kissner, T. Nauser, P. Bugnon, P. G. Lye and W. H. Koppenol, *Chem. Res. Toxicol.*, 1997, **10**, 1285.
107. K. P. Bhabak and G. Mugesh, *Chem. Eur. J.*, 2010, **16**, 1175.
108. K. P. Bhabak, A. A. Vernekar, S. R. Jakka, G. Roy and G. Mugesh, *Org. Biomol. Chem.*, 2011, **9**, 5193.
109. B. J. Bhuyan and G. Mugesh, *Org. Biomol. Chem.*, 2011, **9**, 1356.
110. B. J. Bhuyan and G. Mugesh, *Org. Biomol. Chem.*, 2011, **9**, 5185.
111. K. P. Bhabak, K. Satheeshkumar, S. Jayavelu and G. Mugesh, *Org. Biomol. Chem.*, 2011, **9**, 7343.

CHAPTER 10.3

The Role of Sulfur and Selenium Species in the Thyroid

SURENDAR REDDY JAKKA AND GOVINDASAMY MUGESH*

Department of Inorganic & Physical Chemistry, Indian Institute of Science, Bangalore 560 012, India
*Email: mugesh@ipc.iisc.ernet.in

10.3.1 Introduction

The thyroid is one of the biggest endocrine glands in humans. Anatomically, it is located on the anterior surface of the trachea at the base of the neck (Figure 10.3.1). The butterfly-shaped gland consists of two elongated oval lobes on either side of the trachea connected by an isthmus located just below the larynx, and weighs about 15–25 g.[1–7] The thyroid gland secrets and controls body functions through two major iodine-containing hormones, L-thyroxine (T4, 3,3′,5,5′-tetraiodo-L-thyronine, T4) and 3,3′,5-triiodo-L-thyronine (T3).

Thyroid hormones belong to a class of modified tyrosine derivatives in which T3 is the active form, which exerts its genomic functions by acting as a ligand to a class of nuclear receptors called thyroid receptors (TR).[8] Two isoforms of thyroid receptors are found in human, namely THRA (TRα) and THRB (TRβ). TRα participates in the regulation of cardiac rate and contractility, whereas TRβ controls the cholesterol levels and T3-mediated feedback suppression of TSH. T3 contributes significantly to the meticulous regulation of various genes that are involved in growth, development, metabolism, and energy homeostasis[9] by combining with a dimer of TR with itself or with other

Handbook of Chalcogen Chemistry: New Perspectives in Sulfur, Selenium and Tellurium, 2nd Edition, Volume 2
Edited by Francesco Antonio Devillanova and Wolf-Walther du Mont

Published by the Royal Society of Chemistry, www.rsc.org

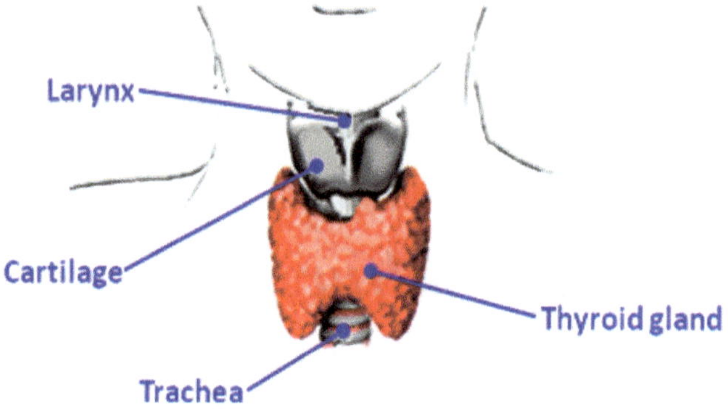

Figure 10.3.1 Thyroid gland.

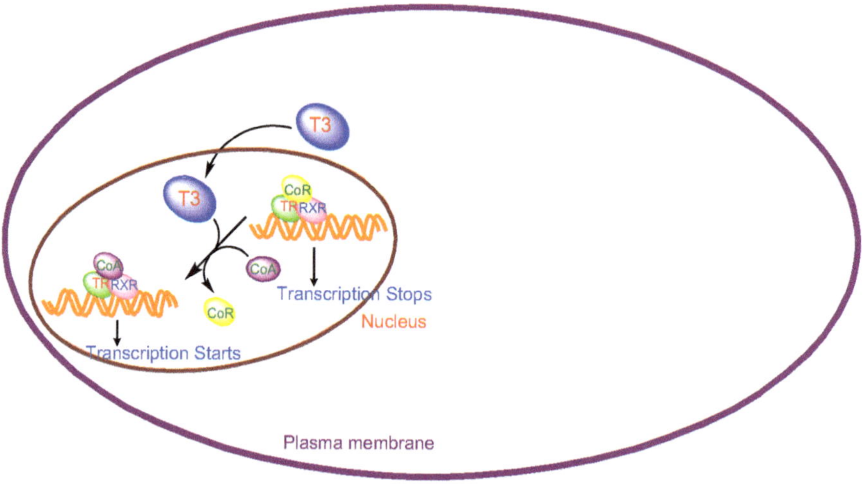

Figure 10.3.2 Schematic representation of thyroid hormone action in regulation of gene expression in the cell nucleus.

nuclear receptors (NR), such as retinoid receptor X and binds to the DNA at the thyroid receptor elements (TREs) in the promoter region of its targeted genes. T3 binding recruits the coactivator protein, which brings out the conformational changes in chromatin and compositional changes in the transcriptional machinery complex (Figure 10.3.2).[10–13]

10.3.2 Thyroid Hormone Homeostasis

Endocytosis-mediated secretion of thyroid hormones from the apical surface of the thyroid follicular cell uses the stored iodinated Tg (thyroglobulin).[14] The proteolytic digestion of monoiodotyrosine (MIT) and diiodotyrosine (DIT) is recaptured in to phagolysomes and releases T4 and T3 into the circulation. T4 is the major form in the released thyroid hormones and is about 40-fold higher

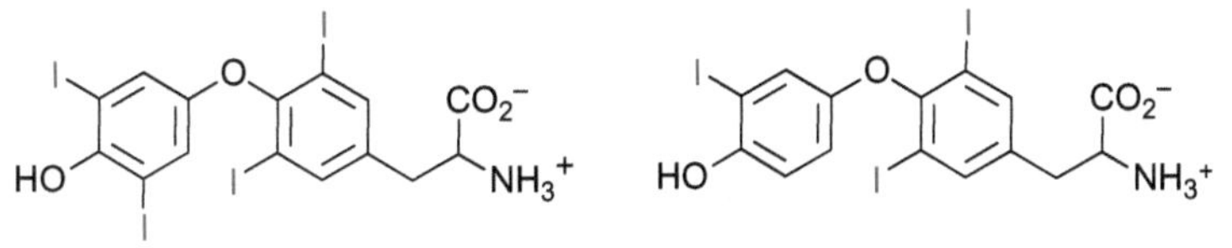

3, 3′, 5, 5′-tetraiodo-L-thyronine (T4) 3, 3′, 5-triiodo-L-thyronine (T3)

Figure 10.3.3 Structures of T4 and T3.

in serum in comparison to T3 (90 *vs* 2 nM). Most of the thyroid hormones bind to carrier proteins, such as thyroxine binding globulin (TBG), serum albumin, and prealbumin, and only ~0.03% is present in its free form. On the other hand, 0.3% of the total T3 is present in its free form, whereas the remaining is in bound state with TBG. The biological functions of thyroid originate from the free serum T3 (Figure 10.3.3).[15]

T4 was first isolated in 1915 by Edward C. Kendall, and after 37 years (in 1952), Gross and Pitt-Rivers found T3 in humans.[16] This observation led to an assumption that T4 may undergo an 'iodothyronine deiodination' during which one of the iodines of T4 is removed and replaced by a hydrogen atom to produce T3. However, 18 years later, Braverman *et al.* provided remarkable and conclusive evidence for the iodothyronine deiodination by administrating radiolabelled thyroid hormones to athyrotic patients. They found both T3 and ^{125}I-T3 in the patient's serum after the administration of only stable ^{125}I-T4 (Scheme 10.3.1).[17] A radioactive scan and histogram of the paper chromatogram of serum of patient after receiving oral stable and radioactive T4 is shown in Figure 10.3.4. This study also suggested that the two iodines in the outer ring of thyroxine are chemically identical and one of them could be removed during the deiodination reaction.

The initial studies suggested that thyroxine deiodination is the first metabolic pathway that confers biological activity to T4. The removal of one iodine atom from T4 results in T3, which has been found to be the most active, short-lived form of thyroid hormone. It should be mentioned that the removal of iodine from the inner ring of thyroxine was not known at that time. It is found that most of the circulating T3 is derived by deiodination of extrathyroidal T4. As discussed in the introduction, T3 modulates the gene expression in virtually almost all cells by thyroid hormone receptors, which are ligand-dependent transcription factors.[18–20] As T4 is a polyiodinated phenoxyphenyl molecule, deiodination must occur to produce T3 from T4 and the reaction is catalysed by two enzymes, type 1 and type 2 iodothyronine deiodinase (D1 and D2 respectively). As the four iodines are attached to two different ring systems in T4, *i.e.* at the 3,5- and 3′,5′-positions (Figure 10.3.3), there is a possibility of the formation of two regioiosomeric products T3 and rT3 during the mono-deiodination. It has been established that both D1 and D2 perform the outer-ring deiodination, *i.e.* removal of iodine from the 3′ or 5′ position of the phenoxyphenyl moiety. Later, it was shown that a type 3 iodothyronine

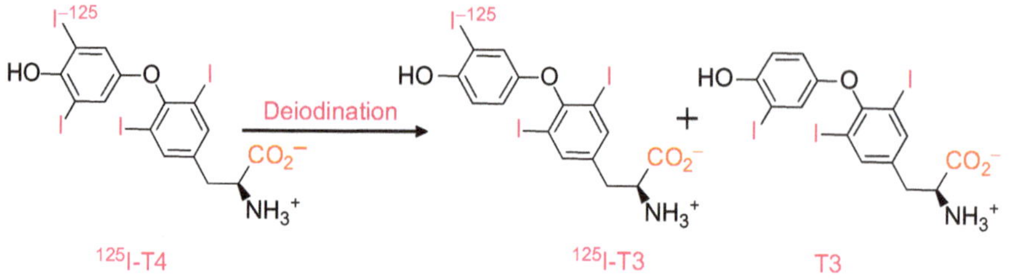

Scheme 10.3.1 Chemical changes observed during the deiodination of radiolabelled T4.

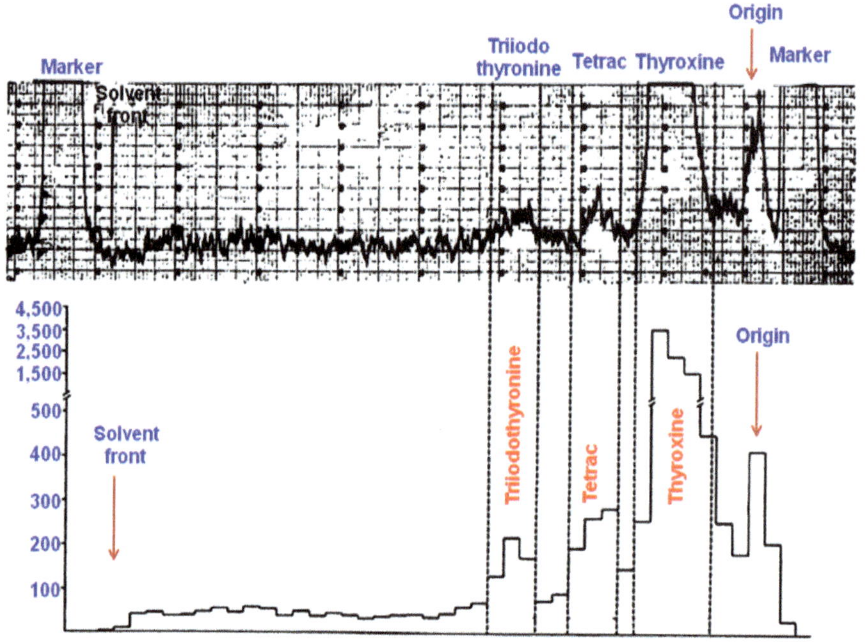

Figure 10.3.4 Radioactive scan and histogram of paper chromatogram of serum of patient after receiving oral stable and radioactive T4. (Modified from Ref. 17).

deiodinase (D3) terminates the function of T4 and T3 by removing iodine from the 3- or 5-positions of T4 and T3 to generate rT3 and T2, respectively. While T3 is highly active in the regulation of genomic profile, rT3 is found to be much less active. Although it took several years to discover all three deiodinases and their catalytic pathways, the homeostasis of thyroid hormone has now been clearly established (Scheme 10.3.2).

At the moment, the function of D1 in humans is not fully understood, except the fact that T4 is converted in large amounts to T3 in hyperthyroidism.[21] However, the widely expressed enzyme D2 exhibits three times higher affinity towards T4 than D1. This led to an assumption that D2 is responsible for the

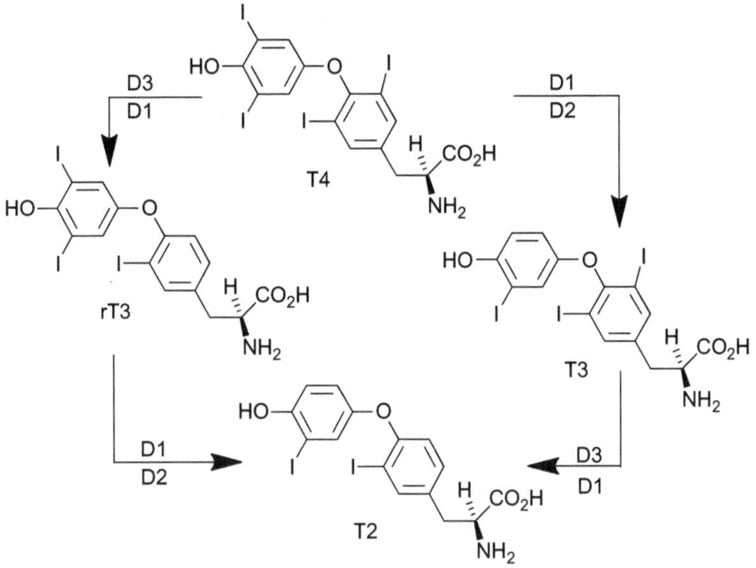

Scheme 10.3.2 Inner- and outer-ring deiodinations of T4 catalysed by three iodothyronine deiodinases (D1, D2, and D3).

predominant pathway of T3 generation. After this report, several new functions of deiodinases in thyroid hormone homeostasis started to emerge.[22] D3, which inactivates the thyroid hormones by removing iodine from the inner ring of the phenoxyphenyl moiety, is normally found in placenta and in the central nervous system, unlike D1 and D2.[23–25] This discovery completed the picture of thyroid hormone homeostasis by coordinated reciprocal changes in their activities and response to iodine availability (Scheme 10.3.2). It should be noted that D1 is the only enzyme that can catalyse both outer- and inner-ring deiodination of T4.

10.3.2.1 Selenoenzymes Modulate the Cell-Specific T3 Concentration

On the basis of the assumption that the deiodinases play a key role in thyroid hormone homeostasis, it was suggested that these enzymes also modulate the local thyroid hormone availability to its receptors and subsequently its function. The major support for this concept is the exciting study on brown adipose tissue (BAT) of rats exposed to cold.[26] In this study, it has been found that the D2 activation in BAT increased and saturated the local TR[27–29] and resulted in significant changes in the transcription of T3-responsive genes in D2-expressing cells.[27,28] Similar to the D2 expression studies, time-dependent expression of D3 has been shown to promote cell-specific hypothyroidism by deactivating the thyroid hormone through inner-ring deiodination. This results in a decrease in the T3-dependent gene transcription in relation to the development and tissue injury.[30,31] Finally, this concept was supported by studies on numerous cellular systems in which deiodinases regulates thyroid hormone signalling.[32–35] Now it has been well established that selenium-containing deiodinases modulate the thyroid hormone concentration in cellular level systemically and temporally, which is reasonably independent of serum thyroid hormone concentrations (Figure 10.3.5).[36]

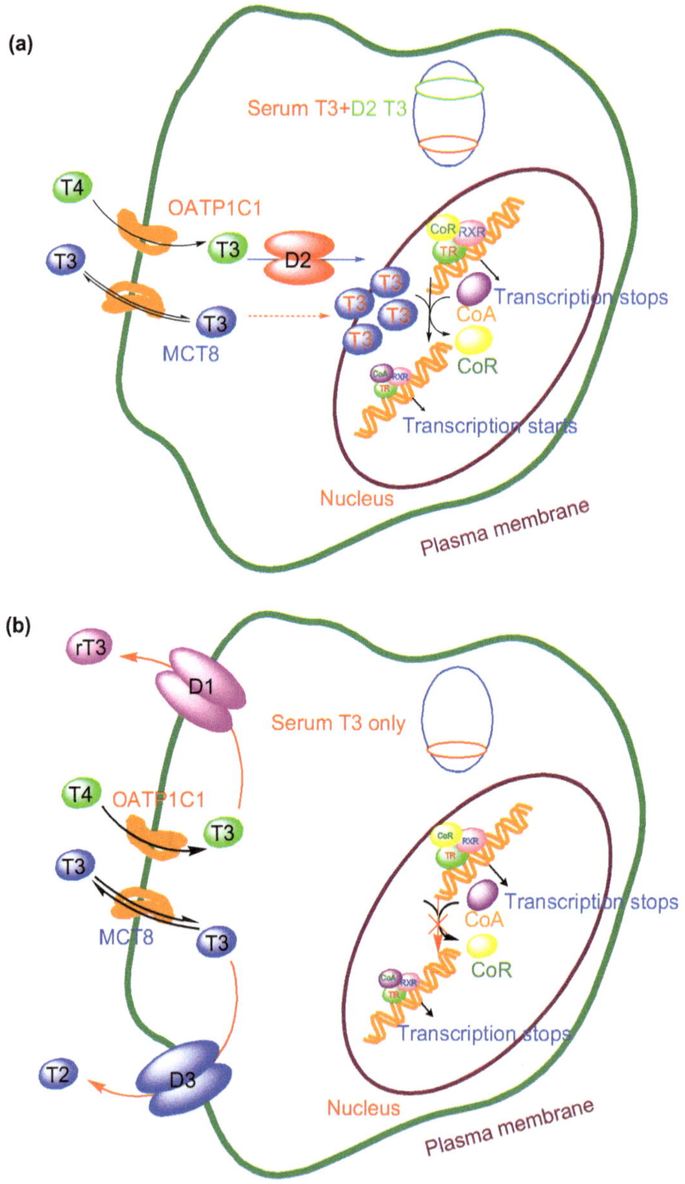

Figure 10.3.5 (A) Schematic representation of increasing local T3 production by enzyme D2. (B) Schematic representation of decreasing local T3 production by enzymes D3 and D1.
(Modified from Ref. 36).

10.3.2.2 Biosynthesis of Selenoproteins, Iodothyronine Deiodinases

Selenoproteins contain one or more residues of the rare amino acid selenocysteine (Sec) in their structure. To date, more than 30 selenoproteins have been identified, and the biological functions of most of these selenoproteins are not well understood. The second eukaryotic selenoprotein is D1 and its cloning led to the discovery of the Sec incorporation mechanisms in higher organisms.[37,38] All the iodothyronine deiodinases (D1, D2, and D3) contain Sec at their active centre, and it is responsible for their activity.[39] A UGA codon encodes for the insertion of Sec during translation. However, in most cases of protein synthesis this codon is normally read as a signal for termination of translation. In the case of selenoproteins, with the help of additional factors, the UGA codon signalling for termination is bypassed and Sec is inserted.[20,40] Typically, the ribosomal selenoprotein synthesis machinery requires an in-frame UGA codon and a SECIS (Sec insertion sequence) element, which is a downstream stem loop that binds SECIS binding protein-2 (SBP-2). SBP-2 also interacts with an elongation factor, EFsec, to mediate the insertion of Sec from a specific tRNA (Sec-tRNA^{Sec}) by the ribosome at the UGA codon (Figure 10.3.6). Ribosomal selenoprotein synthesis is not as efficient as that of its cysteine homologue. It has been shown that the replacement of the UGA codon in D1 by a cysteine codon (CysD1) led to ~400-fold enhancement in the production of D1 protein.[41] As all three deiodinases belong to the selenoprotein family, all the constraints that apply to selenoprotein synthesis are also applicable to the biosynthesis of deiodinases. Furthermore, this concept was substantiated with the discovery of mutated SBP-2 in patients with impaired deiodinase activity.[42]

In prokaryotes, synthesis of Sec synthesis starts from serine (Ser). Initially, Sec-tRNA^{Sec} is misacylated with serine, followed by dehydroxylation by the Sec synthetase (SelA). Subsequently, an activated selenium generated by the selenophosphate synthetase (SelD) is added.[43] In eukaryotes, the corresponding SelD homologues, selenophosphate synthetase 1 and 2 (SPS-1 and SPS-2), had been identified, but a functional SelA candidate remained elusive.[44,45] However, the first piece of this confusion was solved by the identification of a phosphoseryl-tRNA[Ser]^{Sec} kinase, which specifically phosphorylates seryl-tRNASec to form *O*-phosphoseryl-tRNA^{Sec} (Figure 10.3.6).[46] Subsequently, a Sec-tRNASec binding protein of previously unknown function, soluble liver antigen (SLA), has been shown to perform the dephosphorylation of *O*-phosphoseryl-tRNASec in a pyridoxal phosphate-dependent manner.[47,48] An activated selenium species generated by SPS-2 can then be accepted after dephosphorylation to form Sec in eukaryotes.[47]

Recently, the ribosomal protein L30 has been identified as an additional factor that binds SECIS elements and thus increases the incorporation of Sec (Figure 10.3.6).[49] These sequential studies suggested that the SECIS could have either an open or kinked conformation. L30 binds to both conformations, whereas the SBP-2 prefers to bind to the open form. L30 also competes with SBP-2 for SECIS binding, leading to a mechanistic picture in which the SBP-2 initially recruits EFsec and Sec-tRNA^{Sec} to the translating selenoprotein mRNA, and a

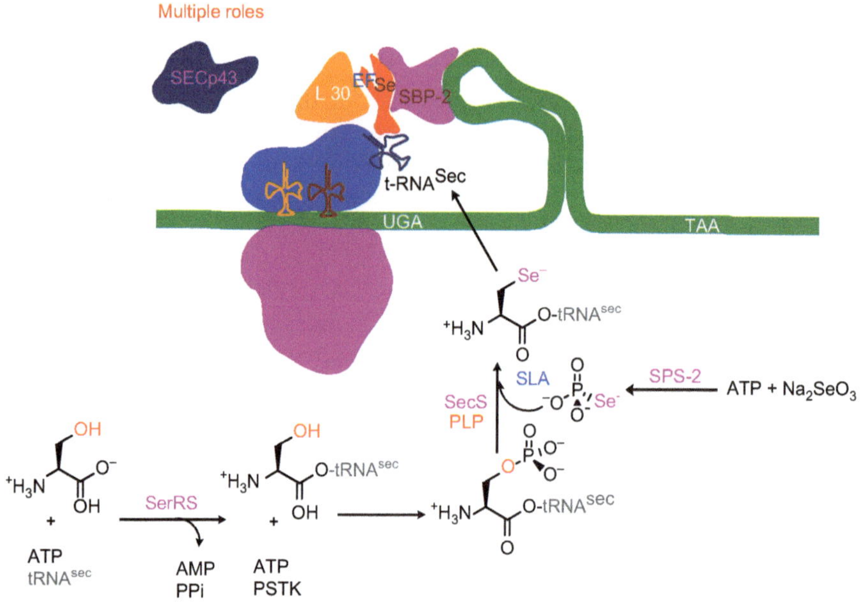

Figure 10.3.6 Schematic representation of selenoprotein biosynthesis.

subsequent shift in the conformation in the SECIS triggers the displacement of SBP-2 and favours the L30 binding. Much of the current research on selenoprotein synthesis and its mechanism reveals the role of subcellular localization and intermolecular interactions in Sec incorporation. Depending on the cellular oxidation state, the SBP-2 is found to have several nuclear localization sequences and shuttles between the cytoplasm and nucleus with EFsec, and is modulated by different redox stress conditions.[50,51] However, SBP-2 is unable to interact simultaneously with the ribosome and SECIS element, but can associate with the ribosome, giving further evidence for a model where SBP-2 binding to the SECIS is displaced by L30.[49,52] Recent findings indicate that a new Sec-tRNASec interacting protein, SECp43, may coordinate multiple steps, leading to selenoprotein synthesis (Figure 10.3.6).[53] SECp43 has been found to help in SLA and SPS1 localization to the nucleus by interacting with them. SECp43 also strengthens the interaction between SBP-2 and EFsec and increases in overall Sec incorporation and selenoprotein mRNA levels.[53,54]

10.3.2.3 Structural Aspects of Iodothyronine Deiodinases

Although there are significant differences in the catalytic functions of three deiodinases, they have some notable similarities. All three enzymes belong to the family of integral membrane proteins with molecular mass 29–33 kDa, and have high sequence similarity around the active centre. Sec is the key residue for the catalytic activity of all the three deiodinases.[20,55,56] However, for D1, some structural and functional data are available, which were derived from basic enzyme kinetics and localization studies. Conformational details, catalytic mechanism, and three-dimensional structural information are greatly limited due to the unavailability of pure proteins and the difficult in synthesizing the enzymes in large quantities in soluble and catalytically active forms for the purpose of structure determination by crystallography or by NMR spectroscopy. This is due to the presence of a transmembrane domain and the inefficient synthesis of selenoproteins owing to the relatively large difference between their synthetic pathways in prokaryotes and eukaryotes.[20,38,55–57]

Hydrophobic cluster analysis[58,59] shows that all the three deiodinases exhibit significant similarity in their sequences and they contain a thioredoxin (TRx) fold. For example, the structure of the three proteins contains a single TM segment in the N-terminal.[60] The model based on the hydrophobic cluster analysis predicts that the active centre of deiodinases is a Sec-containing pocket with β_1-α_1-β_2 motif of TRx fold and the IDUA-like insertion segment around the active centre (Figure 10.3.7). Hydrophobic cluster analysis further predicted that the highly conserved deiodinase-specific segment between amino acids 152 and 166 (according to D1 numbering) has striking similarities with another enzyme, lysosomal-α-L-iduronidase, which belongs to the family of glycoside hydrolases (GH)[61] that remove α-linked iduronic acid residues from the non-reducing end of the glycosaminoglycans heparan sulfate and dermatan sulfate. The local structural similarity partly reflects on the substrate selection. For example, the overall structure of T4 or T3 and α-L-iduronic acids is similar. Notably, in dehalogenases, the helical cap accompanying the βαβ motif along with residues at the end of β strands includes the ligand binding site.[65] In this

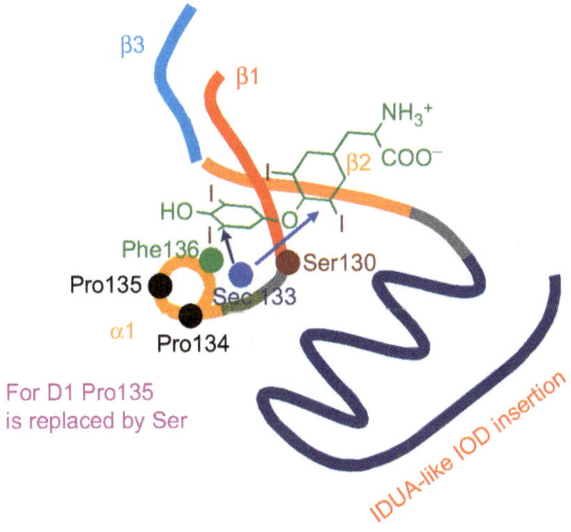

Figure 10.3.7 Schematic representation of proposed model shows the putative active site of deiodinases. Deduced from sequence alignment and by associated modelling.
(Modified figure from Ref. 60).

model, it has been speculated that the IDUA insertion may play a similar role in the deiodinases. However, accurate structural information based on crystal structure or solution-based NMR structural data should enrich our understanding of mechanism by which the deiodination proceeds. Such information would also be useful for inhibitor design.

In the active site, the sequence Thr–Sec–Pro–Pro/Ser–Phe is identical in all the three deiodinases. In the transition from β_1 to α_1 strands, D1 uses an uncharged Ser at position 128, whereas D2 and D3 use a non-polar Pro residue in the corresponding position. Detailed mutagenesis has been carried out at various positions to understand the role of different amino acid residues. The replacement of Phe-136 with Ala in D2 inactivated the enzyme and suggested the possible role of Phe-136 in π-π stacking with substrate T4 or T3. Similarly, mutational studies were carried out to understand the significance of Pro-128/135/146 in all the three deiodinases. The S128P D1, P135S D2, and P146S D3 mutants were used for this purpose (Table 10.3.1). Interestingly, when Pro-135 was replaced by Ser, the enzyme became PTU sensitive. The replacement of Pro-135 to Ser of D2 alters the kinetics and the enzyme may follow a D1-like mechanism. Other mutational studies were performed to understand the effect of different residues in different predicted regions of all the deiodinases and the changes in K_M and PTU sensitivity pattern (Table 10.3.1).

10.3.2.4 Deiodination Mechanism

Although complete structural information is not available for any of the three deiodinases, studies on recombinant and endogenous deiodinases suggested that the D1-mediated catalysis of thyroxine deiodination follows a bisubstrate, ping-pong type of kinetics. On the basis of experiments with histidine specific reagents, it has been postulated that the Sec residue in D1 may form an imidazolium-selenolate ion pair, which is essential for the reaction (Scheme 10.3.3). Further pH-dependent kinetics and site-directed mutagenesis studies[67–69] also supported the formation of an imidazolium-selenolate ion pair at the active site of D1. According to this model, the active site selenolate first attacks at one of the iodines of iodothyronines to produce a putative selenenyl iodide intermediate. Subsequently, the selenenyl iodide species is reduced by an intracellular, endogenous yet unidentified thiol cofactor to regenerate the selenolate. The formation of a selenenyl iodide intermediated was postulated mainly based on the inhibition of D1 activity by 6-n-propyl-2-thiouracil (PTU). It has been shown that PTU may react with the selenenyl iodide intermediate to produce a D1–PTU complex as dead-end product, in which the sulfur of PTU moiety is covalently bound to the selenium of D1. However, the formation of a selenenyl iodide in D1 has not been experimentally proved.

As the enzymatic deiodinations of T4 are very complicated, there have been attempts to develop simple selenium compounds as mimics of the selenium-containing deiodinases. Engman *et al.* have reported that nulcleophilic selenium and tellurium compounds such as PhSeH, PhTeH, and NaHTe are capable of removing iodine from a variety of activated 2,6-diiodophenol

Table 10.3.1 Description of different mutations and their effect on deiodination kinetics.

	Construct	K_m *rT3*	K_m *T4*	K_m*T3*	*PTU*	*kinetics*	*Source*
D1	*Wild type*	300–500 nM	2 μM	2.8 μM	Yes	Ping-pong	63
D2	*Wild type*	8 nM	1–2 nM		No	Sequential	64
D3	*Wild type*		4–5 nM	1–2 nM	No	Sequential	65
Thioredoxin fold							
β1 motif							
D2	F128A		No change				60
D1	C124A	No change				Ping-pong	66
D2	A131C		No change		No	Sequential	77
D2	A131S		No change		No	Sequential	77
α1 motif							
D1	S128P	No change			No	Sequential	60
D2	P134A		No change		No		60
D2	P135S		250 nM		Yes	Ping-pong	60
D3	P146S			10 nM	Yes		
D2	F136A		Inactive				60
β2 motif							
D1	E156A	Inactive					60
D1	E156D	1.8 μM					60
D2	E163A		Inactive				60
D2	E163D		No change				60
D3	E174A		Inactive				60
IDUA-like motif							
D1	H158N	Inactive					69
D1	H158Q	Inactive					69
D2	H165N		Inactive				60
D1	W163A	Inactive					60
D2	W170A		Inactive				60
D1	H174Q	4 μM					69
D1	H174N	20 μM					69
D2	H185Q		No change	No change			60

derivatives.[75] The tellurium compound was particularly effective in performing a complete deiodination reaction. Recently, Goto *et al.* have reported that the synthetic selenol **1** converts the thyroxine derivative **2** (*N*-butyrylthyroxine methyl ester) to the corresponding triiodo derivative (**3**) by an 5′-deiodination (Scheme 10.3.4A). The reaction was carried out in organic solvent ($CDCl_3$) and a relatively high temperature (50 °C) and long reaction time (7 days) were required for about 65% conversion. However, the above model study also led to the conclusion that the two outer-ring iodines of T4 are more reactive than the inner-ring iodines since an enol–keto tautomerism is required for the deiodination (Scheme 10.3.4B).

Interestingly, the report by Goto *et al.* provided experimental evidence for the formation of a selenenyl iodide (R–Se–I) in the deiodination of a thyroxine derivative by an organoselenol. In general, selenenyl iodides react rapidly with

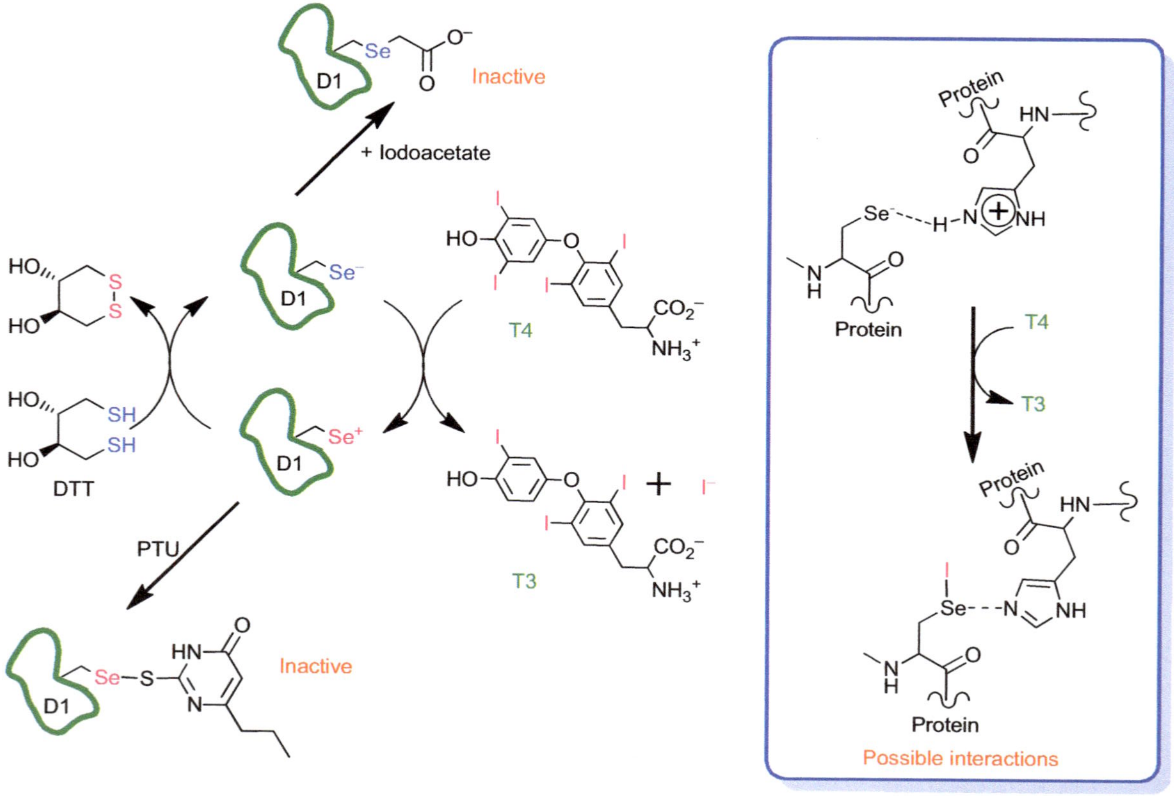

Scheme 10.3.3 Proposed mechanism for the deiodination of T4 by D1.[55,70–74]

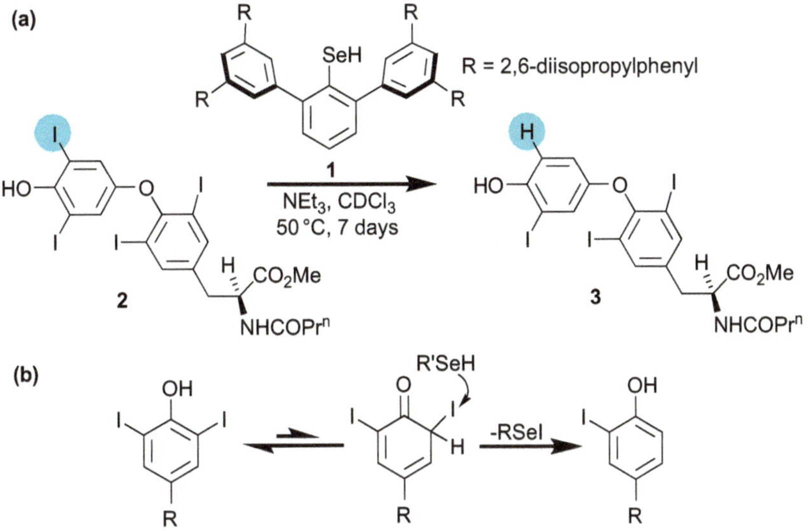

Scheme 10.3.4 (A) Deiodination of thyroxine derivative **2** by selenol **1**. (B) Mechanism for the deiodination *via* enol–keto tautomerism as reported by Goto *et al.*[76]

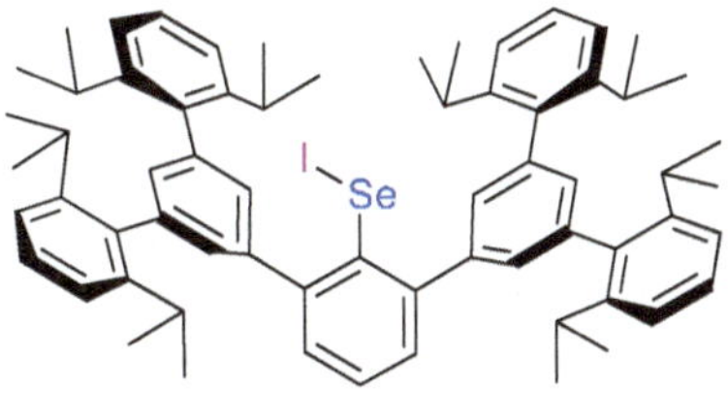

Figure 10.3.8 Structure of bulky aromatic selenenyl iodide, BpqSeI.

selenols to produce the corresponding diselenides. In the case of **1**, the sterically bulky substituents block the formation of diselenides, and therefore, the stable selenenyl iodide (BpqSeI) could be isolated in good yield (Figure 10.3.8). In D1, the amino acid residues present in the active site may stabilize the selenenyl iodide intermediate. However, it is not clear whether D3 uses any additional functional groups for the selective removal of iodine from the inner-ring (5-deiodination) of T4.

Visser and co-workers[77] proposed a completely different mechanism for the outer-ring deiodination of T4 by D2. Accordingly to this mechanism, the selenol (or selenolate) attacks at the 2-position of T4 or rT3 to form a covalent bond between the enzyme selenium moiety and T4 (Scheme 10.3.5). Abstraction of iodonium ion from the enzyme–substrate complex by a thiol cofactor (*e.g.* DTT) leads to formation of a sulfenyl iodide, which may react with another molecule of thiol to produce the corresponding disulfide with elimination of iodide. On the other hand, the protonation of the T3-carbanion

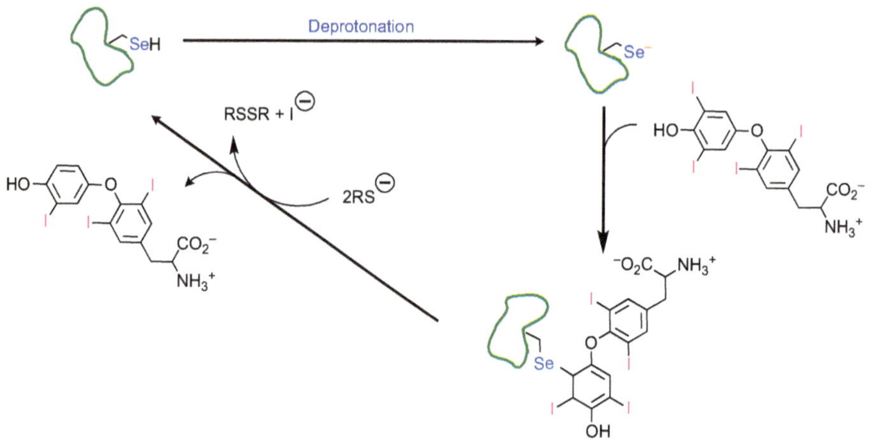

Scheme 10.3.5 Proposed mechanism for the deiodination of T4 by D2.[77]

generated during the reaction produces T3 and unmodified enzyme. As this reaction requires two thiol groups, DTT works efficiently as cofactor. In this model, the activated T4 is directly reduced by the thiol cofactor, which is in contract to the D1 model where a selenenyl iodide has been shown to be an intermediate.

The mechanism proposed for the reductive deiodination of T4 is based on the relative nucleophilicity of selenol and thiol and the K_M values observed for the deiodination. A similar mechanism has been proposed for the reductive dehalogenation of 5-bromo- and 5-iodo-2-deoxyuridylate by thymidylate synthetase.[78–80] As the mutant D2 Sec133Cys is also active in the deiodination, the Cys133 residue, which is suitably positioned at the active site, may attack at the 2-position of T4. This is again similar to thymidylate synthetase in which Cys198 attacks at the C6-position of the pyrimidine in bromodeoxyuridine monophosphate. This mechanism explains the 1000-fold increase in K_M and 10-fold decrease in turnover number for the cysteine mutant as compared to the wild-type D2. This model also explains the relative insensitivity of D2 toward PTU during the deiodination. In D1, the selenenyl iodide (-SeI) intermediate reacts with PTU to produce a selenenyl sulfide.

Recently, a chemical model for the inner-ring deiodination of T4 and T3 by D3 has also been reported.[81–83] It has been shown that the naphthyl-based selenol **4** bearing a thiol group in close proximity to the selenium acts as a model for D3 by selectively deiodinating T4 and T3 to produce rT3 and T2, respectively, under physiologically relevant conditions (Scheme 10.3.6). This study suggested that the nature of substituents may alter its reactivity towards inner- or outer-ring iodine atoms. The presence of an additional inbuilt thiol group adjacent to the selenium is probably essential as this group not only acts as thiol cofactor in the deiodination reactions, but may also assist the selenol in polarizing the C-I bond. The selective removal of iodine from the inner ring of T4 indicates that an enol–keto tautomerism is not required for the

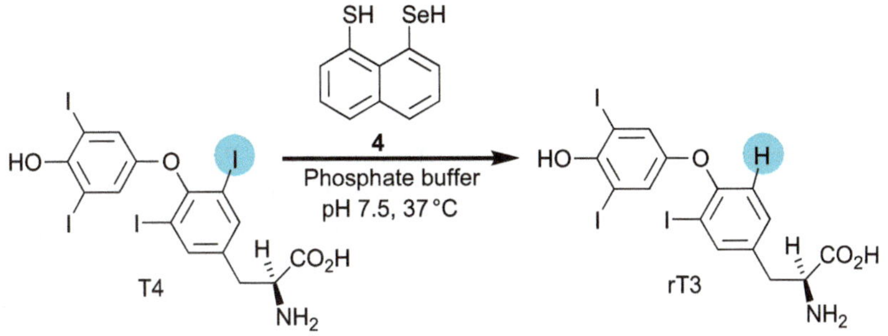

Scheme 10.3.6 A highly selective inner-ring (**5**) monodeiodination of thyroxine (T4) by compound **4**.[81]

deiodination. Although the activation of Sec by histidine is believed to be important for D1-mediated deiodination, such activation is probably not essential for the D3-catalysed reaction.

10.3.3 Conclusions and Future Perspectives

In this chapter, the role of sulfur and selenium species in thyroid gland is described. Thyroxine is produced on thyroglobulin by a TPO/hydrogen peroxide/iodide system and deiodinated to its active and inactive forms by three isoforms of Sec-containing iodothyronine deiodinases (D1, D2, D3). While D1 can deiodinate T4 in both the outer and inner rings, D2 and D3 are selective for either the outer ring (D2) or inner ring (D3). Kinetics and mechanistic studies have revealed that the reaction of Sec residue in D1 with T4 produces a selenenyl iodide intermediate, which can react with a thiol cofactor to produce T3 and unmodified enzyme. In contrast, the mechanism for the reductive deiodination of T4/T3/rT3 by D2 and D3 is still not clear. Some synthetic selenium compounds that functionally mimic the activity of deiodinases have proved useful in understanding the role of selenium in deiodination reactions. However, further studies with both the natural enzymes and synthetic models are required to fully understand the regioselectivity of the deiodination. Such studies may also help in the design of isoform-specific inhibitors and for the development of drugs for thyroid-related problems.

References

1. S. C. Werner, and S. Ingbar, in *The Thyroid: A Fundamental and Clinical Text*, ed. L. E. Braverman, R. D. U. Tiger, Lippincott, Philadelphia, 1991.
2. C. H. Bastomasky, in *Handbook of Physiology*, ed. S.R. Geiger, American Physiological Society, Washington, 1974, vol. 3, pp. 81–89.
3. H. Studer, H. Kohler and H. Burgi, in *Handbook of Physiology*, ed. M. A. Greer, D. H. Solmon, American Physiological Society, Washington, 1974, vol. 3, p. 974.

4. P. R. Larsen, T. F. Davies and I. D. Hay, The thyroid gland, in *Williams Textbook of endocrinology*, ed. J. D. Wilson, D. W. Foster, H. M Kronenberg, and P. R. Larsen, W. B. Saunders, Philadelphia, 9th edn, 1998, pp. 389–515.
5. P. R. Larsen, H. M. Kronenbrg, S. Melmed and K. Polonsky, in *Williams Textbook of endocrinology*, W. B. Saunders, Philadelphia, 10th edn, 2002.
6. G. W. Anderson, C. N. Mariash and J. H Oppenheimer, Molecular actions of thyroid hormone, in *Werner and Ingbar's The Thyroid*, ed. L. E. Braverman, R. D. Utiger, Lippincott, Williams & Wilkins, Philadelphia, 2000, pp. 174–195.
7. V. M. Darras, R. Hume and T. J. Visser, *Mol. Cell Endocrinol.*, 1999, **151**, 37–47.
8. M. A. Lazar, *J. Clin. Invest.*, 2003, **112**, 497–499.
9. D. S. Cooper, F. S. Greenspan and P. W. Ladenson, The thyroid gland, In *Greenspan's Basic and Clinical Endocrinology*, ed. J. Shanahan, McGraw Hill-Medical, New York, 8th edn, 2007, pp. 209–280.
10. P. Sadana, Y. Zhang, S. Song, G. A. Cook, M. B. Elam and E. A. Park, *Mol. Cell. Endocrinol.*, 2007, **267**, 6–16.
11. M. Govindan, X. Meng, C. L Denis, P. Webb, J. D. Baxter and P. G. Walfish, *Proc. Natl. Acad. Sci. U.S.A.* 2009, **106**, 19854–19859.
12. W. Chen, Q Yang and R. G. Roeder, *Mol. Cell.*, 2009, **35**, 755–768.
13. K. B. Horwitz, T. A. Jackson, D. L. Bain, J. K. Richer, G. S. Takimoto and L. Tung, *Mol. Endocrinol.*, 1996, **10**, 1167–1177.
14. A. Taurog, Hormone synthesis. In *Werner and Ingbar's The Thyroid*, ed. L. Bravaman, R. Utiger, Lippincott, Raven Philadelphia, PA, 1996, pp. 47–81.
15. M. Yen, *Physiol. Rev.*, 2001, **81**, 1097–142.
16. J. Gross and R. Pitt-Rivers, *Lancet*, 1952, **1**, 439–41.
17. L. E. Braverman, S. H. Ingbar and K. Sterling, *J Clin Invest.*, 1970, **49**, 855–864.
18. Y. Wu and R. J. Koenig, *Trends. Endocrinol. Metab.*, 2000, **11**, 207–211.
19. P. M. Yen, S. Ando, X. Feng, Y Liu, P. Maruvada and X. Xia, *Mol. Cell Endocrinol.*, 2006, **246**, 121–127.
20. A. C. Bianco, D. Salvatore, B. Gereben, M. J. Berry and P. R. Larsen, *Endocr. Rev.*, 2002, **23**, 38–89.
21. M. Sugawara, R Lau, H. L. Wasser, A. M. Nelson, K. Kuma and J. M. Hershman, *Metabolism*, 1984, **33**, 332–336.
22. P. R. Larsen, J. E. Silva and M. M. Kaplan, *Endocr. Rev.*, 1981, **2**, 87–102.
23. M. M. Kaplan and K. A. Yaskoski, *J. Clin. Invest.*, 1981, **67**, 1208–1214.
24. K. B. Becker, M. J. Schneider, J. C. Davey and V. A. Galton, *Endocrinology*, 1995, **136**, 4424–4431.
25. D. L. St. Germain and V. A. Galton, *Thyroid*, 1997, **7**, 655–668.
26. J. E. Silva and P. R. Larsen., *J. Clin. Invest.*, 1985, **76**, 2296–2305.
27. A. C. Bianco and J. E. Silva., *Endocrinology*, 1987, **120**, 55–62.
28. A. C. Bianco, X. Sheng and J. E. Silva., *J. Biol. Chem.*, 1988, **263**, 18168–18175.

29. S. D. Carvalho, E. T. Kimura, A. C. Bianco and J. E. Silva., *Endocrinology*, 1991, **128**, 2149–2159.
30. N. Marsh-Armstrong, H. Huang, B. F. Remo, T. T. Liu and D. D. Brown., *Neuron*, 1999, **24**, 871–878.
31. W. S. Simonides, M. A. Mulcahey, E. M. Redout, A. Muller, M. J. Zuidwijk, T. J. Visser, F. W. Wassen, A. Crescenzi, W. S. da-Silva, J. Harney, F. B. Engel, M. J. Obregon, P. R. Larsen, A. C. Bianco and S. A. Huang, *J. Clin. Invest.*, 2008, **118**, 975–983.
32. V. A. Galton, *Thyroid*, 2005, **15**, 823–834.
33. D. D. Brown, *Thyroid*, 2005, **15**, 815–821.
34. M. Dentice, A. Bandyopadhyay, B. Gereben, I. Callebaut, M. A. Christoffolete, B. W. Kim, S. Nissim, J. P. Mornon, A. M. Zavacki, A. Zeold, L. P. Capelo, C. Curcio-Morelli, R. Ribeiro, J. W. Harney, C. J. Tabin and A. C. Bianco, *Nat. Cell Biol.*, 2005, **7**, 698–705.
35. M. Dentice, C. Luongo, S. Huang, R. Ambrosio, A. Elefante, D. Mirebeau- Prunier, A. M. Zavacki, G. Fenzi, M. Grachtchouk, M. Hutchin, A. A. Dlugosz, A. C. Bianco, C. Missero, P. R. Larsen and D. Salvatore, *Proc Natl Acad Sci USA*, 2007, **104**, 14466–14471.
36. A. C. Bianco and B.W. Kim, *J. Clin. Invest.*, 2006, **116**, 2571–2579.
37. M. J. Berry, L. Banu and P. R. Larsen, *Nature*, 1991, **349**, 438–440.
38. M. J. Berry, L. Banu, Y. Y. Chen, S. J. Mandel, J. D. Kieffer, J. W. Harney and P. R. Larsen, *Nature*, 1991, **353**, 273–276.
39. G. V. Kryukov, S. Castellano, S. V. Novoselov, A. V. Lobanov, O. Zehtab, R. Guigo and V. N. Gladyshev, *Science*, 2003, **300**, 1439–1443.
40. P. R. Hoffmann and M. J. Berry, *Thyroid*, 2005, **15**, 769–77.
41. M. J. Berry, A. L. Maia, J. D. Kieffer, J. W. Harney and P. R. Larsen, *Endocrinology*, 1992, **131**, 1848–1852.
42. A. M. Dumitrescu, X. H. Liao, M. S. Abdullah, J. Lado-Abeal, F. A. Majed, L. C. Moeller, G. Boran, L. Schomburg, R. E. Weiss and S. Refetoff, *Nat. Genet.*, 2005, **37**, 1247–1252.
43. K. Forchhammer and A. Bock, *J Biol Chem*, 1991, **266**, 6324–6328.
44. M. J. Guimaraes, D. Peterson, A. Vicari, B. G. Cocks, N. G. Copeland, D. J. Gilbert, N. A. Jenkins, D. A. Ferrick, R. A. Kastelein, J. F. Bazan and A. Zlotnik, *Proc. Natl. Acad. Sci. USA*, 1996, **93**, 15086–15091.
45. S. C. Low, J. W. Harney and M. J. Berry, *J. Biol. Chem.*, 1995, **270**, 21659–21664.
46. B. A. Carlson, X. M. Xu, G. V. Kryukov, M. Rao, M. J. Berry, V. N. Gladyshev, D. L. Hatfield, *Proc. Natl. Acad. Sci. U. S. A.* 2004, **101**, 12848–12853.
47. X. M. Xu, B. A. Carlson, H. Mix, Y. Zhang, K. Saira, R. S. Glass, M. J. Berry, V. N. Gladyshev and D. L. Hatfield, *PLoS. Biol.*, 2007, **5**, e4.
48. L. A. de Jesus, P.R. Hoffmann, T. Michaud, E. P. Forry, A. Small-Howard, R. J Stillwell, N. Morozova, J. W. Harney and M. J. Berry, *Mol. Cell. Biol.*, 2006, **26**, 1795–1805.

49. J. Yuan, S. Palioura, J.C. Salazar, D. Su, P. O'Donoghue, M.J. Hohn, A. M Cardoso, W.B. Whitman and D. Soll, *Proc. Natl. Acad Sci. U. S. A.* 2006, **103**, 18923–18927.
50. L. Chavatte, B.A. Brown and D. M. Driscoll, *Nat. Struct. Mol. Biol.*, 2005, **12**, 408–416.
51. L. V. Papp, J. Lu, F. Striebel, D. Kennedy, A. Holmgren and K. K. Khanna, *Mol. Cell. Biol.*, 2006, **26**, 4895–4910.
52. S. A. Kinzy, K. K. Caban and P. R. Copeland, *Nucleic Acids Res.*, 2005, **33**, 5172–5180.
53. A. Small-Howard, N. Morozova, Z. Stoytcheva, E. P. Forry, J. B. Mansell, J. W. Harney, B. A. Carlson, X. M. Xu, D. L. Hatfield and M. J. Berry, *Mol. Cell. Biol.*, 2006, **26**, 2337–2346.
54. X. M. Xu, H. Mix, B. A. Carlson, P. J Grabowski, V. N. Gladyshev, M. J. Berry and D. L. Hatfield, *J. Biol. Chem.*, 2005, **280**, 41568–41575.
55. M. J. Berry, J. D. Kieffer, J. W Harney and P. R. Larsen, *J. Biol. Chem.*, 1991, **266**, 14155–14158.
56. W. Croteau, S. L. Whittemore, M. J. Schneider and D. L. Germain, *J. Biol. Chem.*, 1995, **270**, 16569–16575.
57. C. Buettner, J. W. Harney and P. R. Larsen, *Endocrinology*, 2000, **141**, 4606–4612.
58. C. Gaboriaud, V. Bissery, T. Benchetrit and J. P. Mornon, *FEBS Lett.*, 1987, **24**, 149–155.
59. I. Callebaut, G. Labesse, P. Durand, A. Poupon, L. Canard, J. Chomilier, B. Henrissat and J. P. Mornon, *Cell Mol. Life Sci.*, 1997, **53**, 621–645.
60. I. Callebaut, C. Curcio-Morelli, J. P. Mornon, B. Gereben, C. Buettner, S. Huang, B Castro, T. L. Fonseca, J. W. Harney, P. R. Larsen and A. C. Bianco, *J. Biol Chem.*, 2003, **278**, 36887–96.
61. P. M. Coutinho and B. Henrissat, 1999, *Carbohydrate-active Enzymes* (afmb.cnrs-mrs.fr/_cazy/CAZY/index.html).
62. J. Newman, T. S. Peat, R. Richard, L. Kan, P. E. Swanson, J. A. Affholter, I. H. Holmes, J. F. Schindler, C. J. Unkefer and T. C. Terwilliger, *Biochemistry*, 1999, **38**, 16105–16114.
63. M. J. Berry, L. Banu and P. R. Larsen, *Nature*, 1991, **349**, 438–440.
64. D. Salvatore, T. Bartha, J. W. Harney and P. R. Larsen, *Endocrinology*, 1996, **137**, 3308–3315.
65. D. Salvatore, S. C. Low, M. Berry, A. L. Maia, J. W. Harney, W. Croteau, D. L. St. Germain and P. R. Larsen, *J. Clin. Invest.*, 1995, **96**, 2421–2430.
66. B. C. Sun, J. W. Harney, M. J. Berry. and P. R. Larsen, *Endocrinology*, 1997, **138**, 5452–5458.
67. J. Köhrle, *Methods Enzymol.*, 2002, **347**, 125–167.
68. J. A. Mol, R. Docter, G. Hennemann and T. J. Visser, *Biochem. Biophys. Res. Commun.*, 1984, **120**, 28–36.
69. M. J. Berry, *J. Biol. Chem.*, 1992, **267**, 18055–18059.
70. J. L. Leonard and I. N. Rosenberg, *Endocrinology*, 1978, **103**, 2137–2144.
71. T. J. Visser, I. van der Does-Tobe, R. Docter and G. Hennemann, *Biochem. J.*, 1976, **157**, 479–482.

72. T. J. Visser, D. Fekkes, R. Docter and G. Hennemann, *Biochem. J.*, 1978, **174**, 221–229.
73. J. L. Leonard and I. N. Rosenberg, *Endocrinology*, 1980, **106**, 444–451.
74. I. J. Chopra, *Science*, 1978, **199**, 904–905.
75. A. A. Vasilev and L. Engman, *J. Org. Chem.*, 1998, **63**, 3911–3917.
76. K. Goto, D. Sonoda, K. Shimada, S. Sase and T. Kawashima, *Angew. Chem. Int. Ed.*, 2010, **49**, 545–547.
77. G. J. M. Kuiper, W. Klootwijk and T. J. Visser, *Endocrinology*, 2002, **143**, 1190–1198.
78. Y. Wataya and D. V Santi, *Biochem. Biophys. Res. Commun.*, 1975, **67**, 818–823.
79. C. Garrett, Y. Wataya and D. V Santi, *Biochemistry*, 1979, **18**, 2798–2804.
80. J. T. Kealey., J. Eckstein and D. V. Santi, *Chem. Biol.*, 1995, **2**, 609–614.
81. D. Manna and G. Mugesh, *Angew. Chem. Int. Ed.*, 2010, **122**, 9432–9435.
82. D. Manna and G. Mugesh, *J. Am. Chem. Soc.*, 2011, **133**, 9980–9983.
83. D. Manna and G. Mugesh, *J. Am. Chem. Soc*, 2012, **134**, 4269–4279.

Material Chemistry

CHAPTER 11.1

Stable Chalcogen Radicals

JEREMY M. RAWSON* AND JOHN J. HAYWARD

Department of Chemistry and Biochemistry, The University of Windsor, 401 Sunset Avenue, Windsor ON N9B 3P4, Canada
*Email: jmrawson@uwindsor.ca

11.1.1 Introduction

The word ‘stable’ is open to many different interpretations and should be considered within such contexts as lifetime, and physical and chemical environment. We will focus our attention on those chalcogen radicals whose lifetimes are sufficiently long that they can at least be directly detected by EPR spectroscopy (unlike many biological radicals whose lifetimes are too short and need to be identified by means of spin-traps). In most cases our attention will focus on those radicals that are stable enough to be isolated in the solid state, and we review their solid state properties.

In the first edition of this book,[1] which covered the period up to 2005, we provided an overview of developments in this area since the reports of the first thiazyl radical, $S_3N_2^{+\bullet}$, in 1880.[2] In this second edition we outline developments in stable chalcogen radical chemistry over the last 7 years with a focus on the electronic properties of chalcogen–nitrogen radicals; their conductivity, magnetism, and spin-switching behaviour. In order to put these developments in context, some reference to earlier studies will be made where appropriate.

In the last decade, we have witnessed a blossoming of research in dithiazolyl radical chemistry and the materials properties of these derivatives are explored at length, though this has not precluded coverage of other developments in chalcogen radical chemistry and we also include recent work on the interaction between π-based radicals and *d*-orbital based unpaired electrons. However, the

Handbook of Chalcogen Chemistry: New Perspectives in Sulfur, Selenium and Tellurium, 2nd Edition, Volume 2
Edited by Francesco Antonio Devillanova and Wolf-Walther du Mont

Published by the Royal Society of Chemistry, www.rsc.org

huge area of stable C/S-based radicals such as tetrathiafulvalenes (TTF) and their heavier chalcogenide congeners is left for others to discuss later in this book.

11.1.2 Binary Thiazyl Radicals

Binary thiazyl (S/N) compounds have a long and established history dating back to the 1800s with the discoveries, amongst others, of S_4N_4 in 1835 by Gregory[3] and $(SN)_x$ in 1910 by Burt.[4] Indeed, the area of chalcogen–nitrogen chemistry is now extensive and has been the subject of a recent text.[5] To the best of our knowledge the thiazyl radical $S_3N_2^{+\bullet}$ was the first radical to be reported,[2] pre-dating the 'discovery' of Gomberg's first carbon-based (triphenylmethyl) radical[6] in 1900 by some 20 years. Indeed, binary thiazyl compounds have posed fundamental problems in terms of our understanding of the structure, bonding, and reactivity of electron-rich rings and cages and, more recently, in relation to their materials properties.

Recent work on binary thiazyl radicals has focused on the electrochemical generation of radicals *in situ* by $1e^-$ redox processes involving closed-shell binary S/N molecules. Elegant studies, including ^{15}N labelling experiments, by Boeré and Chivers have identified the presence of $S_4N_4^{\bullet-}$ as a 9-line multiplet spectrum (Figure 11.1.1) by EPR during spectro-electrochemical studies of both S_2N_2 and S_4N_4.[7,8] In their studies of S_2N_2 the presence of trace water permitted the observation of the longer-lived $S_2N_2H^{\bullet-}$ radical. EPR spectra of the related 8-membered ring anions, $[(RC)_2N_4S_2]^{\bullet-}$ and $[(R_2P)_2N_4S_2]^{\bullet-}$, have also been reported and rate constants for their first-order decay found to be less than $0.1\,s^{-1}$.[9] Several computational papers have probed the agreement between calculated and observed hyperfine interactions in binary thiazyl radicals.[10]

11.1.3 1,2,3,5-Dithiadiazolyl Radicals

The 1,2,3,5-dithiadiazolyl (DTDA) radical was first characterized by EPR spectroscopy in 1982 by Markovski.[11] A number of routes to DTDA radicals have been reported,[12] but the methodology developed by Oakley to access

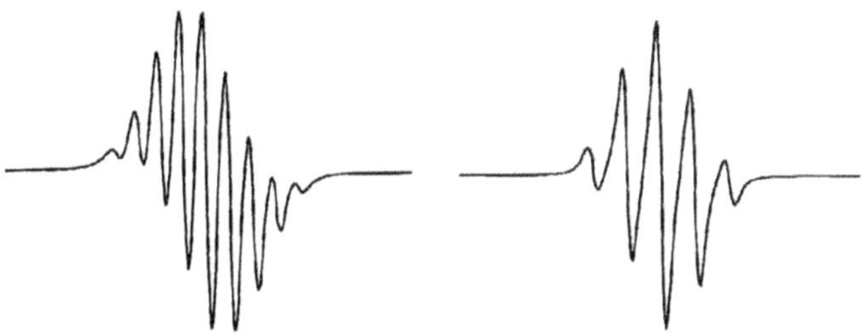

Figure 11.1.1 Solution EPR spectra of $S_4N_4^{\bullet-}$; (left) natural abundance $S_4N_4^{\bullet-}$; (right) 100%-^{15}N-isotopically-enriched $S_4N_4^{\bullet-}$.

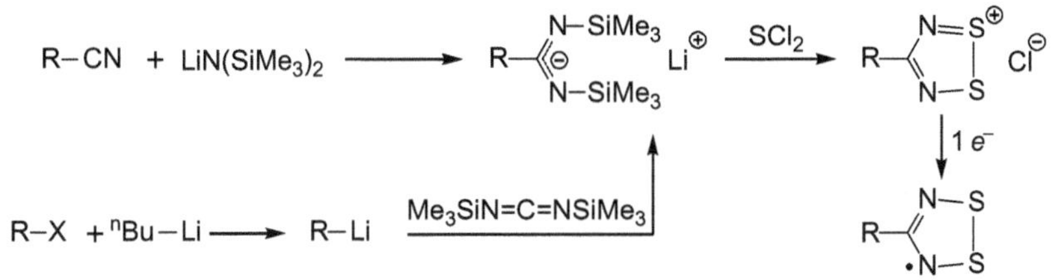

Scheme 11.1.1

benzamidines[13] (Scheme 11.1.1) is remarkably versatile and has provided access to a large number of derivatives from readily accessible aromatic nitriles.[14] The use of $Li[N(SiMe_3)_2]$, however, precludes the use of substrates with acidic protons and does not appear particularly amenable to derivatives where there is steric crowding of the nitrile precursor. In the latter case Rawson has shown[15] that DTDA rings can be accessed *via* Li/H or Li/X exchange methodologies followed by treatment with $Me_3SiNCNSiMe_3$ to access the key benzamidinate intermediate (Scheme 11.1.1).[16] Treatment of the benzamidinates with SCl_2 (2.2 eq) affords the $DTDA^+$ salts which are readily reduced to the radical by $1e^-$ reduction.

Spin density distribution studies on p-$O_2NC_6F_4CNSSN^•$ by polarized neutron diffraction have revealed that the spin density is localized on the CN_2S_2 ring.[17] This is in agreement with electrochemical studies and DFT calculations[18] which indicate that the redox properties and electron distribution are rather insensitive to substituent changes at C. Typically these radicals can be isolated in a crystalline form by vacuum sublimation. In the majority of cases the radicals are found to associate *via* one of several favourable ($\Delta H_{dim} \sim 35$ kJ/mol) π^*–π^* multicentre bonding interactions which are normally, but not exclusively, formed by face-to-face contacts with S···S distances falling in the range 2.9–3.2 Å. This dimerization process is typically considered to generate a diamagnetic 'spin-paired' ground state but recent studies by Passmore[19] and Rawson[20] have revealed that while there is a singlet ground state, these dimers exhibit a thermally accessible triplet excited state in which the singlet-triplet gap seems sensitive to the degree of spatial overlap between radical centres. The triplet state has been characterized by solid state EPR spectroscopy (Figure 11.1.2).

The 1,2,3,5-DTDA radicals have been central to the development of heavier p-block conductors and magnets based on neutral radicals.[21] Oakley's pioneering work on conducting radicals revealed that the propensity for these radicals to dimerize in the solid state opened up a bandgap in the electronic structure, and semiconducting or insulating behaviour was observed for the pristine radicals. However, halogen (Br_2 or I_2) doping has led to improved, albeit activated, conductivities (up to $10^2\,S\,cm^{-1}$), consistent with semiconducting rather than conducting behaviour.[22]

When dimerization can be suppressed through tailoring of the substituents then the radicals are found to be paramagnetic in the solid state. Long-range

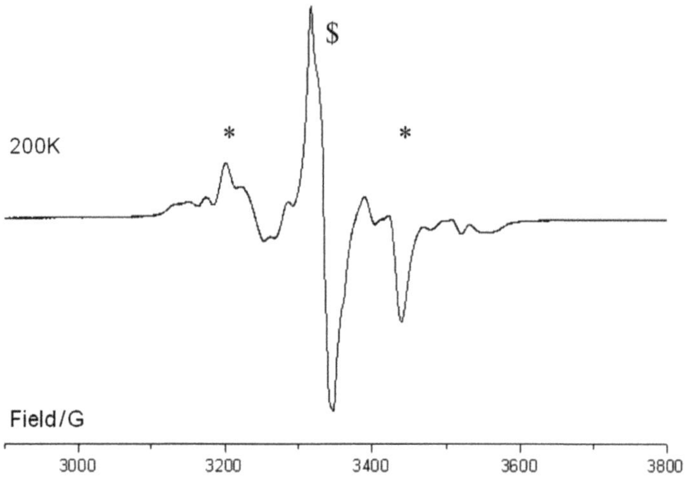

Figure 11.1.2 Solid state EPR spectrum of $(2,5\text{-}Cl_2C_6H_3CNSSN)_2$. Features marked * correspond to the zero field splitting of the S = 1 state while the central intense resonance ($) is attributed to S = $\frac{1}{2}$ defects in the crystal lattice.

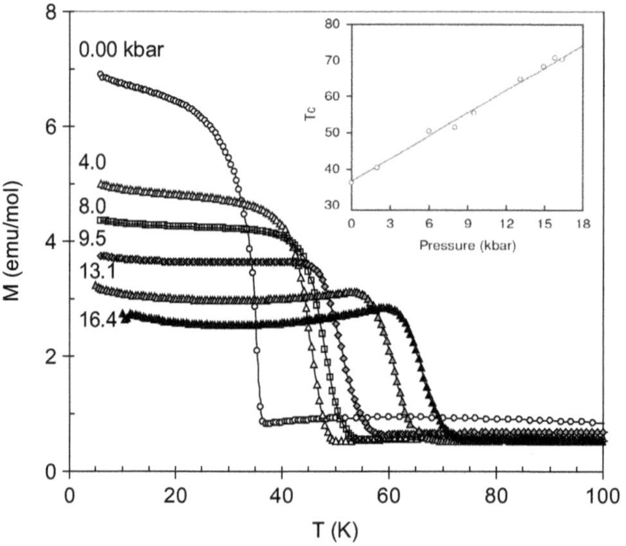

Figure 11.1.3 Temperature dependence of the magnetization of p-NCC_6F_4CNSSN under pressure; the abrupt increase in M is associated with the magnetic ordering.

magnetic order has been observed in two DTDA derivatives; p-$NCC_6F_4CNSSN^{\bullet}$ has been shown to be polymorphic[23] and the β-phase exhibits weak ferromagnetism below 36 K, increasing to 70 K under pressure (Figure 11.1.3).[24] A series of physical studies have been undertaken to probe its magnetic structure and dynamics.[25] In addition, theoretical studies on both α and β-phases have been undertaken to explain their differing magnetic response;[26] the strong magnetic communication in the β-phase is found to occur *via* close intermolecular S···N contacts between DTDA rings which propagate in three dimensions leading to a bulk ordered phase, whereas the α-phase exhibits low-dimensional antiferromagnetic interactions. Structural studies on β-*p*-$NCC_6F_4CNSSN^•$ under pressure reveal the enhancement of T_c towards liquid nitrogen temperature arises from a shortening of the S···N contacts.[24]

Ferromagnetism has also been reported in *p*-$O_2NC_6F_4CNSSN^•$ at the comparatively low temperature of 1.3 K,[27] comparable with many lighter-atom N/O-based radicals. Its magnetism has been probed by DFT which confirm a three-dimensional diamond-like magnetic topology.[28] Notably, the magnetic ordering temperatures generally observed for thiazyl and selenazyl radicals (see later) are typically substantially higher than those observed for lighter-atom C/N- and N/O-based radicals as a consequence of close intermolecular S···S or S···N contacts and the more radially diffuse nature of the orbitals of the heavier chalcogenides.

Clearly the transport and magnetic properties of DTDA radicals are both very sensitive to solid state structure, and work by Haynes has looked at trends in the solid state structures of DTDA radicals in order to establish favourable repeat motifs which can be used to engineer specific architectures, including recent successful co-crystallization studies which auger well for future developments in molecular materials.[14,29] The effects of halogens on crystal packing have attracted considerable attention and prototypical DTDA structures such as $(XCNSSN)_2$ (X = F, Cl, Br) have proved useful grounds to develop more detailed understanding of the factors which control solid state structure.[30] Recent studies by Awaga have shown that a layer of the 4,4′-bis(1,2,3,5-dithiadiazolyl), NSSNC-CNSSN, between Al and ITO (indium tin oxide) exhibit transient photocurrents which may have potential applications in optoelectronic devices. This behaviour has been attributed to a significant imbalance between hole and electron mobilities within the bulk material.[31] Functionalization of commercial cyano-biphenyl liquid crystals ($RC_6H_4C_6H_4CN$) to generate DTDA rings appears to inhibit their mesogenic behaviour and the radicals adopt a π*–π* interaction which renders the bulk material diamagnetic. Nevertheless, variable-temperature EPR studies reveal that radical defects in the diamagnetic host exhibit a thermally activated dynamic disorder *via* rotation about the molecular long axis upon warming.[32]

The coordination chemistry of DTDA radicals has been pioneered through detailed studies by Preuss in recent years.[33] Initial studies on the 2′-pyridyl-DTDA radical (pyDTDA, Scheme 11.1.2) revealed it acts as a chelate ligand towards a range of first-row transition metal ions (Mn, Fe, Co, Ni, Cu)[34] and the solid state structures (Figure 11.1.4) can be tailored by further functionalization of the substituent.[35] Notably, the N atom of the DTDA ring itself appears to be a sufficiently strong donor under certain circumstances to bind to a metal, though this requires some flexibility of the substituent so as not to

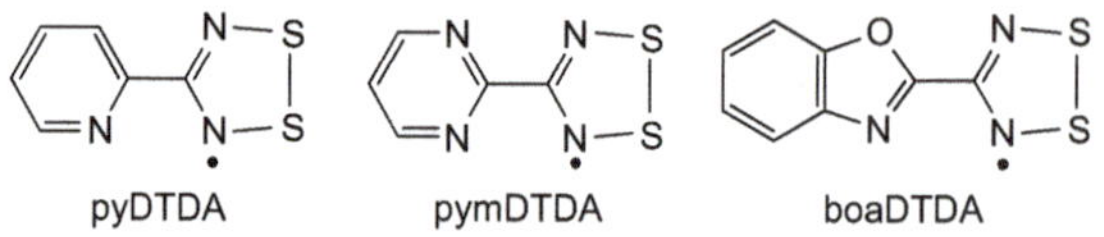

Scheme 11.1.2 Structures of pyDTDA, pymDTDA, boaDTDA.

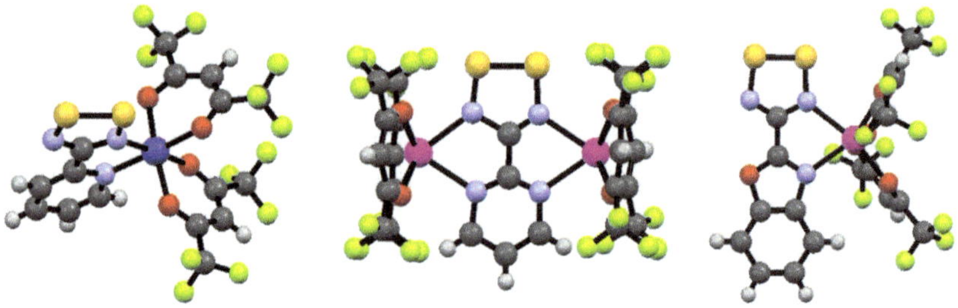

Figure 11.1.4 Structures of (i) $Co(pyDTDA)(hfac)_2$ (ii) $[Mn(hfac)_2]_2(pymDTDA)$ (iii) $Mn(hfac)_2(boaDTDA)$.

hinder coordination.[36] Within these complexes there is very strong magnetic coupling of the radical electron to the paramagnetic metal centre. However, the strong prevalence for the DTDA ring to dimerize sometimes leads to association of these molecular species in the solid state and additional strong antiferromagnetic magnetic interactions between clusters. Use of additional donor atoms on the framework such as pyrimidyl-DTDA (pymDTDA) and benzoxazolyl-DTDA (boaDTDA) offer the potential to bind multiple metal centres; while pyrimidyl-DTDA forms a dimetallic complex $\{Mn(hfac)_2\}_2(pymDTDA)$ (Figure 11.1.4),[37] reaction of boaDTDA with $Mn(hfac)_2$ afforded a 1:1 complex with one of the DTDA N atoms and the O atom of the substituent non-coordinated.[38] In the latter system the $Mn(hfac)_2(boaDTDA)$ complex dimerizes in the solid state through intermolecular S···O interactions which lead to an $S_T = 4$ ground state per pair of $Mn(hfac)_2(boaDTDA)$ complexes.

While the coordination chemistry of DTDA radicals appears centred on the harder N-donor atom, the organometallic chemistry focuses on the softer disulfide bridge. Recent work by Boeré has shown[39] that a range of DTDA derivatives react with $[CpCr(CO)_3]_2$ to form $18e^-$ complexes of the form $CpCr(CO)_2(DTDA)$ in which the DTDA ring acts as a $3e^-$ donor. The S–S bond in the DTDA ring is retained upon coordination and, remarkably, the bond length appears virtually unaltered in relation to the parent DTDA radical ($d_{S\text{-}S} = 2.1$ Å). This is in stark contrast to previous studies on similar Fe and Ni complexes as well as heavier Group 10 phosphines in which oxidative addition of the radical to the metal occurs forming mono-, di-, and trimetallic complexes with cleavage of the S–S bond.[40] Boeré attributed this to a π-type interaction of the DTDA ring with the Cr centre; donation of electron density occurs from a fully occupied low-lying weakly S–S π-bonding MO and one from the S–S antibonding π* SOMO such that the DTDA radical becomes a formal $3e^-$ donor. Removal of electron density from both bonding and antibonding π orbitals has a minimal net effect on the net S-S bond order.

The heavier diselenadiazolyl congeners, $RCN_2Se_2^{\bullet}$ are well known and can be prepared using analogous synthetic methods.[13] However, they appear to have substantially larger dimerization enthalpies than their lighter S-analogues[41] and in all cases reported to date, the structures are dimeric and diamagnetic. The replacement of S by Se appears to lead to small improvements in solid state transport properties, though their tendency to dimerize means they are best described as thermally activated semiconductors.[42] With the exception of their higher dimerization enthalpy, their chemistry otherwise appears to closely resemble that of the dithiadiazolyls, *e.g.* they undergo oxidative addition reactions with a range of low-valent organometallics with Se–Se bond cleavage.[43]

11.1.4 1,3,2,4-Dithiadiazolyl Radicals

The 1,3,2,4-dithiadiazolylium ring system is most commonly generated *via* the reverse electron demand dipolar cycloaddition chemistry of the SNS^+ cation with nitriles pioneered by Passmore.[44] The cycloaddition chemistry is a robust methodology which supports a range of organic nitriles[45,46] as well as inorganic substituents, including covalently bound cyanometallates (Scheme 11.1.3). The corresponding 1,3,2,4-DTDA radicals generated by $1e^-$ reduction of the cations have been shown to be susceptible to thermal or photochemical rearrangement to the 1,2,3,5-DTDA radical both in solution and in the solid state.[47] As a consequence few 1,3,2,4-DTDA radicals have been isolated and fully characterized.[19a,47] Nevertheless, the structural evidence available indicates that they, like the corresponding 1,2,3,5-DTDA radicals, exhibit a marked propensity for dimerization *via* a π*–π* dimerization process. In the case of p-$C_6H_4(CNSNS^{\bullet})_2$ a centrosymmetric *trans*-dimerization mode is adopted whereas in the mixed diradical, •SNSNC.CNSSN•, dimerization occurs *via* an unusual heterodimer in which 1,3,2,4- and 1,2,3,5-radicals dimerize together (Figure 11.1.5). The propensity for these radicals to rearrange in solution has hampered extensive studies of both their reactivity towards metals and their physical properties. Nevertheless, those magnetic measurements which exist are consistent with the formation of essentially diamagnetic dimers in the solid state, though there is evidence for an increase in paramagnetism upon grinding.[19a]

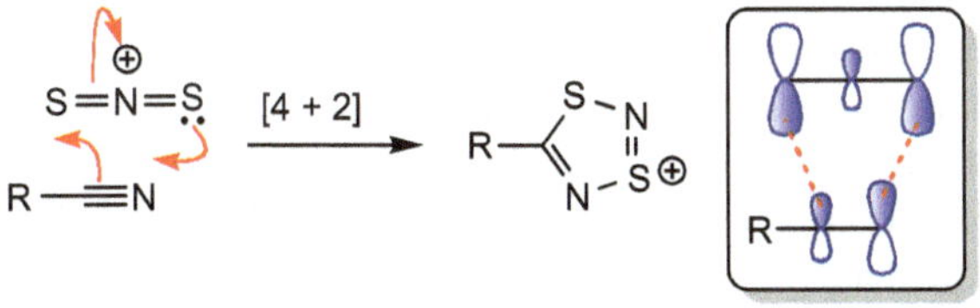

Scheme 11.1.3 Cycloaddition chemistry of SNS^+ with nitriles and (inset) frontier orbitals of SNS^+ and a nitrile.

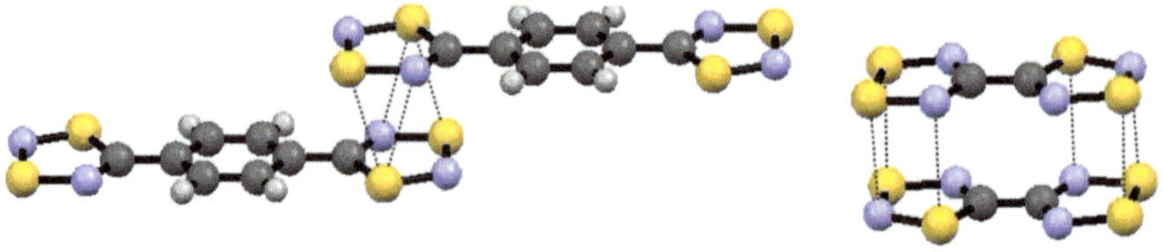

Figure 11.1.5 π^*–π^* dimerization in *p*-$C_6H_4(CNSNS)_2$ (left) and [SNSNC.CNSSN] (right).

11.1.5 1,3,2-Dithiazolyl Radicals

Unlike both the 1,2,3,5- and 1,3,2,4-DTDA radicals described in the last two sections, the 1,3,2-dithiazolyls (1,3,2-DTA) do not exhibit such a strong tendency to dimerize.[48] As a consequence their magnetochemistry is substantially richer and more varied and has been at the centre of a substantial quantity of research in recent years to explore their magnetic properties which include a series of 'high temperature' magnetically ordered phases and spin-transition phenomena.[21]

Non-fused dithiadiazolylium salts $RCSNSCR^+$ are conveniently prepared by the cycloaddition chemistry of SNS^+ with alkynes (Scheme 11.1.3).[1] For fused derivatives, Wolmershäuser developed several routes to 1,3,2-DTA rings starting from appropriate fused 1,2-dithiols.[49] However, only a limited number of 1,2,-dithiols are commercially available and the generation of 1,3,2-DTA radicals has often required tailored syntheses of appropriate dithiol precursors[50] and alternative general synthetic methodologies to access benzo-fused 1,3,2-DTA radicals have been investigated by several groups.

Passmore has recently found that SNS^+ undergoes cycloaddition reactions with quinone and its derivatives. Subsequent migration of H^+, which appears favoured by rearomatization of the benzene ring, affords quinone-fused dithiazolyl derivatives.[51] Reduction of these conjugated systems seems facile using mild reductants (*e.g.* ferrocene and I^-). Using this methodology his group have generated the series of radicals outlined in Scheme 11.1.4. It is clear that other variants on this methodology ought to be possible, provided there is an appropriate H^+ migration route to facilitate rearomatization. The structure of the naphthaquinone derivative has been reported and adopts a *trans*-cofacial π^*–π^* dimer motif with further π-stacking of these dimers (Figure 11.1.6). The intradimer contacts (3.244(2) Å) are sufficient to render the material diamagnetic and are substantially shorter than the interdimer contacts (3.926(2) Å) along the stacking direction. The intradimer distances are comparable with the benzo-1,3,2-dithiazolyl which was also found to be diamagnetic.[48]

Rawson has utilized the S_NAr reaction of [^tBuS]Na with 1,2-dihaloarenes, followed by deprotection of the sulfur by chlorination and finally ring closure with Me_3SiN_3 to generate the 1,3,2-DTA cations. Subsequent 1e^- reduction with Ag powder in polar solvents (THF, MeCN, or *l*.SO_2) afforded the corresponding DTA radicals (Scheme 11.1.5). This methodology appears

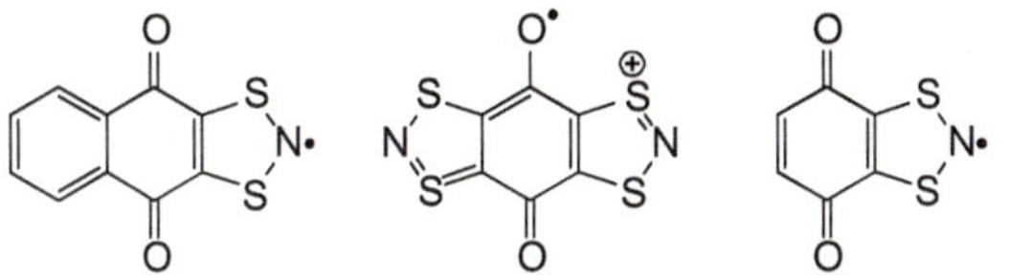

Scheme 11.1.4 Naphthaquinone-DTA, and related systems (ref. 51).

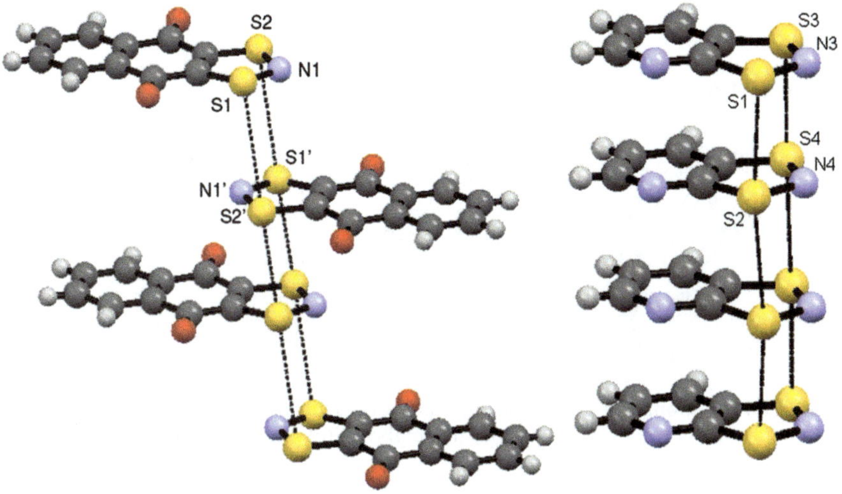

Figure 11.1.6 Structure of NQDTA (left) and pyDTA (right).

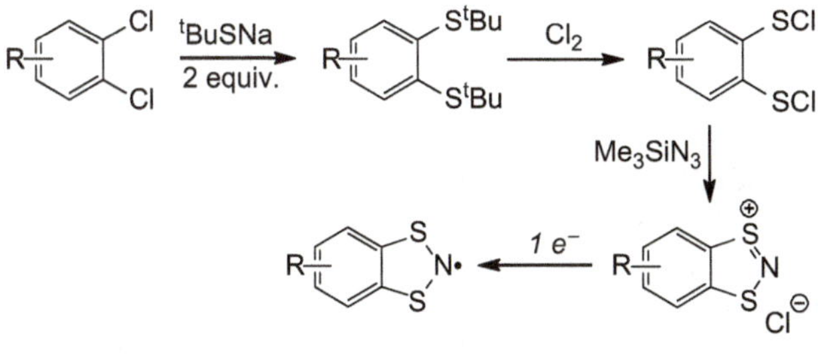

Scheme 11.1.5

relatively versatile and supports electron-withdrawing substituents (CN, CF_3) as well as mildly electron-releasing groups (Me) and heteroatoms (N).[52] The structure of pyDTA adopts a *cis*-cofacial dimer motif (Figure 11.1.6).[52]

An alternative strategy has recently been reported which provides access to selected π-electron-releasing functional groups which is based upon work by Stender *et al.* to generate tetrathiocenes *via* reaction of S_2Cl_2 with 1,2-dialkoxy-benzenes in glacial acetic acid.[53] Rawson has shown that these tetrathiocenes are readily oxidized to the sulfenyl chlorides, and subsequent ring closure with Me_3SiN_3 followed by reduction follows standard protocols

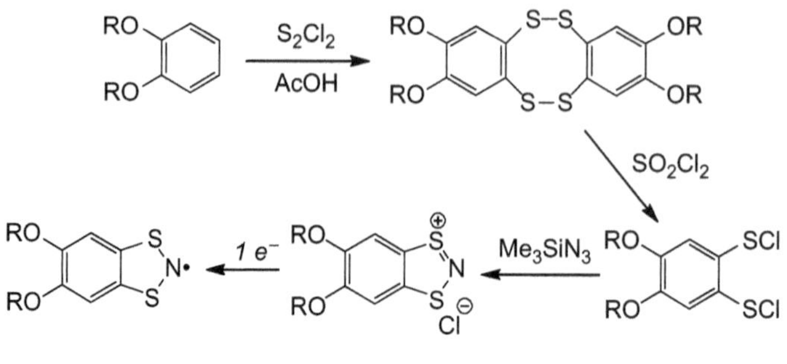

Scheme 11.1.6

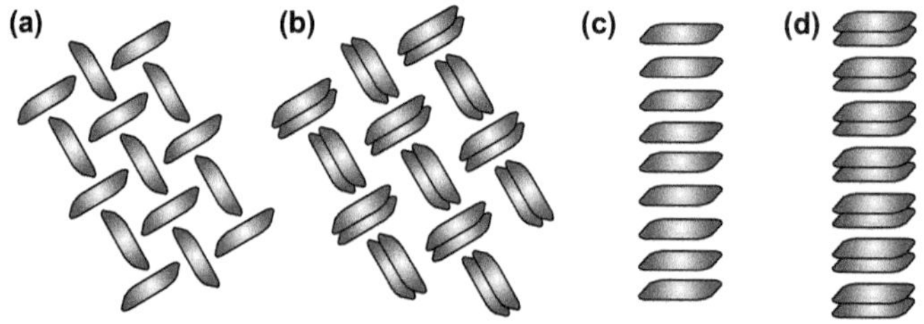

Figure 11.1.7 Common packing motifs for neutral 1,3,2-DTA radicals; (a) herringbone, (b) dimer herringbone, (c) π-stacked, (d) dimer π-stacked.

(Scheme 11.1.6).[54] This methodology ought to be applicable to other π-electron-releasing functional groups such as R_2N.

Computational and experimental studies have probed the electronic structure of the isolated radicals in dilute solution.[55] The solid state properties of 1,3,2-DTA radicals exhibit a rich tapestry of behaviour which is very sensitive to the packing motif; In the absence of electronegative elements these radicals tend to adopt herringbone motifs or 'dimer' herringbone motifs when the radicals form dimers (Figure 11.1.7a,b). While the dimers are essentially diamagnetic consistent with a multicentre spin-paired interaction, the monomeric herringbone structures are paramagnetic and tend to exhibit a network of intermolecular S···N and S···S contacts between DTA radicals. In most cases this appears to propagate low-dimensional antiferromagnetism, characterized by broad rather than λ-type maxima in their χ *vs* T plots. The exception is benzo-1,3,2-dithiazolyl, BDTA (Scheme 11.1.7) which adopts a diamagnetic π*–π* double herringbone motif in the solid state[56] but was found by Awaga to undergo an unusual phase transition upon warming;[57] it melts, then resolidifies at a higher temperature to form a new paramagnetic phase. Upon rapid cooling this material orders as a bulk antiferromagnet at 11 K.

The alternative packing strategies adopted by these neutral 1,3,2-DTA radicals are π-stacked motifs (Figure 11.1.7c,d). This π-stacking, coupled with the weak dimerization enthalpy, confers unusual magnetic properties on many of these 1,3,2-DTA derivatives; At low temperature they often exhibit dimeric

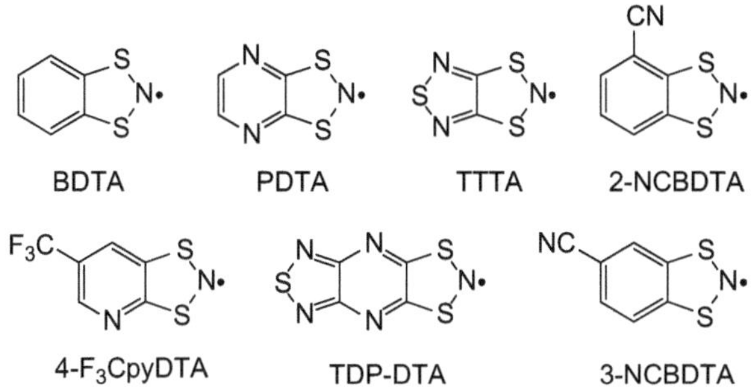

Scheme 11.1.7 Structures of molecules in Table 1; TDP-DTA, TTTA, PDTA, 2-NCBDTA, 3-NCBDTA, 4-F_3CpyDTA.

π-stacked structures, thereby optimizing the small enthalpic stabilization associated with the weak bonding interaction. These dimer phases are diamagnetic and can be considered as a significant Peierl's distortion of a regular π-stack with alternating short (~3.2 Å) and long (~3.9 Å) intermolecular S···S contacts along the stacking direction. However, on warming they typically undergo a solid state phase transition to an entropically favoured regular π-stack (~3.7 Å). These regular π-stacked structures are paramagnetic but exhibit strong antiferromagnetic interactions parallel to the stacking direction. This strong antiferromagnetic exchange is associated with the significant spatial overlap of the π-systems at the interplanar spacings involved. These solid state phase transitions lead to abrupt discontinuities in their magnetic response which can be considered as a spin-switching (diamagnetic–paramagnetic) phenomenon.[58] While the phase transition is reversible, a number of derivatives exhibit thermal hysteresis arising from large activation barriers to the solid state rearrangement. These activation barriers are associated with the cleavage of favourable inter-stack contacts (particularly S···S and S···N) and lead to regions of bistability in which the radicals can exist in either diamagnetic or paramagnetic phases depending upon sample history (Figure 11.1.8). The most thoroughly studied example is TTTA initially prepared by Wolmershäuser.[59] This compound undergoes a reversible phase transition with transition temperatures of 320 K upon warming (diamagnetic to paramagnetic) and 230 K upon cooling (paramagnetic to diamagnetic).[60] Detailed studies by Awaga have shown that this phase transition can not only be driven thermally but can be induced by the application of pressure or light,[61] thereby offering potential for the development of sensors, switches, and data storage applications. Computational studies by Novoa have probed the magnetism of this radical.[62]

These π-stacked, layer-like structures appear to be stabilized by the presence of electronegative groups in the backbone which facilitate in-plane electrostatic interactions. The tailoring of the molecular structures of 1,3,2-DTA derivatives to specifically incorporate heteroatoms and/or structure-directing electronegative groups to promote π-stacking has permitted a range of

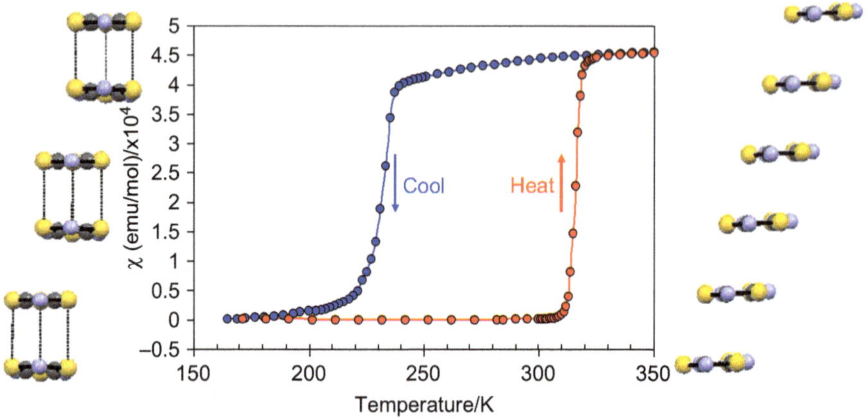

Figure 11.1.8 Temperature dependence of the magnetic susceptibility for TTTA (centre), with diamagnetic low temperature phase (left) and paramagnetic high temperature phase (right) illustrated.

Table 11.1.1 Spin-transition behaviour for 1,3,2-DTA radicals.

Radical	$T_{C(\uparrow)}/K$	$T_{C(\downarrow)}/K$	$\Delta T/K$
TDP-DTA	200	150	50
TTTA	317	225	92
PDTA	343	297	46
2-NCBDTA	250	250	0
3-NCBDTA	304	293	7
4-F_3CpyDTA	50	50	0

spin-switching 1,3,2-DTA to be synthesized, many with transition temperatures near ambient temperature[52,63] (Table 11.1.1).

The conductivities of 1,3,2-DTA radicals have also been explored but tend to exhibit low conductivities ($<10^{-6}$ S cm^{-1}). This can be attributed to (1) a large spin-Peierls distortion for dimeric structures leading to the opening up of a large band gap at the Fermi level; or (2) a very narrow bandwidth arising from the poor spatial overlap associated with the larger interplanar separation, coupled with large on-site Coulomb repulsion leading to a Mott insulating state.[50]

A number of charge-transfer salts of 1,3,2-DTA 'radicals' have been prepared in recent years with a range of acceptors depicted in Scheme 11.1.8.

For BDTA derivatives, the degree of charge transfer can be estimated from the S–N bond length which shortens as electron density is removed from the π* SOMO of BDTA.[64] In a number of cases the extent of electron transfer from the radical is in excess of 0.9 e^- and the DTA heterocycle is best considered as a 1,3,2-DTA$^+$ cation, although it appears that even in its formally cationic state the DTA ring should not be considered as 'innocent' and may provide a pathway for magnetic/electronic exchange. For example, BDTA$^+$ salts of $[M(mnt)_2]^{n-}$ typically comprise mixed π-stacks of cations and anions, yet both

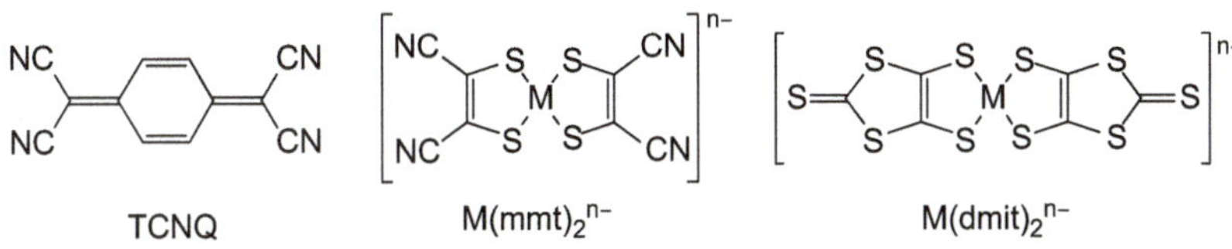

Scheme 11.1.8 TCNQ, $M(mnt)_2^{n-}$, $M(dmit)_2^{n-}$.

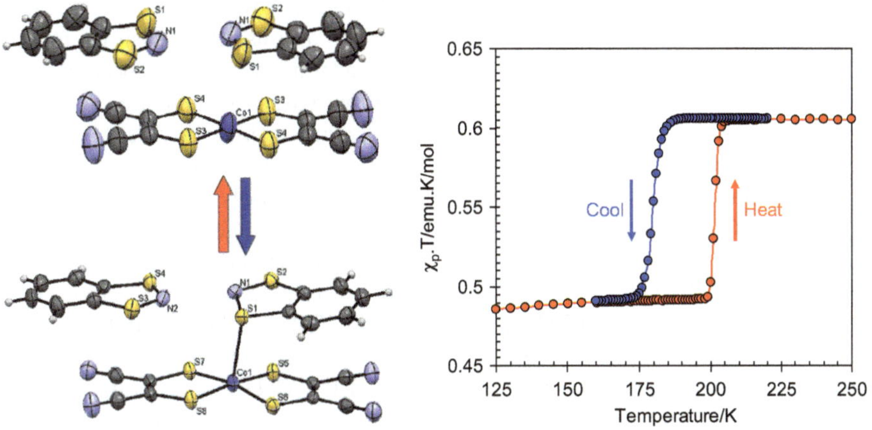

Figure 11.1.9 (left) structural phase transition in $[BDTA]_2[Co(mnt)_2]$ and (right) temperature dependence of χ_pT.

$[BDTA][Ni(mnt)_2]$ and $[BDTA]_2[Cu(mnt)_2]$ exhibit ferromagnetic coupling between the $S=\frac{1}{2}$ metal ions (Ni^{III} and Cu^{II}) which appears mediated *via* the diamagnetic $BDTA^+$ counterions.[65,66] The structure of $[BDTA]_2[Co(mnt)_2]$ also exhibits a mixed-stack structure and, above 190 K, is structurally analogous to the copper complex and best considered as $[BDTA^+]_2[Co(mnt)_2]^{2-}$ containing the low-spin Co^{II} ion ($S=\frac{1}{2}$). However this compound undergoes a phase transition on cooling below 180 K, leading to a weak Co···S coordinate bond to one of the BDTA ring which is associated with charge transfer of $\sim 0.5e^-$ from Co to one BDTA cation, leading to a weak π-d type bonding interaction of the BDTA radical with the Co centre (Figure 11.1.9).[67] Although partial transfer of the unpaired electron from Co to BDTA does not quench the paramagnetism, electron transfer leads to significant contribution from a configuration which can be envisaged as $BDTA^\bullet$ and diamagnetic $Co(mnt)_2^-$. The BDTA radical has a lower g-value due to lack of significant spin-orbit coupling and hence the value of χ_pT is reduced. This process appears reversible and Co–S bond cleavage occurs on warming above 200 K, leading to a 20 K window of hysteresis.

Unlike the $[M(mnt)_2]^{n-}$ salts, $[BDTA][Ni(dmit)_2]_2$ adopts a segregated stacked structure with an estimated charge transfer of 0.24 e^- from BDTA. It is a semiconductor with a room-temperature conductivity of 0.1 S cm^{-1} and an activation energy of 0.19 eV.[68] [BDTA][TCNQ] has been reported[69] to also adopt a segregated π-stacked structure. In this case the estimated degree of charge transfer from BDTA to TCNQ is 0.63 e^- and it is a semiconductor (bandgap $\Delta E = 0.06$ eV) which exhibits a single-crystal conductivity of 2.9×10^{-2} S cm^{-1} at room temperature in agreement with Wolmershäuser's preliminary findings.[49]

The benzo-bis-1,3,2-dithiazolyl ring system BBDTA was independently reported around 1986/7 by the groups of Wolmershäuser and Wudl[70] but the structure of BBDTA was only reported by Oakley a decade later.[71] Recent studies have focused on the radical cation BBDTA$^{+\bullet}$ which forms a diverse range of salts with both diamagnetic main group anions and transition metal salts. The delocalization of the unpaired electron over both ends of the molecule appears to favour three-dimensional networks of magnetic exchange interactions and many complexes have been characterized in recent years a number of which exhibiting long-range magnetic order (Table 11.1.2). The [BBDTA][$GaCl_4$] salt crystallizes in three polymorphic phases; α-[BBDTA][$GaCl_4$] and β-[BBDTA][$GaCl_4$] are diamagnetic due to dimerization of the BBDTA units, whereas γ-[BBDTA][$GaCl_4$], generated by desolvation of [BBDTA][$GaCl_4$].MeCN, orders as a ferromagnet below 6.7 K.[72] The related salts, [BBDTA][MX_4] ($MX_4 = TlBr_4^-$, TlI_4^- and InI_4^-) all adopt two-dimensional layered structures with square-planar lattice of BBDTA$^{+\bullet}$ interspersed by MX_4^- layers. [BBDTA][$TlBr_4$] exhibits canted antiferromagnetism below 11.6 K while [BBDTA][TlI_4] and [BBDTA][InI_4] are metamagnets, *i.e.* possess an antiferromagnetic ground state but undergo a field-induced spin re-orientation to a ferromagnet-like state. Their respective ordering temperatures are 8.9 and 11.3 K respectively.[73] The difference in magnetic response of these [BBDTA][MX_4] salts has been attributed to the relative orientation of the BBDTA$^{+\bullet}$ ions in the organic layer; this leads to antiferromagnetic intralayer interactions in [BBDTA][$TlBr_4$] but ferromagnetic intralayer interactions in [BBDTA][TlI_4] and [BBDTA][InI_4]. In all cases the interlayer interactions are weak and antiferromagnetic. [BBDTA][$FeCl_4$] was originally reported in 1988 by Wolmershäuser and reported to order as a ferrimagnet at 44 K on the basis of Mössbauer studies.[74] Detailed physical studies by Awaga have confirmed[75] these original findings.

The effect of pressure on γ-[BBDTA][$GaCl_4$], which has a ferromagnetic transition of 7.0 K at ambient pressure, leads to an increase in ordering

Table 11.1.2 Magnetic ordering temperatures for BBDTA$^{+\bullet}$ salts.

Complex	*Magnetic order**	T_C/T_N
[BBDTA][$FeCl_4$]	FI	44.0 K
[BBDTA][$TlBr_4$]	CAFM	11.6 K
[BBDTA][InI_4]	MM	11.4 K
[BBDTA][TlI_4]	MM	8.9 K
γ-[BBDTA][$GaCl_4$]	FM	6.7 K
[BBDTA][$FeCl_4$].Me_2CO	AFM	6.4 K

AFM, antiferromagnet, CAFM, canted antiferromagnet (weak ferromagnet), FI, ferrimagnet, FM, ferromagnet, MM, metamagnet.

temperature in excess of 14 K at a hydrostatic pressure of 16.2 kbar (1.62 GPa). This increase in critical temperature is associated with a substantial reduction (88 + %) in magnetic moment which has been attributed to changes in crystal packing and a transformation to diamagnetic α- and β-phases.[76]

The *d*-block Lewis acid $Cu(hfac)_2$ forms a coordination polymer with TTTA in which the heterocyclic N atoms coordinate to Cu forming a one-dimensional coordination polymer with ferromagnetic interactions.[77] More recent studies using the radical cation $BBDTA^{+\bullet}$ as the bridging ligand has led to a range of coordination polymers; $[BBDTA][InBr_4]$ undergoes a spin-Peierls phase transition at 250 K, leading to a marked decrease in magnetic susceptibility below 250 K, consistent with an alternating elongation and contraction of the In–N bonds.[78] The analogous $[BBDTA][InCl_4]$ complex undergoes a similar spin-Peierls transition around 108 K.[79]

The heavier selenium analogues of these 1,3,2-DTA radicals have not been reported, although the selenium analogue of $BDTA^+$, $[C_6H_4Se_2N]^+$, was reported by Wolmershäuser in 1992.[80]

11.1.6 1,2,3-Dithiazolyl Radicals and Their Selenium Analogues

Common routes to 1,2,3-dithiazolylium rings comprise[81] the reaction of aniline derivatives with S_2Cl_2 and the condensation of 2-amino-thiophenols with $SOCl_2$ (this methodology can be implemented to access both Se and Te derivatives also[82]). Subsequent $1e^-$ reduction affords EPR active solutions of the 1,2,3-DTA radicals, though it has only been relatively recently that the first 1,2,3-DTA-based systems have been isolated in the solid state. Non-fused derivatives exhibit delocalization of the electron density around the C_2S_2N ring with significant spin density at C, N, and S and a propensity for σ-dimerization. Derivatives of Appel's salt **1** tend to generate tetrathiadiazafulvalenes **2** *via* initial C-C σ-bond formation at C-5 (Scheme 11.1.9, Figure 11.1.10).[83] Steric protection[84] of C-5 (*e.g.* **3**) or extended π-conjugation (*e.g.* **4**) inhibits this

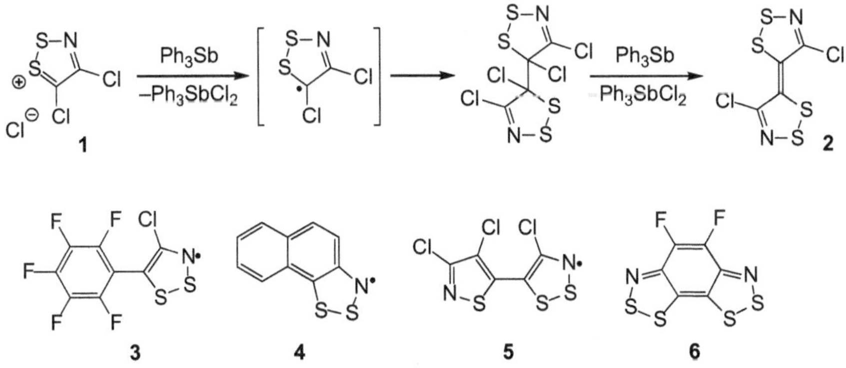

Scheme 11.1.9

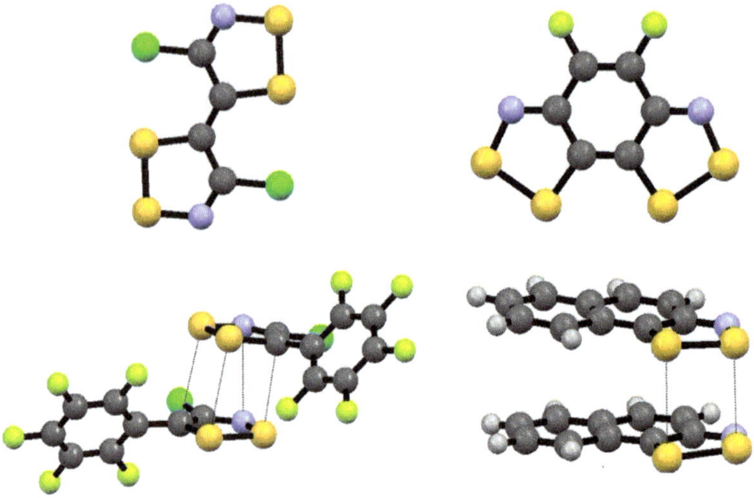

Figure 11.1.10 Crystal structures of (top) quinoidal bis-(1,2,3)-DTAs **2** and **6**; (bottom) π*–π* dimerization in **3** and **4**.

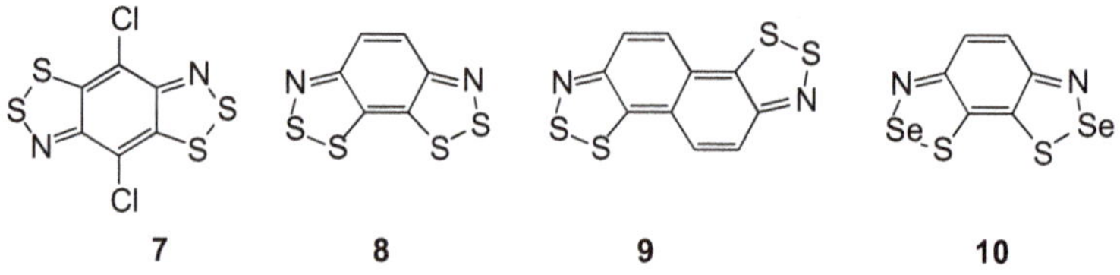

Scheme 11.1.10

σ-bond dimerization process in favour of a π*–π* interaction ($d_{S\cdots S}$ = 3.2–3.3 Å)[85] or, in the case of **5**, a stable solid state 1,2,3-DTA radical.[86] A number of bis-1,2,3-DTAs have been reported such as **6**–**9** which adopt quinoidal closed-shell singlet forms analogous to **2** (Scheme 11.1.10).[87] These 1,2,3-DTA species therefore typically (i) adopt closed-shell quinoidal ground states or (ii) associate in the solid state *via* π*–π* interactions. In both cases, singlet ground states are generated and they are typically insulators or thermally activated semiconductors with bandgaps ≥0.4 eV.[85a]

A number of charge-transfer (CT) salts of quinoidal bis-1,2,3-DTA derivatives have been reported with varying stoichiometries. There is some tendency for these CT salts, like the pristine 1,2,3-DTAs, to stabilize the unpaired electron *via* weakly associated dimeric π*–π* motifs. This includes unusual open-shell 1e^- bridged dimers with long intradimer separations ($d_{S\cdots S}$ = 3.4–3.6 Å)[83b,88] as well as more conventional 2e^- π*–π* dimers such as those seen in the salts [**7**]$_3$[ClO$_4$]$_2$ and [**8**]$_3$[MX$_4$]$_2$ (MX$_4$ = BF$_4$, GaCl$_4$, and FeCl$_4$) which contain dimers of [**7**]$_2^{2+}$ and [**8**]$_2^{2+}$ respectively as well as neutral BT and NT units.[89] The resultant room-temperature pressed pellet conductivities (σ_{RT}) for the 3:2 CT salts are around 10^{-1}–10^{-2} S cm^{-1}. Replacement of

S by Se affords the $[\mathbf{10}]_3^{2+}$ unit which is structurally similar to $[\mathbf{7}]_3^{2+}$ and can be similarly described as discrete $[\mathbf{10}]_2^{2+}$ and neutral **10** units and exhibits an order of magnitude improvement in conductivity ($\sigma_{RT} \sim 10^{-1}\,S\,cm^{-1}$).[90]

The related structures **11–18** have the N atoms positioned *meta* (thereby inhibiting quinone-like closed-shell configurations) but also replace R–C by R–N in the bridging unit, thereby adding an extra electron to the π-system generating resonance-stabilized π-radicals. The substituent at N appears to inhibit π*–π* dimerization and all structures **11–17** (Scheme 11.1.11) adopt slipped π-stacked structures and retain their paramagnetism in the solid state. The slippage leads to the evolution of poorly developed band structures (σ_{RT} 10^{-5}–$10^{-6}\,S\,cm^{-1}$) and more localized electronic structures. Their magnetic responses are sensitive to the degree of slippage, *e.g.* ferromagnetic interactions are observed in **11**[91] and **13**[92] whereas **15** is essentially paramagnetic and **16** exhibits strong antiferromagnetic interactions.[93]

Further molecular tailoring through the inclusion of an additional heteroatom in the framework leads to an unexpected change in which there appears to be a close balance between stable π-stacked radicals and assorted σ-bonded variants depending upon the nature of the substituents. For example, **18** is polymorphic revealing a tendency for dimerization *via* σ-bond formation in the solid state; either *via* C–C σ-bond formation at C-5 (α-**18**) or *via* an S–S σ-bonded dimer (β-**18**) (Scheme 11.1.12),[94] yet the related N-methyl analogue avoids σ-bond formation and adopts a regular π-stack at room temperature.[95]

The selenium analogues exhibit a similar propensity for Se–Se σ-bonded dimers[96] and a rich variety of solid state transformations: for example, conversion from σ-bonded dimers to π-bonded dimers under pressure,[97] and cleavage of the Se–Se σ-bonded dimer to generate two radicals on heating above 380 K (Scheme 11.1.13).[98]

The transport properties of these materials are typically characterized by modest conductivities under ambient pressure but increase with (1) successive replacement of S by Se; (2) increasing temperature; and (3) increasing pressure.[99] The transition from semiconducting to metallic behaviour under pressure is particularly noteworthy and appears correlated with the structural transitions observed.

Resonance-stabilized 1,2,3-DTA radicals with significant π-slippage do not lead to well-evolved band structures and their more localized electronic characteristics have led to the observation of magnetically ordered phases

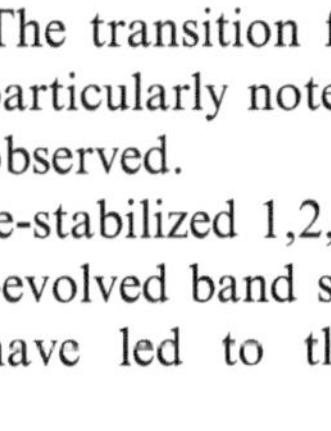

11 R = Me, X = Cl
12 R = Pr, X = Cl
13 R = Me, X = H
14 R = Et, X = H
15 R = Me, X = Me
16 R = Me, X = Ph
17 R = Ph, X = Et

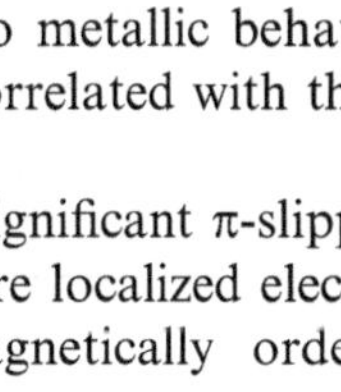

18

Scheme 11.1.11

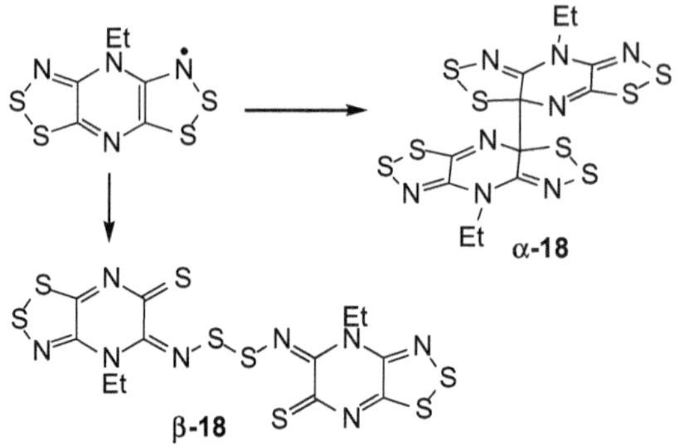

Scheme 11.1.12 Two modes of dimerization observed for **18**.

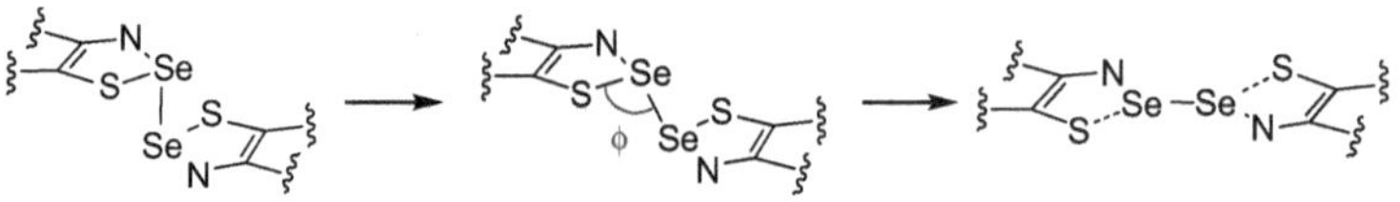

Scheme 11.1.13 Under pressure ϕ distorts from 90° to 180°, moving from a π-bonded dimer to a σ-bonded dimer.

Table 11.1.3 Magnetic ordering in resonance-stabilized bis(1,2,3-dithiazolyls).

Compound	*Magnetic order*	T_C/T_N	H_c/*Oe (2 K)*
19	CAFM	18 K	–
20	CAFM	27 K	–
21	FM	12.8 K	250
22	FM	17.0 K	1370
23	CAFM	14.0 K	–
24	FM	13.6 K	230
25	FM	14.1 K	320
26	MM	5 K	–
27	CAFM	4.5 K	–
28	FM	17.5 K	1600

CAFM, canted antiferromagnet (weak ferromagnet); FM, ferromagnet; MM, metamagnet.

(Table 11.1.3).[100] It is notable that the larger anisotropy afforded by Se (due to greater spin-orbit coupling) leads to remarkably large coercive fields characteristic of 'hard magnets' which is unprecedented in the field of π-based radicals (Figure 11.1.11). The ordering temperatures for these radicals are consistently high for π-based radicals and comparable with those observed for both 1,2,3,5-DTDA and 1,3,2-DTA derivatives (Section 12.1.3 and 12.1.5).

The application of pressure on **22** has been undertaken and leads to an increase in magnetic ordering temperature from 17 K at ambient pressure to

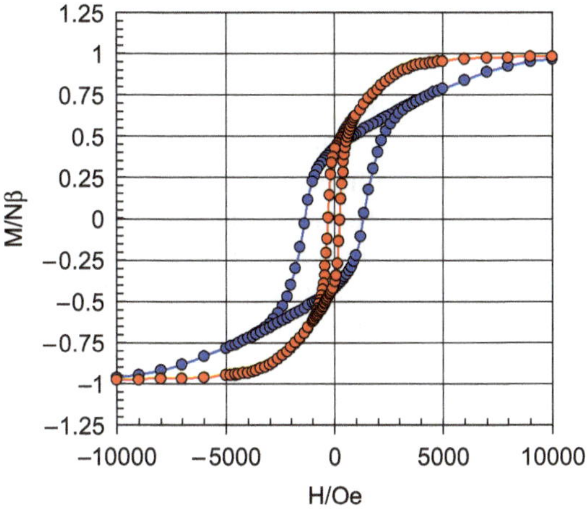

Figure 11.1.11 Hysteresis loops for the ferromagnets **21** (red) and **22** (blue) at 2 K.

21 K at 1 GPa but then decreases again with further pressure, returning to 17 K at 3 GPa.[101] In the case of **28**, pressure increases T_c from 17.5 K to 24 K before decreasing again. The increase in pressure also leads to a weakly metallic state so that these materials seem capable not only of exhibiting long-range magnetic order but also of exhibiting a conducting metallic state.[102]

11.1.7 Trithiazolyl Radical Cations

A 1 : 1 mixture of $[S_4][MF_6]_2$ and $[S_8][MF_6]_2$ (M = As, Sb) effectively acts as source of incipient '$S_3^{+\bullet}$' MF_6^- and undergoes cycloaddition chemistry with dicyanogen, NC–CN, to afford quantitative yields of $[(CNS_3^+)_2][MF_6]_2$.[103] The fluorinated organic substrates $C_2F_4(CN)_2$ and $C_4F_8(CN)_2$ have also been shown to undergo similar chemistry, affording $[C_xF_{2x}(CNS_3)_2][AsF_6]_2$ ($x = 2$, 4).[104] The dication $[(CNS_3)_2]^{2+}$ is isoelectronic with **2**. However, while the latter exhibits a closed-shell quinoidal electronic structure, $[(CNS_3)_2]^{2+}$ exists as an open-shell diradical. The radical centres are essentially disjoint with a very small singlet-triplet gap ($\Delta E_{ST} < 2\,cm^{-1}$) and the presence of a triplet state has been confirmed by EPR spectroscopy. In the solid state intermolecular exchange interactions appear stronger than the very weak intramolecular exchange, with the magnetism reflecting low-dimensional magnetic behaviour.[103] Weak multicentre π*–π* bonding interactions ($d_{S \cdots S} = 3.31$–3.46 Å) between $CNS_3^{+\bullet}$ radical centres fails to lead to a closed-shell configuration but propagates very strong antiferromagnetic interactions between radicals ($\boldsymbol{J} = -500$ to $-900\,cm^{-1}$) with the thermally accessible triplet state characterized by EPR spectroscopy.[104] The radicals $[Cl_3CCNS_3][AsF_6]$ and $[C_2F_5CNS_3][AsF_6]$ were similarly prepared from Cl_3CN and C_2F_5CN and retain their paramagnetism in the solid state and exhibit local antiferromagnetic exchange interactions but no evidence for long-range ordering.[105]

11.1.8 Thiadiazolyl Radical Anions

The redox chemistry of 2,1,3-thiadiazoles has been examined and a number of stable radicals identified by EPR spectroscopy.[106] The 2,1,3-benzothiadiazolidyl radical anion was characterized as its potassium salt $[K(THF)][C_6H_4N_2S]$ by X-ray diffraction.[107] Alkylation of benzo-chalcogenodiazoles, $C_6H_4N_2E$ (E = S, Se Te) with methyl triflate afforded the corresponding alkylated salts (Scheme 11.1.14). In the case of E = S and E = Se these could be reduced to the N-alkylated radical using Cp_2Co and characterized by solution EPR spectroscopy (Scheme 11.1.15).[108]

The [1,2,5]thiadiazolo[3,4c][1,2,5]thiadiazole (**29**) is readily reduced by both organic and inorganic electron donors to generate the resonance-stabilized thiadiazolide radical anion, $[C_2N_4S_2]^{\bullet-}$, isoelectronic with TTTA (Section 11.1.5).[109] Salts of this anion with crown-ether complexes of the alkali earth metals, organometallic cations, and organic donors have all been reported; $[Li(12\text{-Crown-}4)_2][C_2N_4S_2]$, $[Na(15\text{-Crown-}5)][C_2N_4S_2]$, $[K(18\text{-Crown-}6)][C_2N_4S_2]$ all show weak local antiferromagnetic interactions.[110] Variable-temperature magnetic measurements indicate $[Cp^*{}_2Cr]$ orders as an antiferromagnet at 9.7 K,[111] whereas the related salt $[Cp_2^*Co][C_2N_4S_2]$ reveals strong antiferromagnetic interactions between anions, but no evidence for long-range order down to 2 K.[112] The $[C_2N_4S_2]^{\bullet-}$ anion, like other thiazyl radicals, also exhibits some tendency to dimerize. Thus while $[(Me_2N)_3S][C_2N_4S_2]$ is monomeric,[110] $[(Me_2N)_2C{=}C(NMe_2)_2][C_2N_4S_2]_2$ exists as a singlet π^*–π^* dimer ($d_{S\cdots S} = 3.25$ Å) (Figure 11.1.12).[113]

The related selenium analogue, C_2N_4SSe, exhibits similar redox behaviour and its radical anion has been crystallized as $[K(18\text{-Crown-}6)][C_2N_4SSe]$ and exhibits very weak antiferromagnetic exchange between radicals.[114]

19 $E_1 = E_2 =$ Se, X = H, Y = NEt
20 $E_1 = E_2 =$ Se, X = H, Y = NEt
21 $E_1 =$ S, $E_2 =$ Se, X = Cl, Y = NEt
22 $E_1 = E_2 =$ Se, X = Cl, Y = NEt
23 $E_1 =$ Se, $E_2 =$ S, X = Cl, Y = NEt

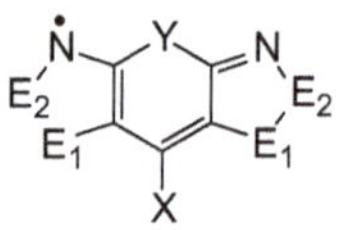

24 $E_1 =$ S, $E_2 =$ Se, X = Me, Y = NEt
25 $E_1 =$ S, $E_2 =$ Se, X = Br, Y = NEt
26 $E_1 = E_2 =$ S, X = H, Y = NMe
27 $E_1 = E_2 =$ S, X = Ph, Y = CO
28 $E_1 = E_2 =$ Se, X = Br, Y = NEt

Scheme 11.1.14

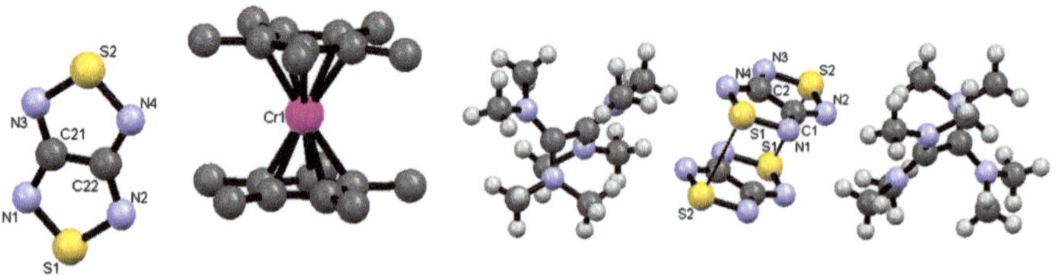

Figure 11.1.12 Crystal structures of $[Cp^*{}_2Cr][C_2N_4S_2]$ (left) containing the radical anion $C_2N_4S_2{}^{\bullet-}$ and $[(Me_2N)_2CC(NMe_2)_2]_2[C_2N_4S_2]_2$ containing the $[C_2N_4S_2]_2{}^{2-}$ π^*–π^* dimer.

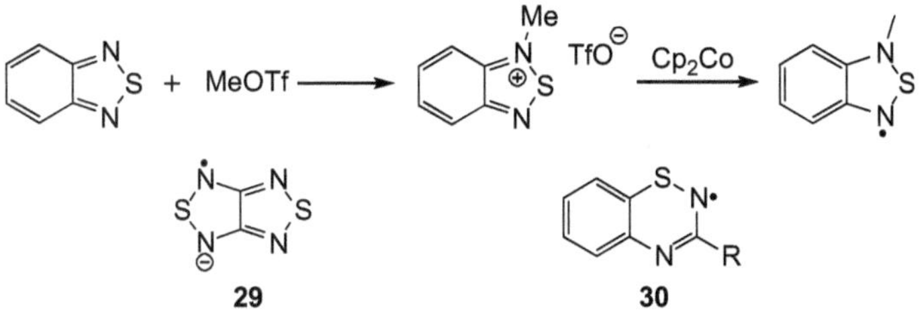

Scheme 11.1.15

11.1.9 Benzothiadiazinyl Radicals

Several benzothiadiazinyl radicals (**30**) have been prepared by oxidation of the 4-H benzothiadiazine using AgO and characterized by EPR spectroscopy. Several derivatives bearing fluorinated substituents were isolated in the solid state.[115] A related benzothiadiazinyl radical generated by oxidation of a mesogenic benzothiadiazine using AgO, PbO_2 or SO_2Cl_2 as oxidant has also been reported.[116]

11.1.10 Conclusions

These studies clearly reveal that π-delocalized chalcogen-nitrogen rings provide a particularly stable framework upon which to stabilize unpaired electrons. These radicals are typically thermally stable, persistent radicals in solution and many can be isolated in the solid state. Their solid state structures reveal a propensity to adopt π*–π* dimer motifs with short intermolecular S···S contacts. When the extent of spatial overlap is high then this dimerization can lead to diamagnetic closed-shell singlet configurations, but recent evidence clearly supports the presence of thermally accessible triplet states across a range of radical types when overlap is less efficient. When dimerization can be suppressed then these materials can offer exceptional properties; these are highlighted by the large range of magnetically ordered phases constructed from 1,3,2- and 1,2,3-dithiazolyl radicals and their heavier selenium congeners. These exhibit some of the highest magnetic ordering temperatures known for π-based radicals and remarkably large coercive fields which are superior to many metal-based magnets. Careful tailoring of the redox properties of these radicals has led to a range of semiconducting materials, a number of which undergo phase transitions to metallic states under pressure. Other radicals exhibit diamagnetic–paramagnetic phase transition behaviour near ambient temperature which has been shown to be driven by pressure and light irradiation. Initial photoconductivity studies indicate that these radicals also have potential applications in optoelectronic devices.

New families of chalcogen-nitrogen radicals are now being reported and will provide new building blocks from which to design new molecule-based materials. Moreover the interplay between paramagnetic radicals and *d*-block (or indeed *f*-block) electrons is clearly being actively pursued and will likely generate more complex multi-property materials as their chemistries become more fully developed. We can undoubtedly look forward to further exciting developments in the coming years.

Acknowledgements

We would like to thank Professors R. T. Boeré (Lethbridge), K. Awaga (Nagoya) and R. T. Oakley (Waterloo) for kindly providing data used to generate Figures 11.1 and 11.9 and 11.12.

References

1. J. M. Rawson and A. Alberola, *Chapter 12.1 in Handbook of Chalcogen Chemistry* (F. A. Devillanove Ed.), RSC, 2007.
2. E. Demarçay, *Compt. Rend.*, 1880, **91**, 854.
3. M. Gregory, *J. Pharm.*, 1835, **21**, 315.
4. F. P. Burt, *J. Chem. Soc.*, 1910, 1121.
5. T. Chivers, "*A Guide to Chalcogen-Nitrogen Chemistry*", World Scientific, 2005.
6. M. Gomberg, *J. Am. Chem. Soc.*, 1900, **22**, 757.
7. T. L. Roemmele, J. Konu, R. T. Boeré and T. Chivers, *Inorg. Chem.*, 2009, **48**, 9454.
8. R. T. Boeré, T. Chivers, T. L. Roemmele and H. M. Tuononen, *Inorg. Chem.*, 2009, **48**, 7294.
9. R. T. Boeré, A. M. Bond, T. Chivers, S. W. Feldberg and T. L. Roemmele, *Inorg. Chem.*, 2007, **46**, 5596.
10. H. M. Tuononen, T. Chivers, A. Armstrong, C. Fedorchuk and R. T. Boeré, *J. Organomet. Chem.*, 2007, **692**, 2705; R. T. Boeré, H. M. Tuononen, T. Chivers and T. L. Roemmele, *J. Organomet. Chem.*, 2007, **692**, 2683.
11. L. N. Markovski, O. M. Polumbrik, V. S. Talanov and Y. G. Shermolovich, *Tetrahedron Lett.*, 1982, **23**, 761.
12. For a detailed review of DTDA chemistry to 1995 see: J. M. Rawson, A. J. Banister and I. Lavender, *Adv. Heterocycl. Chem.*, 1995, **62**, 137
13. P. Del Bel Belluz, A. W. Cordes, E. M. Kristof, P. V. Kristof, S. W. Liblong and R. T. Oakley, *J. Am. Chem. Soc.*, 1989, **111**, 9276; R. T. Boeré, R. G. Hicks and R. T. Oakley, *Inorg. Synth.*, 1997, **31**, 94; R. T. Boeré, R. T. Oakley and R. W. Reed, *Organomet. Chem.*, 1987, **331**, 161.
14. For recent studies see: E. M. Fatila, M. C. Jennings, J. Goodried and K. E. Preuss, *Acta. Crystallogr. Sect. C.*, 2010, **66**, O260; C. S. Clarke, D. A. Haynes, J. N. B. Smith, A. S. Batsanov, J. A. K. Howard, S. I. Pascu and J. M. Rawson, *Cryst.Eng.Comm.*, 2010, **12**, 172; C. Allen, D. A. Haynes, C. M. Pask and J. M. Rawson, *Cryst.Eng.Commun.*, 2009, **11**, 2048.

15. A. Alberola, C. S. Clarke, D. A. Haynes, S. I. Pascu and J. M. Rawson, *Chem. Commun.*, 2005, **37**, 4726.
16. F. T. Edelmann, *Coord. Chem. Rev.*, 1994, **137**, 403.
17. J. Luzon, J. Campo, F. Palacio, G. J. McIntyre, J. M. Rawson, R. J. Less, C. M. Pask, A. Alberola, R. D. Farley, D. M. Murphy and A. E. Goeta, *Phys. Rev. B.*, 2010, **81**, 144429.
18. C. M. Aherne, A. J. Banister, I. B. Gorrell, M. I. Hansford, Z. V. Hauptman, A. W. Luke and J. M. Rawson, *Dalton Trans.*, 1993, 967; R. T. Boeré and T. L. Roemmele, *Coord. Chem. Rev.*, 2000, **210**, 369.
19. A. Decken, T. S. Cameron, J. Passmore, J. M. Rautiainen, R. W. Reed, K. V. Shuvaev and L. K. Thompson, *Inorg. Chem.*, 2007, **46**, 7436; K. V. Shuvaev, A. Decken, F. Grein, T. S. M. Abedin, L. K. Thompson and J. Passmore, *Dalton Trans.*, 2008, 4029.
20. A. Alberola, E. Carter, C. P. Constantinides, D. J. Eisler, D. M. Murphy and J. M. Rawson, *Chem. Commun.*, 2011, **47**, 2532.
21. J. M. Rawson, A. Alberola and A. Whalley, *J. Mater. Chem.*, 2006, **16**, 2560.
22. C. D. Bryan, A. W. Cordes, R. M. Fleming, N. A. George, S. H. Glarum, R. C. Haddon, R. T. Oakley, T. T. M. Palstra, A. S. Perel, L. F. Schneemeyer and J. V. Waszczak, *Nature*, 1993, **365**, 821; C. D. Bryan, A. W. Cordes, R. C. Haddon, R. G. Hicks, D. K. Kennepohl, C. D. MacKinnon, R. T. Oakley, T. T. M. Palstra, A. S. Perel, S. R. Scott, L. F. Schneemeyer and J. V. Waszczak, *J. Am. Chem. Soc.*, 1994, **116**, 1205; C. D. Bryan, A. W. Cordes, J. D. Goddard, R. C. Haddon, R. G. Hicks, C. D. MacKinnon, R. C. Mawhinney, R. T. Oakley, T. T. M. Palstra and A. S. Perel, *J. Am. Chem. Soc.*, 1996, **118**, 330.
23. A. J. Banister, N. Bricklebank, W. Clegg, M. R. J. Elsegood, C. I. Gregory, I. Lavender, J. M. Rawson and B. K. Tanner, *J. Chem. Soc., Chem. Commun.*, 1995, 679; A. J. Banister, N. Bricklebank, I. Lavender, J. M. Rawson, C. I. Gregory, B. K. Tanner, W. Clegg, M. R. J. Elsegood and F. Palacio, *Angew. Chem., Int. Ed. Engl.*, 1996, **35**, 2533.
24. M. Mito, T. Kawae, K. Takeda, S. Takagi, Y. Matsushita, H. Deguchi, J. M. Rawson and F. Palacio, *Polyhedron*, 2001, **20**, 1509; R. I. Thomson, C. M. Pask, G. O. Lloyd, M. Mito and J. M. Rawson, *Chem. Eur. J*, 2012, **18**, 8629.
25. F. Palacio, G. Antorrena, M. Castro, R. Burriel, J. M. Rawson, J. N. B. Smith, N. Bricklebank, J. Novoa and C. Ritter, *Phys. Rev. Lett.*, 1997, **79**, 2336; F. L. Pratt, A. E. Goeta, F. Palacio, J. M. Rawson and J. N. B. Smith, *Physica B*, 2000, **289**, 119; A. Alberola, C. M. Pask, J. M. Rawon, E. J. L. McInnes, J. Wolowska, H. El-Mkami and G. M. Smith, *J. Phys. Chem. B.*, 2003, **107**, 14158.
26. M. Deumal, S. LeRoux, J. M. Rawson, M. A. Robb and J. J. Novoa, *Polyhedron*, 2007, **26**, 1949; M. Deumal, J. M. Rawson, A. E. Goeta, J. A. K. Howard, R. C. B. Copley, M. A. Robb and J. Novoa, *Chem. Eur. J.*, 2010, **16**, 2741; J. M. Rawson, J. Luzon and F. Palacio, *Coord. Chem. Rev.*, 2005, **249**, 2631.

27. A. Alberola, R. J. Less, C. M. Pask, J. M. Rawson, F. Palacio, P. Oliete, C. Paulsen, A. Yamaguchi, D. M. Murphy and R. D. Farley, *Angew. Chem., Int. Ed.*, 2003, **42**, 4782.
28. J. Jornet, M. A. Robb, M. Deumal and J. J. Novoa, *Inorg. Chim. Acta*, 2008, **361**, 3586.
29. D. A. Haynes, *Cryst.Eng.Commun.*, 2011, **13**, 4793.
30. C. Knapp, E. Lork, K. Gupta and R. Mews, *Z. Anorg. Allg. Chem.*, 2005 **631**, 1640; A. D. Bond, D. A. Haynes, C. M. Pask and J. M. Rawson, *J. Chem. Soc., Dalton Trans.*, 2002, 2522; C. S. Clarke, S. I. Pascu and J. M. Rawson, *Cryst.Eng.Commun.*, 2004, 79.
31. L. Hu, A. Iwasaki, R. Suizu, H. Yoshikawa, K. Awaga and H. Ito, *Chem. Phys. Lett.*, 2010, **484**, 177; A. Iwasaki, L. Hu, R. Suizu, K. Nomura, H. Yoshikawa, K. Awaga, Y. Noda, K. Kanai, Y. Ouchi, K. Seki and H. Ito, *Angew. Chem. Int. Ed. Engl.*, 2009, **48**, 4022; K. Kaname, H. Yoshida, Y. Noda, A. Iwasaki, R. Suizu, J. Tsutumi, H. Imabayashi, Y. Ouchi, N. Sato, K. Seki and K. Awaga, *Phys. Chem. Chem. Phys.*, 2009, **11**, 11432.
32. J. M. Rawson, C. S. Clarke and D. W. Bruce, *Magn. Reson. Chem.*, 2009, **47**, 3.
33. K. E. Preuss, *Dalton Trans.*, 2007, 2357.
34. N. G. R. Hearns, K. E. Preuss, J. F. Richardson and S. Bin-Salamon, *J. Am. Chem. Soc.*, 2004, **126**, 9942; J. Britten, N. G. R. Hearns, K. E. Preuss, J. F. Richardson and S. Bin-Salamon, *Inorg. Chem.*, 2007, **46**, 3934; N. G. R. Hearns, E. M. Fatila, R. Clerac, M. Jennings and K. E. Preuss, *Inorg. Chem.*, 2008, **47**, 10330.
35. N. G. R. Hearns, R. Clerac, M. Jennings and K. E. Preuss, *Dalton Trans.*, 2009, 3193.
36. N. G. R. Hearns, K. D. Hesp, M. Jennings, J. L. Korcok, K. E. Preuss and C. S. Smithson, *Polyhedron*, 2007, **26**, 2047.
37. M. Jennings, K. E. Preuss and J. Wu, *Chem. Commun.*, 2006, 341.
38. E. M. Fatila, J. Goodreid, R. Clerac, M. Jennings, J. Assoud and K. E. Preuss, *Chem. Commun.*, 2010, **46**, 6569.
39. H. F. Lau, P. C. Y. Ang, V. W. L. Ng, S. L. Kuan, L. Y. Goh, A. S. Borisov, P. Hazendonk, T. L. Roemmele, R. T. Boeré and R. D. Webster, *Inorg. Chem.*, 2008, **47**, 632; R. T. Boeré, L-Y. Goh, C. Y. Ang, S. L. Kuan, H. F. Lau, V. W. L. Ng, T. L. Roemmele and S. D. Seagrave, *J. Organomet. Chem.*, 2007, **692**, 2697; H. F. Lau, V. W. - L. Ng, L. L. Koh, G. K. Tan, L. Y. Goh, T. L. Roemmele, D. S. Seagrave and R. T. Boeré, *Angew. Chem. Int. Ed. Engl.*, 2006, **45**, 4498.
40. A. J. Banister, I. May, J. M. Rawson and J. N.B. Smith, *J. Organomet. Chem.*, 1998, **550**, 241.
41. A. W. Cordes, C. D. Bryan, W. M. Davis, R. H. de Laat, S. H. Glarum, J. D. Goddard, R. C. Haddon, R. G. Hicks, D. K. Kennepohl, R. T. Oakley, S. R. Scott and N. P.C. Westwood, *J. Am. Chem. Soc.*, 1993, **115**, 7232.
42. J. F. Britten, O. P. Clements, A. W. Cordes, R. C. Haddon, R. T. Oakley and J. F. Richardson, *Inorg. Chem.*, 2001, **40**, 6820.

43. See for example: J. E. Davies, R. J. Less, I. May and J. M. Rawson, *New. J. Chem.*, 1998, **22**, 763; N. Feeder, R. J. Less, J. M. Rawson and J. N.-B. Smith, *Dalton Trans.*, 1998, 4091.
44. S. Parsons and J. Passmore, *Accts. Chem. Res.*, 1994, **27**, 101.
45. T. S. Cameron, A. Decken, M. Gabriel, C. Knapp and J. Passmore, *Can. J. Chem.*, 2007, **85**, 96.
46. J. Passmore and X. Sun, *Inorg. Chem.*, 1996, **35**, 1313; N. Burford, J. Passmore, M. J. Schriver, *Chem. Commun.*, 1986, 140; W. V. F. Brooks, N. Burford, J. Passmore, M. J. Schriver and L. H. Sutcliffe, *Chem. Commun.*, 1987, 69; C. Aherne, A. J. Banister, A. W. Luke, J. M. Rawson and R. J. Whitehead, *Dalton Trans.*, 1992, 1277.
47. A. J. Banister, J. M. Rawson, W. Clegg and S. L. Birkby, *Dalton Trans.*, 1991, 1099.
48. E. G. Awere, N. Burford, R. C. Haddon, S. Parsons, J. Passmore, J. V. Waszczak and P. S. White, *Inorg. Chem.*, 1990, **29**, 4821; E. G. Awere, N. Burford, C. Mailer, J. Passmore, M. J. Schriver, P. S. White, A. J. Banister, H. Oberhammer and L. H. Sutcliffe, *Chem. Commun.*, 1987, 66.
49. G. Wolmershäuser, M. Schnauber and T. Willhelm, *Chem. Commun.*, 1984, 573; G. Heckmann, R. Johann, G. Kraft and G. Wolmershäuser, *Synth. Met.*, 1991, **41–43**, 3287; G. Wolmershäuser and G. Kraft, *Chem. Ber.*, 1990, **123**, 881.
50. G. D. McManus, J. M. Rawson, N. Feeder, F. Palacio and P. Oliete, *J. Mater. Chem.*, 2000 **10**, 2001; A. Alberola, R. D. Farley, S. M. Humphrey, G. D McManus, D. M. Murphy and J. M. Rawson, *Dalton Trans.*, 2005, 3838; T. M. Barclay, A. W. Cordes, N. A. George, R. C. Haddon, M. E. Itkis, M. S. Mashuta, R. T. Oakley, G. W. Patenaude, R. W. Reed, J. F. Richardson and H. Zhang, *J. Am. Chem. Soc.*, 1998, **120**, 352; T. M. Barclay, A. W. Cordes, N. A. George, R. C. Haddon, R. T. Oakley, G. W. Patenaude, R. W. Reed and H. Zhang, *Chem. Commun.*, 1997, 873.
51. A. Decken, A. Mailman, J. Passmore, J. M. Rautiainen, W. Scherer and E. W. Scheidt, *Dalton Trans.*, 2011, 868; A. Decken, A. Mailman and J. Passmore, *Chem. Commun.*, 2009, 6077; A. Decken, A. Mailman, S. M. Mattar and J. Passmore, *Chem. Commun.*, 2005, 2366.
52. A. Alberola, D. J. Eisler, L. Harvey and J. M. Rawson, *Cryst. Eng. Commun.*, 2011, **13**, 1794; A. Alberola, O. P. Clements, R. J. Collis, L. Cubbitt, C. M. Grant, R. J. Less, R. T. Oakley, J. M. Rawson, R. W. Reed and C. M. Robertson, *Cryst. Growth. Des.*, 2008, **8**, 155; A. Alberola, R. J. Collis, R. J. Less and J. M. Rawson, *J. Organomet. Chem.*, 2007, **692**, 2743; A. Alberola, J. Burley, R. J. Collis, R. J. Less and J. M. Rawson, *J. Organomet. Chem.*, 2007, **692**, 2750; A. Alberola, R. J. Collis, S. M. Humphrey, R. J. Less and J. M. Rawson, *Inorg. Chem.*, 2006, **45**, 1903.
53. K.-W. Stender, N. Wölki and G. Klar, *Phos. Sulph. Sil.*, 1989, **42**, 111.

54. A. Alberola, D. Eisler, R. J. Less, E. Navarro-Moratalla and J. M. Rawson, *Chem. Commun.*, 2010, **46**, 6114.
55. S. M. Mattar and J. Durelle, *Magn. Reson. Chem.*, 2010, **48**, S122; S. M. Mattar, *Chem. Phys. Lett.*, 2006, **427**, 438; S. M. Mattar, j. Sanford and A. D. Goodfellow, *Chem. Phys. Lett.*, 2006, **418**, 30; S. M. Mattar, *Chem. Phys. Lett.*, 2005, **405**, 382; A. Alberola, R. D. Farley, S. M. Humphrey, G. D. McManus, D. M. Murphy and J. M. Rawson, *Dalton Trans.*, 2005, 3838.
56. E. G. Awere, N. Burford, C. Mailer, J. Passmore, M. J. Schriver, P. S. White, A. J. Banister, M. Oberhammer and L. H. Sutcliffe, *Chem. Commun.*, 1987, 66.
57. W. Fujita, K. Awaga, Y. Nakazawa, K. Saito and M. Sorai, *Chem. Phys. Lett.*, 2002, **352**, 348.
58. "Reversible Spin-Pairing in Crystalline Organic Radicals" J. M. Rawson and J. J. Hayward in *Spin-Crossover Materials: Properties and Applications* (M. Halcrow, Ed.), J. Wiley, 2012.
59. G. Wolmershäuser and R. Johann, *Angew. Chem., Int. Ed. Engl.*, 1989, **28**, 920.
60. W. Fujita and K. Awaga, *Science*, 1999, **286**, 281; G. D. McManus, J. M. Rawson, N. Feeder, E. J. L. McInnes, J. J. Novoa, R. Burriel, F. Palacio and P. Oliete, *J. Mater. Chem.*, 2001 **11**, 1992.
61. H. Matsuzaki, W. Fujita, K. Awaga and H. Okamoto, *Phys. Rev., Lett.*, 2003, **91**, 017403; W. Fujita and K. Awaga, *Chem. Phys. Lett.*, 2004, **393**, 150.
62. C. S. Clarke, J. Jornet-Somoza, F. Mota, J. J. Novoa and M. Deumal, *J. Am. Chem. Soc.*, 2010, **132**, 17817; C. S. Clarke, J. jornet, M. Deumal and J. J. Novoa, *Polyhedron*, 2009, **28**, 1614.
63. J. L. Brusso, O. P. Clements, R. C. Haddon, M. E. Itkis, A. A. Leitch, R. T. Oakley, R. W. Reed and J. F. Richardson, *J. Am. Chem. Soc.*, 2004, **126**, 8256; J. L. Brusso, O. P. Clements, R. C. Haddon, M. E. Itkis, A. A. Leitch, R. T. Oakley, R. W. Reed and J. F. Richardson, *J. Am. Chem. Soc.*, 2004, **126**, 14692; T. M. Barclay, A. W. Cordes, R. H. deLaat, J. D. Goddard, R. C. Haddon, D. Y. Jeter, R. C. Mawhinney, R. T. Oakley, T. T. M. Palstra, G. W. Patenaude, R. W. Reed and N. P. C. Westwood, *J. Am. Chem. Soc.*, 1997, **119**, 2633.
64. K. Awaga, Y. Umezono, W. Fujita, H. Yoshikawa, H. Cui, H. Kobayashi, S. S. Staniland and N. Robertson, *Inorg. Chim. Acta.*, 2008, **361**, 3761.
65. Y. Umezono, W. Fujita and K. Awaga, *Chem. Phys. Lett.*, 2005, **409**, 139.
66. S. S. Staniland, W. Fujita, Y. Umezono, K. Awaga, P. J. Camp, S. J. Clark and N. Robertson, *Inorg. Chem.*, 2005, **44**, 546.
67. Y. Umezono, W. Fujita and K. Awaga, *J. Am. Chem. Soc.*, 2006, **128**, 1084; M. Kepenekian, B. Le Guennic, K. Awaga and V. Robert, *Phys. Chem. Chem. Phys.*, 2009, **11**, 6066.
68. S. S. Staniland, W. Fujita, Y. Umezono, K. Awaga, S. J. Clark, H. B. Cui, H. Kobayashi and N. Robertson, *Chem. Commun.*, 2005, 3204.

69. Y. Umezono, W. Fujita and K. Awaga, unpublished work, cited in Y. Umezono, W. Fujita and K. Awaga, *Chem. Phys. Lett.*, 2005, **409**, 139.
70. G. Wolmershäuser, M. Schnauber, T. Wilhelm and L. H. Sutcliffe, *Synth. Met.*, 1986, **14**, 239; E. Dormann, M. J. Nowak, K. A. Williams, R. O. Angus and F. Wudl, *J. Am. Chem. Soc.*, 1987, **109**, 594.
71. T. M. Barclay, A. W. Cordes, R. H. de Laat, J. D. Goddard, R. C. Haddon, D. Y. Jeter, R. C. Mawhinney, R. T. Oakley, T. T. M. Palstra, G. W. Patenaude, R. W. Reed and N. P. C. Westwood, *J. Am. Chem. Soc.*, 1997, **119**, 2633.
72. W. Fujita and K. Awaga, *Chem. Phy. Lett.*, 2002, **357**, 385.
73. W. Fujita, K. Takahashi and H. Kobayashi, *Cryst. Growth. Des.*, 2011, **11**, 575.
74. G. Wolmershäuser, G. Wortman and M. Schnauber, *J. Chem. Res.(S)*, 1988, 358.
75. W. Fujita, K. Awaga, M. Takahashi, M. Takeda and T. Yamazaki, *Chem. Phys. Lett.*, 2002, **462**, 97.
76. M. Mito, M. Fujino, Y. Komorida, H. Deguchi, S. Takagi, W. Fujita and K. Awaga, *J. Phys. Soc. Jpn.*, 2008, **77**, 124713; M. Mito, M. Fujino, H. Deguchi, S. Takagi, W. Fujita and K. Awaga, *Polyhedron*, 2005, **24**, 2501.
77. W. Fujita and K. Awaga, *J. Am. Chem. Soc.*, 2001, **123**, 3601.
78. W. Fujita, K. Kikuchi and K. Awaga, *Angew. Chem., Int. Ed. Engl.*, 2008, **47**, 9480.
79. W. Fujita, K. Awaga, R. Kondo and S. Kagoshima, *J. Am. Chem. Soc.*, 2006, **128**, 6016.
80. G. Wolmershäuser, W. Kaim, G. Heckmann and A. Lichtblau, *Z. Naturforsch.*, 1992, **47B**, 675.
81. R. Herz, *Chem. Zentr.*, 1922, **4**, 948.
82. M. Risto, A. Assoud, S. M. Winter, R. Oilunkaniemi, R. S. Laitinen and R. T. Oakley, *Inorg. Chem.*, 2008, **47**, 10100.
83. T. M. Barclay, A. W. Cordes, R. T. Oakley, K. E. Preuss and R. W. Reed., *Chem. Commun.*, 1998, 1039; T.M Barclay, L. Beer, A. W. Cordes, R. C. Haddon, M. E. Itkis, R. T. Oakley, K. E. Preuss and R. W. Reed, *J. Am. Chem. Soc.*, 1999, **121**, 6657; T. M. Barclay, L. Beer, A. W. Cordes, R. T. Oakley, K. E. Preuss, R. W. Reed and N. J. Taylor, *Inorg. Chem.*, 2001, **40**, 2709.
84. T. M. Barclay, L. Beer, A. W. Cordes, R. T. Oakley, K. E. Preuss, N. J. Taylor and R. W. Reed, *J. Chem.Commun.*, 1999, 531.
85. T. M. Barclay, A. W. Cordes, R. C. Haddon, M. E. Itkis, R. T. Oakley, R. W. Reed and H. Zhang, *J. Am. Chem. Soc.*, 1999, **121**, 969; A. W. Cordes, J. R. Mingie, R. T. Oakley, R. W. Reed and H. Zhang, *Can. J. Chem.*, 2001, **79**, 1352; R. T. Oakley, R. W. Reed, C. M. Robertson and J. F. Richardson, *Inorg. Chem.*, 2005, **44**, 1837.
86. L. Beer, A. W. Cordes, R. C. Haddon, M. E. Itkis, R. T. Oakley, R. W. Reed and C. M. Robertson, *Chem. Commun.*, 2002, 1872.

87. T. M. Barclay, A. W. Cordes, J. D. Goddard, R. C. Mawhinney, R. T. Oakley, K. E. Preuss and R. W. Reed, *J. Am. Chem. Soc.*, 1997, **119**, 12136; T. M. Barclay, I. J. Burgess, A. W. Cordes, R. T. Oakley and R. W. Reed, *Chem. Commun.*, 1998, 1939; T. M Barclay, A. W. Cordes, R. T. Oakley, K. E. Preuss and R. W. Reed, *Chem. Mater.*, 1999, **11**, 164; V. V Zhivonitko, A. Y. Makarov, I. Y Bagryanskaya, Y. V. Gatilov, M. M. Shakirov and A. V Zibarev, *Eur. J. Inorg. Chem.*, 2005, 4099.
88. T. M. Barclay, A. W. Cordes, J. R. Mingie, R. T. Oakley and K. E. Preuss, *Cryst.Eng.Comm.*, 2000, **15**, 89.
89. T. M. Barclay, A. W. Cordes, R. T. Oakley, K. E. Preuss and R. W. Reed, *Chem. Mater.*, 1999, **11**, 164; K. Okamoto, T. Tanaka, W. Fujita and K. Awaga, *Phys. Rev. B.*, 2007, **76**, 075328.
90. L. Beer, J. F. Britten, A. W. Cordes, O. P. Clements, R. T. Oakley, M. Pink and R. W. Reed, *Inorg. Chem.*, 2001, **40**, 4705.
91. L. Beer, J. L. Brusso, A. W. Cordes, R. C. Haddon, M. E. Itkis, K. Kirschbaum, D. S. MacGregor, R. T. Oakley, A. A. Pinkerton and R. W. Reed, *J. Am. Chem. Soc.*, 2002, **124**, 9498.
92. L. Beer, J. L. Brusso, A. W. Cordes, E. Godde, R. C. Haddon, M. E. Itkis, R. T. Oakley and R. W. Reed, *Chem. Commun.*, 2002, 2562; L. Beer, J. F. Britten, J. K. Brusso, A. W. Cordes, R. C. Haddon, M. E. Itkis, D. S. MacGregor, R. T. Oakley, R. W. Reed and C. M. Robertson, *J. Am. Chem. Soc.*, 2003, **125**, 14394.
93. L. Beer, J. F. Britten, O. P. Clements, R. C. Haddon, M. E. Itkis, K. M. Matkovich, R. T. Oakley and R. W. Reed, *Chem. Mater.*, 2004, **16**, 1564.
94. A. A. Leitch, C. E. McKenzie, R. T. Oakley, R. W. Reed, J. F. Richardson and L. D. Sawyer, *Chem. Commun.*, 2006, 1088.
95. A. A. Leitch, R. W. Reed, C. M. Robertson, J. F. Britten, X. Yu, R. A. Secco and R. T. Oakley, *J. Am. Chem. Soc.*, 2007, **129**, 7903.
96. L. Beer, J. L. Brusso, R. C. Haddon, M. E. Itkis, A. A. Leitch, R. T. Oakley, R. W. Reed and J. F. Richardson, *Chem. Commun.*, 2005, 1543; A. A. Leitch, X. Yu, C. M. Robertson, R. A. Secco, J. S. Tse and R. T. Oakley, *Inorg. Chem.*, 2009, **48**, 9874.
97. J. S. Tse, A. A. Leitch, X. Yu, X. Bao, S. Zhang, Q. Liu, C. Jin, R. A. Secco, S. Desgreniers, Y. Ohishi and R. T. Oakley, *J. Am. Chem. Soc.*, 2010, **132**, 4876.
98. K. Lekin, S. M. Winter, L. E. Downie, X. Bao, J. S. Tse, S. Desgreniers, R. A. Secco, P. A. Dube and R. T. Oakley, *J. Am. Chem. Soc.*, 2010, **132**, 16212.
99. L. Beer, J. L. Brusso, R. C. Haddon, M. E. Itkis, R. T. Oakley, R. W. Reed, J. F. Richardson, R. A. Secco and X. Yu, *Chem. Commun.*, 2005, 5745; J. L. Brusso, S. Derakhshan, M. E. Itkis, H. Kleinke, R. C. Haddon, R. T. Oakley, R. W. Reed, J. F. Richardson, C. M. Robertson and L. K. Thompson, *Inorg. Chem.*, 2006, **45**, 10958; L. Beer, J. L. Brusso, R. C. Haddon, M. E. Itkis, H. Kleinke, A. A. Leitch, R. T. Oakley, R. W. Reed., J. F. Richardson, R. A. Secco and X. Yu,

J. Am. Chem. Soc., 2005, **127**, 18159; A. A. Leitch, X. Yu, S. M. Winter, R. A. Secco, P. A. Dube and R. T. Oakley, *J. Am. Chem. Soc.*, 2009, **131**, 7112; J. L. Brusso, K. Cvrkalj, A. A. Leitch, R. T. Oakley, R. W. Reed and C. M. Robertson, *J. Am. Chem. Soc.*, 2006, **128**, 15080.
100. A. A. Leitch, J. L. Brusso, K. Cvrkalj, R. W. Reed, C. M. Robertson, P. A. Dube and R. T. Oakley, *Chem. Commun.*, 2007, 3368; C. M. Robertson, D. J. T. Myles, A. A. Leitch, R. W. Reed, B. M. Dooley, N. L. Frank, P. A. Dube, L. K. Thompson and R. T. Oakley, *J. Am. Chem. Soc.*, 2007, **129**, 12688; C. M. Robertson, A. A. Leitch, K. Cvrkalj, R. W. Reed, D. J. T. Myles, P. A. Dube and R. T. Oakley, *J. Am. Chem. Soc.*, 2008, **130**, 8414; C. M. Robertson, A. A. Leitch, K. Cvrkalj, D. J. T. Myles, R. W. Reed, P. A. Dube and R. T. Oakley, *J. Am. Chem. Soc.*, 2008, **130**, 14791; S. M. Winter, K. Cvrkalj, P. A. Dube, C. M. Robertson, M. R. Probert, J. A. K. Howard and R. T. Oakley, *Chem. Commun.*, 2009, 7306; X. Yu, A. Mailman, P. A. Dube, A. Assoud and R. T. Oakley, *Chem. Commun.*, 2011, **47**, 4655.
101. M. Mito, Y. Komorida, H. Tsuruda, J. S. Tse, S. Desgreniers, Y. Ohishi, A. A. Leitch, K. Cvrkalj, C. M. Robertson and R. T. Oakley, *J. Am. Chem. Soc.*, 2009, **131**, 16012; H. Tsuruda, M. Mito, H. Deguchi, S. Takagi, A. A. Leitch, K. Lekin, S. M. Winter and R. T. Oakley, *Polyhedron*, 2011, **30**, 2997.
102. A. A. Leitch, K. Lekin, S. M. Winter, L. E. Downie, H. Tsuruda, J. S. Tse, M. Mito, S. Desgreniers, P. A. Dube, S. Zhang, Q. Liu, C. Jin, Y. Ohishi and R. T. Oakley, *J. Am. Chem. Soc.*, 2011, **133**, 6051.
103. T. S. Cameron, A. Decken, F. Grein, C. Knapp, J. Passmore, J. M. Rautiainen, K. V. Shuvaev, R. C. Thompson and D. J. Wood, *Inorg. Chem.*, 2010, **49**, 7861.
104. A. Decken, M. Ebdah, R. M. Kowalczyk, C. P. Landee, E. J. L. McInnes, J. Passmore, K. V. Shuvaev and L. K. Thompson, *Inorg. Chem.*, 2007, **46**, 7756.
105. A. Decken, S. M. Mattar, J. Passmore, K. V. Shuvaev and L. K. Thompson, *Inorg. Chem.*, 2006, **45**, 3878.
106. E. A. Suturina, N. A. Semenov, A. V. Lonchakov, I. Y. Bagryanskaya, Y. V. Gatilov, I. G. Irtegova, N. V. Vasilieva, E. Lork, R. Mews, N. P. Gristan and A. V. Zibarev, *J. Phys. Chem. A*, 2011, **115**, 4851; A. Y. Makarov, V. V. Zhivonitko, A. G. Makarov, S. B. Zikirin, I. Y. Bagryanskaya, V. A. Bagryansky, Y. V. Gatilov, I. G. Irtegova, M. M. Shakirov and A. V. Zibarev, *Inorg. Chem.*, 2011, **50**, 3017; N. V. Vasilieva, I. G. Irtegova, N. P. Gritsan, A. V. Lonchakov, A. Y. Makarov, L. A. Shundrin and A. V. Zibarev, *J. Phys. Org. Chem.*, 2010, **23**, 536.
107. S. N. Konchenko, N. P. Gritsan, A. V. Lonchakov, U. Radius and A. V. Zibarev, *Mend. Commun.*, 2009, **19**, 7.
108. M. Risto, R. W. Reed, C. M. Robertson, R. Oilunkaniemi, R. S. Laitinen and R. T. Oakley, *Chem. Commun.*, 2008, 3278.

109. A. Y. Makarov, I. G. Irtegova, N. V. Vasilieva, I. Y. Bagryanskaya, T. Borrmann, Y. V. Gatilov, E. Lork, R. Mews, W. D. Stohrer and A. V. Zibarev, *Inorg. Chem.*, 2005, **44**, 7194.
110. V. N. Ikorskii, I. G. Irtegova, E. Lork, A. Y. Makarov, R. Mews, V. I. Ovcharenko and A. V. Zibarev, *Eur. J. Inorg. Chem.*, 2006, 3061.
111. S. N. Kochenko, N. P. Gritsan, A. V. Lonchakov, I. G. Irtegova, R. Mews, V. I. Ovcharenko, U. Radius and A. V. Zibarev, *Eur. J. Inorg. Chem.*, 2008, 3833.
112. N. A. Semenov, N. A. Pushkarevsky, A. V. Lonchakov, A. S. Bogomyakov, E. A. Pritchina, E. A. Suturina, N. P. Gritsan, S. N. Konchenko, R. Mews, V. I. Ovcharenko and A. V. Zibarev, *Inorg. Chem.*, 2010, **49**, 7558.
113. N. P. Gristan, A. V. Lonchakov, E. Lork, R. Mews, E. A. Pritchina and A. V. Zibarev, *Eur. J. Inorg. Chem.*, 2008, 1994.
114. I. Y. Bagryanskaya, Y. V. Gatilov, N. P. Gritsan, V. N. Ikorskii, I. G. Irtegova, A. V. Lonchakov, E. Lork, R. Mews, V. I. Ovcharenko, N. A. Semenov, N. V. Vasilieva and A. V. Zibarev, *Eur. J. Inorg. Chem.*, 2007, 4751.
115. J. Zienkiewicz, A. Fryszkowska, K. Zienkiewicz, F. Guo, P. Kaszynski, A. Janusko and D. Jones, *J. Org. Chem.*, 2007, **72**, 3510.
116. M. J. Sienkowska, J. M. Farrar and P. Kaszynski, *Liq. Cryst.*, 2007, **34**, 19.

CHAPTER 11.2

Chalcogen-Rich Compounds as Electron Donors

DIEGO CORTIZO-LACALLE, PETER J. SKABARA* AND THOMAS D. WESTGATE

WestCHEM, Department of Pure and Applied Chemistry, University of Strathclyde, 295 Cathedral Street, Glasgow G1 1XL, UK
*Email: peter.skabara@strath.ac.uk

11.2.1 Introduction

In 1970, the discovery of a stable electron-conducting cation radical complex based on tetrathiafulvalene (TTF, **1**) and its chloride ($TTF^{+}Cl^{-}$) (1:1 eq), which showed important electron-donor properties, was responsible for the subsequent explosion in interest of chalcogen-based electron-rich molecular conductors.[1] In 1973, metallic conductivity was found in the charge-transfer salt of tetrathiafulvalene and tetracyano-*p*-quinodimethane (TCNQ, **2**).[2] In this material, with a TTF-TCNQ ratio of 1:1, an average of 0.59 electrons are transferred from the TTF molecules to the electron-deficient TCNQ units. The resulting TTF radical cations ($TTF^{+\bullet}$) and acceptor anions ($TCNQ^{-\bullet}$) form adjacent columnar stacks (Figure 11.2.1).[3] Close contacts between sulfur atoms arising from ring-over-bond overlap in the TTF stacks provide the framework for one-dimensional intermolecular delocalization of π-electrons. The polarization of the sulfur atoms was also cited as helping to improve the electron transport properties of the salt. Materials of this donor–acceptor type are viable conductors because transfer of electrons from donor to acceptor generates charge carriers in the oxidized donor and reduced acceptor. Nearly

Handbook of Chalcogen Chemistry: New Perspectives in Sulfur, Selenium and Tellurium, 2nd Edition, Volume 2
Edited by Francesco Antonio Devillanova and Wolf-Walther du Mont

Published by the Royal Society of Chemistry, www.rsc.org

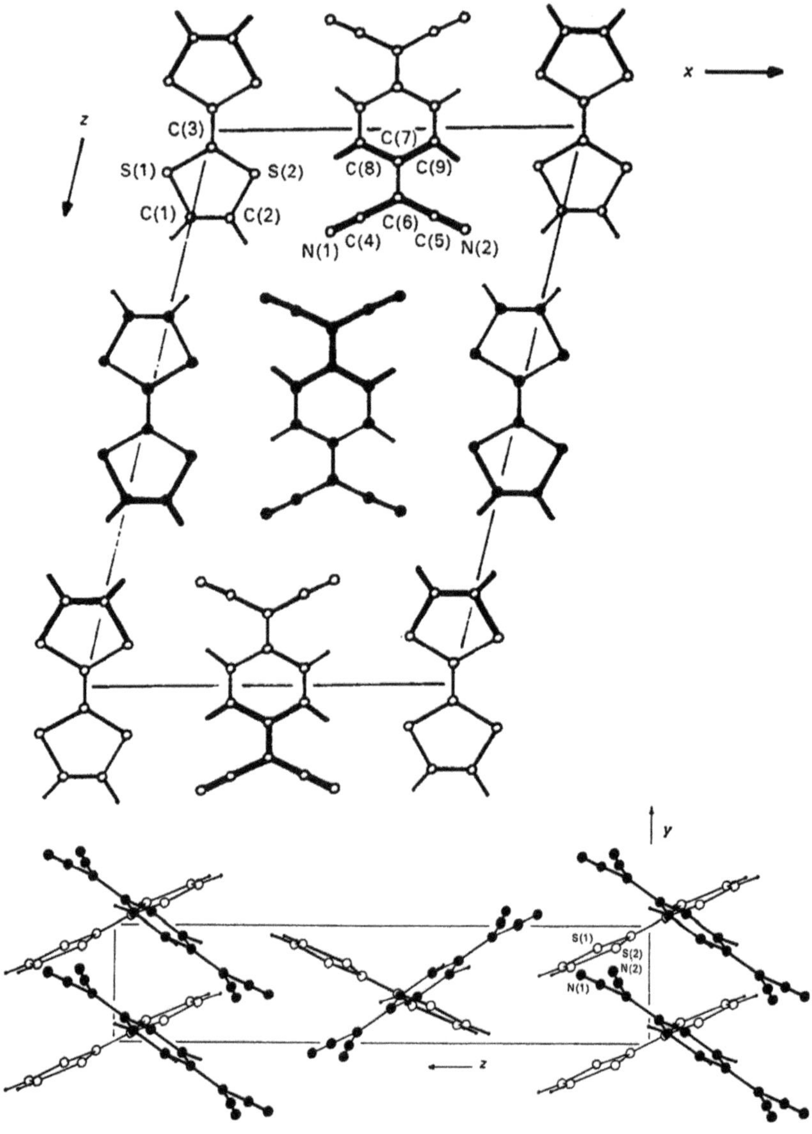

Figure 11.2.1 Top: A view normal to the ac plane of the crystal packing in TTF-TCNQ. The shaded molecules have their centroids at $y = 1/2$. Below: A view along the [100] direction. The shaded molecules have their centroids at $x = 1/2$.
Reproduced from Kistenmacher *et al.*,[3] with permission from IUCr.

four decades after its discovery, this fascinating material is still a matter of study.[4,5]

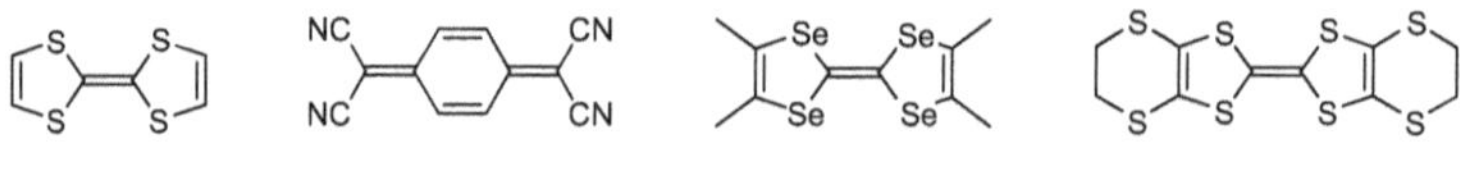

TTF **1** TCNQ **2** TMTSF **3** BEDT-TTF **4**

Vast synthetic effort was subsequently employed to produce new conducting charge-transfer (CT) salts incorporating TTF-based donor structures. The first superconducting complexes, known as the Bechgaard salts, employed the substituted selenium analogue tetramethyltetraselenafulvalene (TMTSF, **3**), which was electrocrystallized with a variety of counter-anions.[6] The best results have originated from the donor *bis*-ethylenedithio-tetrathiafulvalene (BEDT-TTF, **4**), which has four additional sulfur atoms.[7] Although similar to TTF in terms of electrochemical behaviour, the presence of different chalcogen atoms (in the case of TMTSF) or a greater number of chalcogen atoms (BEDT-TTF) in the structure has a profound impact on the crystal packing. Se–Se contacts in TMTSF are remarkably close (3.8–4.0 Å, Figure 11.2.2). The increase in number of chalcogen–chalcogen contacts involving the peripheral sulfur atoms of BEDT-TTF affords new packing modes, characterized by a two-dimensional arrangement of BEDT-TTF dimers.[8–11]

The wide interest in this family of materials is due to the special properties of TTFs as electron-donor compounds which make them suitable for conducting electricity. TTF is a pseudo-aromatic 14 π-electron system in which oxidation to the radical cation and dication occurs reversibly at relatively low potentials ($E^1_{1/2} = 0.37$ V; $E^2_{1/2} = 0.67$ V in dichloromethane *versus* the saturated calomel electrode). The radical cation and the dication possess aromaticity according to Hückel's rule.

The last few decades have seen organic chemists develop hundreds of analogues and derivatives of TTF, including the large family of linearly π-extended TTFs. A diverse range of electron acceptors, including organic, inorganic, and metal coordination species has also been studied in conjunction with this family of donors. Intrinsic physical properties of these counter-anions, such as magnetism[12–15] or chirality,[16,17] can be combined with the conducting nature of the TTF derivative, to provide a multifunctional nanostructured

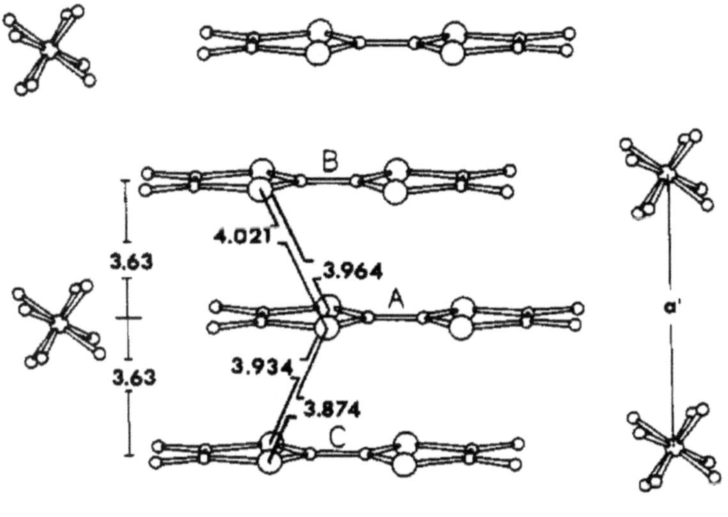

Figure 11.2.2 Side view of the stacks in (TMTSF)ClO_4.
Reprinted with permission from Bechgaard *et al.*[6] Copyright 1981, American Chemical Society.

electroactive material. Many materials with interesting electrochemical properties incorporate a tetrachalcogenafulvalene moiety or an analogue. Their unique electrochemical and structural properties find relevance in areas such as field-effect transistors (FETs),[18–20] molecular memory,[21] proton conduction for fuel cell components,[22] new charge storage devices,[23] mechanical molecular switches,[24] electrochromism,[25] sensitized-solar cells,[26] non-linear optics [27] and multi-stimulus-reponsive gels and micelles.[28,29]

In the 1970s, parallel to the development of these molecular organic conductors of electricity, organic polymers were discovered to be electrical conductors[30] and heterocycles have attracted high interest due to their inherent properties. In that sense, thiophene (**5**), polythiophene (PT), and their derivatives[31] have been widely investigated and employed in a vast number of applications and devices such as organic light-emitting diodes, photovoltaic cells, organic field-effect transistors, electrochromic devices, and sensors. As a consequence of the growing interest in thiophene-containing materials for electronic applications, all these examples are discussed further in the final section of this chapter. By using suitable examples, the electron-donor ability of different types of chalcogen-containing compounds will be discussed.

Thiophene **5**

11.2.2 Structural Diversity

More than 40 years of research has produced great structural diversity in chalcogen-containing electroactive materials. From small molecules to large, elegant macrostructures, a common theme of a hetero-π-delocalized structure is at its heart. In their excellent review of TTF derivatives, Segura and Martín organized these compounds into molecular, macromolecular, polymeric, or supramolecular categories.[32] Here we expand this system beyond TTF derivatives to illustrate the structural range of chalcogen electroactive materials.

11.2.2.1 Molecular Species

Many electroactive molecular systems containing chalcogens are based on CT complexes where tetrachalcogenafulvalene (TEF, Figure 11.2.3) acts as an electron donor. As a means to improve the charge-transport properties of these CT materials, the TEF framework has been altered with a variety of structural motifs, while retaining either the 1,3-dithiole ring or a similar structure. Crystal structure analysis has shown the importance of the packing modes of the donors, as the different phases observed in TEF packing allow for varying

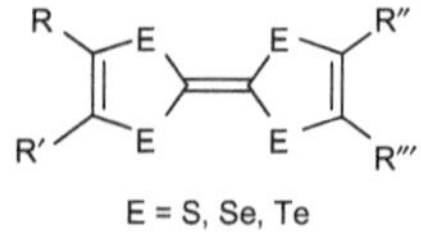

Figure 11.2.3 General substituted tetrachalcogenafulvalene (TEF).

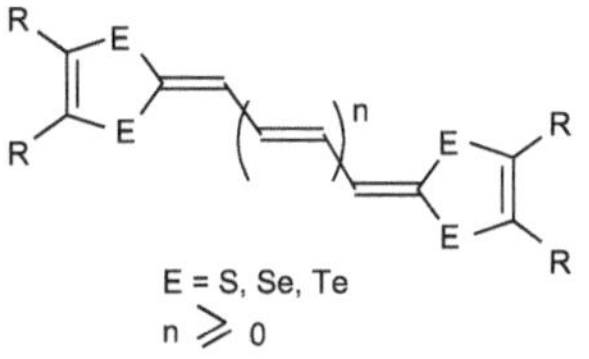

Figure 11.2.4 General linear π-extended TEF.

degrees of interstack interactions through chalcogen–chalcogen contacts. The observation by Laukhina *et al.*[33] of polymorphism in trihalide salts of BEDT-TTF provides a useful example of this principle. This study is the first example of a high-temperature transformation from metallic β″ to semiconducting β′ phase. X-ray data confirms that there are fewer S···S close contacts in the semiconducting phase than in the metallic phase.

Extended TEFs[34] contain a conjugated spacer group between the 1,3-dithiole rings. Upon oxidation, TEF species form cations mainly localized on these heterocycles. Greater spatial separation of these charges is provided in extended analogues. Of course, increased conjugation can also lead to a reduced HOMO–LUMO gap, desirable in semiconductors.[35] Spacers for linear (Figure 11.2.4), annulated (such **6**[36] and **7**[37]), or spiroconjugated[38] types have been varied. The nature of these spacer units dictates the extent of delocalization of π-electrons.[39] The CT complexes of these donors, along with the relationship between solid state structure and electrochemical properties, have been reviewed by Frère and Skabara.[40]

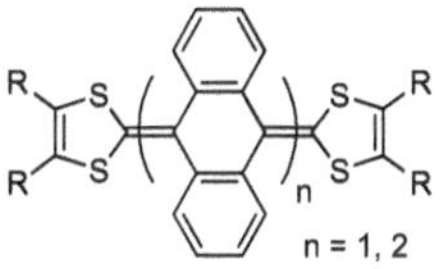

6

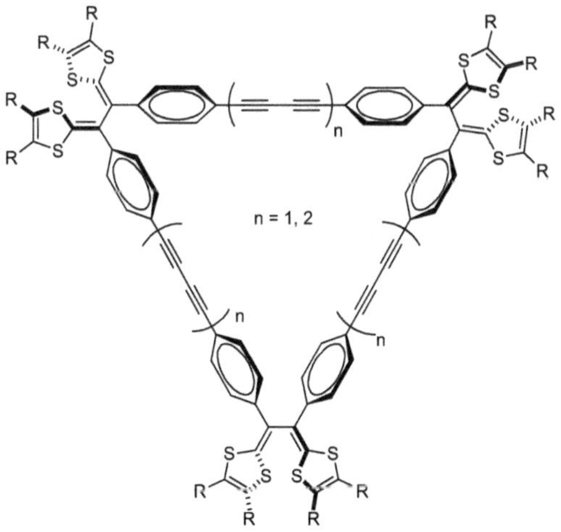

7

Metal complexes based on 1,2-dithione ligands ($M(bdt)_2$, Figure 11.2.5) can be thought of as analogues of TEF, with the central C = C of TTF replaced by a

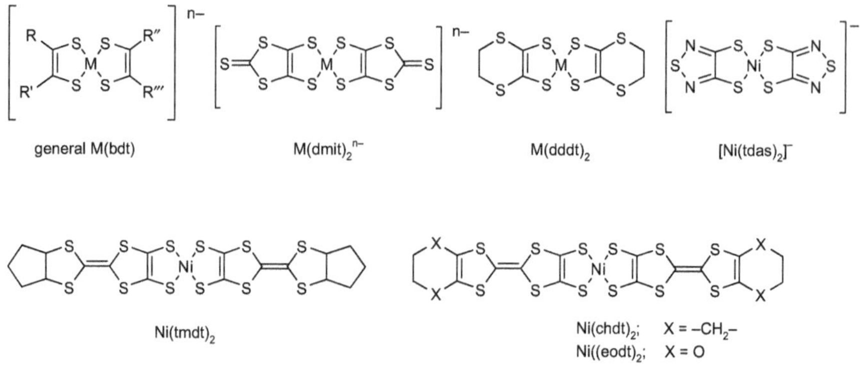

Figure 11.2.5 Metal complexes based on 1,2-dithiolene ligands.

metal ion which is coordinated to bidentate chalcogenide ligands. The conductance band structure is hence augmented by the availability of the metal–ligand CT band. Among these materials, the best studied are the $M(dmit)_2$ and $M(dddt)_2$ complexes (dmit = 1,3-dithiole-2-thione-4,5-dithiolate; dddt = 5,6-dihydro-1,4-dithiin-2,3-dithiolate, Figure 11.2.5). These groups have been reviewed by Cassoux *et al.*[41] and Yagubskii,[42] respectively. $M(dmit)_2$ complexes in the partially reduced state form conducting complexes with donor molecules such as TTF derivatives, while $M(dddt)_2$ complexes form conducting cation salts when partially oxidized and are analogues of BEDT-TTF (**4**). The modes of packing in these conducting salts mirror those seen on the TEF analogues.[43] Variation of metal and counterion gives rise to the diversity in these two groups of materials. Superconducting $M(dmit)_2$ complexes have been prepared, such as $TTF[Ni(dmit)_2]_2$,[44,45] $[Me_4N][Ni(dmit)_2]_2$[46] and $TTF[Pd(dmit)_2]_2$.[47] Many semiconducting complexes of both types are known.[41,42]

Conducting complexes using new classes of chalcogen-rich dithiolate ligands are being continuously developed with interesting results. One example (see Figure 11.2.5) includes a 1:1 semiconducting salt of BEDT-TTF with monomeric $[Ni(tdas)_2]^-$ which also exhibits weak antiferromagnetic coupling.[48] Nickel complexes of ligands incorporating TTF moieties ($Ni(tmdt)_2$,[49] $Ni(chdt)_2$, and $Ni(eodt)_2$[50]) have not only afforded semiconducting salts in their anionic states but have also given rise to single-component molecular metals, *i.e.* complexes that exhibit metallic conductivity in the neutral state. In these cases the HOMO–LUMO gap of the neutral molecules is small enough to allow the bands to overlap and form a partially filled conduction band.[51] In the example of $Ni(tmdt)_2$, X-ray studies account for extensive S···S contacts between molecules in three dimensions, leading to room-temperature conductivity of almost one order of magnitude higher than BEDT-TTF or BETS salts. Furthermore, the high conductivity and metallic behaviour reported for compressed powder samples of the neutral species is consistent with a three-dimensional chalcogen–chalcogen interaction network.

11.2.2.2 Donor-Acceptor Dyad/Triad Materials

Since the preparation of TTF in the early 1970s, the complexity of TTF derivatives has grown stepwise. The development of the macromolecular chemistry of TTF derivatives as electron donors has been of great interest in the last decade, and, by sophisticated synthetic methods, fascinating macromolecular structures have been designed. Macromolecules containing TEF-related donors linked to electron acceptors derived from molecules, such as TCNQ[52,53] and fullerenes,[54] have received much attention in the search for molecular rectifiers and non-linear optical and photovoltaic molecular materials. Similar to the TTF assemblies above, the spacer group between donor and acceptor units can be conjugated (D-π-A) or non-conjugated (D-σ-A), and variable in length, allowing for different intramolecular electron transfer pathways. Some of these molecules, such as **8**,[55] have been proposed as rectifiers.[56]

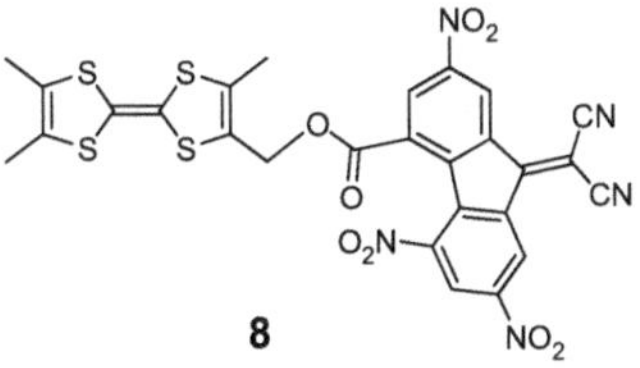

8

Fullerenes have shown particular promise as acceptors in molecular electronics, and numerous interesting TTF/C_{60} ensembles have been reported. Martín and co-workers have been extensively working to enhance the lifetime of the charge-separated states generated on photoinduced electron transfer of TTF/C_{60} systems.[57,58] They managed to mimic the natural photosynthetic process and modulate the electron donor ability by the precise design of the double-exTTF derivative **9**. An efficient charge separation created the radical ion-pair states C_{60}^{-}/exTTF$_1^{+}$/exTTF$_2$ and C_{60}^{-}/exTTF$_1$/exTTF$_2^{+}$ with typical lifetimes of 54–460 ns. The TTF–porphyrin–fullerene triad **10**, prepared by Carbonera *et al.*,[59] showed long-lived photoinduced charge separation.

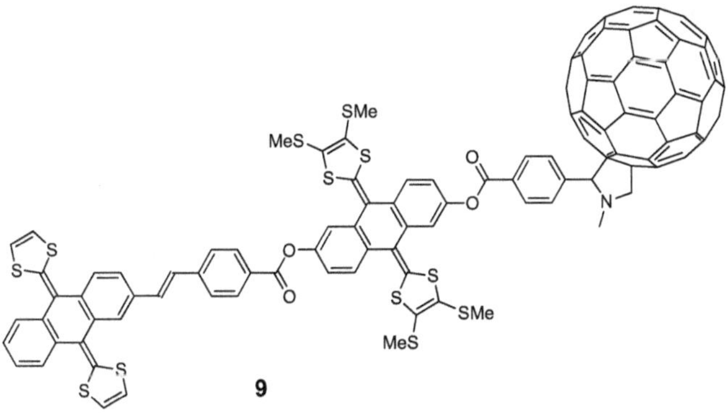

9

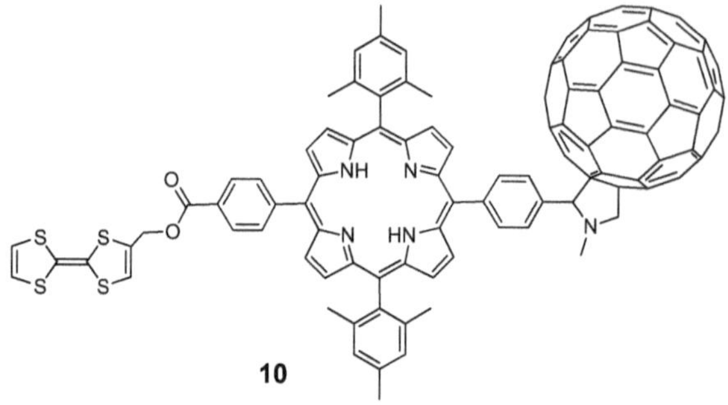

10

The supramolecular versatility of TTF has allowed for the preparation of some large, elegant assemblies such as catenanes and rotaxanes. Neutral TTF acts as a recognition site for the tetracationic cyclophane cyclobis(paraquat-*para*-phenylene) ($CBPQT^{4+}$, **11**) as a result of donor–acceptor interactions.[60–62] These interactions can be 'switched off' when the TTF is electrochemically oxidized, causing the $CBPQT^{4+}$ ring to move to a more thermodynamically favoured location.[63–66] This methodology for construction of TTF-based materials that are capable of doing mechanical work embodies a radical approach to electronic applications and broadens the materials chemistry horizon yet further.[67,68] Likewise, extended TTFs have been described to form further supramolecular assemblies such as head-to-tail supramolecular polymers[69] and clip- and tweezer-like assemblies.[70–72]

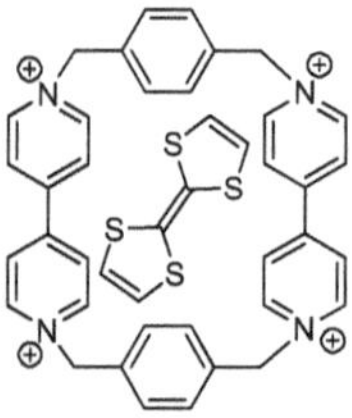

11

11.2.2.3 Polymeric Materials

Undoubtedly, most of the research on chalcogen-containing polymers has been focused on polythiophene and its vast number of derivatives. By strategically tuning the monomeric unit, PT derivatives have been synthesized, to give desired properties. Doped (oxidized) PT is highly conducting, and 3-alkyl-substituted PTs are soluble, easily characterized polymers (see Figure 11.2.6). Self-assembly is an important consideration in these substituted PTs, allowing control of the degree of planarity, π-orbital overlap, and hence optical bandgap.[31] For instance, head-to-head polymerization of 3-alkyl thiophene results in a steric clash between alkyl groups, forcing a twist in the chain between thiophene units. Regioregular (head-to-tail) polymers are not only planar but can also show long-range ordering where interdigitization of

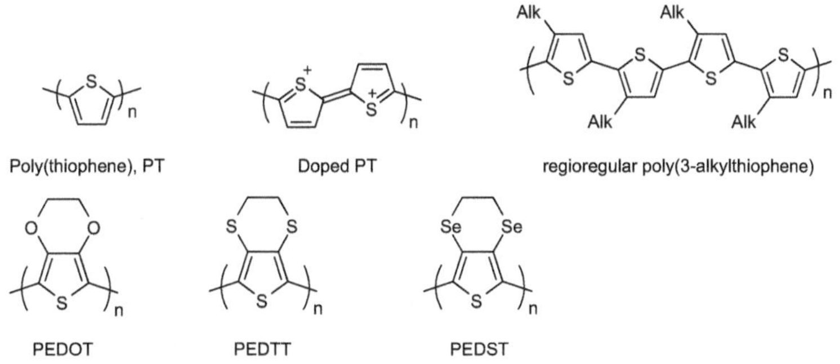

Figure 11.2.6 Poly(thiophene) and some of its common derivatives.

long alkyl chains is possible. Similar to the approach of incorporating additional chalcogens in TEFs, PTs have also been substituted and poly(3,4-ethylenedioxythiophene) (PEDOT), poly(3,4-ethylenedithiathiophene) (PEDTT), and poly(3,4-ethylenediselenathiophene) (PEDST) have been synthesized.[73] Whereas strong intrachain S···O interactions ensure a planar structure in PEDOT, S···S interactions in PEDTT are influenced by twists between the repeat units, bringing the polymer chain out of coplanarity and enlarging the bandgap. On the other hand, PEDST adopts a far more planar structure than PEDTT and therefore its bandgap is closer to that of PEDOT. Poly- and oligothiophenes have been well studied as electron donors in conjunction with acceptors such as the fullerene $PC_{60}BM$;[74] the resulting materials (*e.g.* **12**),[75] have been tested in organic solar cells.[76,77]

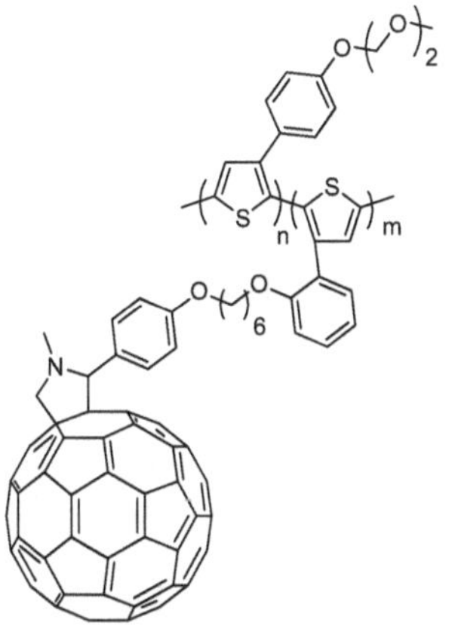

12

Various TEF-containing polymers have been prepared. TEF units have been incorporated into polymer backbones (see for example **13**[78] and **14**[79]) as well as polymers with pendant TEF units.[80,81] The incorporation of both electroactive families—TTF derivatives and polythiophenes—into the polymer backbone has been carried out by Skabara and co-workers. The basic unit is the fused

TTF–thiophene (**15** and **16**),[82,83] although the TTF unit has also been linked to the thiophene through a 1,4-dithiin bridge (**17**).[84] Recently, a thieno-TTF unit has been copolymerized with a diketopyrrolopyrrole derivative to obtain a low-bandgap polymer (**18**) that has been used in the fabrication of organic field-effect transistors (OFET) and OPVs (see Section 11.2.4). Interestingly, the polymer shows good ambient stability and relatively high hole mobility has been obtained. The stability of these devices in ambient atmosphere has been attributed to the partial CT interactions between TTF and DPP units.[85] Intermediate between single molecules and polymers, TEF-containing dendrimers have also been synthesized and studied. These compounds have been reviewed in detail,[32,86] while Iyoda reviewed the TTF dimers, trimers, and oligomers.[87]

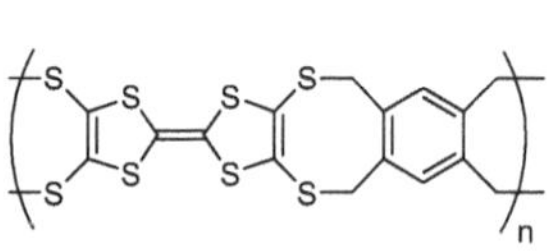
13

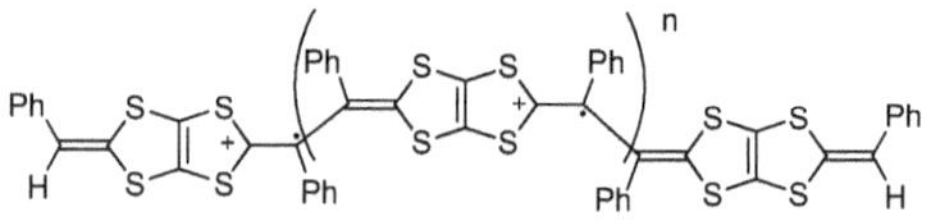
14

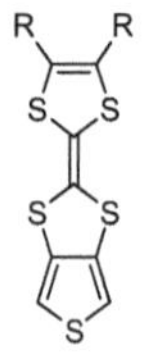
15

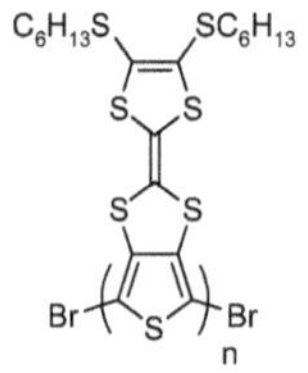
16

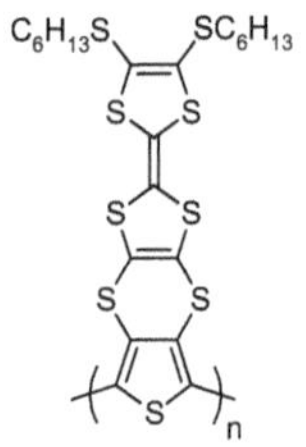
17

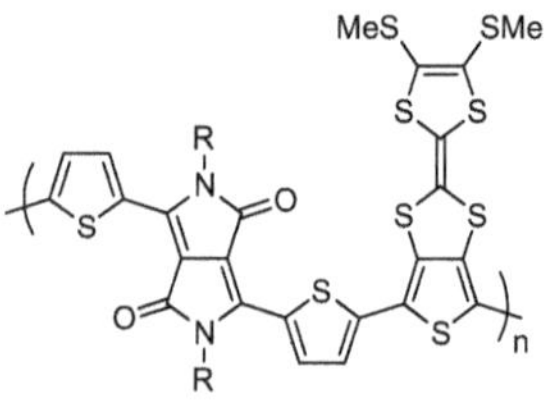
18

11.2.3 The Role of the Chalcogen

In this section, we look more closely at the effect the chalcogen atom has on the properties of the molecular conductors we describe. We do not attempt to review exhaustively all the chalcogen-containing components in electroactive systems; to do so would be a colossal task. Instead, carefully chosen examples and studies illustrate how chalcogen chemistry is used in the design and manipulation of electroactive materials, and ultimately how it affects suitability for molecular device applications.

11.2.3.1 Heterocyclic Structures

Chalcogen-containing heterocycles are ubiquitous in the electroactive structures described in the previous section. A common feature of these structures is sp^2-hybridized chalcogen atoms which impart electron density (from their *p*-orbital lone pair) to delocalized or conjugated π-systems. In some

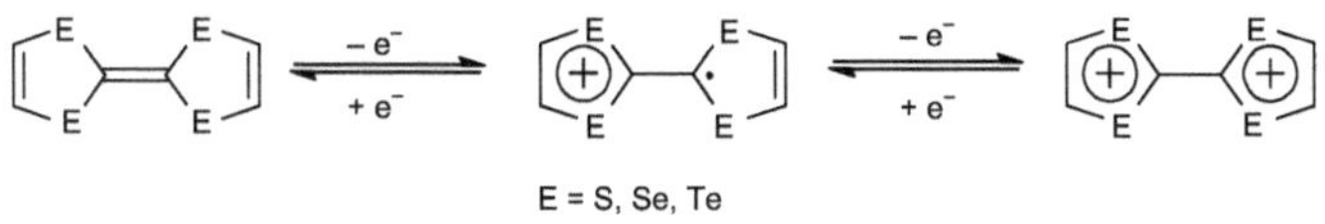

Figure 11.2.7 Redox states of TEF compounds.

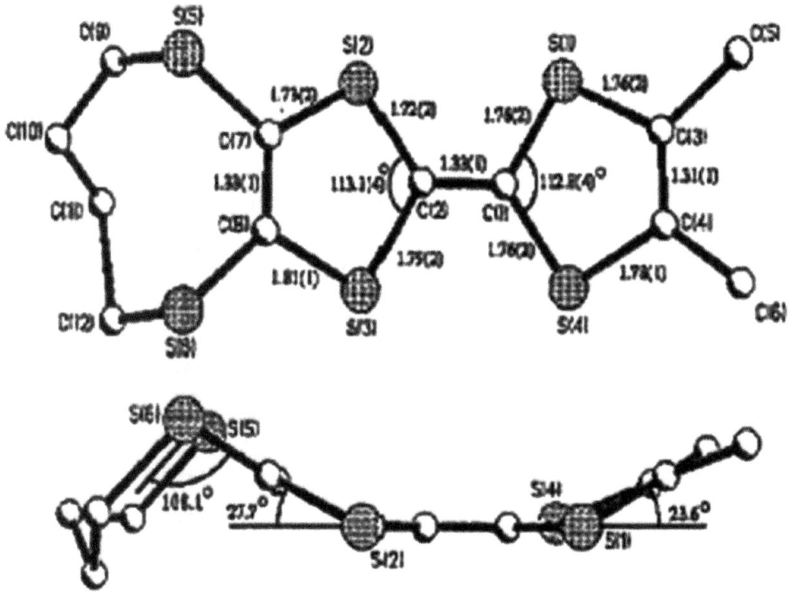

Figure 11.2.8 Molecular structure of tetramethylenedithiodimethyltetrafulvalene (TMDTDM-TTF) showing puckered conformation in the neutral state.
Reprinted from Konarev *et al.*,[88] with permission from Elsevier.

cases this contributes to Hückel aromaticity ($4n+2$ π-electrons), in the thiophene for example. TEF and its derivatives also show interesting aromatic behaviour. Each 1,3-dichalcole ring in TEF can lose one electron (see Figure 11.2.7). The resulting 1,3-dichalcolium cation is 6π aromatic. Where the neutral molecule has a puckered geometry (see for example Figure 11.2.8),[88] aromatization causes a change from puckered to planar in the oxidized radical cation state. Aromatization provides the thermodynamic driving force for TEF to act as an electron donor, while providing a conformational change which is often advantageous to crystal packing.

Short intramolecular contacts between chalcogens and neighbouring chalcogens or other heteroatoms have been shown to influence molecular geometry, particularly planarity, in many structures of electroactive materials. Hence the position of the chalcogen atom in the material can profoundly affect its properties. For example, Crouch *et al.*[89] reported the X-ray crystal structure of compound **19** (Figure 11.2.9), a candidate for an organic field-effect transistor, showing the effect

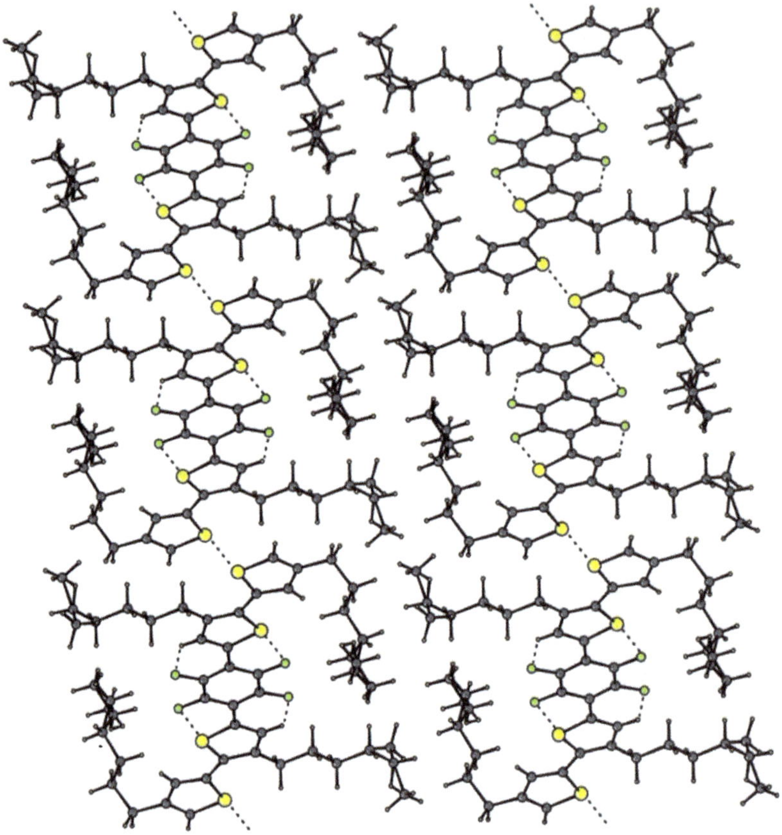

Figure 11.2.9 Packing diagram of compound **19**, viewed along the *a* axis, showing intra- and intermolecular contacts.
Reproduced from Crouch *et al.*,[89] with permission from RSC.

of intramolecular S···F close contacts (in tandem with H···F contacts) on the planarity of the molecule in the solid state. Note also the intermolecular S···S contacts forming chains of molecules. Increased planarity and rigidity generally means better conjugation and π-orbital overlap between molecules, leading to a smaller HOMO–LUMO gap and lower oxidation potential. Planar radical cations benefit from improved chemical stability as a result of delocalization. Stacking in the crystal lattice of the planar molecules also facilitates intermolecular CT. The synthetic chemist who can manipulate these interactions therefore has control over some fundamental properties of the material.

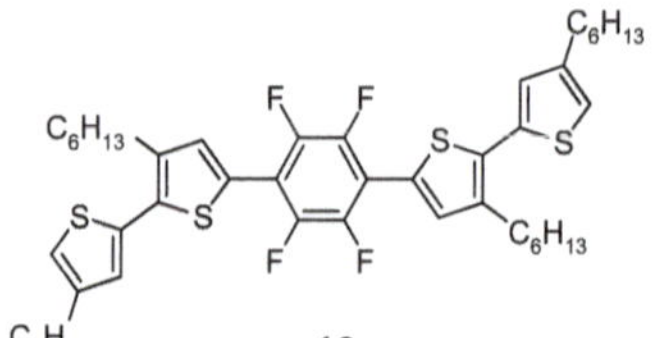

19

In order for a material to act as a useful bulk charge-transport medium, one molecule (be it a polymer or a small molecule in a crystal) must be able to transfer charge to its neighbour, thus establishing a conductance mechanism. Usually this is achieved by intermolecular orbital overlap. The large π-orbitals resulting from the presence of chalcogen atoms provide an excellent basis for co-facial stacking and serve to hold conducting molecules into ordered, self-assembled arrays. The one-dimensional columnar stacks of TTF units in the TTF/TCNQ crystal structure (essentially a 1:1 mixture of radical cation and neutral molecules, Figure 11.2.1) are a useful example. The dimensionality of the structure is reflected in its one-dimensional charge-transport properties, since the only intermolecular close contacts occur between sulfur atoms in co-facial units.[34]

Conduction in these one-dimensional metals is vulnerable to impurities, defects, and thermal vibrations within each stack. Interstack interactions allow for increased dimensionality and, in some cases superconductivity, by suppressing the Peierls distortion, an inherent, disruptive flaw characteristic of one-dimensional conductors. Radical cation salts of BEDT-TTF, for example, employ chalcogen–chalcogen contacts to achieve quasi-two-dimensional conductivity. The flexible ethylene bridge also allows the radical cations to adopt a number of conformations achievable through use of different counterions. Many organic metals have been prepared from salts of BEDT-TTF; this structural family was comprehensively reviewed in 2004 by Geiser and Schlueter,[43] and by Schibaeva and Yagubskii.[11]

Inter- and intrastack chalcogen–chalcogen interactions are important in forming supramolecular structures such as Langmuir–Blodgett films, self-assembled monolayers (SAMs) and liquid crystals. Development of this technology continues to attract attention in order to bridge the gap between promising materials properties and ease of fabrication.[90] Terthiophene (**20**)[91] shows hole and electron mobilities above 1×10^{-2} cm^2 V^{-1} s^{-1} in the smectic G phase at elevated temperatures. Garin and co-workers[92] reported the first discotic liquid crystal built around a TTF-core, namely compound **21**. The same group prepared a CT salt of compound **22**[93] which showed a liquid crystal phase, a promising result in the search for new multi-property materials. The TTF units align in parallel stacks, held by interdigitization of the aryl groups with adjacent stacks.

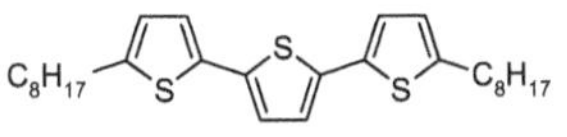

20

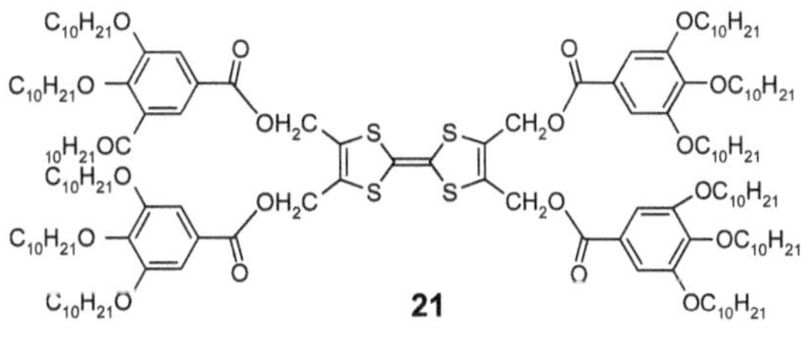

21

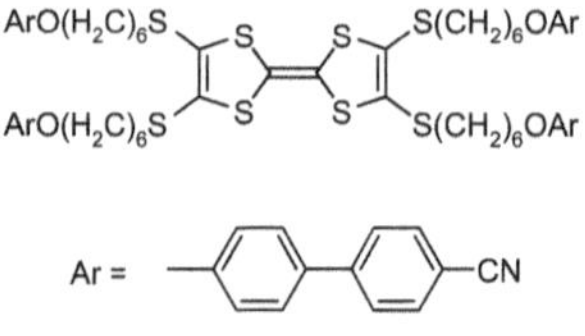

22

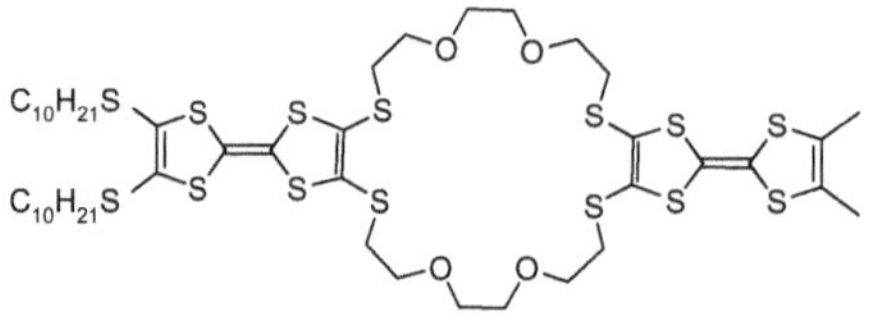

23

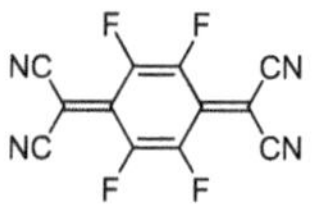

24

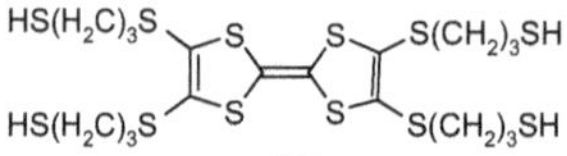

25

Other common techniques have been applied to the assembly of layers or films of TTF-derived molecular conductors. Compound **23** is an example of an amphiphilic TTF derivative. It forms Langmuir–Blodgett CT films with the acceptor TCNQF$_4$ (**24**).[94] Self-assembly of compound **25** on gold by electrochemical methods yielded an electroactive monolayer which was remarkably stable to electrochemical cycling.[95] Recently, an interesting study has been carried out providing the first evidence of the versatility of the electrochemical approach to estimate interaction parameters in SAMs.[96]

11.2.3.2 Use of Chalcogens Heavier than S

We have seen the versatility of chalcogen-containing materials in the previous sections. In most cases, materials properties have been investigated and improved through subtle changes to molecular structure. So far, however, we have not discussed the result of changing the chalcogen atom to form an isostructural material. A change in the substituent group, for example, may change the solid state structure of the material. Substituting a selenium or tellurium atom for a sulfur will, in many cases, cause no change in conformation or packing, so the electronic origins of properties may be probed more insightfully. Synthetic chalcogen chemistry therefore holds the key to interesting variation of physical properties. This approach led to the discovery of the first organic superconductors, the Bechgaard CT salts of tetramethyltetraselenafulvalene (TMTSF, **3**) with various counter-anions: (TMTSF)$_2$PF$_6$ (under pressure) in 1980[97] and (TMTSF)$_2$ClO$_4$ (at ambient pressure) in 1981.[6] Tetraselenafulvalene (TSF, **26**) was first reported by Engler and Patel.[98,99] Otsubo[100] noted that of the 14 TTF-type superconductors, nearly half are

Se-containing. There are 11 tetramethyltetraselenafulvalene (TMTSF) salts with polycyano acceptor anions alone, all showing semiconducting behaviour.[43] Tetratellurafulvalene (TTeF, **27**) was reported by McCullough in 1987.[101] The CT salts of TSF and TTeF with TCNQ followed in 1990.[102] The first tellurium-hybrid TEFs (diselenaditellurafulvalene, DSDTeF, **28** and dimethyldiselenaditellurafulvalene, DMDSDTeF, **29**) were prepared in 1997.[103]

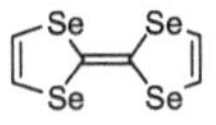
TSF, **26**

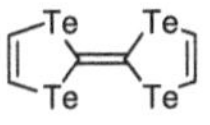
TTeF, **27**

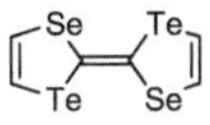
DSDTeF, **28**

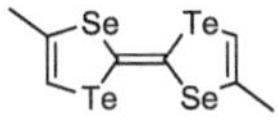
DMDSDTeF, **29**

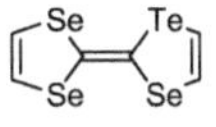
TSTeF, **30**

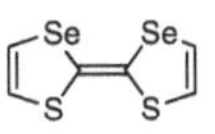
cis-DSDTF, **31**

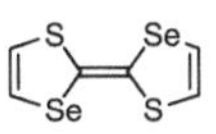
trans-DSDTF, **32**

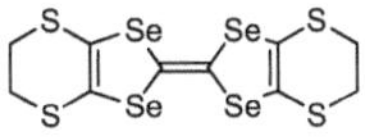
BETS, **33**

Since few examples of isostructural series of chalcogen-containing materials have been reported, this topic is afforded a more detailed treatment here, but this still does not represent an exhaustive list of all reported structures incorporating the heavier chalcogens. The structures discussed here were not chosen necessarily on the basis of outstanding electrical properties but on their value as illustrative comparisons. The reviews by Bryce,[34] Otsubo,[100] and Geiser and Schlueter[43] provide further context in this area.

11.2.3.2.1 Effect of Heavier Chalcogens on Donor Ability

Starting with the simplest material, TTF, it is possible to reveal the effect on the first oxidation potential, $E_{1/2}$ (1) (and hence on electron-donor ability), by insertion of a heavier chalcogen atom into the framework of the molecule. Table 11.2.1 shows the first oxidation potential of a variety of neutral tetrachalcogenafulvalene donors. The first observation to make is that changing the chalcogen from sulfur in TTF to selenium in TSF results in a higher oxidation potential for the neutral molecule, making it a weaker donor, thermodynamically. The Se=C π-bond is weaker than the S=C π-bond, a result of a better size matching of C and S orbitals. Oxidation of the donor generates a radical cation which is stabilized by resonance forms incorporating a C=E double bond (see Figure 11.2.7). It follows that the most thermodynamically stable cation is more easily accessible.[99]

This trend in bond strength continues with TTeF, but the more polarizable Te *p*- and *d*-orbitals allow for greater stabilization of the cationic charge, giving the order TTF<TTeF<TST in $E_{1/2}$ (1). The difference between the first and the second ionization potentials, $\Delta E_{1/2}$, follows the trend TTeF<TSF<TTF. Again, the superior polarizability of Te over Se and S reduces intramolecular coulombic repulsion between the dicationic charges.[101] This behaviour has also been reported in extended TTF derivatives[104] and anthracenediylidenes.[36]

Table 11.2.1 Electrochemical data for tetrathiafulvalene (TTF), tetraselenafulvalene (TSF) and tetratellurafulvalene (TTeF)[101]

Donor	$E_{1/2}(1)$, *V*	$E_{1/2}(2)$, *V*	$\Delta E_{1/2}$, *V*
TTF, **1**	+0.47	+0.81	+0.34
TSF, **26**	+0.62	+0.90	+0.28
TTeF, **27**	+0.59	+0.84	+0.25

11.2.3.2.2 Effect of Heavier Chalcogens on the Properties of CT Salts

We saw in Section 11.2.3.1 that the presence of additional chalcogen atoms in BEDT-TTF/TCNQ promotes interstack interactions, suppressing the Peierls distortion and imparting upon the salt increased dimensionality compared to TTF/TCNQ. The result of including a different chalcogen into the TTF/TCNQ is shown in Table 11.2.2. Despite losing donor efficiency compared to TTF (Table 11.2.1) the TCNQ complexes of *cis*/*trans*-diselenadithiafulvalene (DSDTF, **31**/**32**) and TSF show an improvement in conductivity when two or four selenium atoms are incorporated. The reduced metal–insulator (M–I) transition suggests that this effect is also caused by a suppression of the Peierls distortion. Increased Se–Se interstack contacts add dimensionality to the structure and limit the co-facial dimerization typical of Peierls distortion. Wider conduction bands are afforded from the improved overlap of diffuse orbitals.

Morikami[105] poses the question, 'how many Te atoms in the TSF skeleton are necessary to suppress the M–I transition altogether?' TSF[98] and triselenatellurafulvalene (TSTeF, **30**[103]) form TTF-type one-dimensional stacks with TCNQ in which the Peierls instability is inherent. The crystal structure of DSDTF/TCNQ, on the other hand, shows a completely different relative position of the donor and acceptor columns, brought about by side-by-side Se–Te and Se–Se contacts.[103] No M–I transition is observed down to very low temperatures. This structure is analogous to TTeF, where the Peierls distortion is suppressed by Te–Te interactions. High conductivities accompany these improvements. The findings led Morikami *et al.*[105] to classify TEF/TCNQ complexes into two structural categories: one-dimensional organic metals of the TTTF/TCNQ type, and quasi-one-dimensional organic metals containing multiple side-by-side hetero-contacts between columns of donors.

Saito *et al.*[108] studied the salts of TMTSF and the sulfur analogue tetramethyltetrathiafulvalene (TMTTF), with a polycyano dianion. Although the conductivity of both compounds was low ($\sigma_{rt} = 10^{-5}$ S cm^{-1} for TMTTF) the conductivity of the Se-donor salt was improved by two orders of magnitude. Optical absorption spectroscopy was also used to assess the materials. The electronic transition between radical cations within the segregated donor columns occurred at considerably lower energy (8800 cm^{-1}) in the TMTSF salt than in the TMTTF (11 500 cm^{-1}). A concurrent improvement in inter- and intrastack interactions seems to benefit the heavy chalcogen-containing material in this example.

Table 11.2.2 Electrochemical and structural properties of the TCNQ CT salts of some TEF donors

Donor	*Room-temperature conductivity, σ_{rt}, S cm^{-1}*	*Metal–insulator transition, K*	*Structural classification, (after Morikami et al.[105])*
TTF, **1**[2]	500	59	1D organic metal
Cis/trans-DSDTF, **31**[106]	550	64	1D organic metal
TSF, **26**[98]	800	40	1D organic metal
TSTeF, **30**[103]	2000	85	1D organic metal
DSDTeF, **28**[103]	1400	Metallic at 4.2 K, no transition	Quasi-1D, strong interstack interactions
TTeF, **51**[102,107]	2200	Metallic at 2 K, no transition	Quasi 1D, strong interstack interactions

The BEDT-TTF skeleton has been modified extensively with alternative chalcogens to sulfur. Much attention has been devoted to salts of bis(ethylenedithio)tetraselenafulvalene (BEDT-TSF, or BETS, **33**), as inclusion of larger Se atomic orbitals led to more isotropic electronic structures, and more stable molecular metals. α-$(BETS)_2TlHg(SeCN)_4$ is isostructural with its BEDT-TTF analogue, but the crystal structure shows almost twice as many Se–Se, Se–S, and S–S contacts. A typical unit cell volume increase of 2.8% is seen in the Se-containing structure.[109] Despite the improved interstack contacts, the structures of these two analogues are too similar to bring about a large change in the electrical properties; both α-$(BETS)_2TlHg(SeCN)_4$ and α-$(ET)_2TlHg(SeCN)_4$ are stable metals down to very low temperatures (1.3 K). κ-$(BETS)_2C(CN)_3$,[109] on the other hand, is not isostructural with κ-$(ET)_2C(CN)_3$,[110] and crystallizes in a different space group. It benefits from 22 close chalcogen–chalcogen contacts between BETS molecule and its six neighbours, of which four are face-to-face dimeric interactions within the stack. The all-S analogue contains only eight such contacts (S···S) to its nearest two molecules, but interestingly has no interstack S···S contacts of less than the sum of the van der Waals radii for two sulfurs. In both cases the ethylene bridge remains disordered. These structural alterations are enough to bring about a significant difference in properties between the two complexes. The BETS salt is also a stable metal down to 1.3 K but the ET complex undergoes a M-I transition at about 180 K, indicating that the S···S contacts do not give a sufficiently rigid framework to suppress Peierls transition.

11.2.3.2.3 Higher Chalcogenophenes: Selenophene and Tellurophene and Their Polymers

Whereas thiophene and its related polymers have attracted the vast majority of attention, the higher chalcogenophenes have not been widely investigated yet.

As stated above, the conductivity of some of the Se- and Te-bearing TEF derivatives are higher than that of their sulfur TEF analogues and therefore better properties are also expected for the higher chalcogenophenes. In that sense, since the publication of the first edition of this book, great advances have been achieved in the synthesis and characterization of not only selenophene and poly(selenophene), but also in the chemistry of tellurophene and its polymers.

Selenium and tellurium are well known to have strong Se···Se and Te···Te intermolecular interactions. These interactions should be favourable to enhance close interchain packing of poly(selenophene) (PSe) and poly(tellurophene) (PTe). Se and Te are more polarizable atoms than S, and its replacement by heavier chalcogenophenes is an interesting route to reduce oxidation potentials, and consequently for fine tuning of the electrochemical and optical properties of the corresponding polymer.[111]

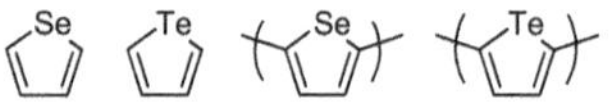

34 **35** **PSe** **PTe**

Although both selenophene (**34**)[112] and tellurophene (**35**)[113,114] are well known, the electrochemical synthesis of their polymers has been more complicated. One of the main reasons is the high monomer oxidation potentials in conventional solvents. In 1985, Sugimoto *at al.*[115] obtained PSe by treatment of selenophene with $FeCl_3$. The optical bandgap of this polymer was 2.0 eV and in the undoped state the polymer showed electrical conductivity as low as 10^{-10} S cm^{-1}, whereas in the doped state a conductivity of 10^{-3} S cm^{-1} was achieved. Semiconductor-type temperature/conductivity dependence was also seen in PSe. The bandgap can be decreased by introducing vinylene groups between the rings (1.6 eV).[116] Glenis *et al.*[117] electropolymerized selenophene and observed a negligible contribution to the π-electron system from the Se atom in the ground state, but an important effect on the conjugation length and conductivity. The PSe instability over oxidation is one of the main drawbacks when electropolymerization is attempted. The best room-temperature conductivity so far reported for a polyselenophene film is 10^{-1} S cm^{-1}.[118] This was prepared in neat boron trifluoride diethyl etherate, in which the oxidation potential of selenophene is significantly lower than in acetonitrile, for example. This result represented an improvement of an order of magnitude over the best previously reported PSe conductivities.[117,119]

Substituted PSe polymers have been also synthesized and characterized.[120] One interesting compound, 3,4-ethylenedioxyselenophene (EDOS, **36**), the Se analogue of EDOT, was synthesized by Aqad *et al.*[111] and showed a moderate enhancement of electron-donating ability compared to EDOT. PEDOS was also obtained electrochemically and chemically.[121] Bendikov's group reported a new synthetic route to prepare EDOS, and PEDOS was prepared chemically (oxidative polymerization using $FeCl_3$), by solid state polymerization from the brominated derivative, and electrochemically. The polymer bandgap of 1.4 eV is lower than that of PEDOT (1.6 eV). Dedoped polymer exposed to air showed conductivity as high as 0.1–0.5 S cm^{-1}, measured in pressed pellet form.

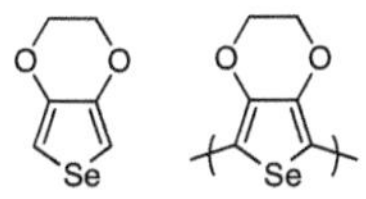

36 **PEDOS**

PTe has proved harder to prepare, but there are some reports[122] of successful electropolymerization, particularly from Otsubo's group where bitellurophene **37** and tertellurophene **38** have been used as monomers.[123,124] It is significant that the tellurophene oligomers have lower oxidation potentials and absorb at longer wavelengths than the corresponding oligoselenophenes and oligothiophenes. Electropolymerization of the dimer and trimer can therefore be performed at lower potentials and better-quality films result. Electropolymerization of tellurophene, bitellurophene, and tertellurophene yielded black films,[123] indicating that the materials are obtained in their doped states. Low stability or extensive branching of the polymers is thought to be responsible for the low conductivities; the highest conductivity so far reported for a tellurophene derivative was 1.3×10^{-5} S cm^{-1}, in the case of electropolymerized tertellurophene.

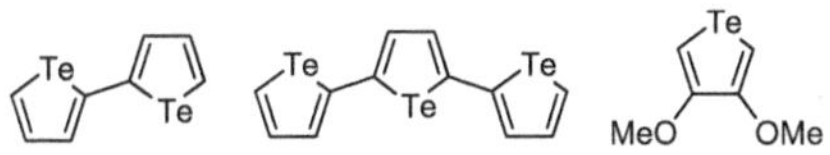

37 **38** **39**

Substituted PTe derivatives have also been reported.[116] Compound **39** [125] underwent electropolymerization experiments, but no evidence to strongly support its formation was reported. It is worth noting that although the crystal structure of **39** did not indicate π–π interactions, it revealed Te···Te close interactions. The interactions are shorter than the Se···Se interactions in the Se analogue.

The best conductivities achieved for PSe and PTe are still much lower than for PT (2000 S cm^{-1}).[31] However, the evidence suggests that the large difference does not originate from the inherent bandgap, bandwidth, or oxidation potential of the polymers. The samples of PSe so far reported may suffer from a significantly shorter effective conjugation length, resulting in charge trapping at chain ends.[126] Indeed, there are numerous other examples of the theoretical superiority of PSe and PTe over PT for electronic materials applications. Inoue *et al.*[127] prepared a series of oligoselenophenes (monomer through to hexamer) and found, as expected, that the higher oligomers showed lower oxidation potentials than the shorter oligomers.

A thorough theoretical comparison of PT, PSe, and PTe by Salzner and co-workers[128] lends further weight to the argument that it is the quality of the samples prepared, not an intrinsic property of the materials, which results in low conductivities. The study predicts very similar values of bandgap, ionization potential, electron affinity, HOMO, and conductance bandwidth for PT and PSe. PTe is predicted to have the lowest bandgap (0.34 eV lower than PT), in agreement with the work on tellurophene oligomers by the Otsubo group. It has a slightly smaller HOMO bandwidth and slightly larger conductance bandwidth than PT. Ionization potentials increase in the direction PTe < PSe < PT, in agreement with the trend in donor strength. The authors assert that there is therefore no reason to suggest that PT is superior to higher chalcogenophenes, unless the increased heteroatom size leads to larger interchain distance and hence less favourable charge 'hopping' between chains. Another theoretical study[129] has shown that the contribution of the quinoidal resonance structure of oligoselenophenes is larger than in the same oligothiophenes due to the lower aromaticity of the selenophene ring. This and a higher degree of planarity lead to lower bandgaps. It is predicted that optimization of PSe and PTe will result in improved conductivities; these materials, while demanding to prepare, still offer great potential.[130,131]

11.2.4 Chalcogen-Rich Compounds as Electron Donors in Materials Chemistry

The field of organic electronics has developed greatly in the last two decades. New materials are constantly synthesized and devices prepared. In that sense, materials based on sulfur-containing rings, and recently on selenium and tellurium (as we have seen above), represent a large percentage of the whole spectrum of materials. In this section, we want to illustrate briefly, by choosing specific examples, the behaviour of chalcogen-rich compounds as electron donors in the materials and in the devices prepared.

Organic photovoltaic (OPV) cells have been mainly fabricated using conjugated polymers. Poly(3-hexylthiophene) (P3HT) has been widely used in combination with soluble derivatives of fullerenes to fabricate OPVs. Efficiencies as high as 5% have been obtained so far from this polymer.[132] In the devices, an electron from the HOMO of P3HT is photoexcited and promoted to the LUMO level. The high electron affinity of the fullerene derivatives leads to the electron transfer from the LUMO of P3HT to the fullerene. One of the main disadvantages of P3HT is its relative high bandgap (1.9 eV). The selenium analogue of P3HT, P3HSe, has a lower bandgap,[133] and OPVs have been also fabricated using these polymers.[134] The efficiencies for P3HT are higher than those reported for P3HSe, but this can be attributed to the vast research and development that P3HT has received in terms of device engineering. To overcome the high bandgap of these homopolymers, and therefore low coverage of the solar spectrum, the use of donor–acceptor units along the backbone of the polymer has been applied. In these polymers an electron-rich unit (Figure 11.2.10) is alternated with an electron-deficient unit. This alternation leads to internal CT along the monomeric units and decreases the bandgap of the polymers.

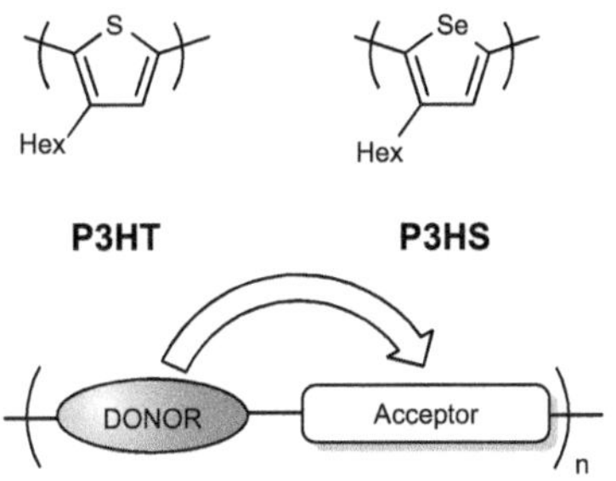

Figure 11.2.10 Donor–acceptor approach for low-bandgap conjugated polymers.

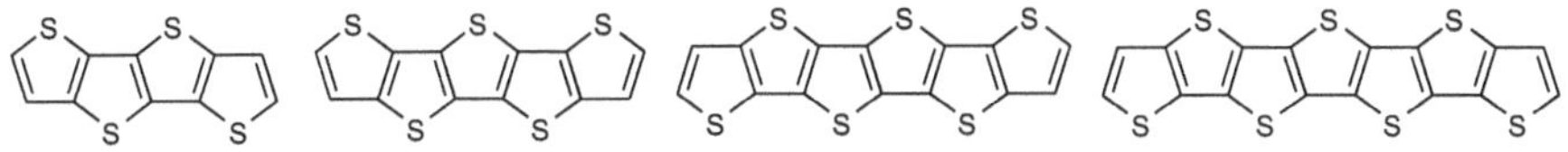

Figure 11.2.11 Fused thienoacenes units used in OFETs.

OFETs are another type of device where sulfur-containing compounds have been widely used. The diversity of compounds that have been used in OFETs is very wide. Many of the TEF derivatives mentioned in the previous sections have been used for OFETs by vacuum deposition or from solution.[20] Fused thienoacenes (Figure 11.2.11) have also been used in OFETs, in many cases obtaining mobilities of around 1 $cm^2 V^{-1} s^{-1}$ and maxima of around 10 $cm^2 V^{-1} s^{-1}$.[135] Polymers such as P3HT have also been used in OFETs,[136] as well as P3HS.[133] Although some materials have shown high electron mobilities,[137] due to the intrinsic electron-donating nature of these compounds, the majority of these compounds act as hole transporters (p-channel) in these devices.

Derivatives of thiophene such as EDOT and proDOT (**40**) have been widely used in polymers for electrochromic devices.[138,139] Upon oxidation the colour of these materials in the doped state changes in comparison with the colour in the neutral state. Ideally, the colour change would be reversible with large coloration efficiencies. PEDOS[121] and PproDOS (**41**)[140] have also shown colour change upon oxidation. Likewise, some copolymers of selenophene derivatives have also been synthesized and used in electrochromic devices showing colour changes in the visible region of the spcctrum, as well as significant contrast (60%) in the near infrared.[141]

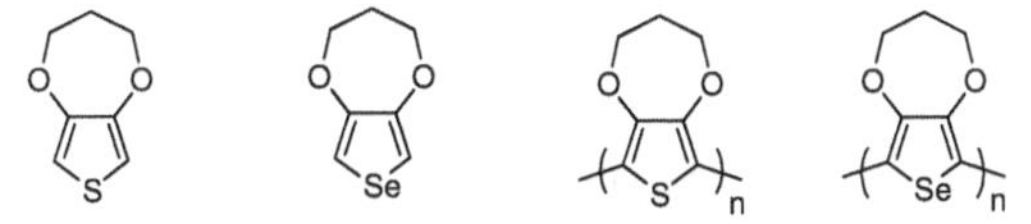

40, proDOT **41, proDOS** **PproDOT** **PproDOS**

Molecular sensors have been studied with chalcogen-rich materials.[70] Thc strong covalent and non-covalent interactions between chalcogens and certain metals are exploited in so-called 'host–guest' and supramolecular chemistry.

Crown ethers containing either S or O are known to bind alkali metal ions, for instance. When 'host' structures are linked to a redox active or electroactive moiety, a change in the physical properties of the molecule is often detected when a 'guest' is bound. In structures such as **42**,[142] complexation of metal ions (in this case Na^+) causes a small positive change (80 mV) in the first ionization potential of the TTF moiety. This is assigned to a through-space inductive effect caused by the presence of the cation, which can be observed by cyclic voltammetry (CV). The second ionization potential is usually unchanged, and can be used as an internal reference. Guest binding can also be detected by NMR, UV-vis spectroscopy or mass-spectrometry.

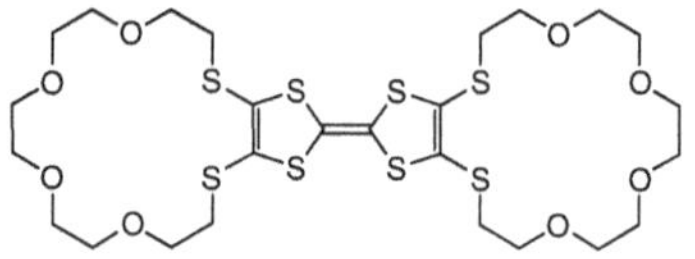

42

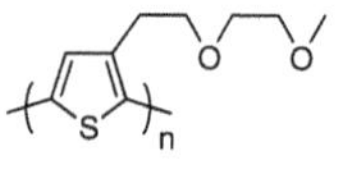

43

TTF is not the only chalcogen-containing motif to be incorporated as the 'signaller' in cation sensors—PT has also been widely studied in this context. Polymer films containing groups capable of molecular recognition are desirable in terms of electrode modification and device fabrication. Roncali *et al.*[143,144] reported the first example of molecular recognition by a conjugated polymer in 1989, in a simple PT system (**43**) containing flexible polyether side chains. Coordination of Li^+ was detected by CV.

References

1. F. Wudl, D. Wobschall and E. J. Hufnagel, *Chem. Comm.*, 1970, 1453.
2. J. Ferraris, V. Walatka, J. H. Perslteln and D. O. Cowan, *J. Am. Chem. Soc.*, 1973, **95**, 948.
3. T. J. Kistenmacher, T. E. Phillips and D. O. Cowan, *Acta Crystallogr. Sec.*, 1974, **B30**, 763.
4. Y. Takahashi, K. Hayakawa, T. Naito and T. Inabe, *J. Phys. Chem. C*, 2012, **116**, 700.
5. S. A. Odom, M. M. Caruso, A. D. Finke, A. M. Prokup, J. A. Ritchey, J. H. Leonard, S. R. White, N. R. Sottos and J. S. Moore, *Adv. Funct. Mater.*, 2010, **20**, 1721.
6. K. Bechgaard, K. Carneiro, F. B. Rasmussen, M. Olsen, G. Rindorf, C. S. Jacobsen, H. J. Pedersen and J. C. Scott, *J. Am. Chem. Soc.*, 1981, **103**, 2440.
7. A. M. Kini, U. Geiser, H. H. Wang, K. D. Carlson, J. M. Williams, W. K. Kwok, K. G. Vandervoort, J. E. Thompson, D. L. Stupka, D. Jung and M. H. Whangbo, *Inorg. Chem.*, 1990, **29**, 2555.
8. H. Urayama, H. Yamochi, G. Saito, K. Nozawa, T. Sugano, M. Kinoshita, S. Sato, K. Oshima, A. Kawamoto and J. Tanaka, *Chem. Lett.*, 1988, 55.

9. C. Rovira and J. J. Novoa, *Chem.-A Eur. J.*, 1999, **5**, 3689.
10. S. A. French and C. R. A. Catlow, *J. Phys. Chem. Solids*, 2004, **65**, 39.
11. R. P. Shibaeva and E. B. Yagubskii, *Chem. Rev.*, 2004, **104**, 5347.
12. T. Enoki and A. Miyazaki, *Chem. Rev.*, 2004, **104**, 5449.
13. E. Coronado and P. Day, *Chem. Rev.*, 2004, **104**, 5419.
14. X. Ribas, M. Mas-Torrent, A. Perez-Benitez, J. C. Dias, H. Alves, E. B. Lopes, R. T. Henriques, E. Molins, I. C. Santos, K. Wurst, P. Foury- Leylekian, M. Almeida, J. Veciana and C. Rovira, *Adv. Funct. Mater.*, 2005, **15**, 1023.
15. B. Zhang, Z. M. Wang, H. Fujiwara, H. Kobayashi, M. Kurmoo, K. Inoue, T. Mori, S. Gao, Y. Zhang and D. B. Zhu, *Adv. Mater.*, 2005, **17**, 1988.
16. E. Coronado, J. R. Galan-Mascaros, C. J. Gomez-Garcia, A. Murcia-Martinez and E. Canadell, *Inorg. Chem.*, 2004, **43**, 8072.
17. C. Rethore, N. Avarvari, E. Canadell, P. Auban-Senzier and M. Fourmigue, *J. Am. Chem. Soc.*, 2005, **127**, 5748.
18. J. I. Naraso, D. Nishida, S. Kumaki, Tokito and Y. Yamashita, *J. Am. Chem. Soc.*, 2006, **128**, 9598.
19. C. Moreno, R. Pfattner, M. Mas-Torrent, J. Puigdollers, S. T. Bromley, C Rovira, J. Veciana and R. Alcubilla, *J. Mater. Chem.*, 2012, **22**, 345.
20. M. Mas-Torrent and C. Rovira, *Chem. Soc. Rev.*, 2008, **37**, 827.
21. C. W. Chu, J. Ouyang, H. H. Tseng and Y. Yang, *Adv. Mater.*, 2005, **17**, 1440.
22. A. Akutsu-Sato, H. Akutsu, S. S. Turner, P. Day, M. R. Probert, J. A. K. Howard, T. Akutagawa, S. Takeda, T. Nakamura and T. Mori, *Angew. Chem.-Int. Edit.*, 2005, **44**, 292.
23. S. Fukuzumi, K. Ohkubo, Y. Kawashima, D. S. Kim, J. S. Park, A. Jana, V. M. Lynch, D. Kim and J. L. Sessler, *J. Am. Chem. Soc.*, 2011, **133**, 15938.
24. T. Avellini, H. Li, A. Coskun, G. Barin, A. Trabolsi, A. N. Basuray, S. K. Dey, A. Credi, S. Silvi, J. F. Stoddart and M. Venturi, *Angew. Chem. Int. Ed.*, 2012, **51**, 1611.
25. C. S. Wang, A. S. Batsanov and M. R. Bryce, *Chem. Commun.*, 2004, 578.
26. S. Wenger, P. A. Bouit, Q. Chen, J. I Teuscher, D. Di Censo, R. Humphry-Baker, J. E. Moser, J. L. Delgado, N. Martín, S. M. Zakeeruddin and M. Grätzel, *J. Am. Chem. Soc.*, 2010, **132**, 5164.
27. I. Fuks-Janczarek, J. Luc, B. Sahraoui, F. Dumur, P. Hudhomme, J. Berdowski and I. V. Kityk, *J. Phys. Chem. B*, 2005, **109**, 10179.
28. C. Wang, Q. Chen, F. Sun, D. Zhang, G. Zhang, Y. Huang, R. Zhao and D. Zhu, *J. Am. Chem. Soc.*, 2010, **132**, 3092.
29. J. Bigot, B. Charleux, G. Cooke, F. Delattre, D. Fournier, J. Lyskawa, L. Sambe, F. Stoffelbach and P. Woise, *J. Am. Chem. Soc.*, 2010, **132**, 10796.
30. C. K. Chiang, M. A. Druy, S. C. Gau, A. J. Heeger, E. J. Louis, A. G. MacDiarmid, Y. W. Park and H. Shirakawa, *J. Am. Chem. Soc.*, 1978, **100**, 1073.

31. J. Roncali, *Chem. Rev.*, 1997, **97**, 173.
32. J. L. Segura and N. Martin, *Angew. Chem.-Int. Edit.*, 2001, **40**, 1372.
33. E. Laukhina, V. Tkacheva, A. Chekhlov, E. Yagubskii, R. Wojciechowski, J. Ulanski, J. Vidal-Gancedo, J. Veciana, V. Laukhin and C. Rovira, *Chem. Mater.*, 2004, **16**, 2471.
34. M. R. Bryce, *J. Mater. Chem.*, 1995, **5**, 1481.
35. Y. Misaki, N. Higuchi, H. Fujiwara, T. Yamabe, T. Mori, H. Mori and S. Tanaka, *Angew. Chem.-Int. Edit.*, 1995, **34**, 1222.
36. A. J. Moore and M. R. Bryce, *J. Chem. Soc.,-Perkin Trans*, 1991, **1**, 157.
37. G. Chen, I. Mahmud, L. N. Dawe, L. M. Daniels and Y. Zhao, *J. Org. Chem.*, 2011, **76**, 2701.
38. P. Sandin, A. Martinez-Grau, L. Sanchez, C. Seoane, R. Pou-Amerigo, E. Orti and N. Martin, *Org. Lett.*, 2005, **7**, 295.
39. H. Brisset, S. LeMoustarder, P. Blanchard, B. Illien, A. Riou, J. Orduna, J. Garin and J. Roncali, *J. Mater. Chem.*, 1997, **7**, 2027.
40. P. Frère and P. J. Skabara, *Chem. Soc. Rev.*, 2005, **34**, 69.
41. P. Cassoux, L. Valade, H. Kobayashi, A. Kobayashi, R. A. Clark and A. E. Underhill, *Coord. Chem. Rev.*, 1991, **110**, 115.
42. E. B. Yagubskii, *J. Solid State Chem.*, 2002, **168**, 464.
43. U. Geiser and J. A. Schlueter, *Chem. Rev.*, 2004, **104**, 5203.
44. L. Brossard, M. Ribault, M. Bousseau, L. Valade and P. Cassoux, *Comptes Rendus De L Academie Des Sciences Serie II*, 1986, **302**, 205.
45. L. Brossard, M. Ribault, L. Valade and P. Cassoux, *Physica B & C*, 1986, **143**, 378.
46. R. Kato, T. Mori, A. Kobayashi, Y. Sasaki and H. Kobayashi, *Chem. Lett.*, 1984, 1.
47. L. Brossard, M. Ribault, L. Valade and P. Cassoux, *Journal De Physique*, 1989, **50**, 1521.
48. S. Curreli, P. Deplano, M. L. Mercuri, L. Pilia, A. Serpe, J. A. Schlueter, M. A. Whited, U. Geiser, E. Coronado, C. J. Gomez-Garcia and E. Canadell, *Inorg. Chem.*, 2004, **43**, 2049.
49. H. Tanaka, Y. Okano, H. Kobayashi, W. Suzuki and A. Kobayashi, *Science*, 2001, **291**, 285.
50. E. Fujiwara, A. Kobayashi, H. Fujiwara and H. Kobayashi, *Inorg. Chem.*, 2004, **43**, 1122.
51. A. Kobayashi, H. Tanaka and H. Kobayashi, *J. Mater. Chem.*, 2001, **11**, 2078.
52. D. F. Perepichka, M. R. Bryce, C. Pearson, M. C. Petty, E. J. L. McInnes and J. P. Zhao, *Angew. Chem. Int. Ed.*, 2003, **42**, 4636.
53. F. Otón, V. Lloveras, M. Mas-Torrent, J. Vidal-Gancedo, J. Veciana and C. Rovira, *Angew. Chem. Int. Ed.*, 2011, **50**, 10902.
54. F. G. Brunetti, J. L. López, C. Atienza and N. Martín, *J. Mater. Chem.*, 2012, **22**, 4188.
55. G. Ho, J. R. Heath, M. Kondratenko, D. F. Perepichka, K. Arseneault, M. Pezolet and M. Bryce, *Chem.-A Eur. J.*, 2005, **11**, 2914.

56. G. J. Ahwell, J. R. Sambles, A. S. Martin, W. G. Parker and M. Szablewski, *J. Chem.Soc.-Chem. Comun.*, 1990, 1374.
57. N. Martín, L. Sánchez, M. A. Herranz, B. Illescas and D. M. Guldi, *Acc. Chem. Res.*, 2007, **40**, 1015.
58. L. Sánchez, I. Pérez, N. Martín and D. M. Guldi, *Chem. Eur. J.*, 2003, **9**, 2457.
59. M. Di Valentin, A. Bisol, G. Agostini, P. A. Liddell, G. Kodis, A. L. Moore, T. A. Moore, D. Gust and D. Carbonera, *J. Phys. Chem. B*, 2005, **109**, 14401.
60. V. Balzani, A. Credi, G. Mattersteig, O. A. Matthews, F. M. Raymo, J. F. Stoddart, M. Venturi, A. J. P. White and D. J. Williams, *J. Org. Chem.*, 2000, **65**, 1924.
61. M. B. Nielsen, J. O. Jeppesen, J. Lau, C. Lomholt, D. Damgaard, J. P. Jacobsen, J. Becher and J. F. Stoddart, *J. Org. Chem.*, 2001, **66**, 3559.
62. D. Philp, A. M. Z. Slawin, N. Spencer, J. F. Stoddart and D. J. Williams, *J. Chem. Soc.-Chem. Commun.*, 1991, 1584.
63. M. A. Olson, A. B. Braunschweig, L. Fang, T. Ikeda, R. Klajn, A. Trabolsi, P. J. Wesson, D. Benítez, C. A. Mirkin, B. A. Grzybowski and J. F. Stoddart, *Angew. Chem.*, 2009, **121**, 1824.
64. E. H. Witlicki, S. S. Andersen, S. W. Hansen, J. O. Jeppesen, E. W. Wong, L. Jensen and A. H. Flood, *J. Am. Chem. Soc.*, 2010, **132**, 6099.
65. S. Nygaard, S. W. Hansen, J. C. Huffman, A. H. Frank Jensen, Flood and J. O. Jeppesen, *J. Am. Chem. Soc.*, 2007, **129**, 7354.
66. J. M. Spruell, A. Coskun, D. C. Friedman, R. S. Forgan, A. A. Sarjeant, A. Trabolsi, A. C. Fahrenbach, G. Bari, W. F. Paxton, S. K. Dey, M. A. Olson, D. Benítez, E. Tkatchouk, M. T. Colvin, R. Carmielli, S. T. Caldwell, G. M. Rosair, S. Gunatilaka Hewage, F. Duclairoir, J. L. Seymour, A. M. Z. Slawin, W. A. Goddard, M. R. Wasielewski, G. Cooke and J. F. Stoddart, *Nature Chem.*, 2010, **2**, 870.
67. S. Saha, A. H. Flood, J. F. Stoddart, S. Impellizzeri, S. Silvi, M. Venturi and A. Credi, *J. Am. Chem. Soc.*, 2007, **129**, 12159.
68. E. H. Witlicki, C. Johnsen, S. W. Hansen, D. W. Silverstein, V. J. Bottomley, J. O. Jeppesen, E. W. Wong, L. Jensen and A. H. Flood, *J. Am. Chem. Soc,.*, 2011, **133**, 7288.
69. G. Fernández, E. M. Pérez, L. Sánchez and N. Martín, *Angew. Chem. Int. Ed.*, 2008, **47**, 1094.
70. D. Canevet, M. Sallé, G. Zhang, D. Zhang and D. Zhu, *Chem. Commun.*, 2009, 2245.
71. S. Goeb, S. Bivaud, P. I. Dron, J. Y. Balandier, M. Chas and M. Sallé, *Chem. Commun.*, 2012, **48**, 3106.
72. M. Hardouin-Lerouge, P. Hudhomme and M. Sallé, *Chem. Soc. Rev.*, 2011, **40**, 30.
73. H. Pang, P. J. Skabara, S. Gordeyev, J. J. W. McDouall, S. J. Coles and M. B. Hursthouse, *Chem. Mater.*, 2007, **19**, 301.
74. T. Benincori, E. Brenna, F. Sannicolo, L. Trimarco, G. Zotti and P. Sozzani, *Angew. Chem., Int. Ed.*, 1996, **35**, 648.

75. A. Cravino, G. Zerza, M. Maggini, S. Bucella, M. Svensson, M. R. Andersson, H. Neugebauer, C. J. Brabec and N. S. Sariciftci, *Monatshefte Fur Chemie*, 2003, **134**, 519.
76. J. Roncali, *Chem. Soc. Rev.*, 2005, **34**, 483.
77. M. Li, P. Xu, J. Yang and S. Yang, *J. Mater. Chem.*, 2010, **20**, 3953.
78. R. L. Meline, I. T. Kim, J. Chen, S. Basak and R. L. Elsenbaumer, *Abstracts Of Papers Of The American Chemical Society*, 1998, **215**, U374.
79. H. Müller, F. Salhi, B. Divisia-Blohorn, F. Genoud, T. Narayanan, M. Lorenzen and C. Ferrero, *Chem. Commun.*, 1999, 1407.
80. Y. Hou, Y. Chen, Q. Liu, M. Yang, X. Wan, S. Yin and A. Yu, *Macromolecules*, 2008, **41**, 3114.
81. Y. Liu, C. Wang, M. Li, G. Lai and Y. Shen, *Macromolecules*, 2008, **41**, 2045.
82. P. J. Skabara and K. Müllen, *Synth. Met.*, 1997, **84**, 345.
83. P. J. Skabara, R. Berridge, E. J. L. McInnes, D. P. West, S. J. Coles, M. B. Hursthouse and K. Müllen, *J. Mater. Chem.*, 2004, **14**, 1964.
84. P. J. Skabara, D. M. Roberts, I. M. Serebryakov and C. Pozo-Gonzalo, *Chem. Commun.*, 2000, 1005.
85. Diego Cortizo-Lacalle, S. Arumugam, S. E. T. Elmasly, A. L. Kanibolotsky, N. J. Findlay, A. Regis Inigo and P. J. Skabara, *J. Mater. Chem.*, 2012, **22**, 11310.
86. J. Yamada and T. Sugimoto, *TTF Chemistry: Fundamentals and Applications of Tetrathiafulvalene*, Kodansha, 2004.
87. M. Iyoda, M. Hasegawa and Y. Miyake, *Chem. Rev.*, 2004, **104**, 5085.
88. D. V. Konarev, E. F. Valeev, Y. L. Slovokhotov, Y. M. Shulga, O. S. Roschupkina and R. N. Lyubovskaya, *Synth. Met.*, 1997, **88**, 85.
89. D. J. Crouch, P. J. Skabara, M. Heeney, I McCulloch, S. J. Coles and M. B. Hursthouse, *Chem. Commun.*, 2005, 1465.
90. M. O'Neill and S. M. Kelly, *Adv. Mater.*, 2003, **15**, 1135.
91. M. Funahashi and J. Hanna, *Appl. Phys. Lett.*, 2000, **76**, 2574.
92. R. Andreu, J. Garin, J. Orduna, J. Barbera, J. L. Serrano, T. Sierra, M. Salle and A. Gorgues, *Tetrahedron*, 1998, **54**, 3895.
93. R. Andreu, J. Barbera, J. Garin, J. Orduna, J. L. Serrano, T. Sierra, P. Leriche, M. Salle, A. Riou, M. Jubault and A. Gorgues, *J. Mater. Chem.*, 1998, **8**, 881.
94. T. Akutagawa, K. Kakiuchi, T. Hasegawa, T. Nakamura, C. A. Christensen and J. Becher, *Langmuir*, 2004, **20**, 4187.
95. H. Fujihara, H. Nakai, M. Yoshihara and T. Maeshima, *Chem. Commun.*, 1999, 737.
96. P. Y. Blanchard, O. Alévêque, S. Boisard, C. Gautier, A. El-Ghayoury, F. Le Derf, T. Breton and E. Levillain, *Phys. Chem. Chem. Phys.*, 2011, **13**, 2118.
97. D. Jerome, A. Mazaud, M. Ribault and K. Bechgaard, *Journal De Physique Lettres*, 1980, **41**, L95.
98. E. M. Engler and V. V. Patel, *J. Am. Chem. Soc.*, 1974, **96**, 7376.

99. E. M. Engler, F. B. Kaufman, D. C. Green, C. E. Klots and R. N. Compton, *J. Am. Chem. Soc.*, 1975, **97**, 2921.
100. T. Otsubo and K. Takimiya, *Bull. Chem. Soc. Jpn.*, 2004, **77**, 43.
101. R. D. McCullough, G. B. Kok, K. A. Lerstrup and D. O. Cowan, *J. Am. Chem. Soc.*, 1987, **109**, 4115.
102. D. O. Cowan, M. D. Mays, T. J. Kistenmacher, T. O. Poehler, M. A. Beno, A. M. Kini, J. M. Williams, Y. K. Kwok, K. D. Carlson, L. Xiao, J. J. Novoa and M. H. Whangbo, *Mol. Cryst. Liq. Cryst.*, 1990, **181**, 43.
103. K. Takimiya, A. Morikami, Y. Aso and T. Otsubo, *Chem. Commun.*, 1997, 1925.
104. T. Sugimoto, H. Awaji, I. Sugimoto, Y. Misaki, T. Kawasi, S. Yoneda, Z. Yoshida, T. Kobayashi and H. Anzai, *Chem. Mater.*, 1989, **1**, 535.
105. A. Morikami, K. Takimiya, Y. Aso and T. Otsubo, *J. Mater. Chem.*, 2001, **11**, 2431.
106. E. M. Engler and V. V. Patel, *J. Chem. Soc.-Chem. Commun.*, 1975, 671.
107. D. O. Cowan, R. McCullough, A. Bailey, K. Lerstrup, D. Talham, D. Herr and M. Mays, *Phosphorus Sulfur Silicon Relat. Elem.*, 1992, **67**, 277.
108. G. Saito, S. Sekizaki, A. Konsha, H. Yamochi, K. Matsumoto, M. Kusunoki and K. Sakaguchi, *J. Mater. Chem.*, 2001, **11**, 364.
109. L. V. Zorina, B. Z. Narymbetov, S. S. Khasanov, R. P. Shibaeva, N. D. Kushch, E. B. Yagubskii, A. Kobayashi and H. Kobayashi, *Synth. Met.*, 1999, **102**, 1735.
110. M. A. Beno, H. H. Wang, L. Soderholm, K. D. Carlson, L. N. Hall, L. Nunez, H. Rummens, B. Anderson, J. A. Schlueter, J. M. Williams, M. H. Whangbo and M. Evain, *Inorg. Chem.*, 1989, **28**, 150.
111. E. Aqad, M. V. Lakshmikantham and M. P. Cava, *Org. Lett.*, 2001, **3**, 4283.
112. S. Gronowitz, T. Frejd, A. Mobergogard and L. Trege, *J. Heterocycl. Chem.*, 1976, **13**, 1319.
113. G. Marino, *Chem. Scr.*, 1975, **A8**, 23.
114. K. Irgolic, *Organotellurium Compounds*, D. Klamann (ed), Thieme, Hamburg, 1990.
115. R. Sugimoto, K. Yoshino, S. Inoue and K. Tsukagosh, *Jpn. J. Appl. Phys.*, 1985, **24**, 425.
116. S. C. Ng, H. S. O. Chan, T. T. Ong, K. Kumura, Y. Mazaki and K. Kobayashi, *Macromolecules*, 1998, **31**, 1221.
117. S. Glenis, D. S. Ginley and A. J. Frank, *J. Appl. Phys.*, 1987, **62**, 190.
118. S. Z. Pu, J. Hou, J. K. Xu, G. M. Nie, S. S. Zhang, L. Shen and Q. Xiao, *Mater. Lett.*, 2005, **59**, 1061.
119. K. Yoshino, Y. Kohno, T. Shiraishi, K. Kaneto, S. Inoue and K. Tsukagoshi, *Synth. Met.*, 1985, **10**, 319.
120. C. Mahatsekake, J. M. Catel, C. G. Andrieu, M. Ebel and Y. Mollier, *Phosphorous, Sulfur, and Silicon*, 1990, **47**, 35.

121. A. Patra, Y. H. Wijsboom, S. S. Zade, M. Li, Y. Sheynin, G. Leitus and Michael Bendikov, *J. Am. Chem. Soc.*, 2008, **130**, 6734.
122. S. C. Ng, H. Ding and H. S. O. Chan, *Chem. Lett.*, 1999, 1325.
123. S. Inoue, T. Jigami, H. Nozoe, T. Otsubo and F. Ogura, *Tetrahedron Lett.*, 1994, **35**, 8009.
124. T. Otsubo, S. Inoue, H. Nozoe, T. Jigami and F. Ogura, *Synth. Met.*, 1995, **69**, 537.
125. A. Patra, Y. H. Wijsboom, G. Leitus and Michael Bendikov, *Org. Lett.*, 2009, **7**, 1487.
126. A. Bourahla, J. M. Saiter and C. Vautier, *Mater. Chem. Phys.*, 2001, **69**, 163.
127. S. Inoue, H. Nakanishi, K. Takimiya, Y. Aso and T. Otsubo, *Synth. Met.*, 1997, **84**, 341.
128. U. Salzner, J. B. Lagowski, P. G. Pickup and R. A. Poirier, *Synth. Met.*, 1998, **96**, 177.
129. S. S. Zade, N. Zamoshchik and M. Bendikov, *Chem. Eur. J.*, 2009, **15**, 8613.
130. A. Patra and M. Bendikov, *J.* Mater. Chem., 2010, **20**, 422.
131. A. A. Jahnke and D. S. Seferos, *Macromol. Rapid. Commun.*, 2011, **32**, 943.
132. W. Ma, C. Yang, X. Gong, K. Lee and A. J. Heeger, *Adv. Funct. Mater.*, 2005, **15**, 1617.
133. M. Heeney, W. Zhang, D. J. Crouch, M. L. Chabinyc, S. Gordeyev, R. Hamilton, S. J. Higgins, I. McCulloch, P. J. Skabara, D. Sparrowe and S. Tierney, *Chem. Comm.*, 2007, **43**, 5061.
134. W. H. Lee, S. K. Lee, S. K. Son, J. Choi, W. S. Shin, K. Kim, S. Moon and I. Kang, *J. Polym. Sci. A*, 2012, **50**, 551.
135. K. Takimiya, S. Shinamura, I. Osaka and E. Miyazaki, *Adv. Mater.*, 2011, **23**, 4347.
136. H. Yan, Z. Chen, Y. Zheng, C. Newman, J. R. Quinn, F. Dötz, M. Kastler and A. Facchetti, *Nature*, 2009, **457**, 679.
137. L. Tan, Y. Guo, G. Zhang, Y. Yang, D. Zhang, G. Yu, W. Xu and Y. Liu, *J. Mat. Chem*, 2011, **21**, 18042.
138. C. M. Amb, A. L. Dyer and J. R. Reynolds, *Chem. Mater*, 2011, **23**, 397.
139. P. M. Beaujuge and J. R. Reynolds, *Chem. Rev.*, 2010, **110**, 268.
140. B. Kim, J. Kim and E. Kim, *Macromolecules*, 2011, **44**, 8791.
141. B. A. Trofimov, E. Y. Schmidt, A. I. Mikhaleva, C. Pozo-Gonzalo, J. A. Pomposo, M. Salsamendi, N. I. Protzuk, N. V. Zorina, A. V. Afonin, A. V. Vashchenko, E. P. Levanova and G. G. Levkovskaya, *Chem. Eur. J*, 2009, **15**, 6435.
142. T. K. Hansen, T. Jorgensen, P. C. Stein and J. Becher, *J. Org. Chem.*, 1992, **57**, 6403.
143. J. Roncali, R. Garreau, D. Delabouglise, F. Garnier and M. Lemaire, *J. Chem. Soc.-Chem. Commun.*, 1989, 679.
144. J. Roncali, L. H. Shi and F. Garnier, *J. Phys. Chem.*, 1991, **95**, 8983.

CHAPTER 11.3

1,2-Dichalcogenolene Ligands and Related Metal Complexes

MASSIMILIANO ARCA,* M. CARLA ARAGONI AND ANNA PINTUS

Dipartimento di Scienze Chimiche e Geologiche, Università degli Studi di Cagliari, S.S. 554 bivio per Sestu, 09042 Monserrato – Cagliari, Italy
*Email: marca@unica.it

11.3.1 Introduction

A concise account on the chemistry of 1,2-dithiolenes, 1,2-diselenolene, and 1,2-ditellurolene ligands was included in the first edition of this handbook.[1] In this chapter selected aspects of 1,2-dichalcogenolene chemistry are reviewed, with the aim of illustrating the general properties, trends, and perspectives of this branch of ligands and the coordination chemistry of their complexes with transition and main group metals. In particular, the synthesis and the structural, electrical, magnetic, and optical properties of the title complexes will be concisely discussed here. Other aspects, of no less importance, such as the role of 1,2-dichalcogenolene complexes in biology and in enzyme modelling,[2–28] along with analytical chemistry applications[29–31] and purely theoretical investigations of the title compounds[32–47] will not be dealt with in this chapter.

11.3.2 Historical Background

The most largely explored sulfur terms of the family, 1,2-dithiolenes, were the subject of study as early as the mid-1930s by Clarck and co-workers.[48–50]

Handbook of Chalcogen Chemistry: New Perspectives in Sulfur, Selenium and Tellurium, 2nd Edition, Volume 2
Edited by Francesco Antonio Devillanova and Wolf-Walther du Mont

Published by the Royal Society of Chemistry, www.rsc.org

In 1957, Bähr and Schleitzer first reported a Pd complex deriving from the sodium salt of the ligand maleonitrile-1,2-dithiolate, Na_2mnt (Scheme 11.3.1).[51] In 1959, dianionic metal complexes derived from quinoxaline-2,3-dithiolate were reported.[52] Nevertheless, it was only in the mid-1960s with the seminal

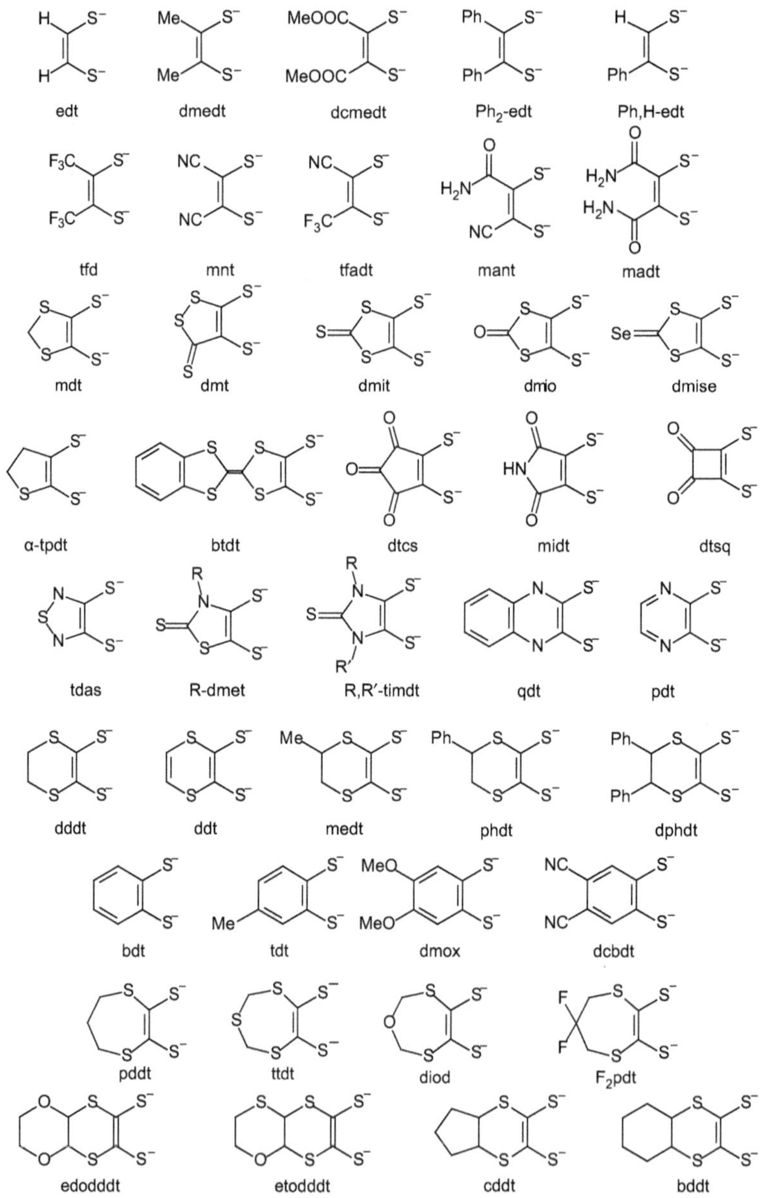

Scheme 11.3.1 Selected 1,2-dithiolene ligands and commonly used acronyms. All ligands have been drawn in their 1,2-dithiolate form.

works by Schrauzer and Mayweg on [Ni(Ph_2-edt)$_2$] (Ph_2-edt^{2-} = diphenylethylene-1,2-dithiolate, also known as dpedt^{2-}),[53] and by Gray on nickel maleonitrile-1,2-dithiolato complexes,[54] that the chemistry of 1,2-dithiolene complexes started to attract the interest of researchers. In 1965, the first example of a structurally characterized trigonal-prismatic tris(1,2-dithiolene) complex, namely [Re(Ph_2-edt)$_3$], was described by Eisenberg and Ibers.[55] Since then, a very large variety of different 1,2-dithiolene ligands and their complexes with numerous metal ions have been reported. The interest in 1,2-dithiolene chemistry was further enhanced by the discovery of a large number of applications, in fields as varied as conductivity, magnetism, linear and non-linear optics, and of the role of 1,2-dithiolene systems in biology. Therefore, the number of publications has increased year by year (Figure 11.3.1): along with papers and communications, several reviews have appeared, dealing with single classes of 1,2-dithiolene complexes, on their reactivity,[56–73] or on their applications.[74–89]

In 2004 an entire volume of *Progr. Inorg. Chem.* was published on the topic,[90] while in 2010 issues 12–13 of volume 254 of *Coord. Chem. Rev.* were entirely dedicated to 1,2-dithiolenes and non-innocent redox-active ligands.[91]

Passing to 1,2-diselenolene ligands (Scheme 11.3.2), it was only in 1967 that Davison and Shawl succeeded in synthesizing several 1,2-diselenolene transition metal complexes.[92,93] The number of publications regarding 1,2-diselenolene complexes remains much smaller than that regarding 1,2-dithiolene complexes, both due to synthetic difficulties and because the properties of these complexes are very similar to those of the fully sulfured congeners. Descending along the group, only very few publications have dealt with the synthesis and characterization of 1,2-ditellurolene complexes.

In general, 1,2-dichalcogenolene ligands can be classified into the three main categories of homoleptic, heteroleptic, and organometallic dithiolene

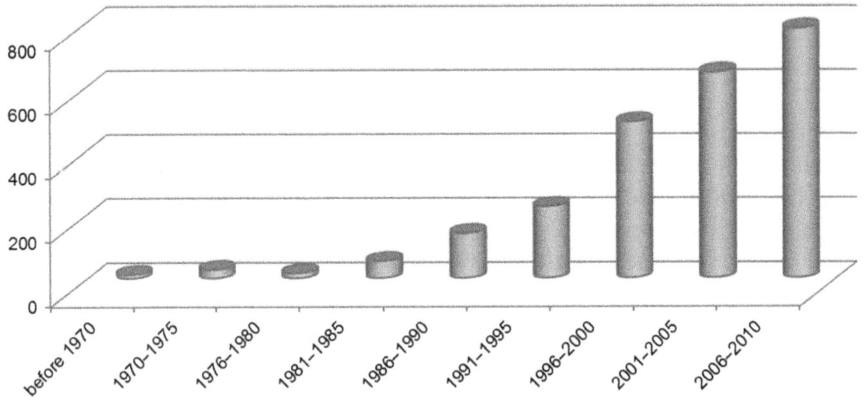

Figure 11.3.1 Occurrences of 1,2-dichalcogenolene metal complexes structurally characterized up to 2010 retrieved from the Cambridge Structural Database (CSD, v.5.33, updated Feb 2012). The number of structures relative to 2011 (108) is not included in the count.

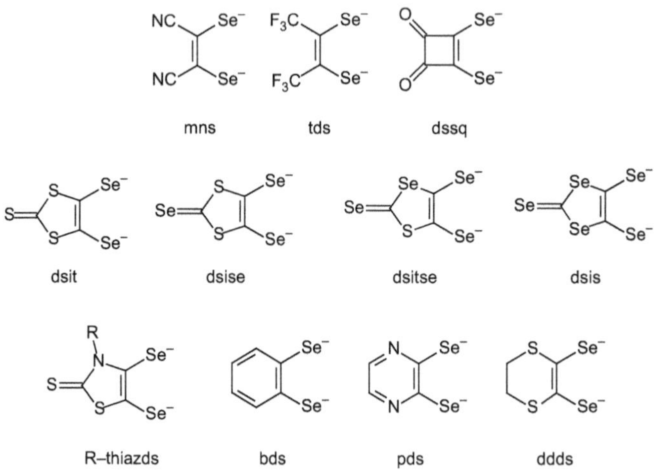

Scheme 11.3.2 Selected 1,2-diselenolene ligands and commonly used acronyms. All ligands have been drawn in their 1,2-diselenolate form.

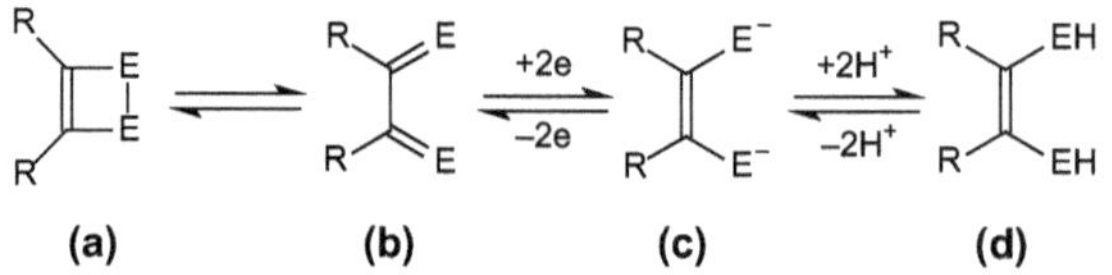

Scheme 11.3.3 Equilibria between 1,2-dichalcogenolene precursors: dichalcogenete (a), 1,2-dichalcogenone (b), ene-1,2-dichalcogenolate (c), and ene-1,2-dichalcogenol (d). E = S, Se, Te.

complexes, where the ligands can be wholly or partly of 1,2-dichalcogenolene type.

11.3.3 Nomenclature

The term dithiolene[94] and the more general expression 1,2-dichalcogenolene are conventionally assigned to metal complexes featuring ligands formally deriving from 1,2-disubstituted alkene- or arene-1,2-dichalcogenolates or the corresponding oxidized species 1,2-dichalcogenones (Scheme 11.3.3).

Although terms such as alkene- or arene-1,2-dithiolate may be useful for referring to the free ligands, they are often difficult to apply to the corresponding coordination metal complexes, as a consequence of the non-innocence shown by these ligands in the complexes, a topic which has been the subject of much debate.[19,39,95–101] As a consequence, the term 'dichalcogenolene' does not refer to a specific molecular charge or to the oxidation state of the central metal ion. This makes it difficult to assign clearly separated *formal* charges to the metal and to the ligands, as illustrated in Scheme 11.3.4 for neutral square-planar bis(1,2-dithiolene) complexes of d^8 metal ions.

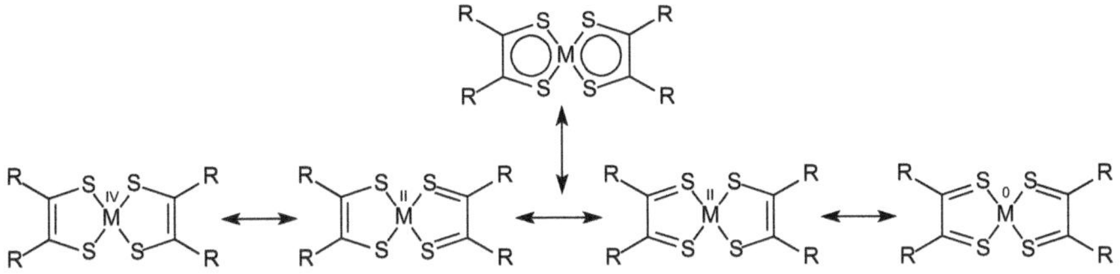

Scheme 11.3.4 Lewis structures and formal oxidation state of the central metal in neutral bis(1,2-dithiolene) complexes of d^8 metal ions.

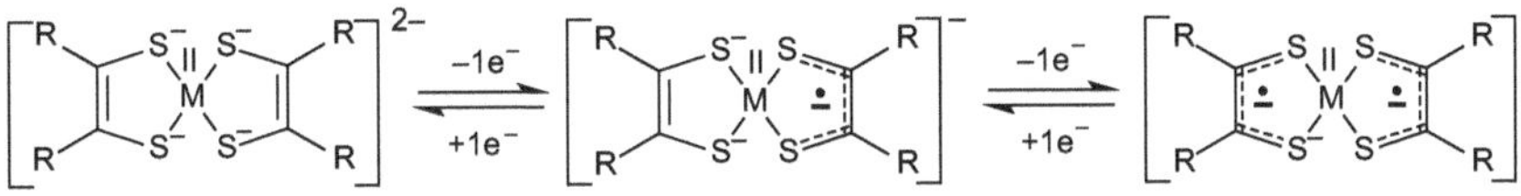

Scheme 11.3.5 Redox electron transfer series in bis(1,2-dithiolene) complexes of d^8 metal ions.

One of the peculiar properties of the title class of complexes is the ability of the molecular entity to carry a charge which can vary considerably, also assuming fractional values in non-integral oxidation state (NIOS) salts. The different molecular oxidation states are reversibly accessible by chemical or electrochemical means. A good example is the case of bis(1,2-dithiolene) complexes of d^8 metal ions (such as Ni^{II}, Pd^{II}, Pt^{II}, or Au^{III}),[102] whose charge can assume values typically ranging between –2 and 0.

Recent studies, also supported by resonance Raman,[103–105] sulfur K-edge X-ray absorption spectroscopy,[14,19,25,97,106–110] and DFT calculations tend to indicate that the three members of the electron transfer series typical of d^8 metal ions containing bis(1,2-dithiolene) complexes are interrelated by redox steps centred on the ligands rather than on the metal, which should be considered to maintain a +II oxidation state throughout (Scheme 11.3.5).

Similar conclusions have been drawn from studies, also supported by multifrequency EPR measurements, carried out on several classes of 1,2-dithiolene complexes such as $[V(L)_3]^z$ (L = pdt^{2-}, mnt^{2-}, bdt^{2-}, tdt^{2-}, edt^{2-}, z = –4, –3, –2, –1, 0, +1),[111] $[Fe(NO)(L)_2]^z$ [L = mnt^{2-}, $\{S_2C_2(p\text{-tolyl})_2\}^{2-}$, z = –3, –2, –1, 0, +1],[33] $[Re(L)_3]^z$ (L = mnt^{2-}, $Ph_2\text{-}edt^{2-}$, z = –1, 0, +1; bdt^{2-}, z = –3, –2, –1, 0, 1),[112,113] $[M(Ph_2\text{-}edt)_3]^{0/-1}$ (M = W, Mo) and $[M(mnt)_2]^z$ (M = W, Mo, z = –2; M = Fe–Zn, Pd, Pt, z = –2, –1),[114–116] and $[Cr(L)_3]^z$ (L = $tbbdt^{2-}$; z = –3, –2, –1, 0), just to cite few examples reported recently.[108]

11.3.4 Synthesis

11.3.4.1 Synthesis of 1,2-Dichalcogenolene Ligands or Precursors

The synthesis of metal complexes deriving from ene-1,2-dichalcogenolato ligands is generally complicated by the tendency of these ligands to

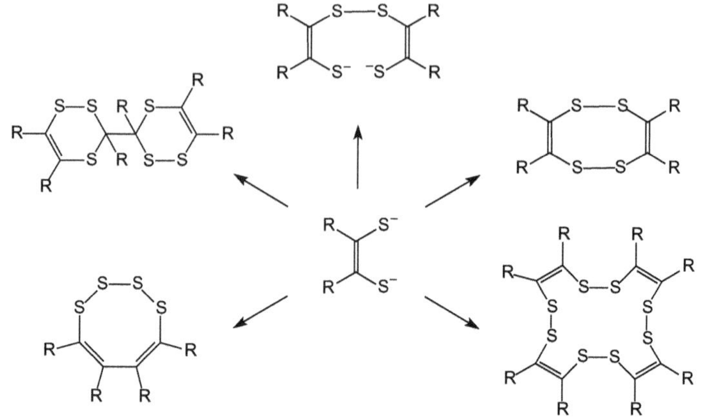

Scheme 11.3.6 Selected compounds obtained by oxidation of 1,2-dithiolene species.

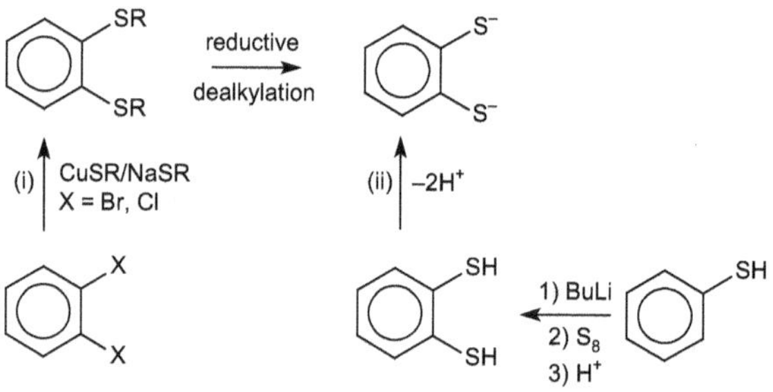

Scheme 11.3.7 Strategies for the synthesis of 1,2-benzenedithiolato ligands.

polymerize or decompose, to give a variety of sulfur-rich products (Scheme 11.3.6).[117–123]

A plethora of different and specific synthetic routes to 1,2-dichalcogenolene ligands has been reported. Nonetheless, some general routes can be identified for the synthesis of most 1,2-dithiolene ligands,[124] often extendible to 1,2-diselenolene and 1,2-ditellurolene species.

11.3.4.1.1 Arene-1,2-dichalcogenolates and Related Ligands

Arene-1,2-dichalcogenolene ligands are generally stable as 1,2-dichalcogenols or 1,2-dichalcogenolate sodium salts.[124,125] Along with aromatic compounds featuring two vicinal thiol groups acting as a dithiolene ligand, such as in the case of benzene-1,2-dithiolate (bdt^{2-}), systems with up to three dithiolene donor groups (benzenehexathiolate) have been reported.[126] Aromatic 1,2-dithiolates can be prepared by following two different synthetic approaches, exemplified for bdt^{2-} in Scheme 11.3.7, which start either from (i) dihalo-substituted arenes or from (ii) mercapto- or vicinal dimercaptoarenes.[127–132]

In the former case, the dibromo- or dichloro-substituted arene is reacted with cuprous or sodium thiolates to give vicinal thioethers which can be reductively dealkylated, for example with Na/NH_3, leading to the corresponding dithiols.[133]

Alternatively, benzenethiol is reacted with *n*-BuLi, and the resulting lithiated product is made to react with elemental sulfur to give the 1,2-dithiolate salt. The corresponding diselenolene ligand can be obtained by reacting 1,2-dibromobenzene with Na_2Se in DMF to give poly(*o*-phenylenediselenide), which can be reduced by $NaBH_4$ or $LiBEt_3H$ to obtain the bds^{2-} anion.[134,135] This can be directly used to obtain the corresponding complexes, protected by conversion to 4,5-benzo-1,3-diselenole-2-thione,[136,137] or stabilized as a Zn complex.[138]

Synthetic strategies similar to those outlined in Scheme 11.3.7 have not been limited to benzene derivatives, but have been exploited to sulfurate different heterocyclic systems. Pyrazine-2,3-dithiol (H_2pdt), for example, was obtained by reacting the 2,3-dichloropyrazine with NaSH.[139] The corresponding selenium species was isolated similarly as the sodium salt Na_2pds.[140] Analogously, the reaction of *o*-carborane, *n*-butyllithium, and selenium was exploited for preparing the corresponding 1,2-diselenolate.[141] The reaction of *N*-substituted 1,3-thiazole with lithium diisopropylamide (LDA), followed by the reaction with sulfur or selenium, was exploited to prepare zinc, nickel, and palladium complexes of the R-$thiazdt^{2-}$ and R-$thiazds^{2-}$ ligands (R-$thiazdt^{2-}$ = R-$dmet^{2-}$ = *N*-substituted 2-thioxo-1,3-thiazoline-4,5-dithiolate; R-$thiazds^{2-}$ = *N*-substituted 2-thioxo-1,3-thiazoline-4,5-diselenolate).[142–144] In addition, the reactions of the Li_2(R-thiazdt), $TEA_2[Zn(R\text{-thiazdt})_2]$, and $TEA_2[Zn(R\text{-thiazds})_2]$ salts (TEA^+ = tetraethylammonium) with $Br(CH_2)_2CN$ were exploited to obtain the corresponding *N*-substituted 4,5-bis(2′-cyanoethylchalcogeno)-1,3-thiazol-2-thiones, stable precursors of the corresponding ligands (see below).[142,145]

Dilithium arene-1,2-dichalcogenates have been used as starting materials for the synthesis of ditellurolene, selenotellurolene, thiotellurolene, and thioselenolene complexes of the ligands having general formula $(1,2\text{-}C_6H_4EE')^{2-}$ (E, E′ = S, Se, Te).[12,146–150] 1,2-Dithiosquarate and 1,2-diselenosquarate ($dtsq^{2-}$ and $dssq^{2-}$) have been isolated as potassium salts, obtained by reaction of 1,2-diethoxysquarate and H_2E (E = S[151,152] and Se,[153,154] respectively), although the selenium compound is highly sensitive to air and moisture.

Notably, contrary to vicinal thioethers, the reductive dealkylation of diseleno- and ditelluroethers has been only marginally tapped,[155] while dichalcogenoethers have been directly exploited as ligands in homoleptic[156] and heteroleptic complexes.[157–159]

11.3.4.1.2 Stable Alkene-1,2-Dichalcogenolato Ligands

Compared to arene-1,2-dithiols, alkene-1,2-dithiols and the corresponding dithiolates are generally much less stable, and must be protected until reacted with the metal salt. In fact, the sodium salts of alkene-1,2-dithiolates are strongly reducing species, highly reactive towards water or halogenated

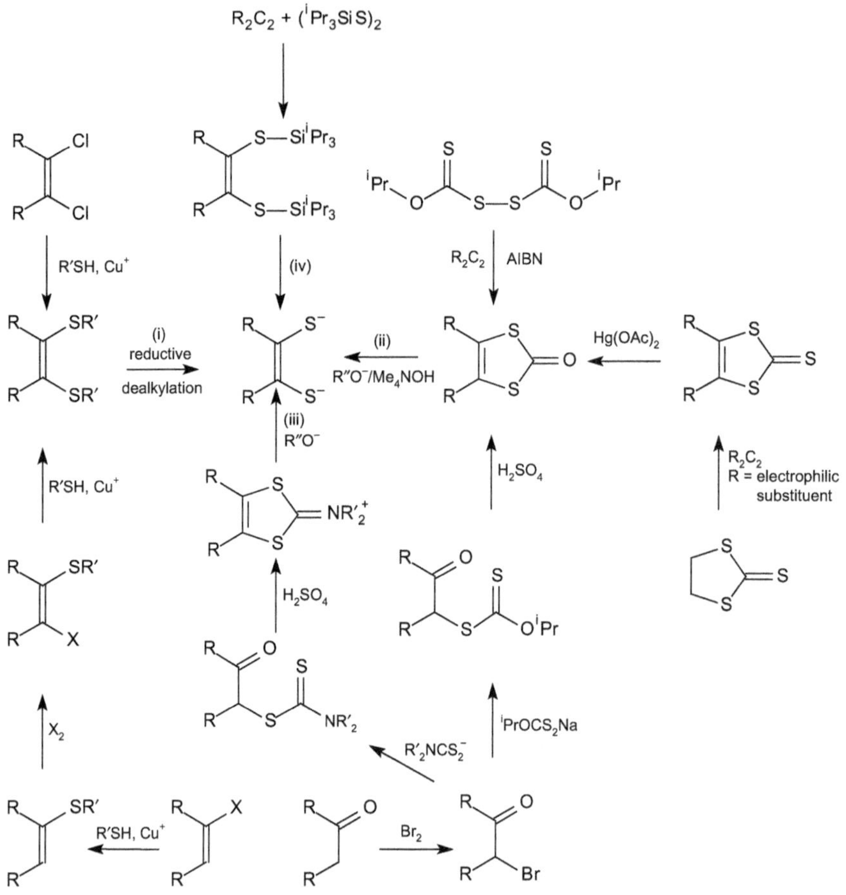

Scheme 11.3.8 Strategies for the synthesis of ethylene-1,2-dithiolato ligands.

solvents. In this case too several synthetic routes, summarized in Scheme 11.3.8, are available.

The first method (i in Scheme 11.3.8), analogous to that described for arene-1,2-dithiolates, starts from dithioethers, typically *cis*-1,2-di(benzylthio)alkenes ($R'=CH_2Ph$ in Scheme 11.3.8), to give the *cis*-alkene-1,2-dithiolates by reaction with strongly reducing agents. Dithioethers can be obtained starting from mono- or 1,2-dihaloalkenes. In the second method, α-bromoketones, obtainable by halogenation of ketones, are reacted with alkylxanthate or dialkyldithiocarbamate salts [typically $^iPrOCS_2Na$ and $(^iPr)_2NCS_2Na$, respectively][160] to give the corresponding α-ketoxanthate or α-ketodithiocarbamate esters, which undergo cyclization in concentrated sulfuric acid to give the vinylene dithiocarbonate or iminiumdithiocarbonate. The reactions of both types of compounds with strong bases (reactions ii or iii in Scheme 11.3.8, respectively) lead to the desired symmetrically or unsymmetrically substituted alkene-1,2-dithiolates.[160–165] When the final 1,2-dithiolene ligand features acidic moieties, such in the case of madt^{2-}, different agents, such as NaSMe, can be used to promote the opening of the vinylene dithiocarbonate.[166]

Alternatively, 1,3-dithiol-2-ones have been obtained by reacting diisopropylxanthogen disulfide with unsaturated substrates, such as disubstituted alkynes R_2C_2, in the presence of AIBN (azoisobutyrylnitrile), a radical initiator.[167] Finally, vinylene dithiocarbonates can be obtained by reacting mercury(II) acetate with the corresponding trithiocarbonates, generated by reaction of electrophilic alkynes with ethylene trithiocarbonate.[168] Analogously, disubstituted vinylene triselenocarbonates have been obtained by reacting the corresponding ethylenetriselenocarbonates with activated alkynes R_2C_2.[169]

It should be noted that the synthetic pathway (ii) has been extended to cyclic 1,2-dithiolene and 1,2-diselenolene ligands, such as dddt^{2-}, ddds^{2-}, ddt^{2-}, phdt^{2-}, medt^{2-}, and so on.[170]

A further form of protected ene-1,2-dithiolate, proposed for obtaining diethylethylene-1,2-dithiolate (iv in Scheme 11.3.8; R = Et), is represented by 3,4-bis(triisopropylsilanylsulfanyl)-hex-3-ene[171] and related compounds, obtained by a Pd^0-catalysed cross-coupling reaction between the alkyne Et–C≡C–Et and bis(triisopropylsilyl)disulfide.[172]

11.3.4.1.3 mnt^{2-}, dmit^{2-} and Related Compounds

Specific procedures have been used for the synthesis of the two most commonly encountered 1,2-dithiolene ligands, mnt^{2-} and dmit^{2-}, and their selenium isologues. Na_2mnt has been prepared by reacting NaCN with CS_2: the intermediate $[S_2CCN]^-$ species undergoes a coupling reaction accompanied by loss of sulfur.[173] Pullen and Olk reviewed the synthesis and coordination properties of dmit^{2-} and isologues in 1999.[77] The chemical or electrochemical reduction of CS_2 and CSe_2 gives the dmit^{2-} and dsis^{2-} ligands, along with the corresponding trichalcogenocarbonate.[174–179]

The dmit^{2-}, dsis^{2-}, and dsit^{2-} ligands can be isolated as air-stable quaternary ammonium or arsonium salts of the corresponding Zn complexes (Scheme 11.3.9 for dmit^{2-}).[138,176,180–182]

The addition of PhC(O)Cl to the Zn complexes leads to the thioesters dmit[C(O)Ph]$_2$, dsis[C(O)Ph]$_2$, and dsit[C(O)Ph]$_2$, whose hydrolysis with NaOMe or Me_4NOH yields the sodium or tetramethylammonium salts of the corresponding 1,2-dichalcogenolates. Metathesis reactions starting from these salts and the desired metal salt (typically halides), followed by addition of a source of a quaternary cation, such as Ph_4PCl, have been extensively used to obtain a large number of different compounds containing dmit^{2-} complexes and related selenated species.[183] In addition, the reactions of the alkylammonium salts of the dmit^{2-} and dsit^{2-} ligands with dichlorophenylstibine[184] have been used to synthesize the compounds [PhSb(dmit)] and [PhSb(dsit)], stable sources of the corresponding dithiolene and diselenolene ligands, respectively (Scheme 11.3.10).[185]

Replacement of the terminal sulfur of the dmit^{2-} ligand with oxygen, performed by reaction with $Hg(OAc)_2$, has been used to prepare the ligand

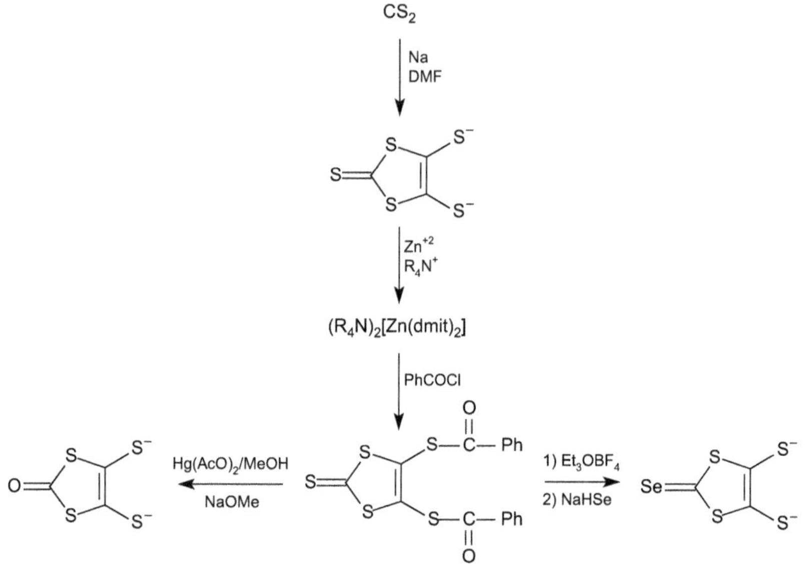

Scheme 11.3.9 Strategies for the synthesis of $dmit^{2-}$, $dmio^{2-}$, and $dmise^{2-}$ ligands.

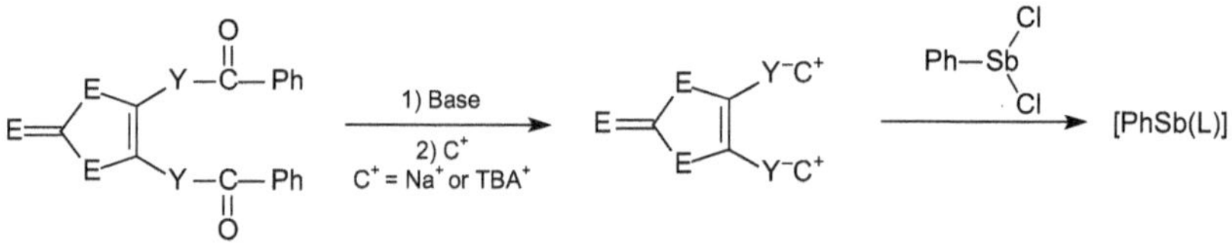

Scheme 11.3.10 Synthesis of the stable salts of the $dmit^{2-}$ (E = Y = S), $dsis^{2-}$ (E = Y = Se), and $dsit^{2-}$ (E = S, Y = Se) ligands.

$dmio^{2-}$ (also known as $dmid^{2-}$).[186] The corresponding 2-selone analog, $dmise^{2-}$, can be prepared from dmit[C(O)Ph]$_2$ by alkylating the terminal thiocarbonyl group with Et_3OBF_4 and then reacting the formed dithiolium salt with NaHSe.[187]

Analogously to what described above for the ligands R-thiazdt^{2-} and R-thiazds^{2-}, lithiation of various vinylene trichalcogenocarbonates followed by reaction with elemental sulfur or selenium has been used to synthesize various $dmit^{2-}$ isologues,[77] stabilized as Zn^{II} complexes. This synthetic strategy, originally proposed by Nigrey for the synthesis of $dsit^{2-}$,[188,189] has been extended to the synthesis of $dsise^{2-}$, $dsis^{2-}$, and $dsitse^{2-}$.[190–196]

Oligo(1,3-dithiole-2,4,5-trithione), $(C_3S_5)_n$,[197] has been also used as a precursor of dmit-related ligands. In particular, its reaction with dipyridyl-acetylene and related compounds yielded 5,6-di(pyrid-2-yl)[1,3]dithiolo[4,5-*b*][1,4]dithiin-2-thione.[198]

The reactions of the $dmit^{2-}$ anion with bis-alkylating reagents, for example 3,6,9-trioxaundecane-l,11-bis(toluene-*p*-sulfonate),[199] were used to prepare dithiolene ligands with the dmit ring condensed with macrocycles of various type. The reaction with $Hg(OAc)_2$, carried out to replace the exocyclic S-atom with oxygen, provides protected forms of ethylene-1,2-dithiolates fused with the selected macrocycles.[200,201]

During the past few years, some novel dmit-related compounds containing peripheral nitrogen donors have been prepared.[59] The reactions of dmit$[C(O)Ph]_2$ with picolyl chloride afforded 4,5-bis(2-pyridylmethylsulfanyl)-1,3-dithiole-2-thione (bpmdmit), while the direct reaction of the TEA^+ salt of the Zn complex afforded 4,5-bis(2-pyridylethylsulfanyl)-1,3-dithiole-2-thione (bpedmit).[202] Analogously, by starting from $[Zn(dmio)_2]^{2-}$, 4,5-bis(2-pyridylethylsulfanyl)-1,3-dithiole-2-one was isolated.[203]

Finally, cross-coupling reactions involving the exocyclic C=S and C=O groups of dmit and dmio derivatives, respectively, have been used to obtain extended planar ligands, as in the case of the reaction of 4,5-dicyanoethyl-1,3-dithiole-2-thione with 4,5-diazafluoren-9-one.[204] Analogously, starting from 4,5-dicyanoethyl-1,3-dithiole-2-one and bpmdmit, 4,5-bis(2-cyanoethylthio)-4′,5′-bis(2-pyridylethylsulfanyl)tetrathiafulvalene was prepared, which is a precursor of the bis(2-pyridylethylsulfanyl)tetrathiafulvalenedithiolene ($pesdt^{2-}$) ligand.[205]

11.3.4.1.4 Synthesis of 1,2-Dichalcogenolene Complexes

1,2-Dichalcogenolene ligands form complexes with main group and *d* transition metal ions.[206] Bis(1,2-dithiolene) complexes have been obtained for transition metals such as Cr, Mn, Ni, Cu, Zn, for the first row, Rh, Pd, Ag, and Cd for the second row, and Pt, Au, and Hg for the third row. Homoleptic tris(1,2-dithiolene) complexes have been obtained for Ti, V, Zr, Nb, Mo, Tc, Ru, Hf, Ta, W, Re, and Os. Cr, Fe, and Co have been found both in bis and tris(1,2-dithiolene) complexes, although tris(1,2-dithiolene) complexes containing these metal ions are more common. In addition, mixed-ligand mono(1,2-dithiolene) complexes of Ir were synthesized. Group *p* metals were also employed for the synthesis of bis (Te, Bi) and tris (Tl, In, Sn) 1,2-dithiolene complexes. During the past decade, lanthanide (Ce, La, Nd, Sm, Gd, Er, Yb) and actinide (U) 1,2-dithiolene complexes have been reported.[196,207–211] Among these, the first examples of tetrakis(1,2-dithiolene) complexes have been characterized in the compounds $[Na_4(THF)_8U(dddt)_4]_\infty$[209] and $[Na_5(THF)_{10}Ce(mnt)_4]_\infty$.[212] There are far fewer reports on complexes containing 1,2-diselenolene ligands, with metals such as Ti, V, Co, Ni, Cu, Zn, Ge, Zr, Rh, Pd, Ag, Cd, In, Sb, W, Ir, Pt, Au, Hg, Tl, and Pb.

Bis and tris(1,2-dichalcogenolene) complexes can exist in different molecular oxidation states (Scheme 11.3.4 for bis(1,2-dithiolene) complexes), but with only a few exceptions, they are obtained as anionic species, counterbalanced by a large variety of organic and inorganic cations.[213–215] These include enantiomerically pure chiral species,[216] crown-ether complexes, for example in the compounds $Cs_2(18\text{-crown-6})_3[Ni(dmit)_2]_2$[217] and [K(18-crown-6)][Ni(α-tpdt)$_2$],[218,219] and several cationic metal complexes,[220] such as $[M(Cp^*)_2]^+$

(M = Fe, Mn)[221] or spin-crossover complexes of the type $[Fe(L)_2]^{2+}$ [L = bpp = 2,6-bis(pyrazol-3-yl)pyridine;[222] 3-R-salEen = 3-R-*N*-(2-ethylamino)-ethyl-salicylaldimine, R = H, CH_3O;[223,224] qsalH = *N*-(8-quinolyl)-salicylaldimine].[225]

As far as 1,2-dithiolene complexes are concerned, it has been established that less negatively charged species can be obtained by air (for couples with approximately $E_{1/2} > 0.00$ V *vs* SCE) or I_2, I_3^-, or Br_3^- oxidation ($E_{1/2} > 0.40$ V *vs* SCE),[132,226] depending on the nature of the ligand and on the electronic structure of the resulting metal complex.[227] On the other hand, oxidized species can be reduced to more negatively charged complexes by several agents, ranging from weakly basic solvents (such as ketones or alcohols, when $E_{1/2} > 0.20$ V) to strong reducing agents, such as hydrazine, sodium amalgam, $NaBH_4$, or $LiAlH_4$.[227]

Stable 1,2-dichalcogenolene salts, such as those of mnt^{2-} or arene-1,2-dithiolates, directly yield the corresponding homoleptic bis or tris(1,2-dichalcogenolene) complexes when reacted with metal halides,[132] such as $NiCl_2$,[95,132,137,228,229] $FeCl_3$,[96] $CrCl_3(THF)_3$,[108] $RuCl_3$,[230] $MoCl_5$ or WCl_6,[133,231] $[AuCl_4]^-$,[140] or – more rarely – different metal salts, such as $[Cr(MeCN)_4](BF_4)_2$ and $[Rh_2(MeCN)_{10}](BF_4)_4$.[232] Similarly, the reaction of 1,2-dithiolates with metal complexes containing halides and one or more ligands of different nature make it possible to obtain mixed-ligand heteroleptic complexes, such as [Pt(dppe)(dmit)] [dppe = 1,2-bis(diphenylphosphino)ethane)].[233] This synthetic route has been largely adopted for the synthesis of the well-known class of diimine-dithiolate mixed-ligand complexes, usually synthesized as neutral species starting from the corresponding metal diimine-dichloro-complexes (diimine = bipyridine, phenathroline and derivatives), by displacement of the two chlorides with the desired 1,2-dithiolate ligand, as first reported by Vogler and co-workers,[234] and more recently by Eisenberg.[235–237] An alternative method consists of the reaction of the homoleptic *bis*(1,2-dithiolene) complex with the appropriate diimine ligand.[238,239]

The synthesis of mono(1,2-ditellurolene) complexes, such as $[Pt(1,2\text{-}Te_2C_5H_6)(PPh_3)_2]$[147] and $[Zr(1,2\text{-}TeEC_6H_4)(t\text{BuCp})_2]$ (E = S; *t*BuCp = *tert*-butylcyclopentadienyl) has been accomplished accordingly.[150]

The isolation of $dmit^{2-}$ and isologues in the form of quaternary ammonium salts of the Zn complex, $(R_4N)_2[Zn(dmit)_2]$, has made it possible to obtain a large number of different anionic complexes through ligand exchange reactions.[240] This strategy was extended to $(PPh_4)_2[Zn(bds)_2]$, used to synthesize the mono(1,2-diselenolene) complex [CpNi(bds)].[138] Also Ni bis(1,2-dithiolene) complexes have been used as starting materials for the synthesis of several types of heteroleptic mono(1,2-dichalcogenolene)-arene nickel complexes (arene = cyclopentadienyl, η^4-C_4H_4; dithiolene = $dmit^{2-}$, $dsit^{2-}$, $dmio^{2-}$, $dddt^{2-}$, $ddds^{2-}$, bdt^{2-}, bds^{2-}, $\{S_2C_2S_2(CH_2)_2X\}^{2-}$; X = CH_2, CF_2, C = CH_2, S).[241–243]

As regards 1,2-dithiolene and 1,2-diselenolene complexes, commonly used precursors are the already cited substituted vinylene di- or sometimes tri-chalcogenocarbonates (1,3-dithiole-2-ones and 1,3-dithiole-2-thione, respectively), or iminiumdichalcogenocarbonates, which are prone to hydrolysis in alkaline media, yielding the corresponding complexes in the presence of suitable metal salts, often halides.[162–165,169,170,188–205,244–246] This route has been very widely applied for the synthesis of a plethora of 1,2-dithiolene complexes,[160] including di- and trimetallic compounds,[247–250] oxometal 1,2-dithiolene complexes, and heteroleptic mono(1,2-dithiolene) complexes, such as those of the type [(arene)M(dcmedt)] [for example arene = cyclopentadienyl, pentamethylcyclopentadienyl (Cp^*), or η^4-C_4H_4],[89,241,242,251,252] [M(dppe)(R,H-edt)] (R,H-edt^{2-} = monosubstituted ethylenedithiolate; R = *p*-cyanophenyl, 2- and 3-thienyl; M = Ni, Pd),[253] and platinum diimine-dithiolate complexes [diimine = bipy, phen; dithiolate = phenyl- (sdt^{2-}) and naphthyl-substituted ethylene-1,2-dithiolate,[254] $(CH_3OOC)_2C_2S_2{}^{2-}$,[236] and few others].[255–257]

The same method has also been applied for the synthesis of 1,2-diselenolene complexes, such as $[Ni(tds)_2]^{2-}$,[258] $[Ni(Se_2C_2Me_2)_2]^{2-}$, and $[Ni(ddds)_2]^{2-}$.[169,170] Also the complex $[Ni(mns)_2]^{2-}$ was synthesized according to this route,[245,259] although an alternative synthesis, extended to other metal ions, was reported.[260–263] Remarkably, the pseudo-cross-coupling synthetic method, which exploits the reaction of a vinylene dichalcogenocarbonate with (1) a second differently substituted vinylene dichalcogenocarbonate or (2) 4,5-bis(benzoylchalcogeno)-1,3-dithiole-2-thione, is an effective method for the synthesis of mixed-ligand heteroleptic bis(1,2-dithiolene) or (1,2-dithiolene)(1,2-diselenolene) complexes.[264–266]

Also bis(triisopropylsilanyl)alkenes were proposed as precursors of dithiolene ligands (see Scheme 11.3.8). In the case of 3,4-bis(triisopropylsilanyl)-hex-3-ene,[171] the ready cleavage of Si–S bonds made it possible to obtain Ni, Co, Fe, Co, Mn, V, Nb, Mo, W, and Re complexes containing the 1,2-diethyl-ethylene-1,2-dithiolate ligand.[172]

As regards ligands tfd^{2-} or tds^{2-},[74,92,93,267,268] use has commonly been made of 1,2-dithietes or 1,2-diselenete, four-membered $R_2C_2E_2$ rings formally deriving from the oxidation of the corresponding 1,2-alkenedichalcogenolates (E = S, Se; Scheme 11.3.3a),[269] as starting materials for the synthesis of mono,[242] bis and tris(1,2-dichalcogenolene) complexes,[124] though their preparation requires severe synthetic conditions.[74,92,93]

Correlated is the sulfuration of α-hydroxyketones (acyloins),[124,270–272] and 1,2-diketones, carried out with various sulfurating agents spanning from P_4S_{10} to Lawesson's reagent,[273] in the presence of metal salts or elemental metals (Scheme 11.3.11).[78,124] In the case of the sulfuration of acyloins with P_4S_{10} in dioxane, the dithiolene thiophosphoryl thiolate intermediate $(R_2C_2S_2)P(S)(SR')$ could be isolated (R = *p*-anisyl, R′ = Me, Bz; R = Ph, R′ = Me; R = Et, R′ = Bz) and identified spectroscopically or structurally.[272] The use of sulfurating agents proved successful to prepare variously functionalized diaryl-substituted 1,2-dithiolene complexes (including Ni and Au long-chain substituted discotic complexes)[274–277] and members of the classes $[M(R,R'\text{-timdt})_2]$.[118,278–280] Notably, complexes containing the R-$dmet^{2-}$ (R-$thiazdt^{2-}$) ligand, intermediate between $dmit^{2-}$ and R_2-$timdt^{2-}$, were obtained both by sulfuration of N-substituted 2-thioxo-1,3-thiazoline-2-thione

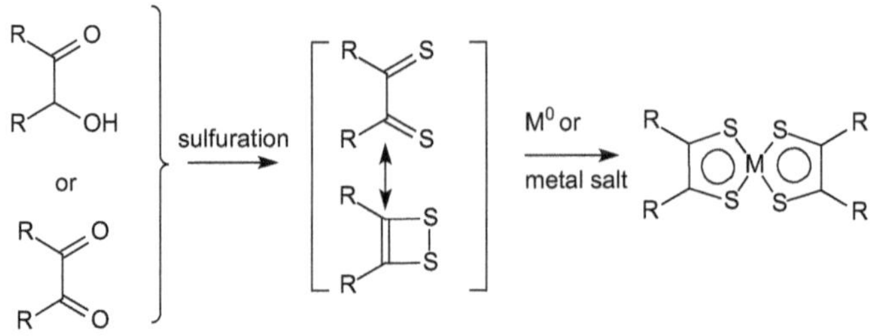

Scheme 11.3.11 Synthesis of bis(1,2-dithiolene) complexes by sulfuration of acyloins or vicinal diketones and complexation in situ.

in the presence of the metal, as in the case of the neutral complexes [M(Et-dmet)$_2$] (M = Ni, Pd, Pt),[281] or starting from the *in situ* generated Li_2(R-thiazdt) salt or from a protected form of the ligand, 3,3′-[(2-thioxo-2,3-dihydro-1,3-thiazole-4,5-diyl)bis(thio)]dipropanenitrile, in the case of TEA[Au(Et-dmet)$_2$][145] and (TEA)$_2$[M(Me-dmet)$_2$] (M = Ni, Pd), as mentioned above.[143]

Sulfuration with P_4S_{10} was also exploited to obtain 1,2-dithiolene complexes with alkylsulfanyl RS substituents (R = ^{i}Pr) starting from S,S′-dialkyldithioxalate, in turn obtained by reacting oxalyl chloride with the required RSH thiol.[282] Closely related are the complexes obtained starting from bpmdmit,[202] bpedmit,[202] and 4,5-bis(2-pyridylethylsulfanyl)-1,3-dithiole-2-one[203] precursors, discussed above. The corresponding nickel complexes, as well as the complex [Au(pesdt)$_2$]$^-$,[205] feature peripheral N-pendants, available to further interact to give extended crystal networks.[59,283,284]

The reactions of electrophilic alkynes, such as dimethyl acetylenedicarboxylate (DMAD), with metal per- and polychalcogenido complexes have been exploited for the synthesis of homoleptic and heteroleptic 1,2-dithiolene, 1,2-diselenolene, and 1,2-ditellurolene complexes.[124,285–290] For example, the reaction of $[WSe_9]^{2-}$ or $[WSe_4]^{2-}$ with DMAD afforded $[W(Se_2C_2COOMe_2)_3]^{2-}$,[288,291] while the reactions of DMAD with $[Cp_2TiSe_5]$[292] and $[Cp^*IrSe_4]_2$[286] yielded the corresponding heteroleptic 1,2-diselenolene complexes. Similarly, the reactions of $[Mo(O)(Te_4)_2]^{2-}$ or $[MTe_7]^{n-}$ (M = Hg, n = 2; M = Au, n = 3) anions with DMAD yielded Mo, Hg, and Au complexes containing the $Te_2C_2(COOMe)_2$ 1,2-ditellurolene ligand.[293,294]

Related are the reactions between $(\mu\text{-EE}')_2Fe_2(CO)_6$ or $(\mu\text{-E})_2Fe_3(CO)_9$ clusters (E, E′ = S, Se, Te) and R–C≡C–R′ alkynes (R = R′ = H;[295] R = H, R′ = Ph;[296,297] R = H, R′ = C_3H_3;[298] R = Me, R′ = C≡C–H)[299] or alkynyl Fischer carbene complexes $(CO)_5M$ = C(OEt)(C≡C–Ph) (M = Cr, W),[300–303] and those between the mixed-metal disulfide compounds $CpMoMn(CO)_5(\mu\text{-}S_2)$ and unsaturated hydrocarbons.[304] Notably, the reaction of $Mo_2(SCH_2CH_2S)_2Cp_2$ with an excess of variously disubstituted alkynes R–C≡C–R′ was reported to give the corresponding bis(1,2-dithiolene) complexes $Mo_2(\mu\text{-SCR}=\text{CR}'\text{S})(SCH_2CH_2S)Cp_2$ and $Mo_2(\mu\text{-SCR}=\text{CR}'\text{S})_2Cp_2$.[305]

The synthesis of bis(1,2-dithiolene) complexes starting from sulfur-rich precursors such as 3,3′-dithio-bis[pyridine-2(*1H*)-thione] in the case of

$[M(pydt)_2]^{2-}$ (M = Ni, Pd, Pt),[306] or $\{1,2\text{-}S_2\text{-}3,6\text{-}(R\text{-}CONH)_2C_6H_2\}_2$ in the case of $[MoO\{1,2\text{-}S_2\text{-}3,6\text{-}(R\text{-}CONH)_2C_6H_2\}_2]^{2-}$ (R = Me, CF_3, *t*-butyl, -C-Ph_3) has also been reported.[307]

4,5-Disubstituted 1,2,3-selenadiazoles react with sulfur or selenium in the presence of organotransition or tetraalkyl/arylphosphino metal complexes to yield the corresponding heteroleptic 1,2-dithiolene or 1,2-diselenolene complexes.[286,308–311] In addition, mono- and dinuclear Pd and Pt 1,2-diselenolene complexes were obtained from bis(cycloalkeno)1,4-diselenins, in their turn obtained from 1,2,3-selenadiazoles.[311–314]

The synthetic pathway to TTeF (tetratellurafulvalene), starting from (*Z*)-1,2-bis(trimethylstannyl)ethane,[315] has been exploited for synthesizing heteroleptic Pt complexes containing the only examples of the Te and Se analogues of edt^{2-}.[316]

Specific synthetic routes have been developed for Mo and W 1,2-dithiolene complexes.[2–22,317,318] Molybdenum oxo-1,2-dithiolene complexes can be typically obtained through routes, reviewed in 2004,[319] starting from $[MoO_2(CN)_4]^{4-}$ or $[MoO(SAr)_4]^{2-}$ (by reaction with H_2bdt and related compounds)[320] or from $[MoOCl(MeCN)_4]^+$ (by reaction with Li_2bdt).[321,322] Dinuclear oxo-bis-μ-sulfido-benzenedithiolene complexes $[M_2O_2(\mu\text{-}S)_2(bdt)_2]^{2-}$ have been synthesized by reacting $[MO_2S_2]^{2-}$ with the 1,2-dithiolate (M = Mo, W).[323] Tungsten monoxodithiolene complexes can be obtained through various synthetic methods,[324] for example by reacting $[W(S_2C_2R_2)_2(CO)_2]$ with Et_4NOH.[319] The corresponding monosulfido- and monoselenido-dithiolene species were obtained by using an alkaline chalcogenide in place of the hydroxide.[325] Bis-oxodithiolene species of the type $[M(O)_2(L)_2]^{z-}$ and $[M(O)_2(L)(L')]^{z-}$ ($L = S_2C_6R_4{}^{2-}$; $L' = (MeO(O)C)_2C_2S_2{}^{-}$) were obtained by oxidation of oxo-1,2-dithiolene complexes with Me_3NO.[162,240,319,322,326,327] Mono(1,2-dithiolene) molybdenum complexes have been synthesized both by following a procedure of stepwise introduction of 1,2-dithiolene ligands, and through removal of a 1,2-dithiolene ligand from bis or tris(1,2-dithiolene) complexes.[11] Bis(η^5-cyclopentadienyl) mono(1,2-dithiolene) complexes of Mo and W have been prepared by reacting $[Cp_2MCl_2]$ complexes with the previously mentioned suitably substituted 1,3-dithiol-2-one pro-ligands.[328] A similar route, starting from $[Cp_2TiCl_2]$, has been reported for the synthesis of bis(η^5-cyclopentadienyl) mono(1,2-diselenolene) titanium complexes.[134] Alternatively, the *in situ* thermolysis of R'_2N_2 azo compounds ($R' = CMe_2CN$ in AIBN or $C_6H_{10}CN$ in 1,1′-azobis(cyclohexane-1-carbonitrile), ACCN), which initiates the reaction between $[Cp_2MS_4]$ complexes and a H–C≡C–R (R = phenyl, 2-pyridyl, or 2-quinoxalyl) was reported to give the corresponding $[Cp_2M(dithiolene)]$ complexes.[329]

Finally, it should be noted that tellurium can not only act as ligand donor site in 1,2-ditellurolene ligands, but can itself be bound by 1,2-dithiolene ligands, as in the case of compounds containing $[Te(mnt)_2]^{2-}$ and $[Te(dtsq)_2]^{2-}$ dianionic 1,2-dithiolene complexes.[330,331]

Whatever the synthetic procedure adopted, the isolated mono, bis, or tris(1,2-dithiolene) or (1,2-diselenolene) complexes can be further exploited: (1) as σ- or η^x (x > 1) π-donors,[332] in complexes such as $[(\eta^5\text{-Cp*})\text{-Fe-}\mu\text{-}\eta^5,\eta^5\text{-}((\text{dmedt})_2\text{Ni})\text{Fe-}\eta^5\text{Cp*}]^{0/+}$;[333] (2) as starting materials for ligand exchange reactions, involving both redox and non-redox processes;[71–73,181,242,334–336] (3) in the presence of suitable ligands, such as those containing thiophene pendants [for example bis(*n*-thienyl)-ethylene-1,2-dithiolate (*n* = 2,3); 4,6-bis(2-thienyl)-thieno-[3,4-d]-1,3-dithiolate][337–339] or indole [in 1-(*N*-methylindol-5-yl)-ethene-1,2-dithiolate],[340] dithiolene complexes have been electropolymerized to give materials with peculiar electrochemical, optical, and conductive properties depending on the nature of the starting materials and the conjugation mode, which can involve the dithiolene core or can occur through the ethylene backbone or the pendants only.[341,342]

Charge-transfer (CT) compounds containing 1,2-dichalcogenolene complexes and NIOS salts, discussed in Section 11.3.6.1, were often prepared using electrocrystallization techniques.[343]

11.3.5 General Structural Features of 1,2-Dichalcogenolene Complexes

Diffraction techniques play a primary role in the characterization of 1,2-dichalcogenolene complexes, not only in view of the importance of solid state properties of metal 1,2-dithiolene and 1,2-diselenolene complexes, but also due to their recent use as building blocks for coordination polymers[204,344] and supramolecular structures.[283,284,345–347] Since the first structural reports on homoleptic nickel complexes deriving from the bdt^{2-} and mnt^{2-} ligands were published,[53,348–350] several reviews have focused on the structural features and packing effects of 1,2-dichalcogenolene complexes.[74,75,89,166,206,207,351,352] On the other hand, apart from metallorganic complexes containing the $Fe_2(CO)_6$ common core,[295–299] only a few papers have dealt with 1,2-dichalcogenolene systems containing different chalcogen donor atoms.[147–149]

As previously pointed out, 1,2-dichalcogenolene ligands possess the ability to carry variable molecular charges due to the considerable electronic flexibility of the planar five-membered 1,2-dichalcogenolene rings, and in this context interatomic distances within the chelate rings have been used as indicators to elucidate the electronic configuration of the complexes, and to distinguish between 1,2-dichalcogenolate and 1,2-dichalcogenone forms (Scheme 11.3.3), not only in monomeric homoleptic 1,2-dichalcogenolene complexes $[M(L)_2]$ and push–pull heteroleptic complexes $[M(L)(L')]$ (L, L′ = 1,2-dichalcogenolene],[353–357] but also in different heteroleptic complexes[358] or more complex polymeric[283,284] or homo- or heterometallic cluster[71–73,182,218,332,359–364] structures.

An examination of the structural reports has allowed deducing some features common to a large variety of compounds:

- In bis(1,2-dichalcogenolene) complexes the central metal ion adopts square-planar, near tetrahedral, or square-pyramidal coordination geometries, also in dependance on the intermolecular interactions established in the solid state.[215] Several structural parameters (such as the dihedral angle between the two E–M–E planes, the angle c_1–M–c_2, where c_1 and c_2 represent the midpoints between the two coordinating chalcogen atoms of the two ligand units, or the bend angle between the E–M–E and E–C–C–E planes) have been used to better define the coordination geometry of the metal and quantify distortions with respect to idealized geometries (E = S, Se).

- Square-planar bis(1,2-dithiolene) and bis(1,2-diselenolene) complexes exhibit a tendency to form dimeric[67,132,170,216,229,266,268,365] or trimeric structures, especially when the transition metal has a formal configuration between d^5 and d^8.[366,367] Different dimers or trimers can interact with each other,[229,266,368] and intermolecular interactions, such as van der Waals, π–π, and π–d interactions are able to address the solid state properties of the resulting molecular materials.[162,215,258,283,369–373] It is worth noting that the tendency to dimerization can be preserved in heteroleptic mono(1,2-dithiolene) complexes. For example, diimine-dithiolate complexes featuring both square-planar and tetrahedral coordination geometries at the metal ions exhibit a tendency to form dyads or dimers held together by intermolecular M–S bonds between layered complexes.[254,374–378]
- In the case of charged complexes, the nature and size of the counterions can play a crucial role in the formation of the stacks[76,139] and in the phase transitions in dependence of the temperature.[379] Stacking formation becomes more common as the π-systems of the complex increase, as observed in dmit^{2-} bis(1,2-dithiolene) complexes.[380] The stacks are often connected by neutral or charged bridging units to give rise to complex supramolecular architectures.[278,381,382] No examples of stacking motifs have been reported so far for mono or bis(1,2-ditellurolene) complexes.
- In tris(1,2-dithiolene) and tris(1,2-diselenolene) complexes the central metal ion adopts an octahedral or a distorted trigonal-prismatic coordination geometry.[112,288,383,384] A number of geometrical parameters have been proposed to discriminate between these two possibilities: (1) the twist angle between E atoms in the two-dimensional projection along the threefold axis, and (2) the dihedral angle between the ligand E–M–E plane and the E–E–E plane formed by three coordinating chalcogen atoms from the three units lying on the same side with respect to the metal centre, or (3) the dihedral angle between the two E–E–E planes defined as described above (E = S, Se).[288,385,386]
- In general, the different ligands feature typical geometrical parameters over a wide range of 1,2-dichalcogenolene complexes.[206]

11.3.6 Applications and Properties of 1,2-Dichalcogenolene Complexes

Metal 1,2-dithiolene and, to a much lesser extent, 1,2-diselenolene complexes have been suggested for a large variety of diverse applications,[217,223,387–391] for example as components of nanolights when doped with carbon quantum dots (CQD),[392] catalysts for the driven reduction of protons from water,[393] components for air-stable ambipolar or unipolar organic field-effect transistors (OFET),[277,394–398] bulk heterojunction photovoltaic devices,[340] *p*-dopants for organic semiconductors[399,400] and in the field of metal-organic frameworks (MOFs).[401] In the following paragraphs, attention will be focused on electrical, magnetic, and optical applications of bis(1,2-dichalcogenolene) complexes. This notwithstanding, it is worth recalling that also complexes of different nature, mono(1,2-dithiolene) complexes for instance, can display features analogous to those of bis(1,2-dichalcogenolene) species. Omary, for example, recently reported on some CT products of mono(1,2-dithiolene) complexes [M(tBu$_2$-bipy)(dmio)] (M = Pd, Pt; tBu$_2$-bipy = 4,4′-di(*tert*-butyl)-2,2′-bipyridine) with organic acceptors such as TCNQ, TCNQF, and TCNE, and demonstrated their magnetic susceptibilities and conductivities are nonzero, suggesting that these systems could be exploited as multifunctional molecular materials.[402]

11.3.6.1 Electrical Properties

The solid state conducting and magnetic properties of 1,2-dithiolene and 1,2-diselenolene complexes have been the subject of a vast number of publications and reviews.[78,80–83,403–405] In all cases, conducting properties (insulating, semiconducting, metal-like, or even superconducting) have been rationalized in terms of their crystal structure, which in turn is affected by temperature and applied pressure (including chemical pressure).[404] For conducting molecular one-dimensional (1D) systems, some criteria have been proven to be necessarily satisfied in order to make the observation of solid state conductivity possible:[403] (1) the existence of stacking structures, (2) a spatial extension of the electronic system so as to allow for a significant overlap between stacked units, and (3) the partial filling of the conducting band, achieved by means of non-integral oxidation or CT processes. In order to avoid Peierls instability,[406] first reported in (TTF)[M(tfd)$_2$] CT complexes (M = Cu, Au; TTF = tetrathiafulvalene, Chapter 11.2),[403,407,408] much effort has been devoted to enhancing the structural dimensionality beyond 1D. BEDT-TTF [bis(ethylenedithio)tetrathiafulvalene] has indeed provided a variety of 2D-metallic systems thanks to the possibility of intermolecular interactions guaranteed by the peripheral sulfur atoms.[404,409] Metal 1,2-dichalcogenolene complexes are a very suitable source of open-shell stable systems (generally radical anions, but also radical cations, *e.g.* in the case of dddt^{2-} complexes) and are able to satisfy these criteria provided that a sufficiently extended planar system is present in the ligands. In particular, open-shell systems deriving from both bis(1,2-dithiolene) and bis(1,2-diselenolene) systems of d^8 metal ions feature a marked tendency to form intermolecular stack interactions with short metal–metal distances (see above). However, in the first 1,2-dichalcogenolene complex showing metal-like conductivity, (Per)$_2$[Pt(mnt)$_2$] (Per = perylene),[410,411] the stacking of perylene units was responsible for the conducting properties, while the first compound containing a 1,2-dichalcogenolene complex responsible for the metal-like behaviour was $(H_3O)_{0.33}Li_{0.82}[Pt(mnt)_2]\cdot 1.67H_2O$.[412] Several 1,2-dichalcogenolene complexes can behave as molecular acceptors in CT complexes with organic donors such as TTF and related species, or can originate NIOS compounds. In principle, the possibility of forming transverse short E···E (E = S, Se) contacts between different stacks could avoid the monodimensionality of the stacking structure.[413,414] Moreover, due to the greater orbital spatial extension of Se compared to S, the replacement of sulfur with selenium has been proposed with the aim of achieving greater interanionic interactions.[267,415]

The complex $TBA_2[Ni(dmit)_2]_7\cdot 2MeCN$ was the first reported conducting $dmit^{2-}$ salt (TBA^+ = tetrabutylammonium).[416] A 2D structure was present in α-$(Et_2Me_2N)[Ni(dmit)_2]_2$,[417] while the first superconductor containing a dithiolene system, $(TTF)[Ni(dmit)_2]$,[418,419] was reported in 1986. To date, about a dozen compounds based on 1,2-dithiolene complexes and exhibiting superconducting properties have been reported.[404,420–432] Remarkably, all of these compounds contain the $dmit^{2-}$ ligand. More common are 1,2-dichalcogenolene-based compounds exhibiting metal-like behaviour. The acknowledgement of the role of TTF in organic conductors (Chapter 11.2) led to the development of ligands embedding the TTF moiety, including 1,2-dithiolene ligands in complexes of various metals.[433] This notwithstanding, most of the complexes deriving from such ligands, recently reviewed by Shatruk and Ray,[434] showed a semiconducting behaviour. Indeed, even if metallic behaviour has been suggested also in single-component 1,2-dithiolene molecular crystals,[87,336] such as $[Au(\alpha\text{-tpdt})_2]$,[435] $[Ni(tmdt)_2]$ (conductivity at room temperature 400 S cm^{-1}), $[Au(Et\text{-dmet})_2]$,[145] or $[Ni(dmdt)_2]$ ($tmdt^{2-}$ = trimethylene-tetrathiafulvalenedithiolate; $dmdt^{2-}$ = dimethyl-tetrathiafulvalenedithiolate),[436–439] metal-like compounds are generally either NIOS salts or CT complexes. More than one-half are again Ni-, Pd-, or Au- complexes based on the $dmit^{2-}$ ligand,[403,440] but recent literature also reports many examples of bis(1,2-dithiolene) complexes deriving from mnt^{2-} (with metals such as Au, Co, Cu, Fe, Ni, Pd, Pt), tfd^{2-}, $dddt^{2-}$, $dmise^{2-}$, $dcit^{2-}$, $dcbdt^{2-}$ (Au, Cu),[441] and $dmbit^{2-}$ (Ni and Pd).[77,83,84,403,404,442]

A few examples of CT salts containing 1,2-diselenolene complexes with metal-like conductivity have also been reported, such as $(TMTSF)_2[M(tds)_2]$ (M = Ni, Pt, Cu; TMTSF = tetramethyltetraselenafulvalene).[415,443] In addition, some examples of CT-compounds containing heteroleptic bis(1,2-dithiolene) complexes, such as (TTF)[Ni(dmit)(mnt)] or (EDT-TTF)[Ni(dmit)(tdas)] (EDT-TTF = ethylenedithiotetrathiafulvalene), have been reported.[444]

In the case of NIOS salts containing cationic radicals, such as $[M(dddt)_2]$, the charge can be counterbalanced by inorganic anions (PF_6^-, HSO_4^-) as well as by different metal compounds, such as $AuBr_2^-$.[445–447] Nevertheless, much more common is the case of NIOS salts containing anionic 1,2-dithiolene or 1,2-diselenolene complexes counterbalanced by different cations, such as alkaline ions, tris- or tetra-alkylammonium, cyclic saturated or unsaturated organic cations, or different spin-crossover transition metal complexes.[223] As previously mentioned, the cation exerts a crucial role in determining the solid state structure of the (cation)(dichalcogenolene) salt: in the context of this book, it seems important to stress the effect that chalcogen–chalcogen interactions between cations and anions can have in determining the conducting properties of the resulting salts. In some salts of [M(dmit)$_2$] radical anions, the secondary bonds between Te-containing cations and the sulfur atoms of the 1,2-dithiolene complex have been reported to be determing factors in increasing the conductivity.[404] An interesting comparison between the solid state properties of compounds containing [M(dmit)$_2$] anions counterbalanced by telluronium cations, such as (Me$_3$Te)[Ni(dmit)$_2$]$_2$, β-(MeTeC$_4$H$_8$O)[Ni(dmit)$_2$]$_2$, (Me$_3$Te)[Pd(dmit)$_2$]$_2$, and (Et$_2$-MeTe)[Pd(dmit)$_2$]$_2$, has been discussed by Kato.[404,448–450] Finally, it it noteworthy that the electrical properties of mono(1,2-dithiolene) complexes have also been investigated. Matsubayashi and co-workers for example demonstrated that Pt^{II} planar sulfur-rich diimine-dithiolate complexes, in their cationic form, can form columnar molecular stackings and/or layered interactions in the solid state through metal–metal and/or sulfur–sulfur contacts, giving rise to high electrical conductivities.[451,452]

11.3.6.2 Magnetic Properties

As mentioned in the previous section, the solid state magnetic properties of 1,2-dithiolene and 1,2-diselenolene complexes have been dealt with in several reviews.[74,80–83,85,403–405] The most common magnetic behaviour is antiferromagnetism, not only in compounds containing non-diamagnetic bis(1,2-dichalcogenolene) complexes,[170,220,229,403,453–456] but also in heteroleptic systems, such as those of the type [Cp$_2$M(dichalcogenolene)], [CpM(dichalcogenolene)$_2$] (M = Ti, Zr, Hf, V, Nb, Ta, Mo, W), or [CpM(dichalcogenolene)] (M = Co, Rh, Ni).[241,371,457,458]

Spin-Peierls (SP) transitions are temperature-dependent magnetoelastic transitions involving quasi-1D insulating systems, and were cited in the previous section.[406] Typically reported systems featuring a Peierls transition are the complexes (TTF)[M(tfd)$_2$], with M = Cu and Au,[407] featuring SP transitions below 12 and 2 K, respectively, resulting in a dramatic decrease in magnetic susceptibility as a consequence of the spin 'dimerization'. Recently, a series of 1D SP-like complexes assembled from [Ni(mnt)$_2$]$^-$ anions with variously substituted Λ-shaped 1-(4′-R-benzyl)pyridinium cations has been reviewed.[459] Among 1,2-diselenolene complexes, (TTF)[Cu(tfs)$_2$] shows a SP transition below 6 K.[460]

The definition of spin-ladder (SL) systems is related to crystalline frameworks resulting from the interactions between a finite number of S = ½ strongly magnetically coupled chains arranged each next to one other.[403] The number of coupled chains (odd or even) and the strength of their interaction produce different magnetic behaviours. Among the most important systems, it is worth mentioning (*p*-EPYNN)[Ni(dmit)$_2$] (*p*-EPYNN = *p*-*N*-pyridinium α-nitronyl nitroxide, a radical cation),[461] which exhibits short-order ferromagnetic interactions below 40 K, and (DT-TTF)$_2$[Au(mnt)$_2$] (DT-TTF = dithiopheno-tetrathiafulvalene).[462–465] In the compound (BEDT-TTF)[Co(mnt)$_2$] the dithiolene anion is diamagnetic and only the organic cation is responsible for the spin system.[466] Also in the compounds (BEDT-TTF)[M(tfadt)$_2$] (M = Ni, Au; tfadt^{2-} = 1-trifluoromethyl-2-cyano-1,2-dithiolate) the antiferromagnetic interaction are established within the chains formed by the radical cations.[467]

Members of the heteroleptic class [Cp$_2$M(L)](TCNQF$_4$), such as [Cp$_2$M(dmio)](TCNQF$_4$)] (M = Mo, W; TCNQF$_4$ = tetrafluoro-tetracyanoquinodimethane),[85,468] and (AsF$_6$)[Cp$_2$W(dsit)][469] have shown SL behaviour as well.

Short-range order ferromagnetic interactions have been reported for several compounds containing 1,2-dithiolene and 1,2-diselenolene complexes, with various types of ligands and metal ions, such as Ni,[258,463,470–480] Pd,[472] Pt,[258,472,481–483] Cu,[484,485] and Mo.[486] Also atypical are long-range ferromagnetic interactions resulting in bulk ferromagnetism,[487–492] a behaviour reported for example in the case of the two homoleptic 1,2-dithiolene complexes (NH$_4$)[Ni(mnt)$_2$]·H$_2$O[84,492,493] and (Cp$_2$*Mn)[Ni(dmit)$_2$].[494]

Only during the past few years the first examples have been reported of single molecular magnets (SMMs) based on different Mn^{II}/Mn^{III} complexes, electrochemically co-crystallized with conducting dithiolenes, such as [Pt(mnt)$_2$]$^{z-}$ or [Ni(dmit)$_2$]$^{z-}$ complexes to give hybridized materials: [{Mn^{II}$_2$Mn^{III}$_2$(hmp)$_6$ (MeCN)$_2$} {Pt(mnt)$_2$}$_4$][Pt(mnt)$_2$]$_2$ (hmp$^-$ = 2-hydroxymethylpyridinate),[493] [Mn$_2$(5-MeO-saltmen)$_2$(acetone)$_2$][Ni(dmit)$_2$]$_7$·4acetone [5-MeO-saltmen^{2-} = *N*,*N*′-(1,1,2,2-tetramethylethylene)bis(5-methoxy-salicylideneiminate)],[494] [Mn$_2$(5-MeO-saltmen)$_2$(MeCN)$_2$][Ni(dmit)$_2$]$_7$·4MeCN,[496] [Mn(5-R-saltmen){M(dmit)$_2$}]$_2$ (R = MeO, M = Ni, Au; R = Me, M = Au),[495] and [Mn(5-MeO-saltmen)(acetone)]$_2$[Ni(dmit)$_2$]$_6$.[496,497]

11.3.6.3 Optical Properties

One of the most striking properties of 1,2-dithiolene[74,78,244,403,498] and 1,2-diselenolene[229,268,500] complexes is their molecular absorption in the Vis-NIR region.[88] For bis(1,2-dithiolene) complexes, this falls at wavelengths (λ_{max}) ranging between 700 and roughly 1750 nm, depending on the nature of the substituents at the carbon atoms of the 1,2-dithiolene system, on the metal ion, on the coordination geometry, on the global charge of the complex (electrochromism), and on the solvent (solvatochromism). Though electronic absorption has been studied also for mono[241,499] and tris(1,2-dithiolene) and (1,2-diselenolene) complexes,[136,162,310] the attention of researchers has focused mainly on bis(1,2-dithiolene) complexes of d^8 metal ions.[102] For M = Ni, Pd, and Pt, the absorption intensity, with ε between 15 000 and 120 000 M^{-1} cm^{-1},[279] is unmatched in any other transition metal compound, whose low-energy absorptions are generally ascribed to weak *d–d* transitions. On the contrary, Vis-NIR absorption has been interpreted as an electronic π–π^* transition between the HOMO and the LUMO, and its nature has been studied at different levels of theory, ranging from EHT to DFT.[38,95,500] In a restricted approach, calculations qualitatively agree in the description of the two molecular orbitals mainly involved. Since only the LUMO has a contribution from the *d* orbitals of the central metal ion, the Vis-NIR transition can be considered as a partial LMCT process.[38,95,106–111]

As regards neutral 1,2-dithiolene complexes, as compared to the simplest 1,2-dithiolene nickel complex ([Ni(edt)$_2$], $\lambda_{max} = 720$ nm),[501] donor substituents cause a bathochromic shift in the λ_{max} value, while acceptor substituents do not markedly affect the energy of the Vis-NIR absorption band. According to the qualitative interpretation by Mueller-Westerhoff,[78] acceptor substituents lower the energies of HOMO and LUMO by similar amounts of energy, not resulting in drastic shifts in transition energies, while donor substituents raise the energy of the filled orbital more than that of the LUMO, thus reducing the HOMO-LUMO energy gap and therefore causing a bathochromic shift of the π–π^* transition towards the NIR region. Worthy of note, by inducing a lowering in the LUMO energy, acceptor substituents tend to stabilize the anionic forms of bis(1,2-dithiolene) complexes, while donor substituents tend to stabilize neutral or, rarely, cationic forms.[78]

The substituent effect has been exploited to fine-tune the position of absorption in order to lower its energy as much as possible. In particular, Mueller-Westerhoff and co-workers reckoned that in Ni bis(1,2-dithiolene) complexes this aim could be achieved by introducing in the 1,2-dithiolene ligand aromatic substituents planar with the dithiolene system, in order to enhance extension of the π-system, and containing amino-donor groups: examples of such 1,2-dithiolene ligands are JUL1^{2-}, JUL2^{2-}, or DETHQ^{2-}, whose nickel complexes exhibited absorptions falling at 1180, 1270, and 1370 nm, respectively in CH_2Cl_2.[403] Further neutral bis(1,2-dithiolene) complexes have been reported to feature low-energy NIR absorptions around 1000 nm, for example the neutral nickel complex of the [4′,5′:5,6][1,4]dithiino[2,3-b]quinoxaline-1′,3′-dithiolate,[502] [Ni(edodddt)$_2$] (edodddt^{2-} = 2,3,4α,8α-tetrahydrodithiine[2,3–b][1,4]dioxo-6,7-dithiolate),[503] [Ni(dmit)$_2$],[281] [Ni(phdt)$_2$],[504] [Ni(bddt)$_2$],[505] [Ni(medt)$_2$],[506] and [Ni(cddt)$_2$],[244] those belonging to the classes [M(R,R′-timdt)$_2$][278,279,506,507] and [Ni(R-dmet)$_2$][281] (M = Ni, Pd, Pt), and a few others.[508–510] In addition, the low-energy absorption bands of neutral 1,2-dithiolene complexes containing ferrocenyl substituents have been discussed.[511] It is worth noting that in the case of neutral bis(1,2-dithiolene) complexes with very narrow HOMO–LUMO gaps, such as single-component molecular conductors, electronic absorption is dramatically shifted into the IR region.[440]

Monoanionic bis(1,2-dithiolene) complexes of Group 10 metal ions feature electronic absorptions that are generally bathochromically shifted and less intense[512] as compared to those of the corresponding neutral species, and in the case of nickel and platinum complexes they fall between 900 and 1450 nm.[281,505,506,513,514] It is interesting to note that salts formed by different 1,2-dithiolene complexes acting both as anions and as cations can present low-energy absorption bands in the solid state, attributed to intermolecular

CT transitions.[515,516] Square-planar neutral Au^{III} bis(1,2-dithiolene) complexes, which can be obtained by chemical[517] or electrochemical[145,205] oxidation of the corresponding non-IR-absorbing anionic species, are formally isoelectronic with monoanionic Group 10 congeners and show an intense absorption in the region between 1400 and 1500 nm (*e.g.* $\lambda_{max} = 1443$ nm for $[Au(4\text{-pedt})_2]$ in DMF,[518] 1408 nm for $[Au(Ph,H\text{-edt})_2]$ in CH_2Cl_2,[164,165] and 1485 nm in CS_2 for $[Au(Ph_2\text{-edt})_2]$.[226,276,519] Also neutral copper bis(1,2-dithiolene) complexes were reported to show NIR absorption, which was investigated in the electrodeposited films ($\lambda_{max} = 1270$ nm for $[Cu(mi\text{-}5edt)_2]$ and $[Cu(mi\text{-}5hedt)_2]$; mi-5edt^{2-} = 1(*N*-methylindol-5-yl)-ethylene-1,3-dithiolate; mi-5Hedt^{2-} = 1(*N*-methylindol-5-yl)-hex-1-ene-1,3-dithiolate).[519]

Only seldom has the effect of pH, in the case of ligands showing protonable sites, such as ppdt^{2-} (ppdt^{2-} = pyrido[2,3–b]pyrazine-2,3-dithiolate) or qdt^{2-}, been investigated as a tool to tune the absorption/emission properties of dithiolene complexes.[520,521]

The strong NIR absorption peculiar to neutral or monoanionic bis(1,2-dithiolene) complexes has been exploited for several applications, summarised in the following paragraphs 11.3.6.3.1–11.3.6.3.3.

11.3.6.3.1 Q-Switching

Laser emission consists of a continuous output of coherent radiation, that can be compressed into ultra-short pulses of extremely high energy[522] by using Q-switching dyes,[78] provided that: (1) the dye has a significant absorption close to the laser emission energy; (2) the dye has no higher excitations at the same energy; (3) the excited state lifetime is long enough to allow for bleaching of the laser; (4) the dye is thermally and photochemically stable.

Bis(1,2-dithiolene) complexes are generally thermally and photochemically very stable, and their Vis-NIR absorption can be tuned in order to reach the wavelength ranges of interest for NIR lasers, such as Nd:YAG, Nd:YLF, and Er:Glass solid state lasers (emission wavelengths $\lambda_{em} = 1064$, 1053, 1540 nm, respectively). In this context, many groups of researchers have devoted their efforts to synthesizing 1,2-dithiolene complexes for this type of application,[504] and indeed some of them have been patented and are commercially available, such as bis(4-dimethylaminodithiobenzil)nickel.[500,523,524]

11.3.6.3.2 Non-Linear Optics

Metal 1,2-dithiolene complexes have been studied for applications in the field of second- and third-order non-linear optical properties.[525] Crystalline second-order NLO materials, possessing large molecular hyperpolarizabilities reflected by their intense low-energy CT absorption,[526] can give rise to the phenomenon of frequency doubling, where an input wave generates another wave with twice the frequency (second-harmonic generation, SHG). Since these materials must not be centrosymmetric,[527,528] heteroleptic bis(1,2-dithiolene) complexes are mainly involved in this field,[403] although homoleptic bis(1,2-dithiolene) complexes as counterions of the hemicyanine dye have been reported to enhance its SHG properties.[529] Several heteroleptic bis(1,2-dithiolene) complexes of the type [M(L)(L′)], featuring both ligands L and L′ of 1,2-dithiolene type, have been reported, such as the monoanion $[Ni(dmit)(mnt)]^-$ isolated as $BzQl^+$ salt ($BzQl^+$ = 1-benzylquinolinium)[530] along with measurements, generally carried out by means of the electric-field-induced second-harmonic generation (EFISH) technique, and theoretical calculations regarding their hyperpolarizability tensors.[531,532] Of particular interest are those systems where the two 1,2-dithiolene ligands have opposite donor/acceptor behaviour, the so-called push–pull systems. In these complexes, the HOMO is mainly localized on the dithiolene ligand with higher acceptor character, while the LUMO is centred on the ligand with higher donor character, so that the relevant low-energy transition possesses an intramolecular CT dithione-dithiolato nature, resulting in high hyperpolarizability values.[353,533] Several neutral complexes of this type, such as $[Pt(Bz_2pipdt)(dcbdt)]$[355] (Bz_2pipdt = 1,4-dibenzylpiperazine-2,3-dithione; $dcbdt^{2-}$ = 4,5-dicyanobenzene-1,2-dithiolate) have been reported.[335,353,356,403,535] Among heteroleptic complexes featuring ligands of different types, the systems most frequently reported are those of the type [M(diimine)(dithiolate)],[529] with different metals M, such as Ni, Pd, Pt, and Zn.[534,535] Due to their optical properties, and in particular to their solvatochromism, this class of complexes has been the subject of a large number of theoretical and experimental studies.[45–47,536–540] Molecular first hyperpolarizabilities (β) of a restricted number of Pt^{II} and Ni^{II} diimine-dithiolate complexes were evaluated from EFISH measurements, using the two-state approximation.[541–544] The values of β determined for these complexes were found to reach as much as $-34\ 10^{-30}$ esu,[545] a value comparable to that of other NLO-active materials, so that these complexes represent one of the most promising class of complexes containing 1,2-dithiolene ligands for NLO applications. An influence of the nature of both the diimine and the dithiolate on the entity of β was found,[544] and, interestingly, the Pt^{II} complexes were identified as the most promising candidates for these applications.[544,545] Other heteroleptic complexes featuring donor and acceptor donors have been reported. An interesting case is that of the neutral complex resulting from the assembly of two fullerenes (acceptors), two cobaltadithiolenes and one TTF bridge (donor) in the single donor–acceptor array C_{60}–Co–TTF–Co–C_{60}, featuring an absorption at 1100 nm in toluene ($\varepsilon = 30\ 000\ M^{-1}\ cm^{-1}$).[546]

Third-order NLO effects, of interest for potential applications, such as in optical limiting[244,527,547–550] impose no symmetry constraints. Thus, studies on this type of non-linearity, carried out for example by means of the Z-scan technique, have been reported on both homoleptic and heteroleptic 1,2-dithiolene complexes.[510,512,519,551–555] Among the members of the former class, studies on neutral nickel bis(1,2-dithiolene) complexes,[556] such as $[Ni(etodddt)_2]$ [$etodddt^{2-}$ = 4,5-(1,4-oxathiane-2,3-diyldithio)-1,3-dithiole][555] or salts of monoanionic species with ligands such as

$dmit^{2-}$,[556] $phdt^{2-}$, $dddt^{2-}$, $pddt^{2-}$,[506] $cddt^{2-}$,[244] and *p*-substituted diphenylethylene-1,2-dithioate[557] have been published. More recently, third-order non-linear susceptibility $|\chi^{(3)}|$ values (in the range $1–8 \times 10^{-13}$ esu) have also been reported for gold complexes, such as $[Au(dmit)_2]$, $[Au(Ph_2\text{-}edt)_2]$[519] and alkylammonium salts of $[AuL_2]^-$ species ($L = dmit^{2-}$, [4′,5′:5,6][1,4]dithiino[2,3-b]quinoxaline-1′,3′-dithiolate, and $Ph_2\text{-}edt^{2-}$).[504,558,559]

Underhill and co-workers have reported on the non-linear properties of unsymmetrically substituted bis(1,2-dithiolene) complexes, such as bis[1-butyl-2-phenylethene-1,2-dithiolato(2^-)-S,S′]nickel(II), embedded in PMMA-based composites (PMMA = polymethylmethacrylate),[560] or in sol–gel materials.[561] As regards the latter class, the above-mentioned array C_{60}–Co–TTF–Co–C_{60} showed a $|\chi^{(3)}|$ value of 9.3×10^{-12} esu.[548] In addition, some works have dealt with complexes containing mnt^{2-} and $dmit^{2-}$ ligands.[403]

11.3.6.3.3 Optical Technology

Several publications and patents have dealt with application of 1,2-dithiolene systems to the various branches of optical information technology. Thus, 1,2-dithiolene complexes have been exploited in materials used for optical data storage, such as CD-ROM or LD-ROM media, in copiers, or as optical switches and wavelength-sensitive NIR-photoconducting devices.[279,562–567] In addition, only citing some recent examples, due to their linear and non-linear optical properties 1,2-dithiolene complexes have been investigated as innovative NIR-absorbing liquid crystals[88] and metallogelators,[568] components of dye-sensitised solar cells,[569,570] for photo-imaging media,[571,572] as Vis-NIR optical filters,[573–579] or as ink components.[580]

11.3.7 Conclusions

1,2-Dithiolene chemistry has been investigated in depth over the last 40 years. The electrical, magnetic, and optical properties of 1,2-dithiolene complexes, in addition to their general thermal and photochemical stabilities, make them useful for a great number of potential applications, thus justifying the increasing attention of more and more researchers. The intrinsic interest in their electronic structure makes this class of complexes appealing for both experimental and theoretical chemists.

Surprisingly, notwithstanding the general interest in 1,2-dithiolene complexes, and the plethora of different applicative fields in which these compounds have been exploited, only a limited number of research groups have systematically investigated the chemistry of 1,2-diselenolene complexes, and only sporadic publications have reported on 1,2-ditellurolenes. In particular, while the electrical and magnetic features of compounds containing 1,2-diselenolene complexes have been studied, their optical properties have been tapped only marginally. While papers dealing with Vis-NIR absorption properties of 1,2-diselenolene generally indicate that their spectroscopic properties are very similar to those of the sulfured congeners, only few systematic explorations of their electrochromic behaviour have been reported, and investigations on NLO properties seem to be at a very early stage. This is clearly shown by the simple comparison of the number of patents dealing with 1,2-diselenolene and 1,2-dithiolene ligands and complexes. The chemistry of 1,2-ditellurolenes is in an even more embryonic phase, and their spectroscopic, electrochemical, electrical and magnetic properties are almost wholly unexplored.

We can foresee that during the next years the chemistry of 1,2-diselenolene complexes, but even more that of 1,2-ditellurolenes, will be investigated in depth, achieving the richness of knowledge and applicative significance that is nowadays a hallmark of 1,2-dithiolene complexes.

Abbreviations

1D	one-dimensional
D	two-dimensional
AcO^-	acetate
α-tpdt[a]	2,3-thiophenedithiolate
ACCN	1,1′-azobiscyclohexanecarbonitrile
AIBN	2,2′-azobisisobutyrylnitrile
$bddt^{2-}$	4*a*,5,6,7,8,8*a*-hexahydro-1,4-benzodithiin-2,3-dithiolate
bds^{2-} [b]	benzene-1,2-diselenolate
bdt^{2-} [a]	benzene-1,2- dithiolate
BEDT-TTF	bis(ethylenedithio) tetrathiafulvalene
bpedmit	4,5-bis(2-pyridylethylsulfanyl)-1,3-dithiole-2-thione
bpmdmit	4,5-bis(2- pyridylmethylsulfanyl)-1,3-dithiole-2-thione
bpp	2,6-bis(pyrazol-3-yl)pyridine
$btdt^{2-}$ [a]	benzotetrathiafulvalenedithiolate
$BrBzPy^+$	4- bromobenzylpyridinium
bz	benzene
Bz_2pipdt	1,4-dibenzylpiperazine-2,3-dithione
$BzQl^+$	1- benzylquinolinium
$cddt^{2-}$ [a]	4*a*,6,7,7*a*-5*H*- cyclopenta[*b*]-1,4-dithiin-2,3-dithiolate
ChSTF	2,3-cyclohexylenedithio-1,4-dithia-5,8-diselenafulvalene
Cp	cyclopentadienyl
Cp^*	pentamethylcyclopentadienyl
CQD	carbon quantum dots
CT	charge-transfer
$dcbdt^{2-}$	4,5-dicyanobenzene-1,2-dithiolate
$dcit^{2-}$	3,4-dimercapto-5-cyanoisothiazole
$dcmedt^{2-}$	1,2-bis(carboxymethyl)-ethylene-1,2-dithiolate
ddt^{2-} [a]	1,4-dithiin-2,3-dithiolate

[a]Depicted in Scheme 11.3.1.
[b]Depicted in Scheme 11.3.2.

$ddds^{2-}$	5,6-dihydro-1,4-dithiin-2,3-diselenolate
$dddt^{2-}$	dihydro-1,4-dithiin-2,3-dithiolate
$DETHQ^{2-}$	bis(3-*N,N′*-diethyltetrahydroquinazolyl)-ethylene-1,2-dithiolate
DFT	density functional theory
$diod^{2-}$ [a]	1,4-dithia-6-oxa-2,3-dithiolate
DMAD	dimethyl acetylenedicarboxylate
$dmbit^{2-}$	2-thione-1,3-dithiole-benzo[*d*]-4,5-dithiolate
$dmbt^{2-}$	dimethyltetrathiafulvalenedithiolate
$dmdt^{2-}$	dimethyltetrathiafulvalenedithiolate
$dmedt^{2-}$	1,2-bis(methyl)ethylenedithiolate
$dmid^{2-}$	see $dmio^{2-}$
$dmio^{2-}$	2-oxo-1,3-dithiole-4,5-dithiolate
$dmise^{2-}$ [a]	2-selenoxo-1,3-dithiole-4,5-dithiolate
$dmit^{2-}$ [a]	2-thioxo-1,3-dithiole-4,5-dithiolate
$dmox^{2-}$ [a]	4,5-dimethoxybenzene-1,2-dithiolate
dmt^{2-} [a]	3-thioxo-1,2-dithiole-4,5-dithiolate
$dpedt^{2-}$	see Ph_2-edt^{2-}
$dphdt^{2-}$ [a]	5,6-diphenyl-1,4-dithiin-2,3-dithiolate
dppe	1,2-bis(diphenylphosphino)ethane
$dsis^{2-}$ [b]	2-selenoxo-1,3-diselenole-4,5-diselenolate
$dsise^{2-}$ [b]	2-selenoxo-1,3-dithiole-4,5-diselenolate
$dsit^{2-}$ [b]	2-thioxo-1,3-dithiole-4,5-diselenolate
$dsitse^{2-}$ [b]	2- selenoxo-1,3-thiaselenole-4,5-diselenolate
$dssq^{2-}$ [b]	1,2-diselenosquarate
$dtcs^{2-}$ [a]	3,4,5-trioxocyclopent-1-ene-1,2-dithiolate
$dtdt^{2-}$ [a]	1,4,6-trithia-2,3-dithiolate
$dtsq^{2-}$ [a]	1,2-dithiosquarate
DT-TTF	dithiopheno-tetrathiafulvalene
edt^{2-} [a]	ethylene-1,2-dithiolate
EDT-TTF	ethylenedithiotetrathiafulvalene
EHT	extended Hückel theory
$edodddt^{2-}$	2,3,4α,8α-tetrahydro-dithiine[2,3–b][1,4]dioxo-6,7-dithiolate
$etodddt^{2-}$	4,5-(1,4-oxathiane-2,3-diyldithio)-1,3-dithiole
F_2pdt^{2-} [a]	2,2-difluoro-1,3-propanediyldithioethylene-1,2-dithiolate
hmp^{-}	2-hydroxymethylpyridinate
HMTTF	bis(trimethylene)-tetrathiafulvalene
HOMO	highest occupied molecular orbital
$JUL1^{2-}$	julolidinyl-ethylene-1,2-dithiolate
$JUL2^{2-}$	1,2-bis(julolidinyl)-ethylene-1,2-dithiolate
LDA	lithium diisopropylamide
LLCT	ligand to ligand charge transfer
LMCT	ligand to metal charge transfer
LUMO	lowest unoccupied molecular orbital
$madt^{2-}$ [a]	maleamide-1,2-dithiolate

$medt^{2-}$ [a]	5,6-dihydro-6-methyl-1,4-dithiin-2,3-dithiolate
5-MeO-saltmen^{2-}	*N*,*N*′-(1,1,2,2-Tetramethylethylene) bis(5-methoxy-salicylideneiminate
mdt^{2-} [a]	2*H*-1,3-dithiole-4,5-dithiolate
mi-5edt^{2-}	1(*N*-methylindol-5-yl)-ethylene-1,3-dithiolate
mi-5Hedt^{2-}	1(*N*-methylindol-5-yl)-hex-1-ene-1,3-dithiolate
$midt^{2-}$ [a]	maleimide-1,2-dithiolate
mns^{2-} [b]	1,2-maleonitrile-1,2-diselenolate
mnt^{2-} [a]	1,2-maleonitrile-1,2-dithiolate
MOF	metal-organic framework
NIOS	non-integral oxidation state
NIR	near infrared
NLO	non-linear optics
OFET	organic field-effect transistors
$pddt^{2-}$ [a]	6,7-dihydro-5*H*-1,4-dithiepinin-2,3-dithiolate
p-EPYNN	*p*-*N*-pyridinium α-nitronyl nitroxide
pds^{2-} [b]	pyrazine-2,3-diselenolate
pdt^{2-} [a]	pyrazine-2,3-dithiolate
4-pedt^{2-}	4-pyridylethylene-1,2-dithiolate
$phdt^{2-}$ [a]	5,6-dihydro-5-phenyl-1,4-dithiin-2,3-dithiolate
Ph,H-edt^{2-}	phenylethylene-1,2-dithiolate
Ph_2-edt^{2-}	diphenylethylene-1,2-dithiolate
Per	perylene
PMMA	polymethylmethacrylate
$pydt^{2-}$	pyridine-2,3-dithiolate
qdt^{2-} [a]	quinoxaline-2,3-dithiolate
qsalH	*N*-(8-quinolyl)salicylaldimine
R-dmet^{2-} [a]	*N*-substituted 2-thioxo-thiazoline-4,5-dithiolate
3-R-salEen	3-substituted *N*-(2-ethylamino)-ethyl)-salicylaldimine
R-thiazds^{2-}	*N*-substituted 2-thioxo-thiazole-4,5-diselenolate
R-thiazdt^{2-}	see R-dmet^{2-}
R,R′-timdt^{2-} [a]	disubstituted 2-thioxoimidazoline-4,5-dithiolate
sdt^{2-}	see Ph,H-edt^{2-}
SHG	second-harmonic generation
SMM	single molecular magnet
SORCI	spectroscopy oriented configuration interaction
SP	spin-Peierls
TBA^+	tetrabutylammonium
$tbbdt^{2-}$	3,5-*tert*- butylbenzene-1,2-dithiolate
*t*BuCp	*tert*-butylcyclopentadienyl
tBu_2-bipy	4,4′-di(*tert*-butyl)-2,2′-bipyridine
$TCNQF_4$	tetrafluorotetracyanoquinodimethane
$tdas^{2-}$ [a]	1,3,5-thiadiazole-3,4-dithiolate
tds^{2-} [b]	1,2-bis(trifluoromethyl)ethylenediselenolate
tdt^{2-} [a]	toluene-3,4-dithiolate
TEA^+	tetraethylammonium

$tfadt^{2-}$	2-(trifluoromethyl)acrylonitrile-1,2-dithiolate
tfd^{2-} [a]	1,2-bis(trifluoromethyl)ethylenedithiolate
$tmdt^{2-}$	trimethylenetetrathiafulvalenedithiolate
TMTSF	tetramethyltetraselenafulvalene
TTeF	tetratellurafulvalene
TTF	tetrathiafulvalene
Vis	visible
YAG	yttrium aluminium garnet
YLF	yttrium lithium fluoride
ZORA	zeroth-order regular approximation

References

1. M. Arca and M. C. Aragoni, in *Handbook of Chalcogen Chemistry: New Perspectives in Sulfur, Selenium, and Tellurium*, ed. F. A. Devillanova, Elsevier, Oxford, 2007, p. 797.
2. S. Leimkühler, M. M. Wuebbens and K. V. Rajagopalan, *Coord. Chem. Rev.*, 2011, **255**, 1129.
3. A. Majumdar and S. Sarkar, *Coord. Chem. Rev.*, 2011, **255**, 1039.
4. F. J. Hine, A. J. Taylor and C. D. Garner, *Coord. Chem. Rev.*, 2010, **254**, 1570.
5. W. Kaim and B. Schwederski, *Coord. Chem. Rev.*, 2010, **254**, 1580.
6. F. J. Hine, A. J. Taylor and C. D. Garner, *Coord. Chem. Rev.*, 2010, **254**, 1570.
7. R. H. Holm, E. I. Solomon, A. Majumdar and A. Tenderholt, *Coord. Chem. Rev.*, 2011, **255**, 993.
8. A. Begum and S. Sarkar, *Eur. J. Inorg. Chem.*, 2012, 40.
9. S. Metz and W. Thiel, *Coord. Chem. Rev.*, 2011, **255**, 1085.
10. L. E. Bevers, P.-L. Hagedoorn and W. R. Hagen, *Coord. Chem. Rev.*, 2009, **253**, 269.
11. P. V. Bernhardt, *Chem. Commun.*, 2011, **47**, 1663.
12. P. Kalimuthu, S. Leimkühler and P. V. Bernhardt, *J. Phys. Chem. B.*, 2011, **115**, 2655.
13. R. P. Mtei, G. Lyashenko, B. Stein, N. Rubie, R. Hille and M. L. Kirk, *J. Am. Chem. Soc.*, 2011, **133**, 9762.
14. K. G. V. Havelius, S. Reschke, S. Horn, A. Döring, D. Niks, R. Hille, C. Schulzke, S. Leimkühler and M. Hausmann, *Inorg. Chem.*, 2011, **50**, 741.
15. M. J. Pushie, C. J. Doonan, K. Moquin, J. H. Weiner, R. Rothery and G. N. George, *Inorg. Chem.*, 2011, **50**, 732.
16. K. G. Matz, R. P. Mtei, R. Rothstein, M. L. Kirk and S. N. Burgmayer, *Inorg. Chem.*, 2011, **50**, 9804.
17. M. Bose, G. Moula, A. Begum and S. Sarkar, *Inorg. Chem.*, 2011, **50**, 3852.
18. T. D. Rapson, U. Kappler, G. R. Hanson and P. V. Bernhardt, *Biochim. Biophys. Acta*, 2011, **1807**, 108.

19. P.P Samuel, S. Horn, A. Döring, K. G. V. Havelius, S. Reschke, S. Leimkühler, M. Haumann and K. Schulzke, *Eur. J. Inorg. Chem.*, 2011, 4387.
20. S. Groysman and R. H. Holm, *Biochem.*, 2009, **48**, 2310.
21. S. Groysman, J.-J. Wang, R. Tagore, S. C. Lee and R. H. Holm, *J. Am. Chem. Soc.*, 2008, **130**, 12794.
22. R. L. McNaughton, B. S. Lim, S. Z. Knottenbelt, R. H. Holm and M. L. Kirk, *J. Am. Chem. Soc.*, 2008, **130**, 4628.
23. S. J. N. Burgmayer, M. Kim, R. Petit, A. Rothkopf, A. Kim, S. BelHamdounia, Y. Hou, A. Somogyi, D. Habel-Rodriguez, A. Williams and M. L. Kirk, *J. Inorg. Biochem.*, 2007, **101**, 1601.
24. S. Groysman and R. H. Holm, *Inorg. Chem.*, 2007, **46**, 4090.
25. A. L. Tenderholt, R. K. Szilagyi, R. H. Holm, K. O. Hodgson, B. Hedman and E. I. Solomon, *J. Inorg. Biochem.*, 2007, **101**, 1594.
26. H. Sugimoto, M. Tarumizu, H. Miyake and H. Tsukube, *Eur. J. Inorg. Chem.*, 2007, 4663.
27. J.-J. Wang, C. Tesslier and R. H. Holm, *Inorg. Chem.*, 2006, **45**, 2979.
28. A. Mujamdar, K. Pal and S. Sarkar, *J. Am. Chem. Soc.*, 2006, **128**, 4196.
29. K. A. Van Houten and R. S. Pilato, *Progr. Inorg. Chem.*, 2004, **52**, 369.
30. Y. Ji, R. Zhang, Y.-J. Li, Y.-Z. Li, J.-L. Zuo and X.-Z. You, *Inorg. Chem.*, 2007, **46**, 866.
31. Y. Ji, R. Zhang, X.-B. Du, J.-L. Zuo and X.-Z. You, *Dalton Trans.*, 2008, 2578.
32. N. J. Wiebelhaus, M. A. Cranswick, E. L. Klein, L. T. Lockett, D. L. Lichtenberger and J. H. Enemark, *Inorg. Chem.*, 2011, **50**, 11021.
33. P. Surawatanawong, S. Sproules, F. Neese and K. Wieghardt, *Inorg. Chem.*, 2011, **50**, 12064.
34. S. Šoralová, M. Breza and M. Gróf, *Polyhedron*, 2011, **30**, 307.
35. P. Romaniello, M. C. D′Andria and F. Lelj, *J. Chem. Phys. A*, 2010, **114**, 5838.
36. S. Meskaldji, L. Belkhiri, T. Arliguie, M. Formigué, M. Ephritikhine and A. Boucekkine, *Inorg. Chem.*, 2010, **49**, 3192.
37. R. Veloso-Bahamonde, R. Ramirez-Tagle and R. Arratia-Pérez, *Chem. Phys. Lett.*, 2010, **491**, 214.
38. L. Serrano-Andrés, A. Avramopoulos, J. Li, P. Labéguerie, D. Bégué, V. Kellö and M. G. Papadopoulos, *J. Chem. Phys.*, 2009, **131**, 134312.
39. J. Conradie, *J. Mol. Struct.-THEOCHEM*, 2009, **915**, 51.
40. H. Jacobsen and J. P. Donahue, *Inorg. Chem.*, 2008, **47**, 10037.
41. Y.-H. Cui, W. Q. Tian, J.-K. Feng, Z.-Z. Liu and W.-Q. Li, *J. Mol. Struct.-THEOCHEM*, 2007, **810**, 65.
42. P. Romaniello, F. Lelj, M. Arca and F. A. Devillanova, *Theor. Chem. Acc.*, 2007, **117**, 621.
43. M. Hofmann, *J. Mol. Struct.-THEOCHEM*, 2006, **773**, 59.
44. M. L. Kirk, R. L. McNaughton and M. E. Helton, *Progr. Inorg. Chem.*, 2004, **52**, 111.
45. C. A. Mitsopoulou, *Coord. Chem. Rev.*, 2010, **254**, 1448.

46. C. Makedonas and C. A. Mitsopoulou, *Eur. J. Inorg. Chem.*, 2006, 590.
47. C. Makedonas and C. A. Mitsopoulou, *Eur. J. Inorg. Chem.*, 2006, 2460.
48. W. H. Mills and R. E. D. Clarck, *J. Chem. Soc.*, 1936, 175.
49. R. E. D. Clarck, *Analyst*, 1936, **60**, 242.
50. R. E. D. Clarck, *Analyst*, 1937, **62**, 661.
51. G. Bähr and H. Schleitzer, *Chem. Ber*, 1957, **90**, 438.
52. D. B. Stevancevic and V. C. Drazic, *Bull. Inst. Nucl. Sci. "Boris Kidrich"*, 1959, **9**, 69.
53. G. N. Schrauzer and V. P. Mayweg, *J. Am. Chem. Soc.*, 1962, **84**, 3221.
54. H. B. Gray, R. Williams, I. Bernal and E. Billig, *J. Am. Chem. Soc.*, 1962, **84**, 3596.
55. R. Eisenberg and J. A. Ibers, *J. Am. Chem. Soc.*, 1965, **87**, 3776.
56. M. Nomura, M. Fujii, K. Fukuda, T. Sugiyama, Y. Yokoyama and M. Kajitani, *J. Organomet. Chem.*, 2005, **690**, 1627.
57. H. Ye, W. Bai, M. Xie, Y. Li and H. Yan, *Eur. J. Inorg. Chem.*, 2011, 2763.
58. M. Nomura, C. Fujita-Takayama, T. Sugiyama and M. Kajitani, *J. Organomet. Chem.*, 2011, **696**, 4018.
59. S. Rabaça and M. Almeida., *Coord. Chem. Rev.*, 2010, **254**, 1493.
60. L. Dang, M. F. Shibl, X. Yang, A. Alak, D. J. Harrison, U. Fekl, E. N. Brothers and M. B. Hall, *J. Am. Chem. Soc.*, 2012, **134**, 4481.
61. M. Nomura, N. Sakamoto, H. Nakajima, C. Fujita-Takayama and T. Sugiyama, *Polyhedron*, 2011, **30**, 2890.
62. M. Nomura, S. Iida, K. Seki, K. Kobayashi, G. Hagino, S. Horikoshi, T. Sugiyama, A. Sugimori and M. Kajitani, *J. Organomet. Chem.*, 2011, **696**, 1723.
63. M. Nomura, K. Terada, A. Onozawa, Y. Mitome, T. Sugiyama and M. Kajitani, *J. Organomet. Chem.*, 2011, **696**, 2720.
64. M. Nomura, H. Oguro, K. Maeshima, S. Iida, S. Horikoshi, T. Sohiyama, A. Sugimori and M. Kajitani, *J. Organomet. Chem.*, 2010, **695**, 1613.
65. M. J. Kerr, D. J. Harrison, A. J. Lough and U. Fekl, *Inorg. Chem.*, 2009, **48**, 9043.
66. D. J. Harrison, A. J. Lough, N. Nguyen and U. Fekl, *Angew. Chem. Int. Ed.*, 2007, **46**, 7644.
67. R. Yu, K. Arumugam, A. Manepalli, Y. Tran, R. Schmehl, H. Jacobsen and J. P. Donahue, *Inorg. Chem.*, 2007, **46**, 5131.
68. A. Mujumdar, K. Pal, K. Nagarajan and S. Sarkar, *Inorg. Chem.*, 2007, **47**, 6136.
69. D. J. Harrison, N. Nguyen, A. J. Lough and U. Fekl, *J. Am. Chem. Soc.*, 2006, **128**, 11026.
70. H. Sugimoto, R. Tajima, T. Sakurai, H. Ohi, H. Miyake, S. Itoh and H. Tsukube, *Angew. Chem. Int. Ed.*, 2006, **45**, 3520.
71. H. Adams, H. C. Gardner, R. A. McRoy, M. J. Morris, J. C. Motley and S. Torker, *Inorg. Chem.*, 2006, **45**, 10967.
72. S. Takemoto, D. Shimadzu, K. Kamikawa, H. Matsuzaka and R. Nomura, *Organometallics*, 2006, **25**, 982.

73. M. Murata, S. Habe, S. Araki, K. Namiki, T. Yamada, N. Nakagawa, T. Nankawa, M. Nihei, J. Mizutani, M. Kurihara and H. Nishihara, *Inorg. Chem.*, 2006, **45**, 1108.
74. J. A. McCleverty, *Progr. Inorg. Chem.*, 1968, **10**, 49.
75. R. Eisemberg, *Progr. Inorg. Chem.*, 1970, **12**, 295.
76. P. I. Clemenson, *Coord. Chem. Rev.*, 1990, **106**, 171.
77. A. E. Pullen and R.-M. Olk, *Coord. Chem. Rev.*, 1999, **188**, 211.
78. U. T. Mueller-Westerhoff, B. Vance and D. I. Yoon, *Tetrahedron*, 1991, **47**, 909.
79. P. Cassoux, L. Valade, H. Kobayashi, A. Kobayashi, R. A. Clark and A. E. Underhill, *Coord. Chem. Rev.*, 1991, **110**, 115.
80. P. Cassoux, L. Valade, in *Inorganic Materials*, ed. D. W. Bruce, D. O'Hare, Wiley, Chichester, 1997, Chapter 1.
81. L. Ouahab, *Chem. Mater.*, 1997, **9**, 1909.
82. E. Canadell, *Coord. Chem. Rev.*, 1999, **186**, 629.
83. P. Cassoux, *Coord. Chem. Rev.*, 1999, **185-186**, 213.
84. N. Robertson and L. Cronin, *Coord. Chem. Rev.*, 2002, **227**, 93.
85. P. Basu, J. F. Stolz and M. T. Smith, *Curr. Sci.*, 2003, **84**, 1412.
86. M. Formigué, *Acc. Chem. Res.*, 2004, **37**, 179.
87. A. Kobayashi, E. Fujiwara and H. Kobayashi, *Chem Rev.*, 2004, **104**, 5243.
88. K. L. Marshall, G. Painter, K. Lotito, A. G. Noto and P. Chang, *Mol. Cryst. Liq. Cryst.*, 2006, **454**, 47.
89. M. Nomura, *Dalton Trans.*, 2011, **40**, 2112.
90. E. I. Stiefel, ed., Dithiolene chemistry: synthesis, properties, and applications, *Progr. Inorg. Chem.*, **52** , 2004.
91. A. Vlcek, *Coord. Chem. Rev.*, 2010, **254**, 1357.
92. A. Davison and E. T. Shawl, *Chem. Commun.*, 1967, 670.
93. A. Davison and E. T. Shawl, *Inorg. Chem.*, 1970, **9**, 1820.
94. J. Locke, J. A. McCleverty and E. J. Wharton, *Chem Commun.*, 1966, 677.
95. K. Ray, T. Weyhermüller, F. Neese and K. Wieghardt, *Inorg. Chem.*, 2005, **44**, 5345.
96. K. Ray, E. Bill, T. Weyhermüller and K. Wieghardt, *J. Am. Chem. Soc.*, 2005, **127**, 5641.
97. K. Ray, T. Petrenko, K. Wieghardt and F. Neese, *Dalton Trans.*, 2007, 1552.
98. G. Periyasamy, N. A. Burton, I. H. Hillier, M. A. Vincent, H. Disley, J. McMaster and C. D. Garner, *Faraday Discuss.*, 2007, **135**, 469.
99. R. Eisenberg and H. B. Gray, *Inorg. Chem.*, 2011, **50**, 9741.
100. R. Eisenberg, *Coord. Chem. Rev.*, 2011, **255**, 825.
101. W. Kaim, *Eur. J. Inorg. Chem.*, 2012, 343.
102. B. Garreau-de-Bonneval, K. I. Moineau-Chan Ching, F. Alary, T.-T. Bui and L. Valade, *Coord. Chem. Rev.*, 2010, **254**, 1462.
103. T. Petrenko, K. Ray, E. Wieghardt and F. Neese, *J. Am. Chem. Soc.*, 2006, **128**, 4422.

104. F. E. Inscore, S. Z. Knottenbelt, N. D. Rubie, H. K. Joshi, M. L. Kirk and J. H. Enemark, *Inorg. Chem.*, 2006, **45**, 967.
105. C. Hemann, P. Ilich, A. L. Stockert, E.-Y. Choi and R. Hille, *J. Phys. Chem. B*, 2005, **109**, 3023.
106. S. Sproules and K. Wieghardt, *Coord. Chem. Rev.*, 2011, **255**, 837.
107. A. L. Tenderholt, R. K Szilagyi, R. H. Holm, K. O. Hodgson, B. Hedman and E. I. Solomon, *Inorg. Chem.*, 2008, **47**, 6382.
108. R. R. Kapre, E. Bothe, T. Weyhermüller, S. DeBeer George, N. Muresan and K. Wieghardt, *Inorg. Chem.*, 2007, **46**, 7827.
109. R. Sarangi, S. De Beer George, D. Jackson Rudd, R. K. Szilagyi, X. Ribas, C. Rovira, M. Almeida, K. O. Hodgson, B. Hedman and E. I. Solomon, *J. Am. Chem. Soc.*, 2007, **129**, 2316.
110. E. I. Solomon, B. Hedman, K. O. Hodgson, A. Dey and R. K. Szilagyi, *Coord. Chem. Rev.*, 2005, **249**, 97.
111. S. Sproules, T. Weyhermüller, S. DeBeer and K. Wieghardt, *Inorg. Chem.*, 2010, **49**, 5241.
112. S. Sproules, T. Weyhermüller, R. Goddard and K. Wieghardt, *Inorg. Chem.*, 2011, **50**, 12623.
113. S. Sproules, F. L. Benedito, E. Bill, T. Weyhermüller, S. DeBeer George and K. Wieghardt, *Inorg. Chem.*, 2009, **48**, 10926.
114. S. Sproules, P. Banerjee, T. Weyhermüller, Y. Yan, J. P. Donahue and K. Wieghardt, *Inorg. Chem.*, 2011, **50**, 7106.
115. T. Waters, H.-K. Woo, X.-B. Wang and L.-S. Wang, *J. Am. Chem. Soc.*, 2006, **128**, 4282.
116. T. Waters, X.-B. Wang, H.-K. Woo and L.-S. Wang, *Inorg. Chem.*, 2006, **45**, 5841.
117. T. Shimizu, H. Murakami, Y. Kobayashi, K. Iwata and N. Kamigata, *J. Org. Chem.*, 1998, **63**, 8192.
118. M. C. Aragoni, M. Arca, F. Demartin, F. A. Devillanova, A. Garau, F. Isaia, F. Lelj, V. Lippolis and G. Verani, *J. Am. Chem. Soc.*, 1999, **121**, 7098.
119. A. Bittermann, B. Bildstein and K. Wurst, *Z. Kristallogr.-New Cryst. Struct*, 2008, **223**, 231.
120. T. Akasaka, M. Nakano, H. Tamura and G. Matsubayashi, *Bull. Chem. Soc. Jpn.*, 2002, **75**, 2621.
121. J. G. Breitzer, A. I. Smirnov, L. F. Szczepura, S. R. Wilson and T. B. Rauchfuss, *Inorg. Chem.*, 2001, **40**, 1421.
122. N. M. Comerlato, L. G. Carreira, R. A. Howie, L. da Silva Jn., and J. L. Wardell, *Z. Kristallogr.*, 2010, **225**, 29.
123. M. C. Aragoni, M. Arca, F. A. Devillanova, F. Isaia, V. Lippolis, A. Mancini, L. Pala, A. M. Z. Slawin and J. D. Woollins, *Chem. Commun.*, 2003, 2226.
124. T. B. Rauchfuss, *Progr. Inorg. Chem.*, 2004, **52**, 1.
125. G. C. Papavassiliou, S. Y. Yiannoupoulos and J. S. Zambounis, *Chem. Scr.*, 1987, **27**, 265.
126. Y. Shibata, B. Zhu, S. Kume and H. Nishihara, *Dalton Trans.*, 2009, 1939.

127. D. M. Giolando and K. Kirschbaum, *Synthesis*, 1992, 451.
128. J. P. Donahue, C. R. Goldsmith, U. Nidiminti and R. H. Holm, *J. Am. Chem. Soc.*, 1998, **120**, 12.
129. E. J. Wharton and J. A. McCleverty, *J. Chem. Soc. (a)*, 1969, 2258.
130. J. Jones and J. Douek, *J. Inorg. Nucl. Chem.*, 1981, **43**, 406.
131. H. Oku, N. Ueyama and A. Nakamura, *Bull. Soc. Chem. Jpn.*, 1996, **69**, 3139.
132. H. Alves, D. Simão, I. C. Santos, V. Gama, R. Y. Henriques, H. Novais and M. Almeida, *Eur. J. Inorg. Chem.*, 2004, 1318.
133. N. D. Lowe and C. D. Garner, *J. Chem. Soc., Dalton Trans.*, 1993, 2197.
134. O. Jeannin, M. Nomura and M. Formigué, *J. Organomet. Chem.*, 2007, **692**, 4113.
135. S. Sproules, A. A. Eagle, M. K. Taylor, R. W. Gable, J. M. White and C. G. Young, *Inorg. Chem.*, 2011, **50**, 4503.
136. S. M. Dibrov and R. E. Bachman, *Inorg. Chim. Acta*, 2004, **357**, 1198.
137. D. J. Sandman, G. W. Allen, L. A. Acampora, J. C. Stark, S. Jansen, M. T. Jones, G. J. Ashwell and B. M. Foxman, *Inorg. Chem.*, 1987, **26**, 1664.
138. P. Grosshans, P. Adkine, H. Sidorenkova, M. Nomura, M. Formigué and M. Geoffrey, *J. Phys. Chem. A*, 2008, **112**, 4067.
139. X. Ribas, J. C. Dias, J. Morgado, K. Wurst, E. Molins, E. Ruiz, M. Almeida, J. Veciana and C. Rovira, *Chem. Eur. J.*, 2004, **10**, 1691.
140. J. Morgado, I. C. Santos, L. F. Veiros, U. Rodrigues, R. T. Henriques, M. T. Duarte, L. Alcacer and M. Almeida, *J. Mater. Chem.*, 2001, **11**, 2108.
141. J. Hu, J. Wen, D. Wu, R. Zhang, G. Liu, Q. Jiang, Y. Li and H. Yang, *Organometallics*, 2011, **30**, 298.
142. S. Eid, T. Roisnel and D. Lorcy, *J. Organomet. Chem.*, 2008, **693**, 2755.
143. S. Eid, M. Formigué, T. Roisnel and D. Lorcy, *Inorg. Chem.*, 2007, **46**, 10647.
144. S. Eid, M. Guerro and D. Lorcy, *Tetrahedron Lett.*, 2006, **47**, 8333.
145. N. Tenn, N. Bellec, O. Jeannin, L. Piekara-Sady, P. Auban-Senzier, J. Íñiguez, E. Canadell and D. Lorcy, *J. Am. Chem. Soc.*, 2009, **131**, 16961.
146. G. D. Figuly, C. K. Loop and J. C. Martin, *J. Am. Chem. Soc.*, 1989, **111**, 654.
147. D. M. Giolando, T. B. Rauchfuss and A. L. Rheingold, *Inorg. Chem.*, 1987, **26**, 1636.
148. S. Ogawa, S. Yoshimura, N. Nagahora, Y. Kawai, Y. Mikata and R. Sato, *Chem. Commun.*, 2002, 1918.
149. M. S. Thomas, J. Darkwa, E. Y. Osei-Twum and L. A. Litorja Jr., *Polyhedron*, 1999, **18**, 2803.
150. P. Tavarès, M. M. Kubicki, P. Meunier and B. Gautheron, *Trans. Met. Chem.*, 1992, **17**, 220.
151. D. Coucouvanis, F. J. Hollander, R. West and D. Eggerding, *J. Am. Chem. Soc.*, 1974, **96**, 3006.

152. D. Coucouvanis, D. G. Holah and F. J. Hollander, *Inorg. Chem.*, 1975, **14**, 2657.
153. U. Drutkowski, B. Wenzel and P. Strauch, *Inorg. Chem Commun.*, 1998, **1**, 449.
154. U. Drutkowski, B. Wenzel, D. Tews and P. Strauch, *Z. Anorg. Allg. Chem.*, 2001, **627**, 1888.
155. T. Kemmitt, W. Levason, M. D. Spicer and M. Webster, *Organometallics*, 1990, **9**, 1181.
156. J. R. Black, N. R. Champness, W. Levason and G. Reid, *Inorg. Chem.*, 1996, **35**, 1820.
157. A. R. J. Genge, W. Levason and G. Reid, *J. Chem. Soc., Dalton Trans.*, 1997, 4549.
158. W. Levason, S. D. Orchard and G. Reid, *Organometallics*, 1999, **18**, 1275.
159. W. Levason, S. D. Orchard, G. Reid and V.-A. Tolhurst, *J. Chem. Soc., Dalton Trans.*, 1999, 2071.
160. E. S. Davies, R. L. Beddoes, D. Collison, A. Dinsmore, A. Docrat, J. A. Joule, C. R. Wilson and C. D. Garner, *J. Chem. Soc., Dalton Trans.*, 1997, 3985.
161. J. M. Tunney, A. J. Blake, E. S. Davies, J. McMaster, C. Wilson and C. D. Garner, *Polyhedron*, 2006, **25**, 591.
162. H. Sugimoto, M. Harihara, M. Shiro, K. Sugimoto, K. Tanaka, H. Miyake and H. Tsukube, *Inorg. Chem.*, 2005, **44**, 6386.
163. O. Jeannin, R. Clérac and M. Formigué, *J. Am. Chem. Soc.*, 2006, **128**, 14649.
164. L. Ambrosio, M. C. Aragoni, M. Arca, F. A. Devillanova, M. B. Hursthouse, S. L. Huth, F. Isaia, V. Lippolis, A. Mancini and A. Pintus, *Chem. – Asian J.*, 2010, **5**, 1395.
165. M. C. Aragoni, M. Arca, F. A. Devillanova, F. Isaia, V. Lippolis and A. Pintus, *Chem. – Asian J.*, 2011, **6**, 198.
166. S. A. Baudron, N. Avarvari and P. Batail, *Inorg. Chem.*, 2005, **44**, 3380.
167. Y. Gareau and A. Beuchemin, *Heterocycles*, 1998, **48**, 2003.
168. R. K. Brown, T. J. Bergendahl, J. S. Wood and J. H. Waters, *Inorg. Chim. Acta*, 1983, **68**, 79.
169. F. Wudl, E. T. Zellers and S. D. Cox, *Inorg. Chem.*, 1985, **24**, 2864.
170. H. Fujiwara, E. Ojima, H. Kobayashi, T. Courcet, I. Malfant and P. Cassoux, *Eur. J. Inorg. Chem.*, 1998, 1631.
171. Y. Gareau, M. Tremblay, D. Gavreau and H. Juteau, *Tetrahedron*, 2001, **57**, 5739.
172. S. Friedle, D. V. Partyka, M. V. Bennet and R. H. Holm, *Inorg. Chim. Acta*, 2006, **359**, 1427.
173. A. Davison and R. H. Holm, *Inorg. Synth.*, 1967, **10**, 8.
174. G. Steimecke, H. J. Sieler, R. Kirmse and E. Hoyer, *Phosphorus Sulfur*, 1979, **7**, 49.
175. R.-M. Olk, W. Dietzsch and E. Hoyer, *Synth. React. Inorg. Met.-Org. Chem.*, 1984, **14**, 915.

176. G. Matsubayashi and K. Akiba, *J. Chem. Soc., Dalton Trans.*, 1990, 115.
177. G. Matsubayashi and A. Yokozawa, *J. Chem. Soc., Dalton Trans.*, 1990, 3013.
178. G. Matsubayashi and A. Yokozawa, *J. Chem. Soc., Dalton Trans.*, 1990, 3535.
179. R.-M. Olk, W. Dietzsh, E. Hoyer, and G. Steimecke, *Patent DD236100*, 1986.
180. M. Kodani, K. Takimiya, Y. Aso, T. Otsubo, T. Nakayashiki and Y. Misaki, *Synthesis*, 2001, 1614.
181. E. Cerrada and M. Laguna, *Can. J. Chem.*, 1998, **76**, 1033.
182. R. Llusar, S. Triguero, V. Polo, C. Vicent, C. J. Gómez-García, O. Jeannin and M. Formigué, *Inorg. Chem.*, 2008, **47**, 9400.
183. R.-M. Olk, W. Dietzsch, R. Kirmse, J. Stack and E. Hoyer, *Inorg. Chim. Acta*, 1987, **128**, 251.
184. M. Nunn, D. B. Sowerby and D. M. Wesolek, *J. Organomet. Chem.*, 1983, **251**, C45.
185. M. Formigué and N. Avarvari, *Dalton Trans.*, 2005, 1365.
186. K. Hartke, T. Kissel, J. Quante and R. Matusa, *Chem. Ber.*, 1980, **113**, 1898.
187. J. P. Cornelissen, B. Pomarede, A. L. Spek, D. Reefman, J. P. Haasnoot and J. Reedijk, *Inorg. Chem.*, 1993, **32**, 3720.
188. P. J. Nigrey, *Synth. React. Inorg. Met. Org. Chem.*, 1986, **15**, 1351.
189. R.-M. Olk, A. Röhr, B. Olk and E. Hoyer, *Z. Chem.*, 1988, **28**, 304.
190. M. V. Lakshmikantham and M. P. Cava, *J. Org. Chem.*, 1980, **45**, 2632.
191. H. Poleschner, *Z. Chem.*, 1986, **26**, 138.
192. B. Olk, R.-M. Olk, J. Sieler and E. Hoyer, *Synth. Met.*, 1991, **42**, 2585.
193. H. Kobayashi, R. Kato and A. Kobayashi, *Synth. Met.*, 1991, **42**, 2495.
194. R.-M. Olk, B. Olk, J. Rohloff and E. Hoyer, *Z. Chem.*, 1990, **30**, 445.
195. H. Poleschner, R. Radeglia and J. Fuchs, *J. Organomet. Chem.*, 1992, **427**, 213.
196. R.-M. Olk, B. Olk, W. Dietzsch and E. Hoyer, *Z. Chem.*, 1989, **29**, 250.
197. N. Svenstrup and J. Becher, *Synthesis*, 1994, 215.
198. A. C. Brooks, P. Day, S. I. G. Dias, S. Rabaça, I. C. Santos, R. T. Henriques, J. D. Wallis and M. Almeida, *Eur. J. Inorg. Chem.*, 2009, 3084.
199. M. L. H. Green, W. B. Heuer and G. C. Saunders, *J. Chem. Soc., Dalton Trans.*, 1990, 3789.
200. A. Famengo, D. Pinero, O. Jeannin, T. Guizouarn and M. Formigué, *Dalton Trans.*, 2012, **41**, 1441.
201. A. Famengo, D. Pinero, O. Jeannin, T. Guizouarn, L. Pickara-Sady and M. Formigué, *New J. Chem.*, 2012, **36**, 638.
202. J. Becher, A. Hazell, C. J. McKenzie and C. Vestergaard, *Polyhedron*, 2000, **19**, 665.
203. S. Rabaça, M. C. Duarte, I. C. Santos, M. Fourmigué and M. Almeida, *Inorg. Chim. Acta*, 2006, **360**, 3797.

204. S. A. Baudron and M. W. Hosseini, *Inorg. Chem.*, 2006, **45**, 5260.
205. S. I. G. Dias, S. Rabaça, I. C. Santos, L. C. J. Pereira, R. T. Henriques and M. Almeida, *Inorg. Chem. Commun.*, 2012, **15**, 102.
206. C. L. Beswick, J. M. Schulman and E. I. Stiefel, *Progr. Inorg. Chem.*, 2004, **52**, 55.
207. T. Arliguie, P. Thuéry, M. Formigué and M. Ephritikhine, *Eur. J. Inorg. Chem.*, 2004, 4502.
208. M. Roger, T. Arliguie, P. Thuéry, M. Formigué and M. Ephritikhine, *Inorg. Chem.*, 2005, **44**, 584.
209. M. Roger, T. Arliguie, P. Thuéry, M. Formigué and M. Ephritikhine, *Inorg. Chem.*, 2005, **44**, 594.
210. M. Roger, L. Belkhiri, P. Thuéry, T. Arliguie, M. Formigué, A. Boucekkine and M. Ephritikhine, *Organometallics*, 2005, **24**, 4940.
211. L. Belkhiri, T. Arliguie, P. Thuéry, M. Formigué, A. Boucekkine and M. Ephritikhine, *Organometallics*, 2006, **25**, 2782.
212. E. M. Weis, C. L. Barnes and P. B. Duval, *Inorg. Chem.*, 2006, **45**, 10126.
213. K. Tomono, Y. Sasaki and K. Miyamura, *Acta Crystallogr.*, 2005, **E61**, m18.
214. X.-Q. Wang, W.-T. Yu, D. Xu, Y.-L. Wang, T.-B. Li, G.-H. Zhang, X.-B. Sun and Q. Ren, *Acta Crystallogr.*, 2005, **E61**, m717.
215. X. M. Ren, Z. P. Ni, S. Noro, T. Akutagawa, S. Nishihara, T. Nakamura, Y. X. Sui and Y. Song, *Cryst. Growth Des.*, 2006, **6**, 2530.
216. J. Lieffrig, O. Jeannin, P. Auban-Senzier and M. Formigué, *Inorg. Chem.*, 2012, **51**, 7144.
217. T. Akutagawa, K. Shitagami, S. Nishihara, S. Takeda, T. Hasegawa, T. Nakamura, Y. Hosokoshi, K. Inoue, S. Ikeuchi, Y. Miyazaki and K. Saito, *J. Am. Chem. Soc.*, 2005, **127**, 4397.
218. A. I. S. Neves, I. C. Santos, L. C. J. Pereira, C. Rovira, E. Ruiz, D. Belo and M. Almeida, *Eur. J. Inorg. Chem.*, 2011, 4807.
219. Q. Ye, T. Akutagawa, T. Endo, S.-I. Noro, T. Takamura and R.-G. Xiong, *Inorg. Chem.*, 2010, **49**, 8591.
220. T. Akutagawa, K. Matsuura, A. Hashimoto and T. Nakamura, *Inorg. Chem.*, 2005, **44**, 4454.
221. D. Belo, L. C. J. Pereira, M. Almeida, C. Rovira, J. Veciana and V. Gama, *Dalton Trans.*, 2009, 4176.
222. E. Coronado, J. C. Dias, M. C. Giménes-Saiz and C. J. Gómez-García, *J. Mol. Struct.*, 2008, **890**, 215.
223. S. Dorbes, L. Valade, J. A. Real and C. Faulmann, *Chem. Commun.*, 2005, 69.
224. C. Faulmann, K. Jacob, S. Dorbes, S. Lampert, I. Malfant, M.-L. Doublet, L. Valade and J. A. Real, *Inorg. Chem.*, 2007, **46**, 8548.
225. J. C. Dias, A. Soriano-Portillo, M. Clemente-Léon, C. Giménes-Saiz, J. R. Galán-Mascarós, C. J. Gómez-García and E. Coronado, *Inorg. Chim. Acta*, 2007, **360**, 3843.

226. G. C. Papavassiliou, G. C. Anyfatis, C. P. Raptopoulou, V. Psycharis, N. Ioannidis, V. Petrouleas and P. Paraskevopoulou, *Polyhedron*, 2009, **28**, 3368.
227. K. Wang, *Progr. Inorg. Chem.*, 2004, **52**, 267.
228. D. Sellmann, H. Binder, D. Häussinger, F. W. Heinemann and J. Sutter, *Inorg. Chem. Acta*, 2000, **300**, 829.
229. X. Ribas, J. C. Dias, J. Morgado, K. Wurst, I. C. Santos, M. Almeida, J. Vidal-Gancedo, J. Veciana and C. Rovira, *Inorg. Chem.*, 2004, **43**, 3631.
230. D. Sellmann, K. Hein and F. W. Heinemann, *Inorg. Chim. Acta*, 2004, **357**, 3739.
231. N. D. Lowe and C. D. Garner, *J. Chem. Soc., Dalton Trans.*, 1993, 3333.
232. F. L. Benedito, T. Petrenko, E. Bill, T. Weyhermüller and K. Wieghardt, *Inorg. Chem.*, 2009, **48**, 10913.
233. K.-S. Shin, Y. J. Jung, S.-K. Lee, M. Formigué, F. Barrière, J.-F. Bergamini and D.-Y. Noh, *Dalton Trans.*, 2008, 5869.
234. A. Vogler and H. Kunkley, *J. Am. Chem. Soc.*, 1981, **103**, 1559.
235. J. A. Zuleta, J. M. Bevilacqua, D. M. Proserpio, P. D. Harvey and R. Eisenberg, *Inorg. Chem.*, 1992, **31**, 2396.
236. J. M. Bevilacqua and R. Eisenberg, *Inorg. Chem.*, 1994, **33**, 2913.
237. S. D. Cummings and R. Eisenberg, *J. Am. Chem. Soc.*, 1996, **118**, 1949.
238. C.-T. Chen, T.-Y. Jeremy Lin, C.-H. Chen and K.-J. Lin, *J. Chin. Chem. Soc.*, 2000, **47**, 197.
239. W.-Y. Guo, Z.-H. Peng, C.-G. Wang and Y.-H. Zhou, *J. Coord. Chem.*, 2006, **59**, 1213.
240. H. Sugimoto, M. Tarumizu, K. Tanaka, H. Miyake and H. Tsukube, *Dalton Trans.*, 2005, 3558.
241. M. Nomura, T. Cauchy and M. Formigué., *Coord. Chem. Rev.*, 2010, **254**, 1406.
242. M. Nomura, T. Cauchy, M. Geoffroy, P. Adkine and M. Formigué, *Inorg. Chem.*, 2006, **45**, 8194.
243. M. Nomura, M. Geoffroy, P. Adkine and M. Formigué, *Eur. J. Inorg. Chem.*, 2006, 5012.
244. Y. Ji, J.-L. Zuo, L. Chen, Y.-Q. Tian, Y. Song, Y.-Z. Li and X.-Z. You, *J. Phys. Chem. Sol.*, 2005, **66**, 207.
245. J. Morgado, I. C. Santos, M. T. Duarte, L. Alcacer and M. Almeida, *Chem. Commun.*, 1996, 1837.
246. S. Rabaça, A. C. Cerdeira, A. I. S. Neves, S. I. G. Dias, C. Mézière, I. C. Santos, L. C. J. Pereira, M. Formigué, R. T. Henriques and M. Almeida, *Polyhedron*, 2009, **28**, 1069.
247. M. Nomura and M. Formigué, *Inorg. Chem.*, 2008, **47**, 1301.
248. K. Arumugam, M. C. Shaw, P. Chandrasekaram, D. Villagrán, T. G. Gray, J. T. Mague and J. P. Donahue, *Inorg. Chem.*, 2009, **48**, 10591.
249. K. Arumugam, R. Yu, D. Villagrán, T. G. Gray, J. T. Mague and J. P. Donahue, *Inorg. Chem.*, 2008, **47**, 5570.
250. M. Nomura and M. Formigué, *Inorg. Chem.*, 2007, **46**, 1301.

251. M. Nomura, S. Kondo, S. Yamashita, E. Suzuki, Y. Toyota, G. V. Alea, G. C. Janairo, C. Fujita-Takayama, T. Sugiyama and M. Kajitani, *J. Organomet. Chem.*, 2010, **695**, 2366.
252. M. Nomura, T. Fujii and M. Kajitani, *Otganometallics*, 2009, **28**, 3776.
253. N. Lardiés, E. Cerrada and M. Laguna, *Polyhedron*, 2006, **25**, 2785.
254. A. Pintus, M. C. Aragoni, N. Bellec, F. A. Devillanova, D. Lorcy, F. Isaia, V. Lippolis, R. A. M. Randall, T. Roisnel, A. M. Z. Slawin, J. D. Woollins and M. Arca, *Eur. J. Inorg. Chem.*, 2012, 3577–3594.
255. Y. Zhang, K. D. Ley and K. S. Schanze, *Inorg. Chem.*, 1996, **35**, 7102.
256. B. W. Smucker, J. M. Hudson, M. A. Omary and K. R. Dunbar, *Inorg. Chem.*, 2003, **42**, 4714.
257. Y. Ji, R. Zhang, X.-B. Du, J.-L. Zuo and X.-Z. You, *Dalton Trans.*, 2008, 2578.
258. S. Rabaça, B. J. C. Vieira, R. Meira, I. C. Santos, L. C. J. Pereira, M. T. Duarte and V. da Gama, *Eur. J. Inorg. Chem.*, 2008, 3839.
259. H. Fujiwara, E. Arai and H. Kobayashi, *Chem. Commun.*, 1997, 837.
260. D. M. Smith, T. E. Albrecht-Schmitt and J. A. Ibers, *Angew. Chem. Int. Ed.*, 1998, **37**, 1089.
261. C. C. McLauchlan and J. A. Ibers, *Inorg. Chem.*, 2000, **39**, 1046.
262. C. C. McLauchlan, S. D. Robowski and J. A. Ibers, *Inorg. Chem.*, 2001, **40**, 1372.
263. C. C. McLauchlan and J. A. Ibers, *Inorg. Chem.*, 2001, **40**, 1809.
264. G. C. Anyfantis, G. C. Papavassiliou, A. Terzis, C. P. Raptopoulou, V. Psycharis and P. Paraskevopoulou, *Polyhedron*, 2010, **29**, 969.
265. G. C. Anyfantis, G. C. Papavassiliou, N. Assimomytis, A. Terzis, V. Psycharis, C. P. Raptopoulou, P. Kyritsis, V. Thoma and I. B. Koutselas, *Solid State Sci.*, 2008, **10**, 1729.
266. G. C. Papavassiliou, G. C. Anyfantis, B. R. Steele, A. Terzis, C. P. Raptopoulou, G. Tatakis, G. Chaidogiannis, N. Glezos, Y. Weng, H. Yoshino and K. Murata, *Z. Naturforsch.*, 2007, **62b**, 679.
267. N. J. Harris and A. E. Underhill, *J. Chem. Soc., Dalton Trans.*, 1987, 1683.
268. W. B. Heuer, A. E. True, P. N. Swepston and B. M. Hoffman, *Inorg. Chem.*, 1988, **27**, 1474.
269. U. Zoller, *Compr. Heterocycl. Chem. II*, 1996, **1B**, 1113.
270. G. N. Schrauzer, V. P. Mayweg and W. Heinrich, *Inorg. Chem.*, 1965, **4**, 1615.
271. G. N. Schrauzer and V. P. Mayweg, *J. Am. Chem. Soc.*, 1965, **87**, 1483.
272. K. Arumugam, J. E. Bolliger, M. Fink and J. Donahue, *Inorg. Chem.*, 2007, **46**, 3283.
273. S. Sheibe, B. J. Pedersen and S.-O. Lawesson, *Bull. Soc. Chim. Belg.*, 1978, **87**, 229.
274. R. Perochon, P. Davidson, S. Rouzière, F. cameral, L. Piekara-Sady, T. Guizouarn and M. Formigué, *J. Mater. Chem.*, 2011, **21**, 1416.

275. T.-T. Bui, O. Thiebaut, E. Grelet, M.-F. Achard, B. Garreau-de Bonneval and K. I. Moineau-Chane Ching, *Eur. J. Inorg. Chem.*, 2011, 2663.
276. R. Perochon, L. Piekara-Sady, W. Jurga, R. Clérac and M. Formigué, *Dalton Trans.*, 2009, 3052.
277. J.-Y. Cho, B. Domercq, S. C. Jones, J. Yu, X. Zhang, Z. An, M. Bishop, S. Barlow, S. R. Marder and B. Kippelen, *J. Mater. Chem.*, 2007, **17**, 2642.
278. F. Bigoli, P. Deplano, F. A. Devillanova, J. R. Ferraro, V. Lippolis, P. J. Lukes, M. L. Mercuri, M. A. Pellinghelli, E. F. Trogu and J. M. Williams, *Inorg. Chem.*, 1997, **36**, 1218.
279. M. Arca, F. Demartin, F. A. Devillanova, A. Garau, F. Isaia, F. Lelj, V. Lippolis, S. Pedraglio and G. Verani, *Dalton Trans.*, 1998, 3731.
280. M. C. Aragoni, M. Arca, T. Cassano, C. Denotti, F. A. Devillanova, R. Frau, F. Isaia, F. Lelj, V. Lippolis, L. Nitti, P. Romaniello, R. Tommasi and G. Verani, *Eur. J. Inorg. Chem.*, 2003, 1939.
281. M. C. Aragoni, M. Arca, F. A. Devillanova, F. Isaia, V. Lippolis, A. Mancini, L. Pala, A. M. Z. Slawin and J. D. Woollins, *Inorg. Chem.*, 2005, **44**, 9610.
282. L. H. Song, C. H. Rhee, S. H. Park, S. L. Lee, D. Grudinin, K. H. Song and J. Choe, *Org. Process Res. Dev.*, 2008, **12**, 1012.
283. R. Wang, L.-C. Kang, J. Xiong, X.-W. Dou, X.-Y. Chen, J.-L. Zuo and X.-Z. You, *Dalton Trans*, 2011, **40**, 919.
284. V. Madhu and S. K. Das, *Dalton Trans.*, 2011, **40**, 12901.
285. C. M. Bolinger and T. B. Rauchfuss, *Inorg. Chem.*, 1982, **21**, 3947.
286. S. Ford, M. R. Lewtas, C. P. Morley and M. Di Vaira, *Eur. J. Inorg. Chem.*, 2000, 933.
287. S. Nagao, H. Seino, T. Okada, Y. Mizobe and M. Hidai, *J. Chem. Soc., Dalton Trans.*, 2000, 3546.
288. M. A. Ansari, C. H. Mahler and J. A. Ibers, *Inorg. Chem.*, 1989, **28**, 2669.
289. T. B. Rauchfuss, *Inorg. Chem.*, 2004, **43**, 14.
290. H. Sugimoto, K. Suyama, K. Sugimoto, H. Miyake, I. Takahashi, S. Hirota and S. Itoh, *Inorg. Chem.*, 2008, **47**, 10150.
291. W.-H. Chiu, Q.-F. Zhang, I. D. Williams and W.-H. Leung, *Organometallics*, 2010, **29**, 2631.
292. O. Jeannin, M. Nomura and M. Formigué, *J. Organomet. Chem.*, 2007, 4113.
293. W. A. Flomer and J. W. Kolis, *Inorg. Chem.*, 1989, **28**, 2513.
294. D. M. Smith, L. C. Roof, M. A. Ansari, J. M. McConnachie, J. C. Bollinger, M. A. Pell, R. J. Salm and J. A. Ibers, *Inorg. Chem.*, 1996, **35**, 4999.
295. P. Mathur, A. K. Dash, Md.M. Hossain and S. B. Umbarkar, *Organometallics*, 1996, **15**, 1356.
296. Th. Fässler, D. Buchholz, G. Huttner and L. Zsolnai, *J. Organomet. Chem.*, 1989, **369**, 297.
297. P. Mathur, Md.M. Hossain and S. Umbarkar, *Orgonometallics*, 1995, **14**, 959.

298. P. Mathur, Md.M. Hossain, S. N. Datta, R.-K. Kondru and M. N. Bhadbhade, *Organometallics*, 1994, **13**, 2532.
299. P. Mathur, S. Ghosh, A. Sarkar and C. Thöne, *Organometallics*, 1998, **17**, 3926.
300. P. Mathur, S. Ghosh, A. Sarkar, C. V. V Satyanarayana and V. G. Puranik, *Organometallics*, 1997, **16**, 4392.
301. P. Mathur, S. Ghosh, S. B. Umbarkar, M. G. Walawalkar and V. G. Puranik, *J. Organomet. Chem.*, 1999, **578**, 150.
302. P. Mathur, S. Ghosh, A. Sarkar, C. V. V Satyanarayana, A. L. Rheingold and L. M. Liable-Sands, *Organometallics*, 1997, **16**, 3536.
303. P. Mathur, S. Hosh, A. Sarkar, A. L. Rheingold and I. A. Guzei, *Tetrahedron*, 2000, **56**, 4995.
304. R. D. Adams and S. Miao, *Inorg. Chem.*, 2004, **43**, 8414.
305. H. Adams, M. J. Morris, A. E. Riddiough, L. J. Yellowless and A. B. P. Lever, *Inorg. Chem.*, 2007, **46**, 9790.
306. S. Shibahara, H. Kitagawa, Y. Ozawa, K. Toriumi, T. Kubo and K. Nakasuji, *Inorg. Chem.*, 2007, **46**, 1162.
307. K. Baba, T.-A. Okamura, C. Suzuki, H. Yamamoto, T. Yamamoto, M. Ohama and N. Ueyama, *Inorg. Chem.*, 2006, **45**, 894.
308. C. P. Morley, R. R. Vaughan and B. J. Wheatley, *J. Organomet. Chem.*, 1988, **353**, C39.
309. C. P. Morley and R. R. Vaughan, *J. Organomet. Chem.*, 1993, **444**, 219.
310. C. P. Morley and R. R. Vaughan, *J. Chem. Soc., Dalton Trans.*, 1993, 703.
311. C. P. Morley, C. A. Webster, P. Douglas, K. Rofe and M. Di Vaira, *Dalton Trans*, 2010, **39**, 3177.
312. S. Ford, P. K. Khanna, C. P. Morley and M. Di Vaira, *J. Chem. Soc., Dalton Trans.*, 1999, 791.
313. S. Ford, C. P. Morley and M. di Vaira, *New J. Chem.*, 1999, **23**, 811.
314. S. Ford, C. P. Morley and M. di Vaira, *Inorg. Chem.*, 2004, **43**, 7101.
315. R. D. McCullough, G. B. Kok, K. A. Lerstrup and D. O. Cowan, *J. Am. Chem. Soc.*, 1987, **109**, 4115.
316. R. D. McCullough, J. A. Belot, J. Seth, A. L. Rheingold, G. P. A. Yap and D. O. Cowan, *J. Mater. Chem.*, 1995, **5**, 1581.
317. S. J. N. Burgmayer, *Progr. Inorg. Chem.*, 2004, **52**, 491.
318. J. McMaster, J. M. Tunney and C. D. Garner, *Progr. Inorg. Chem.*, 2004, **52**, 539.
319. J. H. Enemark, J. J. A. Cooney, J.-J. Wang and R. H. Holm, *Chem. Rev.*, 2004, **104**, 1175.
320. K. Baba, T.-A. Okamura, H. Yamamoto, T. Yamamoto, M. Ohama and N. Ueyama, *Chem. Lett.*, 2005, **34**, 44.
321. H. Oku, N. Ueyama, M. Kondo and A. Nakamura, *Inorg. Chem.*, 1994, **33**, 209.
322. N. Ueyama, H. Oku and A. Nakamura, *J. Am. Chem. Soc.*, 1992, **114**, 7310.
323. A. Döring, C. Shulzke and Q. Zhang, *Inorg. Chim. Acta*, 2010, **363**, 4140.
324. J. Jiang and R. H. Holm, *Inorg. Chem.*, 2004, **43**, 1302.

325. H. Sugimoto, H. Tano, K. Suyuma, T. Kobayashi, H. Miyake, S. Itoh, R. Mtei and M. L. Kirk, *Dalton Trans.*, 2011, **40**, 1119.
326. K. Baba, T. Okamura, H. Yamamoto, T. Yamamoto, M. Ohama and N. Ueyama, *Chem. Lett.*, 2005, **34**, 44.
327. H. Sugimoto, S. Tatemoto, K. Suyuma, H. Miyake, S. Itoh, C. Dong, J. Yang and M. L. Kirk, *Inorg. Chem.*, 2009, **48**, 10581.
328. A. L. Whalley, A. J. Blake, D. Collison, E. S. Davies, H. J. Disley, M. Helliwell, F. E. Mabbs, J. McMaster, C. Wilson and C. D. Garner, *Dalton Trans.*, 2011, **40**, 10457.
329. A. L. Tan, A. J. Blake, E. S. Davies, C. Wilson and C. D. Garner, *Chem. Commun.*, 2011, **47**, 953.
330. L. Golic, W. Dietzsch, K. Kohler, J. Stach and R. Kirmse, *J. Chem. Soc., Dalton Trans.*, 1988, 97.
331. U. Drutkowski and P. Strauch, *Inorg. Chem. Commun.*, 2001, **4**, 342.
332. N. Nakagawa, T. Yamada, M. Sugimoto and H. Nishihara, *Inorg. Chem.*, 2006, **45**, 14.
333. D. J. Harrison, A. G. De Crisci, A. J. Lough, M. J. Kerr and U. Fekl, *Inorg. Chem.*, 2008, **47**, 10199.
334. R.-M. Olk, W. Dietzsch, J. Kahlmeier, P. Jörchel, R. Kirmse and J. Sieler, *Inorg. Chim. Acta*, 1997, **254**, 375.
335. F. Bigoli, P. Cassoux, P. Deplano, M. L. Mercuri, M. A. Pellinghelli, G. Pintus, A. Serpe and E. F. Trogu, *J. Chem. Soc., Dalton Trans.*, 2000, 4639.
336. K. Kubo, A. Nakao, H. M. Yamamoto and R. Kato, *J. Am. Chem. Soc.*, 2006, **128**, 12358.
337. S. Dalgleish, C. A. Morrison, D. S. Middlemiss, A. R. Mount, A. Collins, L. Pilia, A. Serpe, M. L. Mercuri, S. J. Roberts-Bleming, A. Charlton, P. Deplano, P. J. Murphy and N. Robertson, *J. Mat. Chem.*, 2009, **19**, 6194.
338. D. Belo and M. Almeida, *Coord. Chem. Rev.*, 2010, **254**, 1479.
339. T. Anjos, S. J. Roberts-Bleming, A. Charlton, N. Robertson, A. R. Mount, S. J. Coles, M. B. Hursthouse, M. Kalaji and P. J. Murphy, *J. Mater. Chem.*, 2008, **18**, 475.
340. S. Dalgleish, J. B. Labram, Z. Li, J. Wang, C. R. McNeill, T. D. Anthopoulos, N. C. Greenham and N. Robertson, *J. Mater. Chem.*, 2011, **21**, 15422.
341. S. Dalgleish and N. Robertson, *Coord. Chem. Rev.*, 2010, **254**, 1549.
342. P. J. Skabara, C. Pozo-Gonzalo, N. L. Miazza, M. Laguna, E. Cerrada, A. Luquin, B. Gonzáles, S. J. Coles, M. B. Hursthouse, R. W. Harrington and W. Clegg, *Dalton Trans.*, 2008, 3070.
343. P. Batail, K. Boubekeur, M. Formigué and J.-C. P. Gabriel, *Chem. Mater.*, 1998, **10**, 3005.
344. L. N. Dawe, J. Miglioi, L. Turnbow, M. L. Taliaferro, W. M. Shum, J. D. Bagnato, L. N. Zakharov, A. L. Rheingold, A. M. Arif, M. Formigué and J. S. Miller, *Inorg. Chem.*, 2005, **44**, 7530.
345. A. Ranganathan, A. El-Ghavyoury, L. Zorina and P. Batail, *CrystEngComm.*, 2010, **12**, 4268.

346. R. Bolligarla and S. K. Das, *CrystEngComm*, 2010, **12**, 3409.
347. S. Rabaça, D. Belo, A. C. Cerdeira, S. I. G. Dias, M. B. C. Branco, L. C. J. Pereira, I. C. Santos, M. Formigué and M. Almeida, *CrystEngComm*, 2009, **11**, 2154.
348. R. Eisemberg, J. A. Ibers, R. J. H. Clark and H. B. Gray, *J. Am. Chem. Soc.*, 1964, **86**, 113.
349. R. Eisemberg and J. A. Ibers, *Inorg. Chem.*, 1965, **4**, 605.
350. D. Sarain and M. R. Truter, *J. Chem Soc. A*, 1967, 1264.
351. R. P. Burns and C. A. McAuliffe, *Adv. Inorg. Chem. Radiochem.*, 1979, **22**, 303.
352. F.-Y. Dong, J.-M. Dou, D.-C. Li, X.-K. Gao and D.-Q. Wang, *J. Mol. Struct.*, 2005, **738**, 79.
353. P. Deplano, L. Pilia, D. Espa, M. L. Mercuri and A. Serpe, *Coord. Chem. Rev.*, 2010, **254**, 1434.
354. M. C. Aragoni, M. Arca, C. Denotti, F. A. Devillanova, E. Grigiotti, F. Isaia, F. Laschi, V. Lippolis, L. Pala, A. M. Z. Slawin, P. Zanello and J. D. Woollins, *Eur. J. Inorg. Chem.*, 2003, 1291.
355. D. Espa, L. Pilia, L. Marchiò, F. Artizzu, A. Serpe, M. L. Mercuri, D. Simão, M. Almeida, M. Pizzotti, F. Tessore and P. Deplano, *Dalton Trans.*, 2012, **41**, 3485.
356. D. Espa, L. Pilia, L. Marchiò, M. L. Mercuri, A. Serpe, A. Barsella, A. Fort, S. J. Dalgleish, N. Robertson and P. Deplano, *Inorg. Chem.*, 2011, **50**, 2058.
357. L. Pilia, D. Espa, A. Barsella, A. Fort, C. Makedonas, L. Marchiò, M. L. Mercuri, A. Serpe, C. A. Mitsopoulou and P. Deplano, *Inorg. Chem.*, 2011, **50**, 10015.
358. P. Chandrasekaram, K. Arumugam, U. Jayarathne, L. M. Péerez, J. T. Mague and J. P. Donahue, *Inorg. Chem.*, 2009, **48**, 2103.
359. S. Tsukada, Y. Shibata, R. Sakamoto, T. Kambe, T. Ozeki and H. Nishihara, *Inorg. Chem.*, 2012, **51**, 1228.
360. K. Pal, S. Takamizawa and K. Mashima, *Inorg. Chim. Acta*, 2011, **273**, 68.
361. Y. Li, Q. Jiang, Y. Li, X. Shen, H. Yan and V. I. Bregadze, *Inorg. Chem.*, 2010, **49**, 5584.
362. A. Majumdar, J. Mitra, K. Pal and S. Sarkar, *Inorg. Chem.*, 2008, **47**, 3393.
363. B. K. Maiti, K. Pal and S. Sarkar, *Dalton Trans.*, 2008, 1003.
364. P. Ghosh, E. Bill, T. Weyhermüller and K. Wieghardt, *Inorg. Chem.*, 2007, **46**, 2612.
365. A. C. Cerdeira, D. Simão, I. C. Santos, A. Machado, L. C. J. Pereira, J. C. Waerenborgh, R. T. Henriques and M. Almeida, *Inorg. Chim. Acta*, 2008, **361**, 3836.
366. S. Sproules and K. Wieghardt, *Coord. Chem. Rev.*, 2010, **254**, 1358.
367. K. W. Browall, T. Bursh, L. V. Interrante and J. S. Kasper, *Inorg. Chem.*, 1972, **11**, 1800.

368. X. M. Ren, T. Akutagawa, S. Nishihara and T. Nakamura, *Synth. Met*, 2005, **150**, 57.
369. M. L. Mercuri, P. Deplano, L. Pilia, A. Serpe and F. Artizzu, *Coord. Chem. Rev.*, 2010, **254**, 1419.
370. W.-B. Pei, J.-S. Wu, X.-M. Ren, Z.-F. Tian and J. Xie, *Dalton Trans.*, 2012, **41**, 2667.
371. M. Formigué, T. Cauchy and M. Nomura, *CrystEngComm*, 2009, **11**, 1491.
372. O. Jeannin, R. Clérac and M. Formigué, *Chem. Mater.*, 2007, **19**, 5946.
373. Y. Umezono, W. Fujita and K. Awaga, *J. Am. Chem. Soc.*, 2006, **128**, 1084.
374. C. M. Liu, D.-Q. Zhang, Y.-L. Song, C.-L. Zhan, Y.-L. Li and D. B. Zhu, *Eur. J. Inorg. Chem.*, 2002, 1591.
375. Q.-H. Wang, D.-L. Long, H.-M. Hu, Y. Cui and J.-S. Huang, *J. Coord. Chem.*, 2000, **49**, 201.
376. L.-L. Wen, Z.-H. Peng, B.-G. Zhang and Y.-H. Zhou, *J. Mol. Struct.*, 2004, **694**, 223.
377. J. S. Pap, F. L. Benedito, E. Bothe, E. Bill, S. DeBeer George, T. Weyhermuller and K. Wieghardt, *Inorg. Chem.*, 2007, **46**, 4187.
378. J.-L. Zuo, R.-G. Xiong, X.-Z. You and X.-Y. Huang, *Inorg. Chim. Acta*, 1995, **237**, 117.
379. R. Świetlik, D. Jankowski, M. Formigué and K. Yakushi, *Vib. Spectrosc.*, 2011, **55**, 195.
380. T. Akutagawa, K. Shitagami, M. Aonuma, S. Noro and T. Nakamura, *Inorg. Chem.*, 2009, **48**, 4454.
381. M. C. Aragoni, M. Arca, F. Demartin, F. A. Devillanova, F. Lelj, F. Isaia, V. Lippolis, A. Mancini, L. Pala and G. Verani, *Eur. J. Inorg. Chem.*, 2004, 3099.
382. N. Avarvari, M. Formigué and E. Canadell, *Eur. J. Inorg. Chem.*, 2004, 3409.
383. R. B. King, *J. Organomet. Chem.*, 2001, **623**, 95.
384. X. Ribas, J. C. Dias, J. Morgado, K. Wurst, M. Almeida, T. Parella, J. Veciana and C. Rovira, *Angew. Chem. Int. Ed.*, 2004, **13**, 4049.
385. E. I. Stiefel and G. F. Brown, *Inorg. Chem.*, 1972, **11**, 434.
386. G. F. Brown and E. I. Stiefel, *Inorg. Chem.*, 1973, **12**, 2140.
387. L. K. Marshall, B. Shudel and I. A. Lippa, *SPIE – Int- Soc. Opt. Eng.*, 2003, **5213**, 201.
388. C.-T. Chou, Y.-F. Pai, C.-C. Lin, T. K. Misra and C.-Y. Liu, *J. Cromatogr. A*, 2004, **1043**(2), 255.
389. H. Horie, A. Takagi, H. Hasebe, T. Ozawa and K. Ohta, *J. Mater. Chem.*, 2001, **11**, 1063.
390. Z. Ni, X. Ren, J. Ma, J. Xie, C. Ni, Z. Chen and Q. Meng, *J. Am. Chem. Soc.*, 2005, **127**, 14330.
391. W. J. Koro and R. L. Burns, *Patent EP 1468724*, *CA 246541*, *US 2005000899*, 2004.

392. A. Begum, S. K. Sonkar, M. Saxena and S. Sarkar, *J. Mater. Chem.*, 2011, **21**, 19210.
393. W. R. McNamara, Z. Han, P. J. Alperin, W. W. Brennessel, P. L. Holland and R. Eisenberg, *J. Am. Chem. Soc.*, 2011, **133**, 15368.
394. S. P. Tiwari, W. J. Potscavage Jr., T. Sajoto, S. Barlow, S. R. Marder and B. Kippelen, *Org. Electron.*, 2010, **11**, 860.
395. T. D. Anthopoulos, G. C. Anyfantis, G. C. Papavassiliou and D. M. de Leeuw, *Appl. Phys. Lett.*, 2007, **90**, 122105.
396. H. Wada, T. Taguchi, B. Noda, T. Kambayashi, T. Mori, K. Ishikawa and H. Takezoe, *Org. Electron.*, 2007, **8**, 759.
397. T. D. Anthopoulos, S. Setayesh, E. Smits, M. Cölle, E. Cantatore, B. de Boer, P. W. M. Blom and D. M. de Leeuw,, *Adv. Mater.*, 2006, **18**(1900).
398. E. C. P. Smits, T. D. Anthopoulos, S. Setayesh, E. van Veenendaal, R. Coehoorn, P. W. M. Blom, B. de Boer and D. M. de Leeuw, *Phys. Rev. B*, 2006, **73**, 205316–1.
399. Y. Qi, T. Sajoto, M. Kröger, A. M. Kandabarow, W. Park, S. Barlow, E.-G. Kim, L. Wielunski, L. C. Feldman, R. A. Bartynski, J.-L. Brédas, S. R. Marder and A. Kahn, *Chem. Mater.*, 2010, **22**, 524.
400. Y. Qi, T. Sajoto, S. Barlow, E.-G. Kim, J.-L. Brédas, S. R. Marder and A. Kahn, *J. Am. Chem. Soc.*, 2009, **131**, 12530.
401. Y. Kobayashi, B. Jacobs, M. D. Allendorf and J. R. Long, *Chem. Mater.*, 2010, **22**, 4120.
402. B. W. Smucker, J. M. Hudson, M. A. Omary and K. R. Dunbar, *Inorg. Chem.*, 2003, **42**, 4714.
403. C. Faulmann and P. Cassoux, *Progr. Inorg. Chem.*, 2004, **52**, 399.
404. R. Kato, *Chem. Rev.*, 2004, **104**, 5319.
405. R. Llusar and C. Vicent, *Coord. Chem. Rev.*, 2010, **254**, 1534.
406. H. M. McConnell and R. Lynden-Nell, *J. Chem. Phys.*, 1962, **36**, 2393.
407. J. W. Bray, H. R. Hart Jr, L. V. Interrante, I. S. Jacobs, J. S. Kasper, G. D. Watkins, S. H. Wee and J. C. Bonner, *Phys. Rev. Lett.*, 1975, **35**, 744.
408. I. S. Jacobs, J. W. Bray, H. R. Hart Jr, L. V. Interrante, J. S. Kasper, G. D. Watkins, D. E. Prober and J. C. Bonner, *Phys. Rev. B*, 1976, **14**, 3036.
409. J. M. Williams, J. R. Ferrero, R. J. Thorn, K. D. Carlson, U. Geiser, H. H. Wang, A. M. Kini, and M.-H. Wangbo, in *Organic Superconductors (Including Fullerenes)*, Prentice Hall, Englewood Cliffs, NJ, 1992.
410. L. Alcacer, H. Novais, F. Pedroso, S. Flandrois, C. Coulon, D. Chasseau and J. Gaultier, *Solid State Commun.*, 1980, **35**, 945.
411. R. D. McDonald, N. Harrison, J. Singleton, A. Bangura, P. A. Goddard, A. P. Ramirez and X. Chi, *Phys. Rev. Lett.*, 2005, **94**, 106404.
412. M. M. Ahmad, D. J. Turner, A. E. Underhill, C. S. Jacobsen, K. Mortensen and K. Carneiro, *Phys. Rev. B*, 1984, **29**, 4796.
413. See for exampleJ. C. Dias, X. Ribas, J. Morgado, J. Seiça, E. B. Lopes, I. C. Santos, R. T. Henriques, M. Almeida, K. Wurst,

P. Foury-Leylekian, E. Canadell, J. Vidal-Gancedo, J. Veciana and C. Rovira, *J. Mater. Chem.*, 2005, **15**, 3187
414. A. Kobayashi, H. Tanaka, M. Kumasaki, H. Torii, B. Narymbetov and T. Adachi, *J. Am. Chem. Soc.*, 1999, **121**, 10763.
415. W. B. Heuer, P. J. Squattrito, B. M. Hoffman and J. A. Ibers, *J. Am. Chem. Soc.*, 1988, **110**, 792.
416. L. Valade, M. Bousseau, A. Gleizes and P. Cassoux, *J. Chem. Soc., Chem. Commun.*, 1983, 110.
417. R. Kato, H. Kobayashi, H. Kim, A. Kobayashi, Y. Sasaki, T. Mori and H. Inokuchi, *Chem. Lett.*, 1988, 865.
418. L. Brossard, M. Ribault, M. Bousseau, L. Valade and P. Cassoux, *C. R. Acad. Sci., Ser. II*, 1986, **302**, 205.
419. D. de Caro, K. Jacob, C. Faulmann, J.-P. Legros, F. Senocq, J. Fraxedas and L. Valade, *Synth. Metals*, 2010, **160**, 1223.
420. Ten bis(1,2-dithiolene) complexes exhibiting superconducting properties have been reported to date, namely (TTF)[Ni(dmit)$_2$], α-(EDT-TTF)-[Ni(dmit)$_2$], (Me$_4$N)[Ni(dmit)$_2$]$_2$, α′-(TTF)[Pd(dmit)$_2$]$_2$, α-(TTF)[Pd(dmit)$_2$]$_2$, β-(Me$_4$N)[Pd(dmit)$_2$]$_2$, (Et$_2$Me$_2$N)[Pd(dmit)$_2$]$_2$, β′-(Et$_2$Me$_2$P)[Pd(dmit)$_2$]$_2$, β′-(Me$_4$Sb)[Pd(dmit)$_2$]$_2$, β ′-(Me$_4$As)[Pd(dmit)$_2$]$_2$. See refs. 418 and 421–429.
421. H. Tajima, M. Inokuchi, A. Kobayashi, T. Ohta, R. Kato, H. Kobayashi and H. Kuroda, *Chem. Lett.*, 1993, 1235.
422. A. Kobayashi, H. Kim, Y. Sasaki, R. Kato, H. Kobayashi, S. Moriyama, Y. Nishio, K. Kajita and W. Sasaki, *Chem. Lett.*, 1987, 1819.
423. L. Brossard, H. Hurdequint, M. Ribault, L. Valade, L.-P. Legros and P. Cassoux, *Synth. Met*, 1988, **27**, B157.
424. L. Brossard, M. Ribault, L. Valade and P. Cassoux, *J. Phys. (Paris)*, 1989, **50**, 1521.
425. A. Kobayashi, H. Kobayashi, A. Miyamoto, R. Kato, R. A. Clark and A. E. Underhill, *Chem. Lett.*, 1991, 2163.
426. H. Kobayashi, K. Bun, T. Naito, R. Kato and A. Kobayashi, *Chem. Lett.*, 1992, 1909.
427. R. Kato, Y. Kashimura, S. Aonuma, N. Hanasaki and H. Tajima, *Solid State Commun.*, 1998, **105**, 561.
428. See Y. Kashimura, Doctoral Thesis, The University of Tokyo, 1998, and A. Eguchi, M. Tamura, Y. Nishio, and K. Kajita, unpublished results.
429. R. Kato, N. Tajima, M. Tamura and J.-I. Yamaura, *Phys. Rev. B*, 2002, **66**, 020508.
430. A. Tajima, A. Nakao and R. Kato, *J. Phys. Soc. Jpn.*, 2005, **74**, 412.
431. T. Itou, O. Oyamada, S. Maegawa, K. Kubo, H. M. Yamamoto and R. Kato, *Phys. Rev. B*, 2009, **79**, 174517.
432. M. Tamura and R. Kato, *Sci. Technol. Adv. Mater.*, 2009, **10**, 024304.
433. E. Fujiwara, K. Hosoya, A. Kobayashi, H. Tanaka, M. Takumoto, Y. Okano, H. Fujiwara, H. Kobayashi, Y. Fujishiro, E. Nishibori, M. Tanaka and M. Sakata, *Inorg. Chem.*, 2008, **47**, 863.
434. M. Shatruk and L. Ray, *Dalton Trans.*, 2010, **39**, 11105.

435. D. Belo, H. Alves, E. B. Lopes, M.T Duarte, V. Gama, R. T. Henriques, M. Almeida, A. Pérez-Benítez, C. Rovira and J. Veciana, *Chem. – Eur. J*, 2001, **7**, 511.
436. H. Tanaka, Y. Okano, H. Kobayashi, W. Suzuki and A. Kobayashi, *Science*, 2001, **291**, 285.
437. A. Kobayashi, H. Tanaka and H. Kobayashi, *J. Mater. Chem.*, 2001, **11**, 2078.
438. E. Fujiwara, A. Kobayashi, H. Fujiwara and H. Kobayashi, *Inorg. Chem.*, 2004, **43**, 1122.
439. M. Sasa, E. Fujiwara, A. Kobayashi, S. Ishibashi, K. Terakura, Y. Okano, H. Fujiwara and H. Kobayashi, *J. Mater. Chem.*, 2005, **15**, 155.
440. S. S. Staniland, W. Fujita, Y. Umezono, K. Awaga, S. J. Clark, H. Cui, H. Kobayashi and N. Robertson, *Chem .Commun.*, 2005, 3204.
441. E. B. Lopes, H. Alves, I. C. Santos, D. Garf, J. S. Brooks, E. Canadell and M. Almeida, *J. Mater. Chem.*, 2008, **18**, 2825.
442. J. Morgado, L. Alcacer, R. T. Henriques and M. Almeida, *Synth. Met*, 1995, **71**, 1943.
443. W. B. Heuer and B. M. Hoffman, *J. Chem. Soc., Chem. Commun.*, 1986, 174.
444. Y. Kashimura, Y. Okano, J.-I. Yamaura and R. Kato, *Synth. Met*, 1999, **103**, 2123.
445. E. B. Yagubskii, L. A. Kushch, V. V. Gritsenko, O. A. Dyachenko, L. I. Buravov and A. G. Khomenko, *Synth. Met*, 1995, **70**, 1039.
446. L. A. Kushch, V. V. Gritsenko, L. I. Buravov, A. G. Khomenko, G. V. Shilov, O. A. Dyachenko, V. A. Merzhanov, E. B. Yagubskii, R. Rousseau and E. Canadell, *J. Mater. Chem.*, 1995, **5**, 1633.
447. E. B. Yagubskii, *Synth. Met*, 2003, **133-134**, 385.
448. M. Fujiwara and R. Kato, *J. Chem. Soc., Dalton Trans.*, 2002, 3763.
449. R. Kato, M. Fujiwara, Y. Kashimura and J.-I. Yamaura, *Synth. Met*, 2001, **120**, 675.
450. R. Kato, T. Imakubo, H. Yamamoto, R. Maeda, M. Fujiwara and J.-I. Yamaura, *Mol. Cryst. Liq. Cryst.*, 2002, **380**, 61.
451. K. Kubo, M. Nakano, H. Tamura and G.-E. Matsubayashi, *Inorg. Chim. Acta*, 2002, **336**, 120.
452. K. Kubo, M. Nakano, H. Tamura and G. Matsubayashi, *Inorg. Chim. Acta*, 2000, **311**, 6.
453. X. M. Ren, H. Okudera, J. L. Xie and Q. J. Meng, *J. Mol . Struct.*, 2005, **733**, 119.
454. H.-R. Wen, J.-L. Zuo, T. A. Scott, H.-C. Zhou and X.-Z. You, *Polyhedron*, 2005, **24**, 671.
455. S. S. Staniland, W. Fujita, Y. Umezono, K. Awaga, P. J. Camp, S. J. Clark and N. Robertson, *Inorg. Chem.*, 2005, **44**, 546.
456. K. Awaga, Y. Umezono, W. Fujita, H. Yoshikawa, H. Cui, H. Kobayashi, S. S. Staniland and N. Robertson, *Inorg. Chim. Acta*, 2008, **361**, 3761.

457. T. Cauchy, E. Ruiz, O. Jeannin, M. Nomura and M. Formigué, *Chem. – Eur. J*, 2007, **13**, 8858.
458. M. Nomura and M. Formigué, *New J. Chem.*, 2007, **31**, 528.
459. H.-B. Duan, X.-M. Ren and Q.-J. Meng, *Coord. Chem. Rev.*, 2010, **254**, 1509.
460. L. V. Interrante, J. W. Bray, H. R. Hart Jr., I. S. Jacobs, J. S. Kasper, P. A. Piacente and J. C. Bonner, *Lect. Notes Phys*, 1979, **96**, 55.
461. H. Imai, T. Inabe, T. Otsuka, T. Okuno and K. Awaga, *Synth. Met*, 1997, **85**, 1705.
462. D. Arcon, A. Lappas, S. Margadonna, K. Prassides, E. Ribera, J. Veciana, C. Rovina, R. T. Henriques and M. Almeida, *Phys. Rev. B*, 1999, **60**, 4191.
463. X. Ribas, A. Sironi, N. Masciocchi, E. B. Lopes, M. Almeida, J. Veciana and C. Rovira, *Inorg. Chem.*, 2005, **44**, 2358.
464. R. Wesolowski, J. T. Haraldsen, J. L. Musfeldt, T. Barnes, M. Mas-Torrent, C. Rovira, R. T. Henriques and M. Almeida, *Phys. Rev. B*, 2003, **68**, 134405.
465. R. Wesolowski, J. T. Haraldsen, J. L. Musfeldt, T. Barnes, M. Mas-Torrent, C. Rovira, R. T. Henriques and M. Almeida, *Phys. Rev. B*, 2004, **70**, 059901.
466. H. Nakajima, M. Katsuhara, H. Nii, M. Ashizawa, T. Kawamoto and T. Mori, *Synth. Met*, 2004, **145**, 95.
467. O. Jeannin and M. Formigué, *Inorg. Chim. Acta*, 2007, **360**, 3820.
468. M. Formigué, B. Domercq, I. V. Jourdain, P. Molinie, F. Guyon and J. Amaudrut, *Chem. -Eur. J*, 1998, **4**, 1714.
469. B. Domercq, C. Coulon and M. Formigué, *Inorg. Chem.*, 2001, **40**, 371.
470. W. E. Broderick, J. A. Thompson, E. P. Day and B. M. Hoffmann, *Science*, 1990, **249**, 401.
471. J. S. Miller, J. C. Calabrese and A. J. Epstein, *Inorg. Chem.*, 1989, **28**, 4230.
472. S. Zurcher, V. Gramlich, D. Von Arx and A. Togni, *Inorg. Chem.*, 1998, **37**, 4015.
473. M. Uruichi, K. Yakushi, Y. Yamashita and J. Qin, *J. Mater. Chem.*, 1998, **8**, 141.
474. V. Gama, D. Belo, I. C. Santos and R. T. Henriques, *Mol. Cryst. Liq. Cryst.*, 1997, **306**, 17.
475. W. E. Broderick, J. A. Thompson, M. R. Godfrey, M. Sabat, B. M. Hoffmann and E. P. Day, *J. Am. Chem. Soc.*, 1989, **111**, 7656.
476. C. Ni, Y. Li, D. Dang, Y. Song, Z. Ni and Q. Meng, *Inorg. Chim. Acta*, 2005, **358**, 2680.
477. H. Nakajima, M. Katsuhara, M. Ashizawa, T. Kawamoto and T. Mori, *Inorg. Chem.*, 2004, **43**, 6075.
478. W. E. Broderick, J. A. Thompson, M. R. Godfrey, M. Sabat and B. M. Hoffman, *J. Am. Chem. Soc.*, 1989, **111**, 7656.
479. N. C. Schiødt, R. Sessoli and F. C. Krebs, *Inorg. Chem.*, 2004, **43**, 1986.

480. C. Ni, D. Dang, Y. Song, S. Gao, Y. Li, Z. Ni, Z. Tian, L. Wen and Q. Meng, *Chem. Phys. Lett.*, 2004, **396**, 353.
481. M. Fettouhi, L. Ouahab, E. Codjovi and O. Kahn, *Mol. Cryst. Liq. Cryst.*, 1995, **273**, 29.
482. X. M. Ren, H. Okudera, R. K. Kremer, Y. Song, C. He, Q. J. Meng and P. H. Wu, *Inorg. Chem.*, 2004, **43**, 2569.
483. S. Rabaça, R. Meira, L. C. J. Pereira, M. T. Duarte, J. J. Novoa and V. Gama, *Inorg. Chim. Acta*, 2001, **326**, 89.
484. O. Jeannin, R. Clérac, T. Cauchy and M. Formigué, *Inorg. Chem.*, 2008, **47**, 10656.
485. N. Venkatalakshmi, B. Varghese, S. Lalitha, R. F. X. Williams and P. T. Manoharan, *J. Am. Chem. Soc.*, 1989, **111**, 5748.
486. W. B. Heuer, P. Mountford, M. L. H. Green, S. G. Bott, D. O'Hare and J. S. Miller, *Chem. Mater.*, 1990, **2**, 764.
487. J. Xie, X. Ren, Y. Song, W. Tong, C. Lu, Y. Yao and Q. Meng, *Inorg. Chem. Commun.*, 2002, **5**, 395.
488. J. Xie, X. Ren, S. Gao and Q. Meng, *Chem. Lett.*, 2002, 576.
489. N. Robertson and S. Tiwary, *Polyhedron*, 2003, **22**, 2339.
490. M. L. Allan, A. T. Coomber, I. R. Marsden, J. H. F. Martens, R. H. Friend, A. Charlton and A. E. Underhill, *Synth. Met*, 1993, **56**, 3317.
491. A. T. Coomber, D. Beljonne, R. H. Friend, J. L. Bredas, A. Charlton, N. Robertson, A. E. Underhill, M. Kurmoo and P. Day, *Nature*, 1996, **380**, 144.
492. C. Faulmann, A. E. Pullen, E. Riviere, S. Dorbes, F. Senoca, E. Coronado and P. Cassoux, *Eur. J. Inorg. Chem.*, 2003, 2880.
493. H. Hiraga, H. Miyasaka, K. Nakata, T. Kajiwara, S. Takaishi, Y. Oshima, H. Nojiri and M. Yamashita, *Inorg. Chem.*, 2007, **46**, 9661.
494. H. Hiraga, H. Miyasaka, S. Takaishi, T Kajiwara and M. Yamashita, *Inorg. Chim. Acta*, 2008, **361**, 3863.
495. H. Hiraga, H. Miyasaka, R. Clérac, M. Formigué and M. Yamashita, *Inorg. Chem.*, 2009, **48**, 2887.
496. K. Kubo, H. Miyasaka and M. Yamashita, *Phys. B: Condens. Matter*, 2010, **405**, S313.
497. K. Kubo, T. Shiga, T. Yamamoto, A. Tajima, T. Moriwaki, Y. Ikemoto, M. Yamashita, E. Sessini, M. L. Mercuri, P. Deplano, Y. Nakazawa and R. Kato, *Inorg. Chem.*, 2011, **50**, 9337.
498. U. T. Mueller-Westerhoff and R. W. Sanders, *Patent US 2003225296*, 2003.
499. M. Nomura, E. Tsukano, C. Fujita-Takayama, T. Sugiyama and M. Kajitani, *J. Organomet. Chem.*, 2009, **694**, 3116.
500. C. Lauterbach and J. Fabian, *Eur. J. Inorg. Chem.*, 1999, 1995.
501. Z. S. Herman, R. F. Kirchner, G. H. Loew, U. T. Mueller-Westerhoff, A. Nazzal and M. C. Zerner, *Inorg. Chem.*, 1982, **21**, 36.
502. L. Hu, J. Qin, N. Zhou, Y.-F. Meng, Y. Xu, J.-L. Zuo and X.-Z. You, *Dyes Pigm*, 2012, **92**, 1223.

503. G. Soras, N. Psaroudakis, M. J. Manos, A. J. Tasiopoulos, D. G. Liakos and G. A. Mousdis, *Polyhedron*, 2009, **28**, 3340.
504. J.-L. Zuo, T.-M. Yao, F. You, X.-Z. You, H.-K. Fun and B.-C. Yip, *J. Mater. Chem.*, 1996, **6**, 1633.
505. J.-F. Bai, J.-L. Zuo, W.-L. Tan, W. Ji, Z. Shen, H.-K. Fun, K. Chinnakali, I. Abdul Razak, X.-Z. You and C.-M. Che, *J. Mater. Chem.*, 1999, **9**, 2419.
506. P. Romaniello, M. C. Aragoni, M. Arca, T. Cassano, C. Denotti, F. A. Devillanova, F. Isaia, F. Lelj, V. Lippolis and R. Tommasi, *J. Phys. Chem. A*, 2003, **107**, 9679.
507. T. Cassano, R. Tommasi, M. Arca, F. A. Devillanova and F. Lelj, *Recent Res. Devel. Chem. Physics*, 2004, 553.
508. G. C. Anyfantis, G. C. Papavassiliou, P. Aloukps, S. Couris, Y. F. Weng, H. Youshino and K. Murata, *Z. Naturforsch*, 2007, **62b**, 200.
509. G. C. Anyfantis, G. C. Papavassiliou, A. Terzis, C. P. Raptopoulou, Y. F. Weng, H. Yoshino and K. Murata, *Z. Naturforsch*, 2006, **61b**, 1007.
510. P. Aloukos, S. Couris, J. B. Koutselas, G. C. Anyfantis and G. C. Papavassiliou, *Chem. Phys. Lett.*, 2006, **428**, 109.
511. U. T. Mueller-Westerhoff and R. W. Sanders, *Organometallics*, 2003, **22**, 4778.
512. Q. Miao, J. Gao, Z. Wang, H. Yu, Y. Luo and T. Ma, *Inorg. Chim. Acta*, 2011, **376**, 619.
513. T.-M. Yao, J.-L. Zuo, X.-Z. You and X.-Y. Huang, *Polyhedron*, 1995, **14**, 1487.
514. V. Madhu and S. K. Das, *Inorg. Chem.*, 2008, **47**, 5055.
515. F. Bigoli, P. Deplano, M. L. Mercuri, M. A. Pellinghelli, L. Pilia, G. Pintus, A. Serpe and E. F. Trogu, *Inorg. Chem.*, 2002, **41**, 5241.
516. P. Deplano, M. L. Mercuri, L. Marchiò, L. Pilia, M. Salidu, A. Serpe, F. Congiu and S. Sanna, *Eur. J. Inorg. Chem.*, 2005, 1829.
517. C. Chatzikyriakos, I. Papagiannouli, S. Couris, G. C. Anyfantis and G. C. Papavassiliou, *Chem. Phys. Lett.*, 2011, **513**, 229.
518. S. Rabaça, A. C. Cerdeira, S. Oliveira, I. C. Santos, R. T. Henriques, L. C. J. Pereira, J. T. Coutinho and M. Almeida, *Polyhedron*, 2012, **39**, 91.
519. S. Dalgleish, K. Awaga and N. Robertson, *Chem. Commun.*, 2011, **47**, 7089.
520. S. D. Cumming and R. Eisenberg, *Inorg. Chem.*, 1995, **34**, 3396.
521. R. Bolligarla, R. Kishore, G. Durgaprasad and S. K. Das, *Inorg. Chim. Acta*, 2010, **363**, 3061.
522. A. J. Scroggie, G. D'Alessandro, N. Langford and G.-L. Oppo, *Opt. Commun.*, 1999, **160**, 119.
523. K. H. Drexhage and U. T. Mueller-Westerhoff, *US Patent 3743964*, 1973.
524. K. H. Drexhage and U. T. Mueller-Westerhoff, *IEEE J. Quant. Electr*, 1972, **QE-8**, 759.
525. Y. R. Shen, *The Principles of Nonlinear Optics*, Wiley, New York, 1984.
526. J. L. Oudar, *J. Chem. Phys.*, 1977, **67**, 446.

527. K. Base, M. T. Tierney, A. Fort, J. Muller and M. W. Grinstaff, *Inorg. Chem.*, 1999, **38**, 287.
528. H. Le Bozec and T. Renouard, *Eur. J. Inorg. Chem.*, 2000, 229.
529. D. Zhou, G. J. Ashwell and C. Huang, *Chem. Lett.*, 1997, 7.
530. B. Cai, J.-L. Liu, X.-L. Sheng and X.-M. Ren, *Inorg. Chem. Commun.*, 2011, **14**, 1971.
531. P. Romaniello and F. Lelj, *TEOCHEM*, 2003, **636**, 23.
532. P. Romaniello and F. Lelj, *Chem. Phys. Lett.*, 2003, **372**, 51.
533. F. Bigoli, C.-T. Chen, W.-C. Wu, P. Deplano, M. L. Mercuri, M. A. Pellinghelli, L. Pilia, G. Pintus, A. Serpe and E. F. Trogu, *Chem. Commun.*, 2001, 2246.
534. S. D. Cummings, L.-T. Cheng and R. Eisemberg, *Chem. Mater.*, 1997, **9**, 440.
535. C.-T. Chen, S. Y. Liao, K. J. Lin, T. Y. J. Lin, L. L. Lia and C. H. Chen, *Nonlinear Optics*, 1999, **22**, 35.
536. C. Makedonas and C. A. Mitsopoulou, *Eur. J. Inorg. Chem.*, 2007, 110.
537. C. Makedonas and C. A. Mitsopoulou, *Inorg. Chim. Acta*, 2007, **360**, 3997.
538. C. Makedonas, C. A. Mitsopoulou, F. J. Lahoz and A. I. Balana, *Inorg. Chem.*, 2003, **42**, 8853.
539. T. Kusamoto, S. Kume and H. Nishihara, *J. Am. Chem. Soc.*, 2008, **130**, 13844.
540. E. A. M. Geary, K. L. McCall, A. Turner, P. R. Murray, E. J. L. McInnes, L. A. Jack, L. J. Yellowlees and N. Robertson, *Dalton Trans.*, 2008, 3701.
541. K. Base, M. T. Tierney, A. Fort, J. Muller and M. W. Grinstaff, *Inorg. Chem.*, 1999, **338**, 287.
542. S. D. Cummings, L.-T. Cheng and R. Eisenberg, *Chem. Mater.*, 1997, **9**, 440.
543. C.-T. Chen, T.-Y. Jeremy Lin, C.-H. Chen and K.-J. Lin, *J. Chin. Chem. Soc.*, 2000, **47**, 197.
544. C.-T. Chen, S.-Y. Liao, K.-J. Lin and L.-L. Lai, *Adv. Mater.*, 1998, **3**, 334.
545. C. Makedonas and C. A. Mitsopoulou, *Inorg. Chim. Acta*, 2007, **360**, 3997.
546. Y. Matsuo, M. Maruyama, S. S. Gayathri, T. Uchida, D. M. Guldi, H. Kishida, A. Nakamura and E. Nakamura, *J. Am. Chem. Soc.*, 2009, **131**, 12643.
547. W. L. Tan, W. Ji, J. L. Zuo, J. F. Bai, X. Z. You, J. H. Lim, S. S. Yang, D. J. Hagan and E. W. Van Stryland, *Proc. SPIE – Int. Soc. Opt. Eng.*, 1999, **3899**, 475.
548. W. L. Tan, W. Ji, J. L. Zuo, J. F. Bai, X. Z. You, J. H. Lim, S. Yang, D. J. Hagan and E. W. Van Stryland, *Appl. Phys. B*, 2000, **70**, 809.
549. Z. Sun, M. Tong, H. Zeng, L. Ding, Z. Wang, Z. Xu, J. Dai and G. Bian, *Chem. Phys. Lett.*, 2001, **342**, 323.
550. B. Domercq, C. Coulon, P. Feneyrou, V. Dentan, P. Robin and M. Formigué, *Adv. Funct. Mater.*, 2002, **12**, 359.

551. T. Cassano, R. Tommasi, L. Nitti, M. C. Aragoni, M. Arca, C. Denotti, F. A. Devillanova, F. Isaia, V. Lippolis, F. Lelj and P. Romaniello, *J. Chem. Phys.*, 2003, **118**, 5995.
552. T. Cassano, R. Tommasi, M. Arca and F. A. Devillanova, *J. Phys.: Condens. Matter*, 2006, **18**, 5279.
553. G. Soras, N. Psaroudakis, G. A. Mousdis, M. J. Manos, A. J. Tasiopoulos, P. Aloukos, S. Couris, P. Labéguerie, J. Lipinski, A. Avramopoulos and M. G. Papadopoulos, *Chem. Phys.*, 2010, **372**, 33.
554. X. Q. Wang, Q. Ren, H. L. Fan, J. W. Chen, Z. H. Sun, T. B. Li, X. T. Liu, G. H. Zhang, D. Xu and W. L. Liu, *J. Cryst. Growth*, 2010, **312**, 2206.
555. W. F. Guo, X. B. Sun, J. Sun, X. Q. Wang, G. H. Zhang, Q. Ren and D. Xu, *Chem. Phys. Lett.*, 2007, **435**, 65.
556. P. Aloukos, G. Chatzikyriakos, I. Papagiannouli, N. Liaros and S. Couris, *Chem. Phys. Letters*, 2010, **495**, 245.
557. Z. Dai, X. Yue, B. Peng, Q. Yang, X. Liu and P. Ye, *Chem. Phys. Lett.*, 2000, **317**, 9.
558. X.-B. Sun, Q. Ren, X.-Q. Wang, G.-H. Zhang, H.-L. Yang, L. Feng, D. Xu and Z.-B. Liu, *Chin. Phys.*, 2006, **15**, 1009.
559. H. L. Yang, W. Ji, Q. Ren, X. Q. Wang, F. J. Zhang and J. J. He, *Proc. SPIE*, 2009, **7631**, 76310V/1.
560. S. N. Oliver, C. S. Winter, R. J. Manning, J. D. Rush, C. A. S. Hill, and A. E. Underhill, in *Nonlinear Optical Properties of Inorganic Materials V*, ed. D. J. Williams, **110** , 1992.
561. G. J. Gall, T. A. King, S. N. Oliver, C. A. Capozzi, A. B. Seddon, C. A. S. Hill and A. E. Underhill, *Proc. SPIE - Int. Soc. Opt. Eng.*, 1994, **2288**, 372.
562. D. Natali, M. Sampietro, M. Arca, C. Denotti and F. A. Devillanova, *Synth. Metals*, 2003, **137**, 1489.
563. M. C. Aragoni, M. Arca, T. Cassano, C. Denotti, F. A. Devillanova, F. Isaia, V. Lippolis, D. Natali, L. Nitti, M. Sampietro, R. Tommasi and G. Verani, *Inorg. Chem. Commun.*, 2002, **5**, 869.
564. M. C. Aragoni, M. Arca, M. Caironi, C. Denotti, F. A. Devillanova, E. Grigiotti, F. Isaia, F. Laschi, V. Lippolis, D. Natali, L. Pala, M. Sampietro and P. Zanello, *Chem. Commun.*, 2004, 1882.
565. M. Caironi, D. Natali, M. Sampietro, M. Ward, A. Meacham, F. A. Devillanova, M. Arca, C. Denotti and L. Pala, *Synth. Metals*, 2005, **153**, 273.
566. M. C. Aragoni, M. Arca, F. A. Devillanova, F. Isaia, V. Lippolis, A. Mancini, L. Pala, G. Verani, T. Agostinelli, M. Caironi, D. Natali and M. Sampietro, *Inorg. Chem. Commun.*, 2007, **10**, 191.
567. T. Agostinelli, M. Caironi, D. Natali, M. Sampietro, M. Arca, F. A. Devillanova and V. Ferrero, *Synth. Metals*, 2007, **157**, 984.
568. S. Debnath, J.-F. Bergamini, F. Artzner, C. Mériadec, F. Camerel and M. Formigué, *Chem. Commun.*, 2012, **48**, 2283.

569. E. A. M. Geary, L. J. Yellowlees, L. A. Jack, I. D. H. Oswald, S. Parsons, N. Hirata, J. R. Durrant and N. Robertson, *Inorg. Chem.*, 2005, **44**, 242.
570. S. Hattori, Y. Wada, S. Yanagida and S. Fukuzumi, *J. Am. Chem. Soc.*, 2005, **127**, 9648.
571. H. Takahashi, S. Hirano, and I. Kawashima, *Patent JP 2005010742*, 2005.
572. T. Takada, C. Tanaka, M. Watanabe, M. Yamada, and R. Inoue, *Patent JP 2004333784*, 2004.
573. H. Sato, *Patent JP 2004184844*, 2004.
574. H. Nishiguchi, E. Fujisawa, M. Suzuki, S. Yamamura, and Y. Kitayama, *Patent JP 2004182857*, 2004.
575. H. Nishiguchi, T. Takeushi, E. Fujisawa, and M. Suzuki, *Patent JP 2004285314*, 2004.
576. E. Kato, *Patent JP 2004045654*, 2004.
577. I. Sato and Y. Saito, *Patent JP 2005099755*, 2005.
578. S. Kuwabara, *Patent US 2005035336*, 2005.
579. H. Nishiguchi, E. Fujisawa, M. Suzuki, and I. Nakanishi, *Patent JP 2005023104*, 2005.
580. S. A. Deutsch, *Patent WO 2004082956*, 2004.

CHAPTER 11.4

II–VI Semiconductors and Their Device Applications

BIN HE AND WENJUN ZHANG*

Center of Super-Diamond and Advanced Films (COSDAF), and Department of Physics and Materials Science, City University of Hong Kong, 83 Tat Chee Avenue, Kowloon, Hong Kong SAR, China
*Email: apwjzh@cityu.edu.hk

11.4.1 Introduction

II–VI semiconductors are compounds consisting of elements from Groups IIB (Zn, Cd, Hg) and VIA (S, Se, Te) of the periodic table. The combinations of the elements compose various semiconductors which include binary, ternary, and quaternary compounds. II–VI semiconductors have been shown by extensive research to possess distinctive electronic and optoelectronic properties, such as a broad range of direct bandgaps (zero to $\sim$4 eV), large exciton binding energy and high electro-optical coefficient, which make them candidates in a wide range of applications incuding infrared (IR) lasers and detectors, solar cells, blue-green lasers and light-emitting diodes (LEDs), and radiation detectors.

Starting with studies on the luminescence and optical properties of ZnS, the research interest in exploiting the applications of II–VI semiconductors has continued to be a hot topic since the 1920s.[1] Many important applications based on II–VI semiconductors have been realized and are commercially available. For example, HgCdTe (MCT) is a dominant material for IR sensing and imaging and by varying the cadmium concentration, it can cover all of the important IR ranges of interest with negligible variation of lattice constant.[2,3] Actually, HgCdTe is considered the third most studied semiconductor after Si

Handbook of Chalcogen Chemistry: New Perspectives in Sulfur, Selenium and Tellurium, 2nd Edition, Volume 2
Edited by Francesco Antonio Devillanova and Wolf-Walther du Mont

Published by the Royal Society of Chemistry, www.rsc.org

and GaAs.[4] Another example we want to mention here is CdTe which is the first and only thin-film photovoltaic (PV) material to improve on crystalline silicon PV in cost. The direct manufacturing cost for CdTe PV modules has been reduced to US$0.76 per watt since 2010.[5] The highest efficiency of CdTe solar cells reached 17.3% in 2011 and total area efficiency of 14.4% for CdTe PV modules was published in 2012.[6] In the last decade, II–VI compound-based LEDs have suffered from limited lifetimes and diminished gradually due to the rapid development of Group III-nitride based devices. Nevertheless, ZnSe-based white LEDs with lifetime longer than 10 000 hours at room temperature have recently been achieved, which makes the practical use of II–VI light-emitting compounds more competitive and promising.[7]

The advances of nanoscience and nanotechnology in recent years have opened new opportunities for the applications of II–VI semiconductors since the requirements on substrates for growing high-quality, single-crystalline nanostructures is not as strict as for epitaxial thin-film and bulk crystal growth. II–VI nanomaterials, such as nanowires and quantum dots, with high crystalline perfection, reduced defects, and controlled doping, can be achieved more easily.[8] Moreover, wide-bandgap II–VI materials typically have stronger exciton–phonon interactions than III–V semiconductors; their corresponding nanostructures are thus expected to be very useful in fabricating optoelectronic devices and in exploring the exciton nature in low-dimensional structure.[4] The application potential of II–VI nanomaterials has attracted extensive research interests and became one of the hottest topics in material science. Recently it has been shown both theoretically and experimentally that HgTe quantum wells shows a unique new state of matter—the 'topological insulator'.[9] Reviews on the progress of II–VI nanomaterials can be found elsewhere;[8,10] the research output on these hot topics is too wide to be covered in this limited chapter. Therefore, we restrict the review here only to the thin films and bulk crystal materials of II–VI materials and their current applications, in particular the progress in the last decade.

In this chapter, we start with the basic properties of II–VI semiconductors, and then review the synthesis techniques of bulk crystals and epitaxial films of II–VI compounds. Doping and metal–semiconductor contacts, which are essential issues for fabrication of electronic and optoelectronic devices, are also addressed. We finally summarize the current status of applications of II–VI semiconductors in PV conversion (solar cells), LEDs, light and particle detection, phosphors, and the latest progress in commercialized devices by current leading manufacturers.

11.4.2 Basic Properties of II–VI Semiconductors

The foremost binary II–VI semiconductors are ZnS, ZnSe, ZnTe, CdS, CdSe, and CdTe. Basically, II–VI semiconductors exhibit properties intermediate between covalent IV-IV/III–V and ionic I–VII compounds. II–VI semiconductors have comparatively wider bandgap than Ge, Si, and GaAs because of the increasing ionicity in the covalent–ionic mixed bonds.[11] More importantly, most II–VI semiconductors (except HgTe and HgSe) have a direct bandgap, which consequently endows the II–VI materials with great potential for optical and optoelectronic applications. The basic properties of II–VI semiconductors have been described in previous publications.[4,12,13] In Table 11.4.1 we highlight some major structural and semiconductor properties of the main binary II–VI compounds. These binary compounds can be further alloyed to form II–VI ternary semiconductors so that their bandgaps can be tuned continuously.[14] The most studied ternary II–VI semiconductors are HgCdTe, CdZnTe (CZT), and HgZnTe (MZT). Moreover, there are also a few studies on the quaternary II–VI semiconductors, such as $CdTe_{1-x-y}Se_xS_y$, $ZnTe_{1-x-y}Se_xS_y$,[15] $Zn_xCd_{1-x}S_ySe_{1-y}$,[16] $Zn_{1-x}Mg_xS_ySe_{1-y}$,[17,18] *etc.*[19,20] However, due to complex composition, it appeared to be difficult to engineer their bandgaps with high controllability, reproducibility, and stability. We discuss briefly below some interesting properties of selected binary and ternary II–VI semiconductors.

ZnS has the largest bandgap among the II–VI semiconductors (3.72 eV for cubic and 3.58 for hexagonal structures). The cubic structure (zinc blende) is more stable; the hexagonal structure is also known as the mineral wurtzite, although it can be produced synthetically.[21] The transition from the zinc blende to the wurtzite structure occurs at around 1020 °C at atmosphere pressure. ZnS has been used as a scintillator or phosphor material when it is doped with suitable activators. ZnS is also as an important IR optical material, being transparent from the visible region to over 12 μm in wavelength.

ZnSe crystallizes normally in zinc blende structure in ambient conditions. ZnSe is a promising material for fabricating blue-green II–VI LEDs and laser diodes (LDs). ZnSe doped with chromium (ZnSe:Cr) has been used as an IR laser gain medium emitting at about 2.5 μm.[22] In addition, ZnSe is an IR material with a wide transmission wavelength range; and ZnSe activated with tellurium (ZnSe(Te)) is a scintillator with emission peak at 640 nm which is good for photodiode matching.

ZnTe typically has a zinc blende crystal structure, but can be also prepared as hexagonal (wurtzite) crystals. ZnTe is usually a p-type semiconductor and its conductivity can be easily tuned by doping, which make it an attractive semiconducting material with various applications in optoelectronics, such as blue LEDs, LDs, solar cells, components of microwave generators, and generation of pulsed terahertz radiation.[23]

The wurtzite structure of CdS is more stable than its zinc blende structure, and CdS has also a high-pressure form of the NaCl rock salt structure.[24] CdS has been used in manufacturing photoresistors which are sensitive to visible and near IR light. Moreover, CdS has a bandgap of 2.43 eV, which make it an ideal material to be used as the window layer of junction solar cells based on CdTe and $Cu(InGa)Se_2$ (CIGS). High-efficiency thin film CdS/(CdTe or CIGS) solar cells have been obtained by using CdS as an n-partner in a superstrate configuration.[25] Cu^+ and Al^{3+} (CL) doped CdS has also been used as a phosphor.[26]

Table 11.4.1 Physical and chemical properties of II–VI semiconductors

Properties	*ZnS*	*ZnSe*	*ZnTe*	*CdS*	*CdSe*	*CdTe*
Structure, elastic, thermal properties						
Stable phases at 300K	Z/W[a]	Z	Z	Z / W	Z / W	Z
Lattice constant (Z/W) (Å)	$a_0 = 5.405$/ $a_0 = 3.820$, $c_0 = 6.260$	$a_0 = 5.667$/ –	$a_0 = 6.088$/ –	$a_0 = 5.82$/ $a_0 = 4.135$, $c_0 = 6.749$	$a_0 = 6.077$/ $a_0 = 4.300$, $c_0 = 7.011$	$a_0 = 6.46$/ –
Density (g cm^{-3}) (Z/W)	4.088/4.087	5.266/-	5.636/-	4.82/4.82	5.81/5.81	-/5.86
Melting point (K)	1991	1799	1563	1750	1533	1370
Symmetry (Z/W)	$F\bar{4}3m/P6_3mc$	$F\bar{4}3m/-$	$F\bar{4}3m/-$	$F\bar{4}3m/P6_3mc$	$-/P6_3mc$	$F\bar{4}3m/-$
Young's modulus (GPa)	74.5	70	52.8	45	50	37
Thermal conductivity at 300K (W $cm^{-1}K^{-1}$)	0.27 (Z)	0.19	0.18	0.2	0.09	0.01
Linear expansion coefficient (10^{-6} K^{-1}) Z/W	6.6/6.9	7.4/-	8.3/-	3.0/4.5	3.0/7.3	5.1/-
Heat capacity (J $mol^{-1}K^{-1}$)	45	51.88	46.44	53.9	48.46	23.9
Electrical and transport properties						
Bandgap (eV, 300 K) (Z/W)	3.72/3.58	2.64/-	2.35/-	2.5/2.482	1.74/-	1.475
Electron effective mass (m^*/m_0)	–0.40	0.21	0.2	0.25	0.12	0.11
Hole effective mass (m^*/m_0)	–	0.6	0.2	0.8	0.45	0.35
Dielectric constant ($\varepsilon_0/\varepsilon_\infty$)	8.6/5.2	8.6/5.7	10.3/7.28	8.28/5.23	9.3/6.2	10.4/7.1
Breakdown field ($\times 10^5$V cm^{-1})	43	20	13	17	7.1	4.9
Iconicity (%)	62	63	61	69	70	72
Optical properties						
Absorption coefficient 10.6 μm (cm^{-1})	≤0.15	$1–2\times10^{-3}$	-	≤0.007	≤0.0015	≤0.003
Exciton binding energy (meV)	38	21	10	30.5	15	10.5
Refractive index (Z/W)	2.368/2.378	2.5/-	2.72/-	-/2.529	2.5/-	2.72/-
Electro-optical coefficient (m/V) ($\lambda = 10.6$ μm)	2×10^{-12}	2.2×10^{-12}	4×10^{-12}	-	-	6.8×10^{-12}
Thermo-optical coefficient (dn/dT) ($\lambda = 10.6$ μm)	4.7	6.1	-	-	-	11

[a]W, wurtzite; Z, zinc blende.

CdSe has three crystalline structures: zinc blende, wurtzite, and rock salt structures. The sphalerite CdSe structure is unstable and converts to the wurtzite upon moderate heating, and the rock salt structure is only observed under high pressure.[27] The CdSe bulk material crystallizes exclusively in the hexagonal phases, and growing cubic CdSe crystals is considered a difficult task. Several groups have presented the structural and optical properties of zinc blende CdSe layers prepared by various techniques.[28,29] CdSe is transparent to IR light, and has seen limited use in windows for instruments utilizing IR light. Because CdSe shows a size-dependent fluorescence spectrum, much current research on CdSe is focusing on CdSe nanoparticles or quantum dots which have demonstrated applications in optoelectronic devices, LDs, high-efficiency solar cells, nanosensing, and biomedical imaging.[30–33]

CdTe is the only II–VI compound in which both n- and p-type conductivity can easily be controlled by introducing impurities. One of the most important applications of CdTe is in thin-film solar cells. CdTe-based thin-film solar cells accounted for 6% of all solar cells installed in 2010.[34] However, the development of CdTe solar cells may face problems because of the shortage of Te and the toxicity of Cd.[35] CdTe is an IR optical material available for manufacturing optical windows and lenses, but this application is also limited by its toxicity. CdTe has the greatest electro-optic coefficient of the linear electro-optic effect among II–VI compounds, which makes it an ideal electro-optic modulator. CdTe has high atomic number, large bandgap, and high electron mobility ($\sim 1000\ cm^2\ V^{-1}\ s^{-1}$), which result in high intrinsic $\mu\tau$ (mobility-lifetime) product and thus high degree of charge collection. Therefore, CdTe doped with chlorine is used as a radiation detector with excellent spectral resolution for X-rays, gamma-rays, and alpha and beta particles; and CdTe allows the construction of compact detectors which enable operation at room temperature for a wide variety of applications in nuclear spectroscopy.[36]

As for the Hg–VI compounds, HgS is dimorphic with two crystal forms: red cinnabar (α-HgS, hexagonal), is the form in which mercury is most commonly found in nature; black, metacinnabar (*β*-HgS), is less common in nature and adopts the zinc blende ($T_d^2 - F\bar{4}3m$) crystal structure.[37] Mercury metal is produced from the cinnabar ore by roasting in air and condensing the vapour.[38] α-HgS has a direct band gap of 2.1 eV, density of 8.1 g cm^{-3} and melting point of 580 °C. HgSe is a grey-black crystalline solid semimetal with a zinc blende structure and the lattice constant is 0.608 nm.[39] It has a density of 8.3 g cm^{-3}. HgSe can be obtained as the mineral tiemannite. The lattice parameter of HgSe is 0.646 nm in the cubic crystalline form, the bulk modulus is about 42.1 GPa, the thermal expansion coefficient about $5.2 \times 10^{-6}\ K^{-1}$, static dielectric constant 20.8, dynamic dielectric constant 15.1, and low thermal conductivity (2.7 W m K^{-1}). HgTe bonds are weak, leading to low hardness values (2.7×10^7 kg m^{-2}).[40–42] It has a density of 8.1 g cm^{-3}, and it is easily etched by acids, such as hydrobromic acid. Both HgSe and HgTe are zero-gap materials with quadratic energy dispersion.[43,44] Recent observations suggest that HgTe is a high-potential material for applications as optical isolator and modulator in the THz spectral range.[45] Calculations have shown that the cinnabar phase of

HgTe is a promising thermoelectric material which could be used to design devices for both refrigeration and power generation.[46]

HgCdTe (MCT) is an alloy of CdTe and HgTe with variable bandgaps. CdTe has a bandgap of approximately 1.5 eV at room temperature, and HgTe is a semimetal with zero bandgap energy. Mixing these two substances allows one to obtain any bandgap between 0 and 1.5 eV. $Hg_{1-x}Cd_xTe$ remains the main candidate for IR detectors with high quantum efficiency. Due to small changes in lattice constant, the material can be used for bandgap-engineered devices. Practical HgCdTe heterostructures for devices have been grown with advanced molecular-beam epitaxy (MBE) and metal-organic chemical vapour deposition (MOCVD).[47] HgCdTe is a soft material due to the weak bonds of Hg with Te, and the thermal conductivity of HgCdTe is low, which means that it is unsuitable for high-power devices. The refractive index of $Hg_{1-x}Cd_xTe$ is high, reaching nearly 4 for HgCdTe with high Hg content.

CdZnTe (CZT), an alloy of CaTe and ZnTe, is a direct-bandgap semiconductor which is mainly used in radiation detectors. The band gap is tunable from about 1.4 to 2.2 eV, depending on the composition. Radiation detectors made from CZT can operate in photoconductive mode at room temperature, unlike Ge materials which require liquid nitrogen cooling. CZT-based detectors have good performances including high sensitivity for X-rays and gamma-rays due to the high atomic numbers of Cd and Te, and better resolution than scintillator detectors. CZT can also work as substrate for epitaxial growth of HgCdTe since $Cd_{0.96}Zn_{0.04}Te$ has almost perfect lattice matching with LWIR HgCdTe (80% Hg and 20% Cd).

The alloy of HgTe and ZnTe, HgZnTe (MZT), is a narrow-gap semiconductor. HgZnTe was first proposed as an alternative detector material to MCT due to its superior hardness and high energies for Hg vacancy formation and dislocations.[48] MZT has better chemical, thermal, and mechanical stability than MCT, and the bandgap of MZT is more sensitive to composition fluctuations than that of MCT, which is an issue of concern for reproducible device fabrication.

11.4.3 Synthesis of II–VI Semiconductors

The basis for all the semiconducting applications is to obtain high-quality electronic grade materials. For II–VI semiconductors, both bulk crystal and epitaxial film materials have been well established for fabrication of diverse devices. An overall summary of the principles and progresses can be found in the relevant books.[3,4]

11.4.3.1 Bulk Crystal Synthesis Methods

Since the electrical and optical properties of semiconductor compounds depend strongly on impurities and native defects, techniques to grow bulk crystals of II–VI compounds with high purity and high quality are critical to their electronic device applications.[49–51] Bulk crystals of II–VI compounds can be synthesized *via* liquid, vapour and solid phases. Vapour-phase growth includes chemical vapour transport (CVT) and physical vapour transport (PVT) methods; liquid-phase methods includes growth from the melt or solvent.[4] Much effort has been devoted in recent years to improve the uniformity, reproducibility, and impurity and dopant control in bulk crystal growth of II–VI semiconductors.

Growth from the vapour phase has the advantage of being performed at lower temperatures, which can prevent phase transition and undesirable contamination. It has been illustrated that self-seeded approaches by the CVT or PVT techniques enable the growth of single-crystal volume to several centimetres.[3,4] However, grain boundaries and twins form easily during growth. In order to solve these problems, seeded chemical vapour transport (SCVT) and seeded PVT (SPVT), the so-called modified Lely method, have been developed.[52] By employing SPVT and SCVT, Yasuo Namikawa *et al.* from Sumitomo Electronic Industry (SEI) synthesized ZnSe single crystals with large size and low defect density.[7] The most important problem to be solved in the PVT growth of ZnSe single crystal is void formation during crystal growth. In their work, void formation was eliminated by the semi-open free-growth method in which the growing crystal could be kept at the local minimum temperature position during crystal growth. Figure 11.4.1 shows a ZnSe single crystal with the crystal diameter of about 45 mm and the crystal height of about 35 mm grown by the PVT method. The average dislocation density in the wafer was about 5×10^3 cm^{-2}, which meets the requirement for LED substrates.

From the viewpoint of industrial production, melt growth is the most usual method for obtaining large single crystals, and the Bridgman technique is a typical crystal growth method from melt.[4] Thus far, the Bridgman method has been applied to synthesize II–VI compound crystals, including CdZnTe,[53–59] HgCdTe,[60–62]and CdTe,[63,64] with large size and high quality. For CdZnTe room-temperature radiation-detector fabrication, high resistivity is a critical requirement for reducing the bulk leakage current. $Cd_{1-x}Zn_xTe$ crystals with high electrical resistivity, with $x=0.2$, grown unseeded the by Bridgman

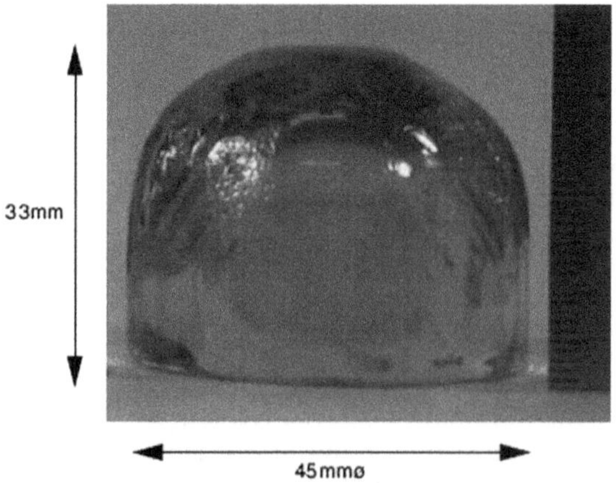

Figure 11.4.1 ZnSe single crystal grown by PVT method.[7] (Reproduced with permission from *SEI Technical Review*, 2011, **72**, 25.)

Figure 11.4.2 (left) CdZnTe boules grown by the modified Bridgman method. These boules are 2.6 and 3.6 inches in diameter (6.7 and 9.2 cm). The large grain structure of the crystals is visible. The boules are sawn into wafers for epitaxial growth of HgCdTe films.[60] (Reproduced from *Opto-Electron. Rev.*, 2002, **10**, 159, with permission from ECS—The Electrochemical Society.) (right) Comparison of the 5.5"-diameter and 8.0"-diameter CdZnTe ingots grown by the Gen-3+ and Gen-5 high-pressure electro-dynamic gradient freeze (HP EDGF) crystal growth systems, respectively.[56] (Reproduced from *Proce. of SPIE*, 2006, **6319**, 191, with permission from SPIE.)

technique under controlled Cd overpressures, have been reproducibly obtained with the proper indium dopant concentration and reservoir temperature. This is achieved by low carrier concentrations, which is the result of balance between the concentrations of intrinsic defects, *i.e.* native point defects such as Cd vacancies, and extrinsic point defects, *i.e.* dopants.[65] By using the Bridgman technique, CdZnTe crystals of very large size can be grown, as shown in Figure 11.4.2.[56,60] These crystals are ideal substrates for epitaxial growth of HgCdTe for IR detector array fabrication because of the perfectly matched lattices.

P. Capper *et al.* have developed an improved process by adding the accelerated crucible rotation technique (ACRT) to the standard Bridgman system. The main effect of this is to stir the melt during growth and cause flatter solid–liquid interfaces, which improves the uniformity of the compositions and enables larger-diameter material (up from 13 to 20 mm), and reduces the densities of grain and subgrain. Based on this modification, HgCdTe crytals with improved radial compositional variations, good axial uniformity, and larger size can be prepared.[62] Figure 11.4.3 shows the FTIR map of a 15 mm diameter (from the conical tip region) short-wavelength slice ($\lambda \sim 1.4$ μm) demonstrating excellent radial uniformity.[62]

The travelling heater method (THM) is an alternative solution growth process particularly useful for the growth of binary and ternary semiconductor alloys, such as CdTe and CZT.[66–68] Recently, a newly developed 100 mm diameter CdTe single crystal, an important II–VI semiconductor for radiation detector, more than 300 mm in length was achieved by the THM method. It is the largest CdTe single crystal grown thus far by THM.[69] For CZT detectors, the thickness for detecting energy less than 200 keV is typically 6 mm or less, which is adequate for applications such as X-ray astronomy or gamma-camera imaging in nuclear medicine. However, applications in gamma spectroscopy require much thicker CZT to ensure sufficient stopping power for high-energy

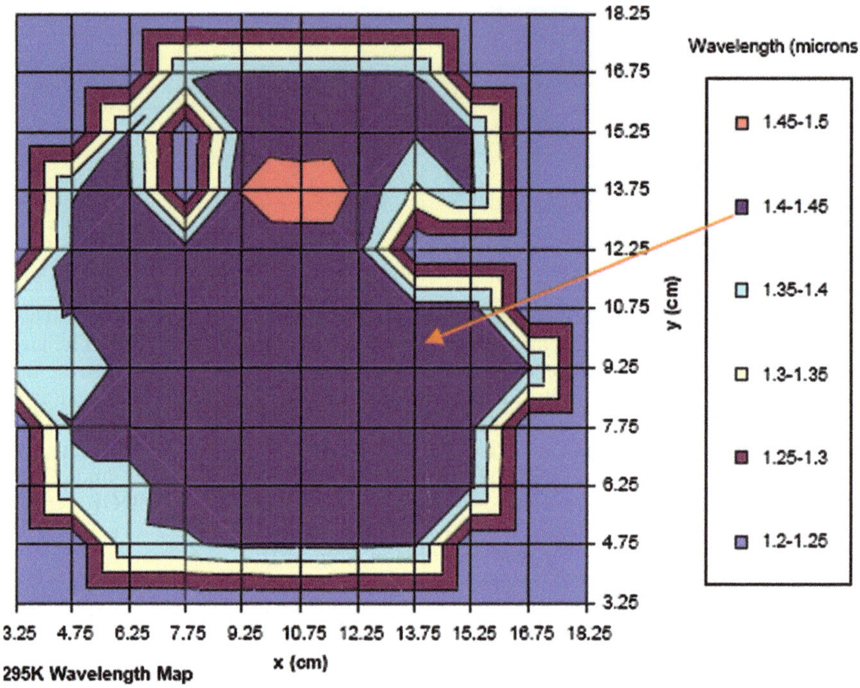

Figure 11.4.3 Radial distribution of wavelength (in mm at room temperature) in high-χ slice.[62]
(Reprinted from *J. Cryst. Growth*, 2005, **275**, 259, with permission from Elsevier.)

gamma photons. Meanwhile, larger-area crystals are needed to provide the required larger volume. Larger CZT crystals which will benefit the commercial deployment of CZT detectors in higher energy gamma application were thus required.[70] THM has been demonstrated as a practically effective approach to crystal growth, supporting volume production of CZT. H. Chen *et al.* synthesized 50 mm and 75 mm diameter CZT ingots by the THM process in 2007 and 2008, respectively. Figure 11.4.4 shows 100% single crystals cut from a 75 mm diameter CZT ingot grown by THM.[71]

The Czochralski method is another technique for growing crystals from melt, and there are some studies on bulk crystal II–VI compound like ZnSe. However, because of the high vapour pressures at melting points, sublimation of II–VI compounds during Czochralski growth needs to be suppressed. It is thus necessary to simultaneously apply high pressure and high temperature, which makes the commercial viability of mass production of these crystals by the Czochralski method uncertain. Therefore, the Czochralski method is not discussed in detail in this chapter.[72]

Besides the growth methods described above, there are also a few reports on alternative techniques for growing single crystals of II–VI wide-bandgap compounds, *e.g.* zone melting,[73] hydrothermal technique,[74] and solid-state

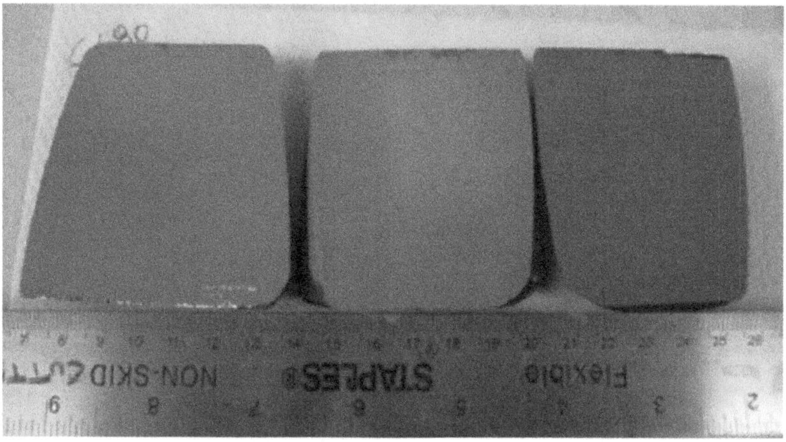

Figure 11.4.4 100% single crystals cut from a 75 mm diameter CZT ingot along the direction of their growth.[71]
(Reproduced with permission from *J. Appl. Phys.*, 2008, **103**, 014903. Copyright 2008, AIP Publishing LLC.)

recrystallization (SSR)[75] for growing bulk crystals of ZnSe, ZnO, ZnS, and CdTe. The methods are not as widely used as those described above, and thus are not included here.

11.4.3.2 Epitaxial Film Growth of II–VI Semiconductors

Epitaxial growth of II–VI compounds has been mainly carried out using liquid-phase epitaxy (LPE) and vapour-phase epitaxy (VPE). In addition to the conventional VPE, various epitaxial growth techniques for thin-film synthesis have been developed since 1960s, including hot-wall epitaxy (HWE),[76] MOCVD,[77] MBE,[78] metal-organic molecular-beam epitaxy (MOMBE),[79] and atomic layer epitaxy (ALE).[80] GaAs wafers with (100) direction are commonly employed as substrate for the II–VI film growth because of the lattice matching and availability of large, high-quality GaAs crystals.[11]

There are two transport mechanisms for the source materials in VPE techniques: physical vapour deposition (PVD) without chemical reaction, and chemical vapour deposition (CVD) in which the formation of the deposited film is the production of chemical reaction of the precursors on the substrate.[4] Thus far, the VPE technique has been widely employed in semiconductor epitaxial growth, *e.g.* high-quality epilayers of ZnS,[81,82] ZnSe,[83–85] ZnTe,[86] CdS,[87] CdSe,[88] and CdTe[89] have been grown on various substrates.

X-ray and gamma-ray detectors are mainly manufactured from bulk crystal CdTe and CZT which have adequate thickness to provide sufficient stopping power for high-energy X-ray or gamma photons. Because large-area bulk crystal CdTe is still expensive, and MBE and MOCVD are not cost effective for thick film growth, the VPE method has thus been used as an economical way to grow thick epitaxial CdTe layers with large area for radiation detector applications. CdTe

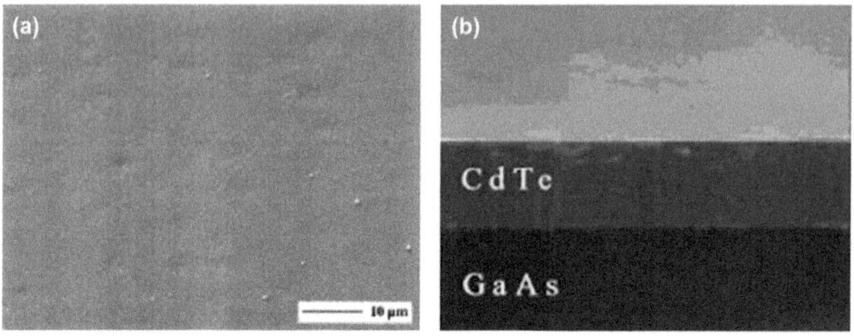

Figure 11.4.5 SEM micrographs of (a) surface plan view and (b) cleaved cross-section of a CdTe epilayer on GaAs substrate.[90]
(Reproduced from *Sensors*, 2009, **9**, 3491, with permission from MDPI.)

films with mirror-like, nearly featureless surfaces (as shown in Figure 11.4.5) have been demonstrated by VPE method.[90] The process is promising for fabricating X-ray detection in the 1–50 keV photon energy range, which requires sensitive material corresponding to CdTe crystals less than 500 μm thick.

MOCVD or MOVPE was introduced in 1968 as an improvement over conventional VPE.[91] The purity and abruptness of grown layers together with the flexibility of the technique make the growth of almost all compound semiconductors possible. To date, many source precursors of II–VI elements have been developed for growth of ZnSe, ZnS, and CdTe.[92–102] The great advantage of using metallo-organics is that they are volatile at moderately low temperatures. Since all constituents are in the vapour phase, precise electronic control of gas flow rates and partial pressures is possible.[4] MOCVD has been employed for growth of p-ZnSe epilayers,[103–105] HgCdTe,[60] thick CdTe,[106–110] ZnS,[111] and ZnTe[112,113].

CdTe and CdZnTe detector crystals are commonly grown by melt-growth techniques and the required high-resistivity crystal is obtained by compensating for background impurities and native defects with external dopants. However, melt-grown ingots presently contain numerous defects, and in some cases exhibit polycrystallinity, poor electrical transport properties, and inhomogeneity, which significantly hampers development of large-area and high-sensitivity, high-energy resolution imaging detectors. The vapour-phase growth of detector-grade single-crystal CdTe or CdZnTe films on large-area substrates (GaAs or Si) offers several advantages. As the growth occurs at a comparatively low substrate temperature, higher-quality crystals with a reduced density of defects can be produced. In addition, low-temperature growth offers stoichiometric crystals as well as high accuracy and easy control of the compensating dopants. As a result, homogeneous large-area crystals with good electrical properties can be obtained. However, growth of detector-grade single-crystal epitaxial films with sufficient thickness, as required to absorb incident X-rays or gamma-rays efficiently, presents a considerable challenge. In addition, the large difference in the lattice constants, thermal expansion coefficients, and chemical bonding characteristics between the film and substrates are other issues of concern.[106]

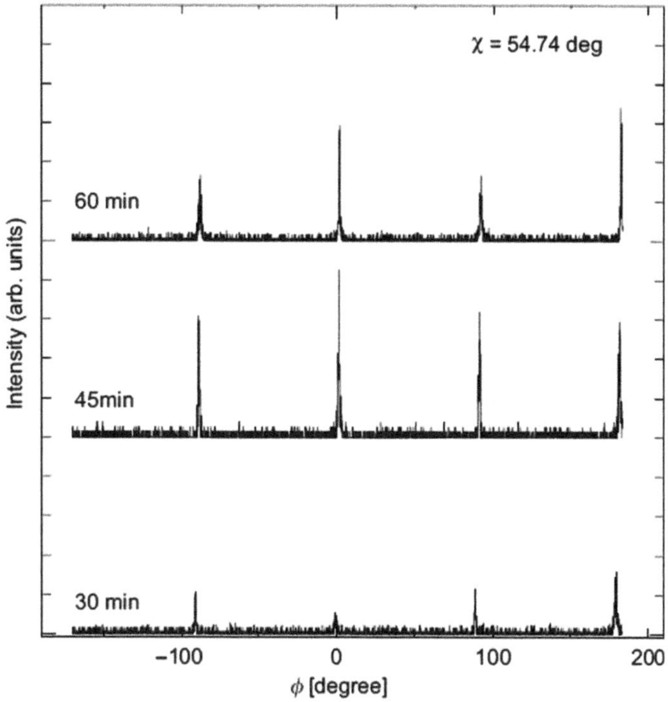

Figure 11.4.6 Line profiles of j scans of CdTe (111) plane corresponding to (100) plane for various TESb treatment time.[107]
(Reproduced from *J. Cryst. Growth*, 2008, **310**, 3440, with permission from Elsevier.)

M. Tahashi *et al.* demonstrated the epitaxial growth of (100)-oriented CdTe films by MOCVD on antimony-treated (100) GaAs substrate by in-plane X-ray measurements which showed fourfold symmetries as depicted in Figure 11.4.6.[107]

By introducing a thin GaAs interfacial layer, M. Niraula *et al.* achieved thick (>260 μm) single-crystal CdTe films on (211) Si substrates. Figure 11.4.7 shows the photoluminescence (PL) spectrum from 100 μm-thick film grown at a substrate temperature of 600 °C and 560 °C. The similarities on the PL results, along with the results from the XRD measurements, imply that crystallinity and film quality remain unchanged with increasing the growth temperature or growth rates. The spectra in Figure 11.4.7 are dominated by a sharp peak at 1.5902 eV, which is usually observed in p-type CdTe samples. This emission peak can be assigned as the recombination of exciton bound to neutral acceptor (A_0, X), with cadmium vacancy or its complex probably acting as neutral acceptors. The small but distinct peak at 1.5937 eV (Figure 11.4.7a, inset) is the donor bound exciton emission (D_0, X), with iodine probably acting as a neutral donor. Weak emissions visible at 1.5958 and 1.5973 eV are due to free exciton emissions. High-intensity and sharp bound excitonic emission line, with a FWHM value of 0.9 meV, together with free exciton emission and weak

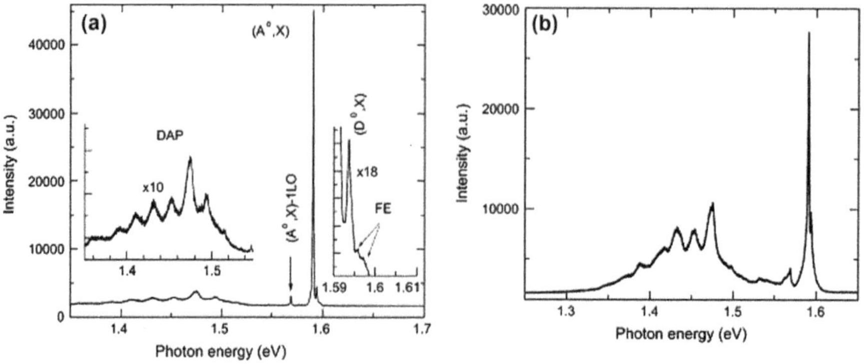

Figure 11.4.7 (a) 4.2 K photoluminescence spectrum of a 100 μm-thick CdTe film on (211) Si substrate grown at 600 °C with a 65 μm/h growth rate. Shown in the inset are the deep level emission band and the edge emission band at a magnified scale. (b) PL spectrum of a 100 μm-thick film grown at 560 °C.[106]
(Reproduced from *IEEE Trans. Nucl. Sci.*, 2009, **56**, 836, with permission from IEEE.)

donor–acceptor pair (DAP) emission band indicate that grown film is of high crystalline quality.[106]

Moreover, quantum wells (QW) of II–VI wide-bandgap compounds have recently become a focus in the field. Successful pulsed laser operation at 77 K in ZnCdSe/ZnSe/ZnMgSSe QW-structure separated-confinement heterostructures has been realized by MOCVD.[114]

MBE was developed at the beginning of the 1970s to grow high-purity high-quality compound semiconductor epitaxial layers.[115–117] Now, it has become one of the most important techniques for growing compound semiconductor epilayers. Large-area epitaxial growth of HgCdTe films by MBE has been the important process for the very large IR focal-plane array (FPA) industry and attracted much research interest.[118–121] R. Sporken *et al.*[122] first demonstrated the successful growth of CdTe films on silicon by MBE which were then used for epitaxial growth of HgCdTe. The (111)*B* oriented layers are made of two domains which are rotated by 90°, and the layer with only one domain can be grown on Si(100) misoriented by 8°. $Hg_{1-x}Cd_xTe$ has been grown by MBE on (111)*B* CdTe and characterized by reflection high-energy electron diffraction (RHEED), which provides information of the surface reconstructions and the growth rate in two-dimensional growth mode.[123]

Epitaxial growth of HgCdTe layer on large-area substrates by MBE has been well developed and has been employed in commercial productions. A uniformly low defect density on epitaxial HgCdTe layers grown on 6 inch (15 cm) diameter Si can be achieved with optimized growth conditions, as shown in Figure 11.4.8.[121]

MBE has also been applied for growth of other II–VI semiconductors such as CdTe,[108,124] CdSe,[125] ZnTe,[126] and ZnSe.[127] D. Greiffenberg *et al.* grew 100 μm thick films of CdTe on semi-insulating (100) GaAs substrates by PVT in a MBE facility for the radiation detector application. The grown layers were highly oriented, as revealed from X-ray pole figure measurements as shown in Figure 11.4.9.[108]

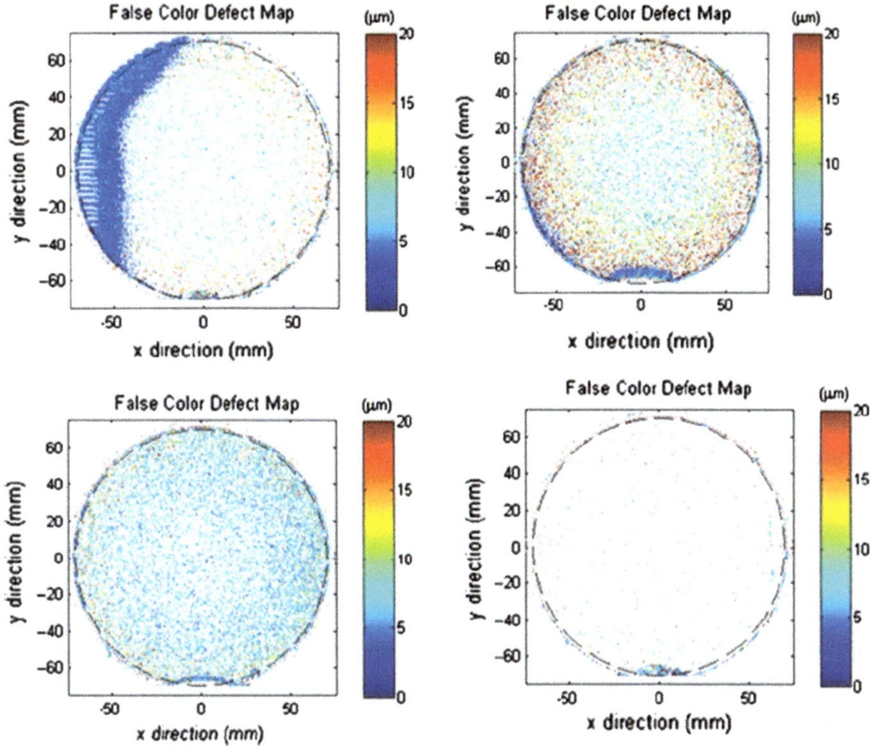

Figure 11.4.8 Defect maps of double-layer heterojunction (DLHJ) HgCdTe layers on 6-inch-diameter Si. In the top-left image, a very high density of small defects is shown on the left side due to a problem in surface preparation. In the top-right image, the defect size increases abruptly near the edge and this is attributed to the larger Te/Hg flux ratio. In the bottom-left image, a relatively high density of smaller size defects are present due to a low Te/Hg flux ratio. In the bottom-right image, a uniformly low defect density is observed when the growth conditions are optimum.[121] (Reproduced from *J. Electron. Mater.*, 2011, **40**, 1706, with permission from Springer.)

Although ZnTe-based green LEDs have been achieved, commercialization of the products is still difficult due to the high cost and small size of current ZnTe bulk substrates. Therefore, larger wafer size ZnTe substrate is highly desirable. The employment of thick, epitaxially grown ZnTe as low-cost substrate has been proposed for various device applications. High-quality thick ZnTe layers have been successfully grown on substrates including GaAs, InP, InAs, and GaSb (001) using MBE. The misfit dislocations present at the various ZnTe/substrate interfaces have been studied extensively.[126] For example, HRTEM images (Figure 11.4.10) indicate that Lomer edge dislocations and 60 partial dislocations are the dominant defects present at the interfaces. The defect densities in films become lower as the ZnTe thickness is increased.[126]

MOMBE is one of the variations of the MBE system.[128,129] The difference is that metal-organic gaseous sources are used as the source materials. Therefore,

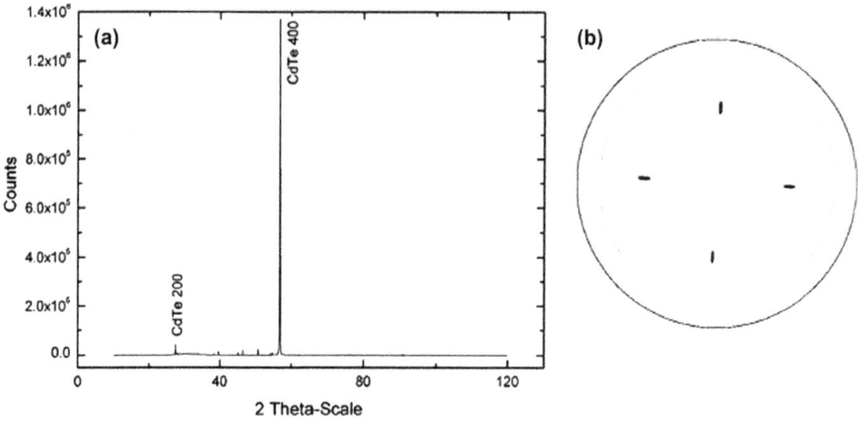

Figure 11.4.9 (a) $\theta/2\theta$ scan of a CdTe layer. The only observable peaks are located at 27.84° and 56.87°, indicating a (100) orientation. (b) Typical 111 pole figure of a CdTe thick film. The four peaks at $\chi = 54.73°$ show a discrete orientation.[108]
(Reproduced from *IEEE Trans. Nucl. Sci.*, 2007, **54**, 773, with permission from IEEE.)

the growth technique has the merits of both MOCVD and MBE. MBE or MOMBE techniques have been used to grow epilayers of almost all II–VI semiconductors.[130–132] In particular, they are very successful in growing superthin layers, such as single quantum wells (SQW), multiple quantum wells (MQW),[133,134] and nanostructures.[135]

The HWE technique has been developed to grow II–VI epilayers, such as CdTe,[136] CdS,[137] CdSe,[138] ZnTe,[139] and also to producing heterostructures for laser and photovoltaic detector fabrication.[140] Since large-area bulk crystal CdTe is expensive, VPE has attracted more and more interest in alternative structures based on the growth of thick (~100 μm) films of CdTe on foreign substrates, *e.g.* GaAs[141] and glass.[142] By using the HWE method, Wang *et al.*[136] and G. M. Lalev *et al.*[143,144] achieved high-quality epitaxial CdTe films, on GaAs and Si respectively, with optimized conditions, although there are large differences in lattice mismatch and thermal expansion. The CdTe epilayers showed good crystal quality as characterized by XRD and PL spectra.

Except for the above techniques, epitaxial growth of II–VI semiconductors can also be realized by ALE and LPE. ALE can grow thin film epitaxially at low temperature and control the deposited layers in superlattice structures with good accuracy and excellent thickness uniformity. ALE has been used to grow II–VI wide-bandgap compounds[145,146] and prepare monolayers (ML).[147] Electrochemical atomic layer epitaxy (E-ALE) is ALE in an electrochemical environment which is developed based on ALE. In recent years, E-ALE has attracted some interest in growing II–VI materials, such as CdTe,[148] HgSe,[149] and CdS[150]. The LPE method, which occurs at near thermodynamic equilibrium conditions, has been used to grow epilayers with low native defect density, such as ZnS,[151] ZnSe,[152] ZnSSe,[153] ZnTe,[154] and HgCdTe.[155] LPE

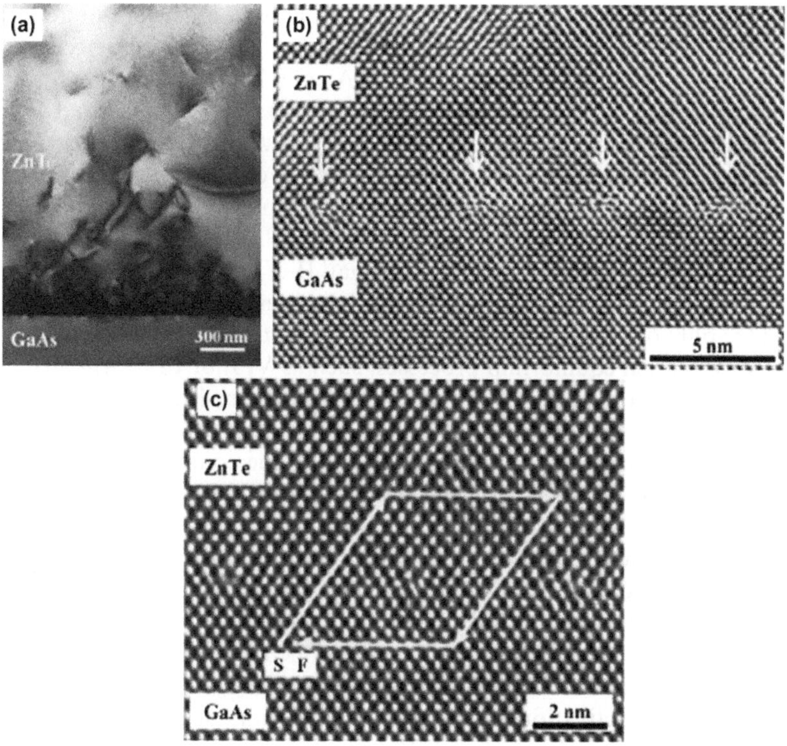

Figure 11.4.10 (a) Low magnification HREM image of ZnTe/GaAs (001) sample. (b) High magnification HREM image. (c) High-resolution image showing ZnTe/GaAs interface with Burgers' circuit. S and F indicate the start and finish points for the Burgers' circuit analysis.[126] (Reproduced from *J. Cryst. Growth*, 2011, **323**, 127, with permission from Elsevier.)

growth technology is now very mature for production of both first- and second-generation tactical HgCdTe detectors.[60] Remarkably, the best current structural quality HgCdTe material is grown using LPE onto lattice-matched substrates of CdZnTe.[3]

In some work, II–VI semiconductor films are not deposited epitaxially but as polycrystalline films by techniques such as pulsed laser deposition (PLD),[156–158] chemical bath deposition (CBD),[159–163] Close-spaced sublimation (CSS),[164–171] sputtering, or other processes.[172–174] However, these deposition techniques have also been used for growing II–VI semiconductor thin films. For instance, CBD-synthesized CdS[175] and CdTe grown by CSS have been widely used in commercial production of CdTe solar panels due to simplicity and cost-effectiveness.[167]

11.4.4 Doping and Metal–Semiconductor Contacts

Achievement of device-grade materials is the first and basic step for the electronic and optoelectronic applications of II–VI semiconductors; however, controlled doping and proper metal–semiconductor contacts are prerequisites for fabricating practical devices, such as LEDs, solar cells, and detectors.

Theoretically, elements from Group III (Al, Ga, In) can substitute for Zn or Cd and act as donors. Halogen donors (F, Cl, Br, I) will normally bond with Zn or Cd vacancies. Group I elements, including Cu, Ag, Au, will diffuse quickly and become interstitial ions, or they can enter the crystal lattice to replace Zn and Cd as a donor. A phosphor (P) impurity normally brings a deep acceptor level in sulfide and selenide.[11]

The first report of doping for wide-bandgap II–VI semiconductors like ZnTe and ZnSe date from the 1950s,[176,177] and it was indicated that they could be doped for only one type of conduction. Similarly, it was demonstrated that, by using conventional equilibrium processes, such as VPE or LPE, wide-bandgap ZnSe, ZnS, CdS could only be n-type doped with low resistivity, whereas ZnTe could only get p-type conduction.[1] In II–VI semicondutors, CdTe is the only exception which can be doped for either n- or p-type conduction. Research on applications of II–VI semiconductors during the 1960s and 1970s was hindered primarily because of the difficulties in doping.[4]

It was well known that the bipolar electrical conduction of II–VI materials was mainly limited by self-compensation by native defects such as vacancies, interstitials and anti-sites, lattice relaxation and related formation of deep levels, and amphoteric dopant incorporation, which induce the conduction type opposite to that induced by the dopants. Moreover, the low solubility of dopants and high activation energy needed for donors or acceptors also cause the degradation of bipolar doping.[11,178]

W. Fashinger *et al.* summarized relative band edge positions of II–VI compounds, as shown in Figure 11.4.11. The experimentally obtained doping type is indicated by different shadings. Their relative band edge position and dopability are compared: compounds with high conduction band edges (*e.g.* MgSe) are not n-dopable, while materials with low valence-band edges (*e.g.* ZnS) are not p-dopable.[179,180]

Ion implantation has also been employed to dope II–VI semiconductors. Ion implantation can provide high concentration doping at low temperature, which benefits restraining the defect that cause self-compensation. However, further studies on ion-implanted ZnSe revealed that ion implantation introduced high concentrations of impurity together with high-density point defects.[11]

With emphasis on wide-bandgap materials, following the advance of crystal growth techniques, such as MBE and MOCVD, the controlled doping of II–VI semiconductors has recently made great progress. P-type ZnSe was achieved by Li doping by MBE and MOCVD.[181,182] However, a net acceptor density of 8×10^{16} cm^{-3} seems to represent the upper limit for Li doping. At higher Li concentrations, strong compensation occurs which renders the ZnSe material highly resistive. However, Li impurities are unstable in ZnSe at temperatures above 300 C. The possibility of heavy diffusion of Li seems to have obstructed the practical applications to optoelectronic devices constituted with thin multilayered structures. P-type behaviour has also been reported, employing the isoelectronic impurity oxygen as a dopant in ZnSe layers grown by MBE.

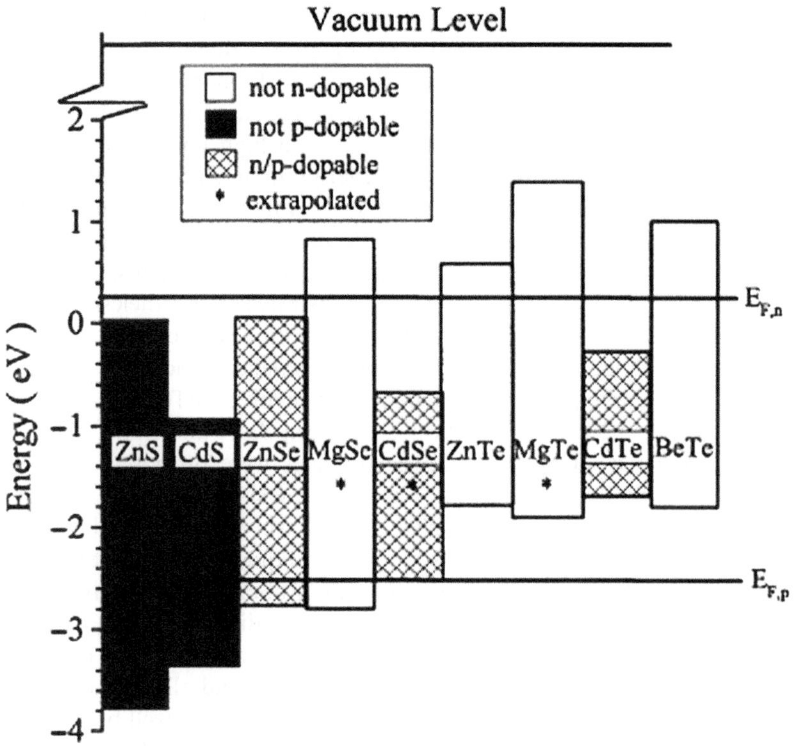

Figure 11.4.11 Relative band edge positions of various II–VI compounds. The experimentally obtained dopability is indicated by different shadings.[179,180] (Reproduced from *J. Cryst. Growth*, 1999, **197**, 557, with permission from Elsevier.)

However, net acceptor concentrations in ZnSe:O layers appear to be low; the largest net acceptor density reported so far is 1.2×10^{16} cm^{-3}.[183]

By introducing nitrogen (N), better p-type doping for ZnS,[184,185] ZnSe,[183,186,187] and CdSe[188] has been achieved. Important progress was made in 1990, when N plasma was introduced into MBE growth as a p-dopant which allowed a high p-doping levels (around 10^{18} cm^{-3}) for ZnSe. Higher rf power and lower substrate temperatures led to increased N incorporation with the maximum hole concentration realized.[183,187] Based on their success in doping, M. A. Haase *et al.* reported blue emitting devices and the first blue-green ZnSe-based laser.[189] The ZnSe-based device was the most promising blue laser before the extensive applications of the gallium nitride (GaN)-based laser. In addition to ZnSe, Cu-doped CdS films were also reported to be p-type.[190–192]

After two decades of research and development, significant progress has been made on controlled doping for II–VI semiconductors and many II–VI-based devices have been made commercially available.[1,4,193,194]

Metal–semiconductor contact is another essential issue in device fabrication, and metal contacts with II–VI semiconductors have been extensively studied.[195,196]

It is normally not easy to obtain good ohmic contacts to wide-bandgap semiconductors, particularly p-type doped materials, due to the low valence-band edges, such as p-type ZnSe.[180] In the first work on ZnSe-based LDs, the room-temperature laser was suppressed due to the large barrier between Au and p-type ZnSe.[189] Besides, the contact caused a high voltage threshold for laser emission. Because of surface Fermi level pinning, ZnSe Schottky barrier heights are nearly independent of metal work functions; and large voltage drops across the resistive contacts are responsible for the high operating voltages of ZnSe-based lasers. To minimize contact problems, researchers constructed laser devices on p-type GaAs substrates. Nevertheless, the large valence-band offset between GaAs and ZnSe continues to cause a significant voltage drop.[197] Since ZnTe exhibits a higher valence-band edge and much higher p-doping levels, p-contacts to ZnSe that involve ZnTe were developed. In order to reduce the valence-band offset between ZnTe and ZnSe, which acts as an additional barrier for holes, two strategies were proposed: Fan *et al.* introduced a ZnSe/ZnTe superlattice with a quasi-continuously varied Te content between the ZnSe and the ZnTe.[198] Ishibashi *et al.* designed a series of ZnTe QW in ZnSe, which was supposed to allow sequential resonant tunnelling of holes from the ZnTe through the QW to the ZnSe.[199] The decisive element for a good contact is believed to be an optimized grading in the first case and a precise design of the resonant tunnelling structure in the second case. Both approaches led to the development of ohmic contacts and allowed a reduction of the laser threshold voltage at room temperature to values well below 5 V.[200]

In CdTe solar cells, due to the high electron affinity of p-type CdTe, an ohmic contact or tunnelling contact is required to increase the cells' efficiency. For this purpose, etching with nitric-phosphoric (NP) acid was proposed to obtain Te-rich layers on the top of CdTe before metal (Cu) deposition. The large amount of holes in the Cu_xTe diffusion layer formed during the annealing led to a good ohmic contact with CdTe layers.[201] An alternative approach is to etch with nitric-acetic (NA) acid which can generate even better ohmic contact with reduced series resistance as compared to the etching by NP acid. ZnTe was also introduced as a buffer layer between the CdTe layer and Cu back-contact, which produces an electrical field to reduce recombination loss of minority carriers at the back-contact surface. Kim *et al.* showed that double-layer structure with Cu-doped ZnTe (ZnTe:Cu) buffer and Cu_xTe back-contact could be a promising configuration for back-contact fabrication, since ZnTe:Cu offered a good alignment of the valence band with CdTe, and Cu_xTe had an excellent contact with CdTe:Cu. Moreover, it was demonstrated that the device stability could be further improved if Cu was replaced by Al as the back-contact metal.[201]

For II–VI semiconductor-based detectors, CdZnTe radiation detectors are usually fabricated by forming ohmic contacts with Pt or Au in metal–semiconductor–metal (MSM) structures.[90] While there are few reported data on electrical characteristics of contacts to HgCdTe in IR detectors, due primarily to the proprietary nature of the work, Blank and Goldberg have published an extensive review on current flow through metal–semiconductor contacts, with references to HgCdTe contacts. The various mechanisms proposed for Schottky barrier formation were reviewed by Rhoderich and Williams.[3]

11.4.5 Optoelectronic and Photovoltaic Applications

The advances in the synthesis and controlled doping techniques of II–VI semiconductors lead to their optoelectronic device applications in a wide range of IR lasers and detectors, solar cells, LDs, and LEDs, non-linear optical materials, and radiation detectors. Many of them such as solar cells, detectors, and LEDs have been available commercially and play an important role in civil and military fields.

11.4.5.1 Light-Emitting Devices

Solid-state lighting is a rapidly evolving technology, and is now virtually certain to some day displace traditional lighting in applications ranging from the low-power spot illuminator to the high-power area illuminator. Moreover, there is a considerable headroom for continued evolution even after this initial displacement.[202] There are several kinds of light-emitting devices such as LEDs, phosphors, light-emitting capacitors, and optically or electron-beam pumped lasers based on II–VI semiconductors.

Most group II–VI semiconductor materials have direct wide bandgaps with high optical absorption and emission coefficients (except HgSe and HgTe which are semimetals), high exciton binding energy, and high electro-optical coefficient, which make them the leading candidates for optoelectronic applications. Particularly, the bandgaps of II–VI semiconductors cover the light spectrum from UV to the visible region, benefiting their diverse applications like medical treatment and diagnostics, spectroscopy, data storage, display, and blue-green light-emitting devices.

11.4.5.1.1 LEDs and LDs

LEDs and LDs are composed of chips of semiconducting materials doped with impurities to create a p-n junction and powered by injected electric current. Enlighten by the preliminary studies of light emission from SiC, ZnO,[203] and III–V compounds,[204–208] II–VI semiconductors are very attractive in light emission since their direct bandgaps cover a broad light spectrum from near-UV to IR.[209–211] However, until the 1980s, despite the considerable research interest, the light-emitting applications based on II–VI semiconductors were not well realized because of the difficulties in obtaining high crystallographic quality materials and stable p-n junctions.

In the late 1970s, the development and application of non-equilibrium growth techniques such as MBE and MOVPE to II–VI compounds enabled the growth of high-quality and high-purity crystals and stimulated intensive new research into the properties of these materials, which promoted the rapid progresses of growth and low-resistivity doping of II–VI semiconductors in the 1980s and early 1990s.[182,183,187,212–218] Based on the success in obtaining device-

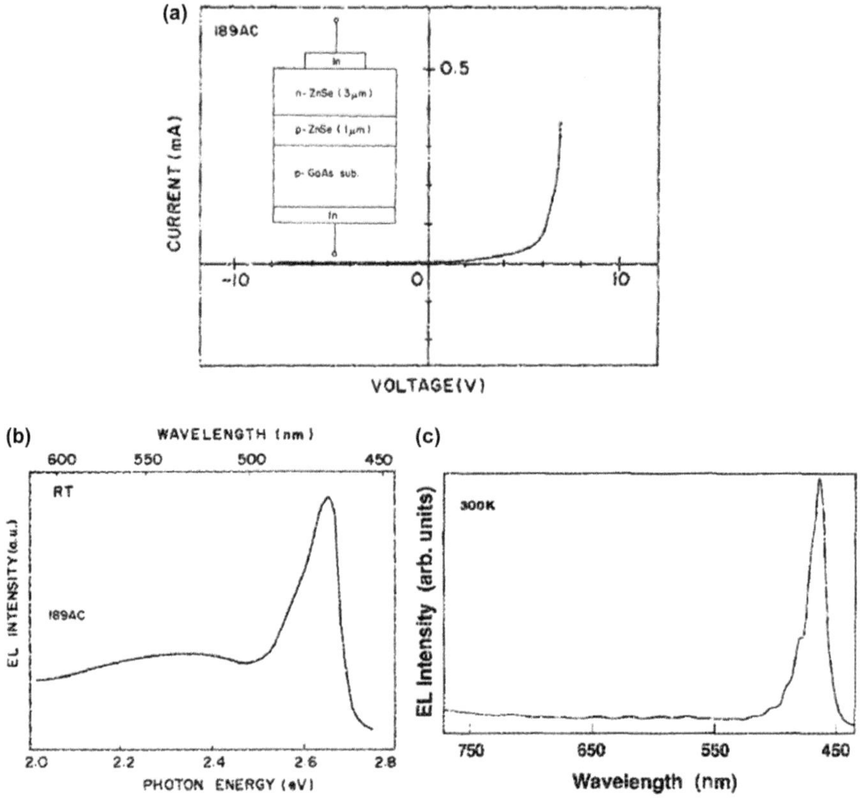

Figure 11.4.12 (a) Current-voltage characteristic of a ZnSe *p-n* diode. Inset is the structure of the p-n diode. (b) Electroluminescence spectrum at room temperature of ZnSe *p-n* diode with Li-doped ZnSe as p-type layer.[182] (Reproduced with permission from *Appl. Phys. Lett.*, 1988, **52**, 57. Copyright 2008, AIP Publishing LLC.) (c) Electroluminescence spectrum at room temperature of ZnSe *p-n* diode with N-doped ZnSe as p-type layer.[183]
(Reproduced with permission from *Appl. Phys. Lett.*, 1990, **57**, 2127. Copyright 2008, AIP Publishing LLC.)

grade materials and controlled doping of II–VI semiconductors, the prototypes of ZnSe-based LEDs have been fabricated and reported.[182,219,220] The typical structure of LEDs with Li-doped p-type ZnSe and its electroluminescence spectrum at room temperature are shown in Figure 11.4.12. The spectrum is composed of a blue emission band peaking at about 467 nm and a broad band located in the longer-wavelength region.[182] As a comparison, the spectrum from the LED with N-doped p-type ZnSe layer is shown in Figure 11.4.12c. Obviously, the latter has a better performance with sharper dominant emission in the blue region and without a broad band located in the longer-wavelength region.

II–VI diode lasers did not emerge until QW structure was developed.[189] In 1986, strained QW heterostructure, which was first proposed in 1963 by Herber

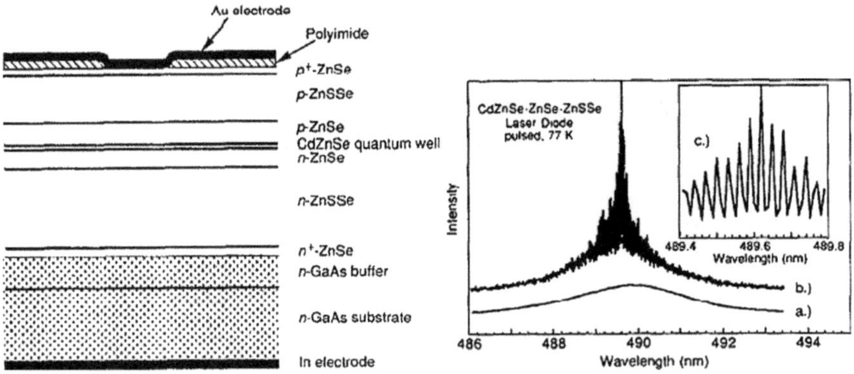

Figure 11.4.13 (left), a cross section of a blue-green laser diode. (right), optical spectra for a blue-green laser diode: (a) below threshold; (b) above threshold; and (c) an expanded view of the lasing spectrum.[189] (Reproduced with permission from *Appl. Phys. Lett.*, 1991, **59**, 1272. Copyright 2008, AIP Publishing LLC.)

Kroemer,[221] was suggested to improve III–V laser performance because the reduced threshold current lowered the in-plane hole effective mass.[222,223] Earlier experience with GaAs and related semiconductor lasers showed the obvious benefit of the two-dimensional density of states in reducing the threshold current density in a QW laser. Following the idea of reducing threshold current by employing a ZnCdSe/ZnSe/ZnSSe QW structures as the active region and the achievements of ZnSe LEDs, the blue-green (490 nm) ZnSe-based pulsed laser operation at 77 K were demonstrated by Haase *et al.* from 3M,[189] which was the first blue-green semiconductor LD and made ZnSe-based materials the most promising material for fabricating efficient blue light-emitting devices before the rapid development of the GaN technique. The typical structure of a LD and its optical spectrum are shown in Figure 11.4.13.[189] At injection currents below threshold, the spontaneouse mission from one of these devices has a FWHM of about 3 nm. Above threshold, the diode operates in many longitudinal modes which are separated by 0.03 nm.

The first demonstration of ZnSe-based LD induced great research interest to improve laser performance, such as room-temperature continuous-wave operation,[224,225] single lateral mode operation, high-power operation, and low threshold current and longer lifetime which were realized on the basis of Mg- and Be-incorporated alloys, proper contacts to p-type materials, the reduction of stacking-fault density, *etc.* The covalent radius of Mg is larger than that of Zn, so Mg chalcogenides have higher ionicity, which means Mg chalcogenides have larger lattice constants and bandgap energy than those of Zn chalcogenides. The structure of LD employing ZnMgSSe alloy is shown in Figure 11.4.14.[225] The LD can operate with RT CW mode. Beryllium chalcogenides have stronger bond strength due to their pronounced covalent bonding, which has impact on the defect generation and propagation, and the lifetime as well.[226,227] The detailed discussion and history can be found elsewhere.[2,228,229]

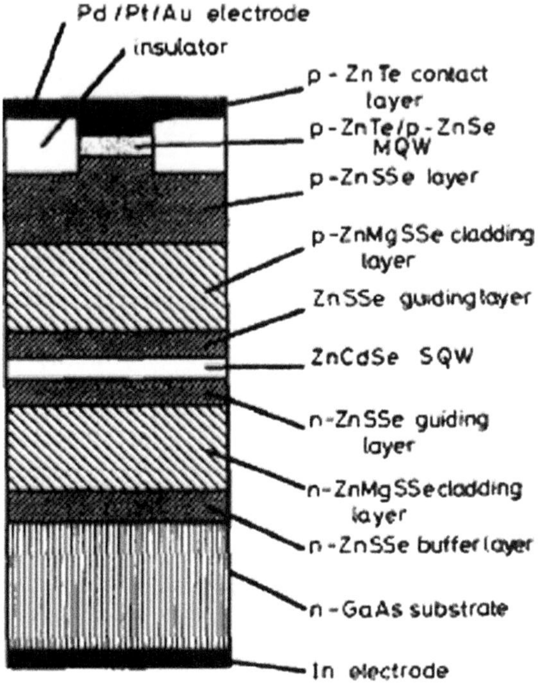

Figure 11.4.14 Schematic structure of ZnCdSe/ZnSSe/ZnMgSSe SQ W-SCH laser diode.[225]
(Reproduced from *Electron. Lett.*, 1993, **29**, 2194, with permission from IET.)

A major focus of research on II–VI light emitters has been to illuminate and mitigate device degradation mechanisms, which is vital for its future commercial applications.[230–235] It has been demonstrated that the degradation is related to crystal defects that originate from pre-existing defects and propagate throughout the active region, acting as non-radiative recombination centres; and room temperature CW basing is probably prevented due to rapid propagation of such defects.[236] Significant progress in device lifetime has been achieved by circumventing the relatively weak bonding and pronounced electrical activity and mobility of defects in ZnSe-based materials. Be-containing alloys have been employed to enhance the bond strength of the material's active region. The implementation of ZnMgBeSe electron-blocking layers has also enabled the reduction of recombination-enhanced defect reaction which is triggered by electron overflow into the p-cladding layer. For example, multi QW (MQW) have enabled 10,000 h and longer lifetimes at SEI. By adopting a double cladding layer, the rapid degradation was suppressed and the lifetime tendency is similar to the LEDs consisting of a III–V semiconductor system.[237,238] It is interesting to note that white LEDs were invented unintentionally at SEI. Through the development of blue LEDs, the researchers found that the impurity doped ZnSe substrate emitted yellow light by absorbing blue

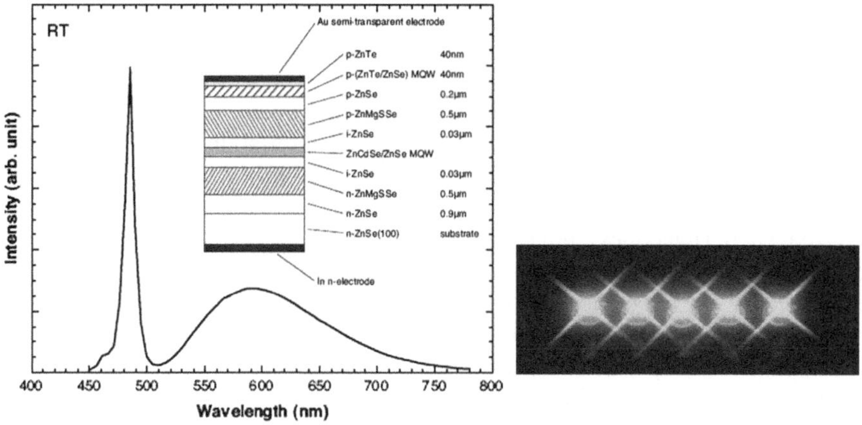

Figure 11.4.15 (left) Emission spectrum of ZnSe-based white LED. Inset is the schematic cross section of the epitaxial structure of ZnSe-based white LED. (right) ZnSe-based white LED actually emitting white light.[7] (Reproduced with permission from *SEI Technical Review*, 2011, **72**, 25.)

light from the active layer. It was then discovered that the white LED could be fabricated by mixing the blue light from the active layer and the yellow emission from the ZnSe substrate. Figure 11.4.15 shows the spectrum and the photo of the white light emission. ZnSe-based white LEDs have advantages such as (1) lower operating voltage, (2) higher controllability of colour attributed to the fluorescent material free process, and (3) simpler structure due to the conductive substrate, as compared with the GaN-based white LEDs on sapphire. However, the devices also have fatal drawbacks, *i.e.* the device efficiency deteriorated easily due to the extension of the crystal defects in the material. The life of the device was a serious problem to be solved.[7]

In its early stage of the development, ZnSe-based LEDs were developed vigorously all over the world. However, in the mid-1990s, as the GaN-based LEDs had made rapid progress in performance, the mainstream of white LEDs has shifted to GaN-based material. Only SEI continued the development of ZnSe-based material.[7] Nobody could have imagined at that time that the popularity of GaN and ZnSe could be completely reversed within a few years.[239] The development of new II–VI alloys and device designs on various substrates have greatly expanded the visible wavelength range for II–VI emitters. Overall, this impressive material's development showcases the flexibility and potential of II–VI alloys as candidate light-emitting materials.[229]

11.4.5.1.2 Light-Emitting Capacitors (LECs)

Thin-film electroluminescent (TFEL) display panels can be used in high contrast displays. The basic alternating current TFEL (AC-TFEL) mechanism is explained by the trapping and recombination of electrons and holes. When a high electric field is applied, electrons and holes are formed by the p–n junction of n-type ZnS and p-Cu_xS. The injected electrons are trapped by shallow donor centres such as Cu_xS-related defects, and the injected holes are trapped by recombination centres such as Cl^- ions. When the field is changed, an emission is generated by the recombination of electrons and holes.[240] The most efficient materials has been the ZnS:Mn phosphor discovered in 1974. AC-TFEL devices with ZnS:Mn as the active layer based on the Inoguchi model are still the most efficient TFEL devices, and panels based on this technology have been utilized where ruggedness, reliability, high contrast, and high viewing angles are required; such as in niche medical, transport, and military applications.[241] For example, AC-TFEL device can be used as a backlight for mobile phones and personal digital assistants (PDAs) because of its flexibility, uniform light emission, and low power consumption.

The device structure of an LEC panel from Ceelite Technologies, the first manufacturer of LEC devices, is shown in Figure 11.4.16.[242] It can be modelled as a capacitor consisting of manganese-activated ZnS as the phosphor and yttrium oxide for the sandwiching insulators. AC-TFEL emission has been only realized in ZnS-based phosphors such as ZnS:Cu,Cl phosphor, which shows blue-green emission depending on the concentration of Cu, and ZnS:Cu, Cl, Mn phosphor, which shows orange or white emission depending on the concentration of Mn and Cu. The energy transfer from the host to the activator has been studied in detail and an excellent review is provided by Daud *et al.*[243,244]

11.4.5.1.3 Optically and Electron-Beam Pumped Lasers

The difficulties in doping II–VI semiconductor materials prevented practical LED and LD applications in the 1960s and 1970s, and thus II–VI wide-bandgap semiconductors had emitted laser only in photopumped[245,246] or electron-beam-excited modes.[228,247–249]

Nakanishi *et al.* showed the photo-pumped lasing of ZnSe/ZnSSe multiple QWs up to 400 K.[245] The threshold at room temperature was 10.5 kW cm^{-2}, which was the lowest value reported for II–VI semiconductor photo-pumped lasers. It should be noted that in the experiment the active region was directly excited by a dye laser at 450 nm and the excitation was more effective than that

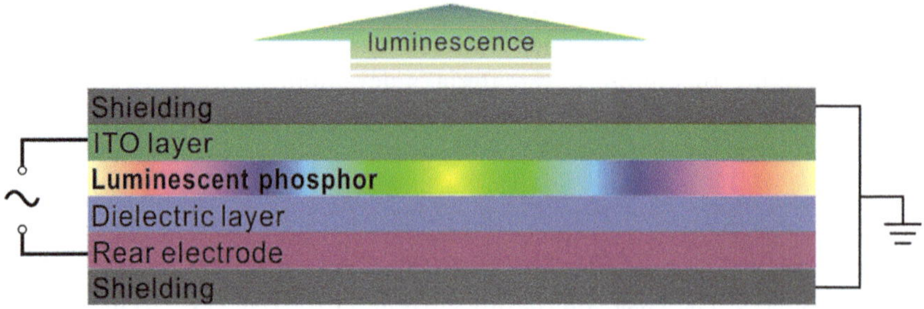

Figure 11.4.16 Schematic of the structure of an LEC device from Ceelite Technologies[242]

in many of the experiments using the 337 nm or 335 nm lines of N, or Nd:YAG lasers, respectively, exciting the cladding layers. For the ZnSe/ZnMgSSe double heterostructure, photo-pumped lasing was reported at room temperature with N, laser excitation at the threshold of 70 kW cm^{-2}.[193] H. Wang *et al.* investigated the light-emitting properties of ZnS/ZnSe and ZnSTe/ZnSe strained II–VI superlattices,[246] and observed room-temperature stimulated emission by photopumping in the deep-blue spectral region. An upper limit of threshold carrier density required to achieve lasing was estimated to be 3×10^{18} cm^{-3}. The peak of stimulated emission was several tens of meV red-shifted from the spontaneous emission peak, suggesting that exciton–exciton scattering was involved in the stimulated emission process.

The current green LEDs are manufactured using GaN-based materials, but the efficiency varies by colour (wavelength) and the performance of green LEDs lags behind other colours. Recently, P. I. Kuznetsov *et al.* reported interesting work in which ZnCdSe/ZnSSe MQW structures fabricated by MOVPE were demonstrated to function as an electron-beam-pumped vertical cavity surface emitting laser (VCSEL) with resonant periodic gain. Based on the grown MQW structure, green VCSEL was fabricated.[250] Figure 11.4.17 shows the main characteristics of the e-beam pumped VCSEL based on the 40 QW structure. Lasing at wavelength of 542 nm with 3 W output power was achieved in longitudinal pumping by a scanning electron beam at RT. The laser efficiency was about 4%.

Based on optical pumping, 3M combined its colour-converting QW structure with a blue LED to produce a green emitter with an efficacy of 181 lm W^{-1} at 350 mA in 2009.[251,252] II–VI compound CdMgZnSe alloys QW

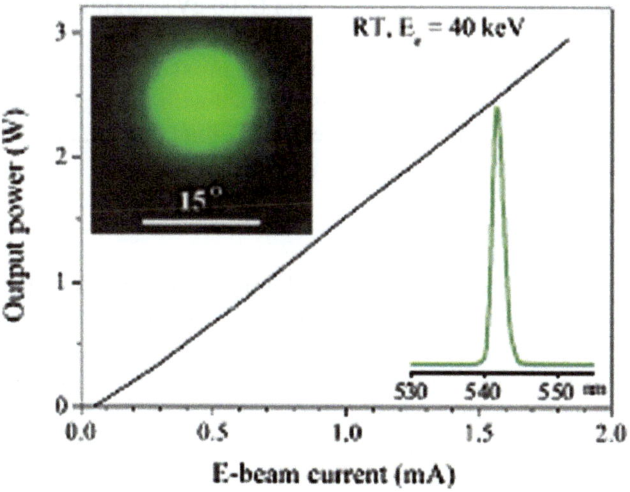

Figure 11.4.17 Output power of VCSEL on the base of MQW structure *vs* e-beam current. Far field pattern (the left top insert) and laser spectrum (the right bottom insert) are shown.[250]
(Reproduced from *Phys. Stat. Sol.* (c), 2006, **3**, 771, with permission from John Wiley and Sons.)

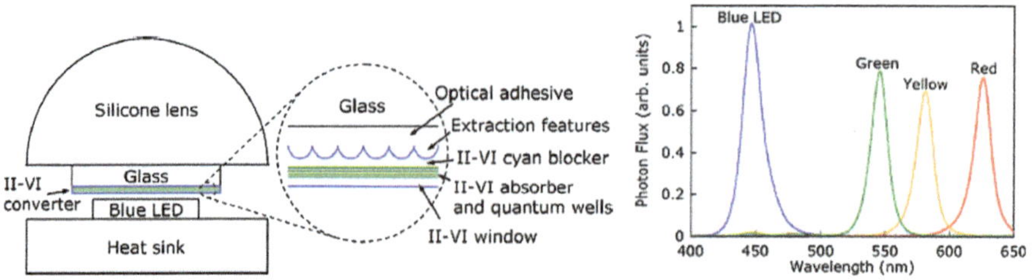

Figure 11.4.18 (left) Diagram of the blue LED and a II–VI colour converter with inset showing details of the converter. (right) Measured spectra of the blue LED, compared to that of three II–VI colour converters when each was placed on top of the LED.[252]
(Reproduced with permission from *Appl. Phys. Lett.*, 2010, **96**, 231116. Copyright 2008, AIP Publishing LLC.)

heterostructures were demonstrated to work as efficient, narrow-spectrum, photoluminescent colour converters to generate green, yellow, and red light when pumped by commercially available blue GaInN LEDs. The QWs were grown on InP substrates by MBE. The emission wavelength can be controlled by the composition and thickness of the QWs. With appropriate light extraction and epitaxial design, the QW colour converters can efficiently provide high-power, narrow-spectrum output at wavelengths in the green through red portion of the visible spectrum.

Figure 11.4.18 (left) shows a diagram of the blue LED and a II–VI colour converter. The spectrum of the LED has a peak emission at 446 nm (2.78 eV). The blue light from the GaInN LED is incident on the CdMgZnSe window, which has a bandgap wavelength of 405 nm (3.06 eV) making it transparent to the blue light, thereby eliminating surface recombination. Figure 11.4.18 (right) shows the spectrum emitted by the blue LED compared to the emission spectra of the of II–VI converters when each was pumped with the LED. The spectra of the converters have peaks at 545 nm (green), 581 nm (yellow), and 626 nm (red) and each has a FWHM of 15 nm.[252]

11.4.5.1.4 II–VI Phosphors

A phosphor is a material that exhibits luminescence, including both phosphorescent materials and fluorescent materials, which show a slow decay in brightness (>1 ms) and the emission decay taking place over tens of nanoseconds respectively. Phosphorescent materials have been used in radar screens and glow-in-the-dark applications, whereas fluorescent materials are common used in cathode ray tube (CRT) and plasma video display screens, sensors, and white LEDs.[244]

Phosphors are usually made from a suitable host material with an added activators (dopants or impurities) which are the luminescence centres. (In rare cases dislocations or other crystal defects can play the role of the impurity.) The wavelength emitted by the emission centre depends on the impurity and the surrounding crystal structure.

Phosphors based on wide-bandgap II–VI compounds have been extensively applied in lighting, displays, scintillators, and medical X-ray imaging. The wide gap nature and the flexibility to incorporate large concentrations of a wide variety of activators make these materials the most efficient phosphors. The best-known type is a Cu-activated ZnS and the Ag-activated ZnS which is the most efficient CRT phosphor.

Extensive work has been carried out to gain an understanding of the impurity-induced luminescence, such as luminescent centres including transition-metal ions and rare-earth ions. The understanding of the interaction of native defects, *e.g.* vacancies, and the impurities incorporated in II–VI materials paved the way for the understanding of impurity-induced luminescence. The most studied phosphor materials include (Zn, Cd)S:Ag, Cl, (Zn, Cd)S:Cu, Cl, ZnS:Cu, Al, ZnS:Ag, Al. The detail data and reviews have been summarized in the relevant literature.[4,244,253]

11.4.5.2 CdTe Solar Cells

The development of solar cells has gone through two generations: a first generation of crystal Si-based solar cells, and a second generation of thin-film solar cells. Currently, third-generation solar cells strive to drastically reduce the cost of solar energy below the current level of around \$1 W^{-1} to less than \$0.2 W^{-1}.[5,254,255]

It has been well accepted that CdTe is the most attractive semiconductor for photovoltaic application among Si, III–V, and II–VI compounds.[4,256,257] CdTe has a direct bandgap energy of about 1.47 eV, which is very close to the theoretically calculated optimum value for solar cells. CdTe has also a high absorption coefficient ($>10^5$ cm^{-1} at a wavelength of 700 nm), so that approximately 90% of the incident light is absorbed by a layer thickness of only 2 μm (compared with around 10 μm for Si). Furthermore, CdTe can be made from polycrystalline thin films on glass substrates, thus avoiding the need for expensive single-crystal substrates. CdTe is one of the most promising photovoltaic materials available for use in low-cost high-efficiency solar cells. The theoretical efficiency for the CdS/CdTe photovoltaic cell is ≈30% at AM 1.5,[4] being the first and only thin-film photovoltaic technology to surpass crystalline silicon PV in cheapness.[258–260] The CdTe PV technique is growing rapidly and CdTe is now the second most utilized solar cell material in the world.[4,34,261]

Research on CdTe can be traced back to the 1950s.[262–264] The development of CdTe solar cells was initially motivated by their potential space application in communications satellites. In 1963, the first thin-film solar cell of the n-CdTe/p-Cu_2Te type was demonstrated based on CdTe single crystals and polycrystalline films by a GE research laboratory.[265] The cell had an efficiency of 6%, but instability of the device (caused by Cu diffusion) led to the quest for an alternative hetero-partner for CdTe. Soon after, the superstrate configuration was recognized to be more efficient as shown in Figure 11.4.19. Interdiffusion between CdS and CdTe, which could easily happen in the

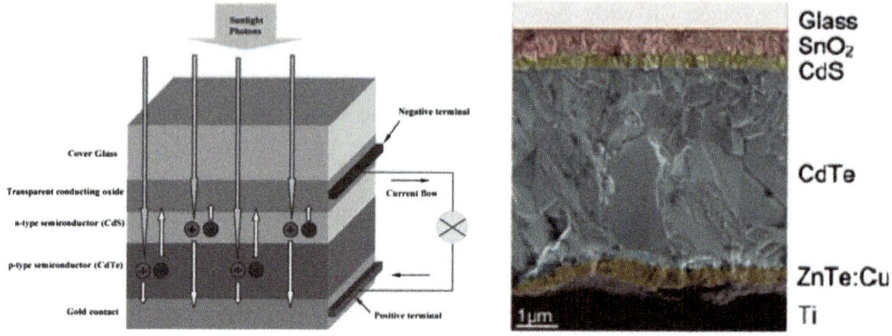

Figure 11.4.19 (Left) A schematic layout of a conventional CdS/CdTe p-n junction thin-film solar cell[270] (Reproduced from Ref. 270.) and (right) the CdTe solar cell structure revealed by cross-sectional SEM image.[269] (Reproduced from IEEE 4th World Conference on Photovoltaic Energy Conversion (WCPEC-4), 2006, **1**, 317, with permission from IEEE.)

process, was believed to be a reason for the superior performance of the superstrate configuration. In 1972, Bonnet and Rabenhorst presented a p-CdTe/n-CdS heterojunction solar cell which also had an efficiency of 6%.[266] Ten years later, Tyan *et al.*[267] presented a superstrate thin-film cell with more than 10% efficiency. The cell was grown by closed-space sublimation—the leading method of later commercial devices—and oxygen was incorporated during film deposition. Encouraged by this achievement, several companies started in-house development. The goal was to find the best low-cost deposition method that allowed high throughput and could deliver more than 10% efficient modules. Cells of about 10% sunlight-to-electricity efficiency were fabricated by the early 1980s at Kodak, Matsushita, Monosolar, and Ametek.[268,269]

Since 1990, it has been known that the activation of CdTe by chlorine treatment could substantially increases the device efficiency. Since then, chlorine activation using specific and critical process parameters has been part of the process. Based on this technique, the efficiency of CdTe solar cells were upgraded greatly. T. L. Chu *et al.*[271] achieved the CdTe solar cells with an efficiency of 14.6%. The increased conversion efficiency appears to be related to the interface reaction between CdS and CdTe during the deposition of CdTe at high temperatures. The formation of CdS_xTe_{1-x} shifts the electrical junction from the metallurgical interface into CdTe, thus improving the electrical and photovoltaic characteristics of the junction. Besides, Chu used resistive tin oxide (SnO_2) as the buffer layer and then thinned the CdS from several micrometres to less than 0.5 μm in thickness to allow more light through. Chu reached 15% efficiency in 1991, the first thin-film cell to do so, as verified at the National Renewable Energy Laboratory (NREL). His achievements made CdTe a primary candidate for commercialization. Ferekides *et al.* also later demonstrated the efficiency reached 15.8%.[272]

In 2001, X. Wu *et al.* from NREL achieved a long-standing record total-area efficiency of 16.5% for CdS/CdTe polycrystalline thin-film solar cell with a high

FF of 77.34% and high J_{sc} of near 26 mA cm^{-2}.[273] Close-space sublimation was used for the absorber preparation, and a modified high mobility window layer of Cd-stannate developed at NREL was employed to improve device performance and reproducibility. The details of the modification are: (1) a high-quality cadmium stannate (Cd_2SnO_4, or CTO) transparent conductive oxide to replace the conventional SnO_2; (2) a high-resistivity ZnSnOx (ZTO) buffer layer located between the transparent conductive oxide (TCO) and the CdS film; (3) an oxygenated, nanocrystalline CdS:O window layer with higher bandgap; (4) a modified CTO/ZTO/CdS/CdTe device structure (Figure 11.4.20 (left)); (5) two new manufacturing processes for producing high-efficiency CdTe modules; (6) CuTe:HgTe-doped graphite paste, followed by a layer of Ag paste was applied as the back-contact layer.[274] The current-voltage and relative quantum efficiency curves for the cell are shown in Figure 11.4.20 (right).

Very recently, an exciting bulletin was issued by First Solar, the leading manufacturer of CdTe solar cells. According to their announcement, the efficiency of CdTe solar cell has been upgraded to 17.3% in 2011 and 14.4% total area efficiency for CdTe PV modules in 2012 which were confirmed by the NREL.[6]

Photovoltaic structures on lightweight and flexible substrates have attracted much interest for their several advantages over heavy glass-based structures in both terrestrial and space applications: (1) the cells, mounted on flexible foils are not fragile; (2) the requirements of the supporting structures are minimal; and (3) they can be wrapped on to any suitably oriented or curved structures. The specific power of the solar cells is an important factor in space applications and hence development of photovoltaic devices on lightweight substrates is desired.[275] In 2009, EMPA, the Swiss Federal Laboratories for Materials Testing and Research, demonstrated a 12.4% efficient solar cell on flexible polyimide substrate.[276]

Although much progress has been achieved, the efficiency of CdTe solar cells is still much lower than the theoretical value. Currently, several challenges may be faced for further development of CdS/CdTe thin-film solar cells:[201] (1) short minority carrier lifetime due to the recombination of electron-hole pairs at the defect centres in CdTe layers and at the interface between CdS and CdTe; (2) insufficient transparency of TCO and CdS window layers; (3) lack of good ohmic contact between CdTe layers and back contacts; and (4) possibility in doping p-type CdTe films in a stable way. Techniques coming up in recent years concentrate mainly on the first three challenges above. Some novel ideas, such as bifacial and tandem structure nanostructured geometries have been proposed for further studies.[201] The opportunity to increase current has been almost fully exploited, but more difficult challenges associated with junction quality, with properties of CdTe, and with contacts have not been as successful. Improved doping of CdTe and increased understanding of key processing steps (*e.g.* cadmium chloride recrystallization and contacting) are keys to progress. Since CdTe has the optimal bandgap for single-junction devices, it may be expected that efficiencies close to or exceeding 20% should be achievable in practical CdTe cells. PV Modules with efficiency of 15% would then be possible.

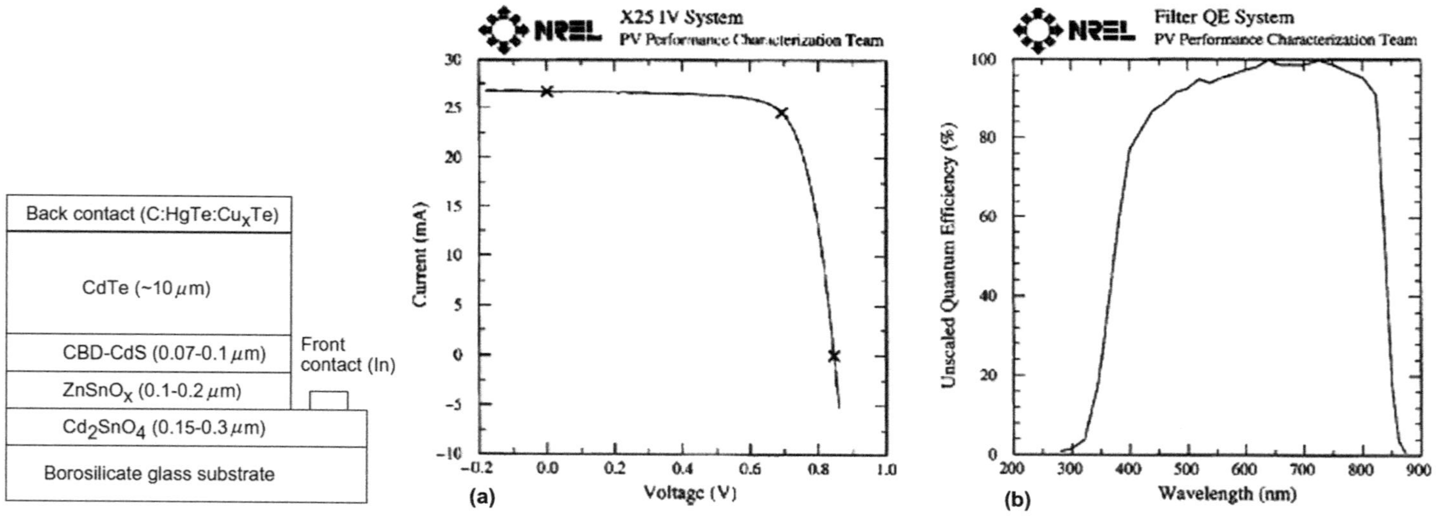

Figure 11.4.20 (left) Modified CTO/ZTO/CdS/CdTe device structure. (right) Current-voltage (a) and relative quantum efficiency (b) curves for 16.5%-efficient CdS/CdTe polycrystalline thin-film solar cell.[274]
(Reproduced from Solar Energy, 2004, **77**, 803, with permission from Elsevier.)

11.4.5.3 Detectors

Semiconductor detectors are devices that use semiconductor materials to detect photons or traversing charged particles. Extensive important military, civil, industrial, and scientific applications depend heavily on semiconductor detectors. Owing to the wide range of physical properties, II–VI semiconductors have been the very important materials for fabrication of various kinks of detectors.

11.4.5.3.1 HgCdTe IR Photodetectors

IR detection based on HgCdTe is the main commercial application for narrow-bandgap II–VI semiconductors.[4] As the Cd composition increases, the direct bandgap for $Hg_{1-x}Cd_xTe$ gradually increases from a negative value for HgTe to a positive value for CdTe. The bandgap tunability results in IR detector applications that span the short wavelength IR (SWIR: 1–3 μm), middle wavelength (MWIR: 3–5 μm), long wavelength (LWIR: 8–14 μm) and very long wavelength (VLWIR: 14–30 μm) ranges.[277] The favourable properties which make HgCdTe ternary alloy an unique and almost ideal IR detector material include: (1) its large optical absorption coefficients that, together with long diffusion lengths, enable high quantum efficiencies (approaching 100% in most cases); (2) the inherent recombination mechanisms that lead to long carrier lifetimes, low thermal generation rates, high operating temperatures, and long diffusion lengths, (3) the band structure of HgCdTe, enabling ideal electron-initiated avalanche photodiodes, with single-carrier multiplication and no excess noise; (4) device-grade films can be grown routinely by MBE and MOVPE on CZT, Si, Ge, and GaAs so that high-quality mesa and planar homojunctions and heterojunctions can be formed; (5) photodiodes with low $1/f$ noise and high radiation tolerance can be achieved by a native CdTe passivation; (6) MCT also has a low dielectric constant for low junction capacitance which benefits high-frequency response.[3,278]

HgCdTe IR detectors have exhibited great progress in recent decades, mainly due to their important applications in remote sensing, IR astronomy research, and military fields. Military technology uses HgCdTe sensors for night vision. In particular, the US Air Force makes extensive use of HgCdTe on all aircraft, and to equip airborne smart bombs. A variety of heat-seeking missiles are also equipped with HgCdTe detectors.

The first report of the synthesis of alloy HgCdTe *via* the semimetal HgTe and the wide-bandgap CdTe was published by the Royal Radar Establishment.[279] This milestone paper reported both photoconductive as well as photovoltaic response at wavelengths extending out to 12 μm. Up to now, HgCdTe has initiated the development of three generations of detector devices. The first generation, linear arrays of photoconductive detectors, was produced in large quantities and is in widespread use today. The second generation, two-dimensional arrays of photovoltaic detectors, is now in high-rate production. Third-generation devices, suggested here to encompass the more exotic device structures embodied in two-colour detectors, avalanche photodiodes, and hyperspectral arrays.[60] The latest comprehensive reviews including growth and properties of HgCdTe, and principles and performance of HgCdTe detectors can be found in monographs.[3,280]

11.4.5.3.1.1 First-Generation Detectors. The first HgCdTe devices were photoconductive because of the simplicity of the technology, and the relative ease of achieving near-ideal IR performance and excellent reliability. The device detects the change in resistance of the element when the photon flux is changed.[3] First-generation HgCdTe detectors consist of linear arrays of photoconductive elements. The diagram of a basic HgCdTe photoconductor is shown in Figure 11.4.21.[60] These detectors are tractable with low-noise preamplifiers, generally having emitter-coupled bipolar front ends.[60] However, the low impedance of the photoconductor makes it unsuitable for coupling to direct injection gates of silicon CT devices or MOSFETs, therefore each element requires a lead-out through the vacuum encapsulation to an off-focal-plane amplifier. The problems of power consumption and cryogenics with too many elements set a practical limit to the size of photoconductor arrays for commercial thermal imaging to a few hundred elements.[3]

The SPRITE (signal processing in the element) detector shown schematically in Figure 11.4.22 is a variation of the standard photoconductive device which

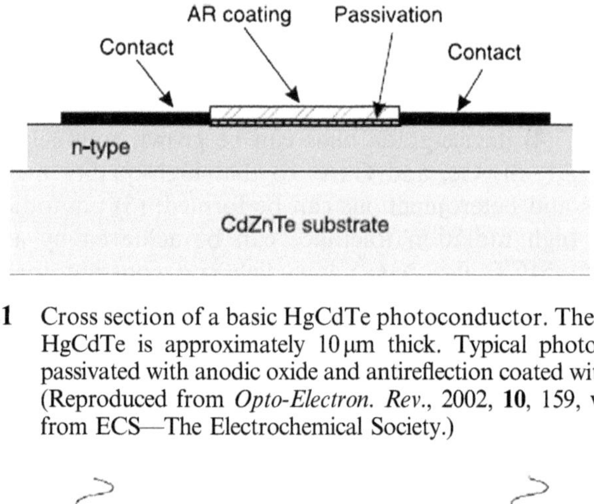

Figure 11.4.21 Cross section of a basic HgCdTe photoconductor. The n-type layer of HgCdTe is approximately 10 μm thick. Typical photoconductors are passivated with anodic oxide and antireflection coated with zinc sulfide.[60] (Reproduced from *Opto-Electron. Rev.*, 2002, **10**, 159, with permission from ECS—The Electrochemical Society.)

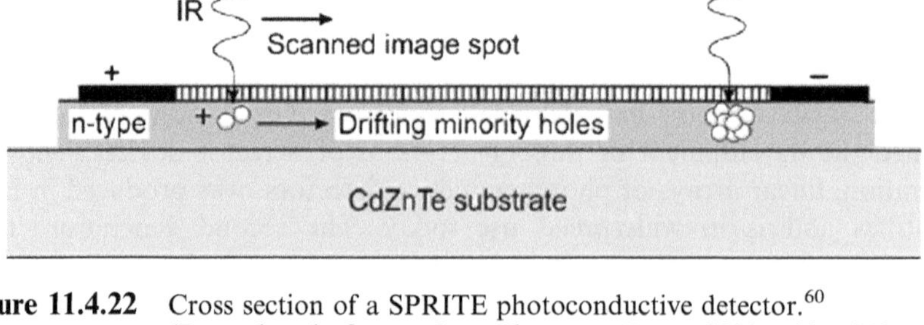

Figure 11.4.22 Cross section of a SPRITE photoconductive detector.[60] (Reproduced from *Opto-Electron. Rev.*, 2002, **10**, 159, with permission from ECS—The Electrochemical Society.)

adopts a smart way to amplify the sensitivity of photoconductive device. The SPRITE detector provides signal averaging of a scanned image spot which is accomplished by synchronization between the drift velocity of minority carriers along the length of a photoconductive bar of material and the scan velocity of the imaging system.[277] The image signal then builds up a bundle of minority charges which is collected at the end of the photoconductive bar, effectively integrating the signal for a significant length of time and thereby improving the signal-to-noise ratio. A family of thermal imaging systems has utilized this configuration; however, a decline in its usage is now observed.[277]

11.4.5.3.1.2 Second-Generation Detectors. Second-generation HgCdTe devices composed of two-dimensional arrays of photovoltaic elements (Figure 11.4.23). Photovoltaic arrays have inherently low power consumption and can be easily connected to a silicon integrated circuit with indium bump bonds at the pixel level to produce a retina-like FPA. Second-generation thermal imaging cameras based on such arrays broke through the performance limits imposed by photoconductive arrays in first-generation cameras.[3] First demonstrated in the mid-1970s, indium bump bonding of readout electronics provides for multiplexing the signals from thousands of pixels onto a few output lines, greatly simplifying the interface between the vacuum-enclosed cryogenic sensor and the system electronics. The signal-to-noise ratio of the sensor was improved upon the increase of the number of detector elements in an array, to the extent that they can collect proportionally more signal from the scene.[60]

The key technology needed to make photovoltaic devices possible was surface passivation. Where the depletion region intercepts the surface, there is often enhanced generation of carriers due to the presence of a high density of interface states. This can be exacerbated if the surface passivation is not properly optimized to give a flat band potential at the junction.[4] With the advent of CdTe passivation, HgCdTe photodiodes could finally be reliably passivated. CdTe passivation is stable during vacuum packaging bake cycles

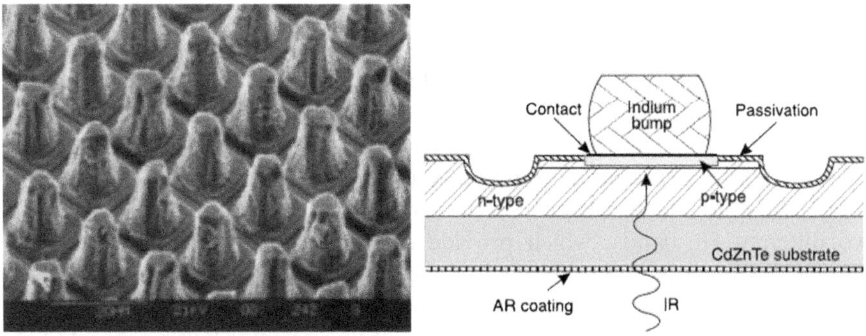

Figure 11.4.23 (left) HgCdTe photovoltaic array with indium bumps. (right) Cross section of a mesa-etched HgCdTe photodiode.[60]
(Reproduced from *Opto-Electron. Rev.*, 2002, **10**, 159, with permission from ECS—The Electrochemical Society.)

Figure 11.4.24 Medium-wavelength infrared image from a 384×288 element array (OSPREY) and the Kenis System. (Courtesy Denel Ltd.)[3] (Reproduced with permission from John Wiley and Sons.)

and shows little effect from the radiation found in space applications. This development has made possible the full-scale production of second-generation devices.[60]

MWIR technologies have now reached maturity in large two-dimentional (2D) arrays with near-ideal performance and few pixel defects. The emphasis for MWIR detectors is to develop wafer-scale processes to provide large, economical 2D arrays for staring thermal imaging cameras. Currently, IR cameras based on HgCdTe arrays are commercially available and are producing remarkable sensitivities down to 10 mK, over an order better than first-generation cameras. Figure 11.4.24 shows a representative image captured by a state of the art HgCdTe 2D array.[3]

11.4.5.3.1.3 Third-Generation Detectors. In recent years, much effort has been devoted to the development of third-generation devices which have substantially enhanced capabilities over an ordinary photodiode.[60,118,121,281–283] It has been commonly accepted that third-generation devices include two-colour or multicolour detectors, avalanche photodiodes, and hyperspectral arrays. Moreover, the devices should have megapixel arrays with high density, higher operating temperature, fast readout rates, very low noise-equivalent temperature difference (NETD), and retina-level signal processing.[4] The technical developments which are keys to third-generation devices include dry etching, VPE, optical coatings, and advanced readout concepts.

The two-colour detector can realize simultaneous readout of flux levels in two different IR bands, which provide enhanced target discrimination and identification, combined with lower false-alarm rates.[280] It is a remarkable solution to the problem of pixel registration in dual-band sensors. $Hg_xCd_{1-x}Te$ two-colour detectors are made with a stack of two detector layers separated by a p-type layer as a common electrode. Figure 11.4.25 illustrates the schematic structure of the device. The Band 1 and Band 2 alloy compositions can be any two x-values as long as Band 1 has a higher x-value than Band 2. Vapour-phase

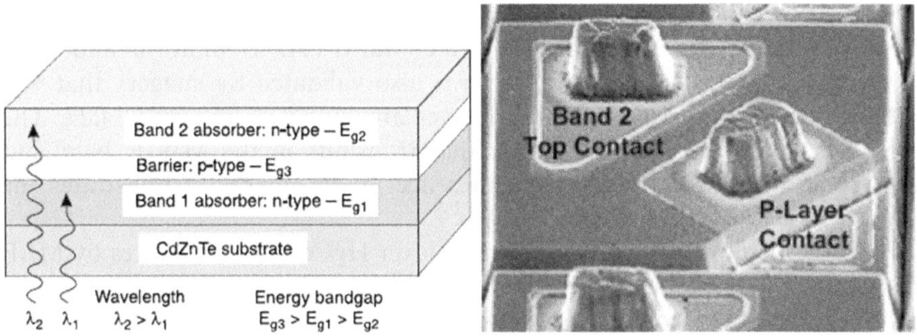

Figure 11.4.25 (left) Structure of a HgCdTe two-colour detector pixel. (right) Structure of a simultaneous two-colour pixel. One indium bump contacts the n-type layer of Band 2. The other bump contacts the p-type layer between Band 1 and Band 2.[60]
(Reproduced from *Opto-Electron. Rev.*, 2002, **10**, 159, with permission from ECS—The Electrochemical Society.)

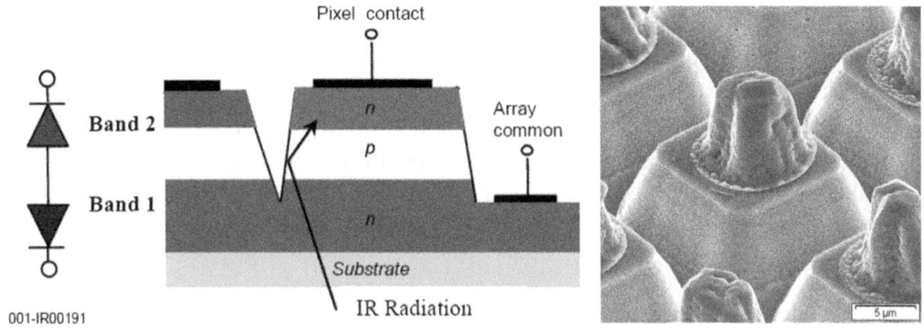

Figure 11.4.26 A cross-section diagram (left) and a SEM photograph (right) of a single-bump 20 μm pitch two-colour detector.[282]
(Reproduced from *Proc. of SPIE*, 2005, **5783**, 331, with permission from SPIE.)

growth is the preferred method for growing this structure. Two-colour detectors have been successfully demonstrated in a variety of spectral combinations.[60]

Raytheon Company, one of the leading manufacturers of IR detectors, also aims to develop the two-colour detector for the U.S. Army. The Raytheon two-colour detector is based on a bias-selectable back-to-back diode structure. Raytheon's single-bump, single-mesa implementation of these two-colour detectors is shown in Figure 11.4.26. This detector is in an n-p-n triple-layer heterojunction structure of HgCdTe grown by MBE. The bottom n-type absorbing layer has a shorter cutoff (MWIR) than the top n-type absorbing layer (LWIR). The mesas are formed using a reactive dry-etch process that produces a smooth and steep sidewall. This structure achieves approximately 100% optical fill factor in each band due to total internal reflection of the incident radiation off the mesa sidewalls.[121,282]

The MWIR and LWIR NETD for Raytheon FPAs have demonstrated that the responsivity and noise of the two-colour FPAs is uniform and well-behaved. This excellent performance was also validated by imagery that was obtained at f/5. Still frames of the images are shown in Figure 11.4.27. The subject is holding a sheet of plastic that transmits in the MWIR band but absorbs in the LWIR band. The difference in the images demonstrates the multispectral capability of these FPAs.[282]

Better control over the growth of multilayer HgCdTe epistructures by MBE has led to the development of a new generation of highly sensitive avalanche photodiodes (APDs).[284] Figure 11.4.28 illustrates an APD structure which consists of six MBE-grown layers. Short-wavelength HgCdTe APDs are able to exploit a very favourable property in the band structure of the alloy when the bandgap is about 0.90 eV. For materials of this alloy composition, the energy required to excite an electron from the top of the valence band to the bottom of the conduction band is identical to the energy for the excitation of an electron from the top of the split-off valence band to the top of the valence band.[60]

The avalanche effect in the high-field region multiplies the number of photoexcited carriers by the avalanche gain. This raises the signal level, which itself may be highly useful for raising low signal levels above the amplifier noise. A second consideration is the amount by which the noise is increased by the avalanche process. To address the problem, it is advantageous to have a large asymmetry between the avalanche gain of holes and electrons. The band structure of HgCdTe gives k-values of 0.1 or less—a highly favourable ratio of hole to electron multiplication during avalanche conditions, resulting in very little noise gain.[60]

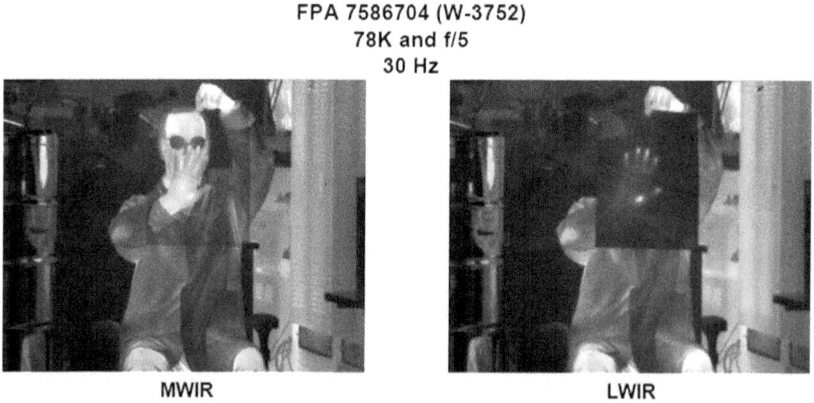

Figure 11.4.27 Example of imagery obtained at 78 K for an Army DBFM 640×480 M/LWIR FPA at f/5 field of view and 30 Hz frame rates. The subject is holding a thin piece of plastic which transmits in the MWIR band but absorbs in the LWIR band.[282]
(Reproduced from *Proc. of SPIE*, 2005, **5783**, 331, with permission from SPIE.)

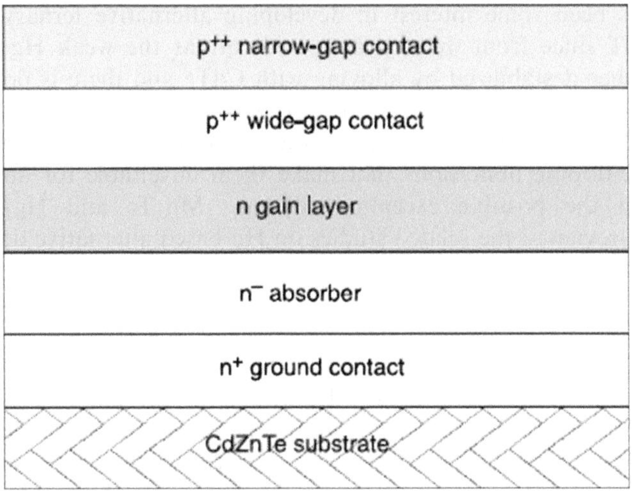

Figure 11.4.28 Avalanche photodiode structure consisting of six HgCdTe layers.[60] (Reproduced from *Opto-Electron. Rev.*, 2002, **10**, 159, with permission from ECS—The Electrochemical Society.)

When a second-generation array is combined in a scanning imager having a means to selectively illuminate each row with a different spectral band, a hyperspectral imager is obtained. Such instruments can image a scene in hundreds of spectral bands per frame, generating a hypercube image. Such a capability is anticipated to revolutionize disciplines such as land resource utilization which today rely on just a handful of spectral bands. Hyperspectral arrays have been built to cover the visible through LWIR spectral regions. HgCdTe and other detector materials such as silicon and InSb have been used in hyperspectral assemblies.[60]

The main limitation of LWIR HgCdTe-based detectors is that they need cooling to temperatures near that of liquid nitrogen (77 K) to enable high sensitivity by reducing noise caused by thermally generated carriers. MWIR HgCdTe cameras can be operated at temperatures accessible to thermoelectric coolers with a small performance penalty. Hence, HgCdTe detectors are relatively heavy compared to bolometers and require maintenance. A number of concepts to improve performance of photodetectors operating at near room temperature have been proposed.[47] Recent considerations of the fundamental detector mechanisms suggest that near-perfect detection can be achieved in the MWIR and LWIR range without the need for cryogenic cooling. A detailed introduction is given by A. Rogalski.[3]

Technology of the HgCdTe alloy continues to develop and improve. This trend is related both to the persistent demand for this material for military and commercial applications as well as the unique properties of HgCdTe that make it possible to use the structures with QWs on the basis of this material in the studies of physical effects and phenomena such as the quantum Hall effect, spin-orbit splitting, and so on.[194,285–287]

There has been some interest in developing alternative ternary alloys to replace MCT since from theoretical considerations the weak Hg–Te lattice bond is further destabilized by alloying with CdTe and there is potential for obtaining materials with increased hardness and detectors with better temperature stability. However, most other compounds have crystal growth problems or doping limitations that make them unsuitable for device fabrication, with the possible exception of $Hg_{1-x}Mn_xTe$ and $Hg_{1-x}Zn_xTe$.[3] A. Rogalski reviewed the related studies on Hg-based alternative detectors.[280]

11.4.5.3.2 Radiation Detectors

A radiation detector is a device that converts a radiation ray into an electrical signal. They can be divided into gas-filled detectors, scintillation detectors, and semiconductor detectors. Since semiconductor radiation detectors have a high spectrometric performance, and can be made portable, they have been applied in nuclear safeguards, computed tomography (CT) and radiographic imaging, industrial safety, non-destructive analysis, security and monitoring, non-proliferation, and astrophysics, as well as in many scientific applications.[4] Si and Ge are traditional semiconductors used for radiation detectors that offer good performance in a wide range of applications. Interests in radiation detectors operating at room temperature gave rise to development of compound semiconductors with wide bandgaps, in comparison to Si and Ge. Moreover, for X-ray and gamma-ray detection, compound semiconductors with high atomic number were preferred in order to emphasize photoelectric interaction.[90]

Radiation detectors using CdTe and CZT can operate in photoconductive mode at room temperature, unlike some other materials (particularly Ge) which require liquid nitrogen cooling.[288] Due to the high atomic number, the high density, and the wide bandgap, CdTe and CdZnTe detectors ensure high detection efficiency, good room temperature performance, and are very attractive for X-ray and gamma-ray applications.[90] The high atomic number and density of CdTe and CZT benefit the strong absorption and high detection efficiency. The wide bandgaps enable highly resistive devices with large depletion depths and low leakage currents, when the material is brought into the semi-insulating state with electrical compensation techniques. The moderately high mobility and lifetime of charge carriers (particularly electrons) allow good charge transport in devices depleted to many mm or even cm thickness.[289]

CdZnTe detectors consist of a full-area electrode on one surface (typically the entrance window of the photons) and a patterned electrode at the opposite surface. Most metals (Pt, Au, Ti or In) form a Schottky barrier at the metal–CdZnTe interface. The surface of the CdZnTe crystal is typically coated with a high-resistivity passivation film to avoid excessive surface leakage currents and detrimental surface charges.[289] The detector arrays are bonded to a substrate or a readout electronics chip with a matching pad distribution using bump bonding or alternative bonding techniques. Figure 11.4.29 shows a typical structure of $Cd_xZn_{1-x}Te$ radiation detector and an actual 256 pixel monolithic CdZnTe detector array ($34\times34\times5$ mm^3 single-crystal).[289]

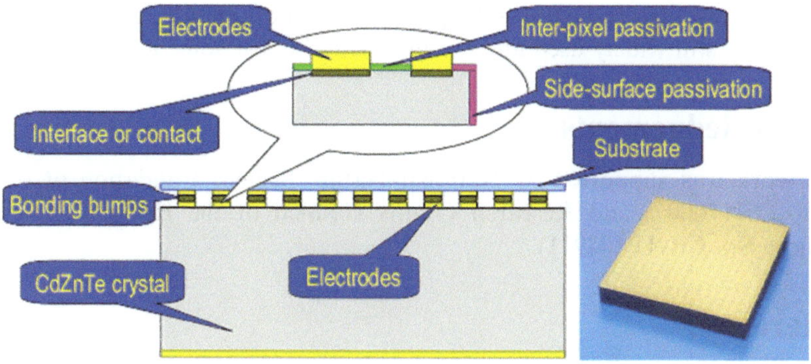

Figure 11.4.29 The typical structure of CdZnTe detectors and photograph of an actual monolithic CdZnTe detector array.[289]
(Reproduced from *Phys. Stat. Sol.* (b), 2004, **241**, 783, with permission from John Wiley and Sons.)

Based on commercial available high-quality crystals, the full potential of CdTe and CZT compounds for high-energy photon detection applications was exploited during the mid-1990s.[70,290–292] In order to obtain highly resistive semi-insulating $Cd_xZn_{1-x}Te$, the native defects and residual impurities need to be electrically compensated. This can be accomplished by deliberate introduction of impurities or native defects that produce charge carriers of the opposite sign to the semiconductor (*i.e.* donors in a p-type material). Column III (Al, Ga, In) and VII (Cl) donors were often proposed to compensate the effect of acceptors such as Cd vacancies in $Cd_xZn_{1-x}Te$. The net carrier concentration in this compensation scheme equals the difference of the concentrations of the donors and the residual acceptors.

A key parameter of the CZT crystal is its thickness, which varies according to the application. The higher the energy to be measured, the thicker the crystal required to detect the excited particles. For low-energy applications, thinner crystals are needed. Typical crystal thickness ranges from 1 mm for low energy to as much as 15 mm for very high-energy capture.[293]

$Cd_xZn_{1-x}Te$ detectors have been studied widely and manufactured commercially.[56,294] Large-area fine-pixel CZT imaging detectors have also been developed at MIT[295] and NASA.[296] However, there are still the severe problems such as polarization effects, long-term stability, and high cost. Further efforts should be focused on the preparation of high-quality materials and improvement of the stability and reliability of detectors.[4]

In addition to $Cd_xZn_{1-x}Te$ radiation detectors, ZnS is used extensively as a scintillator material in ionizing radiation detectors especially for α-ray detection.[297] Scintillators are widely used in neutron and high-energy particle physics experiments, X-ray security, nuclear cameras, CT scanners, gas exploration, and in the petroleum industry as detectors for gamma-ray logs. The U.S. government has employed scintillators as Homeland Security radiation detectors. Although semiconductor ionizing radiation detectors are making increasing inroads into the particle detection market, traditional scintillators are still widely used and are indispensable for some applications.[4]

Acknowledgements

This work was supported by National Nature Science Fundation of China (NSFC 61176007) and Research Grants Council of the Hong Kong SAR (Project No. CityU 104911).

References

1. D. J. Chadi, *Annu. Rev. Mater. Sci.*, 1994, **24**, 45.
2. D. Edwall and J. Bajaj, in *II–VI semiconductor materials and their applications*, ed. M. C. Tamargo, Taylor & Francis, New York, 2002, p. 1.
3. P. Capper and J. Garland, ed., *Mercury Cadmium Telluride: Growth, Properties and Applications*, Wiley, Chichester, 2011.
4. S. Kasap and P. Capper, ed., *Handbook of Electronic and Photonic Materials*, Springer, New York, 2006.
5. M. Osborne, Manufacturing cost per watt at First Solar falls to US$0.76 cents: module faults hit earnings, *PV-Tech.Org*, 29 July 2010.
6. S. Ali-Oettinger, New world record for cadmium-telluride, *PV Magazine*, 17 January 2012.
7. Y. Namikawa, *SEI Technical Review*, 2011, **72**, 25.
8. J. S. Jie, W. J. Zhang, C. S. Igor Bello, Lee and S. T. Lee, *Nano Today*, 2010, **5**, 313.
9. M. König, S. Wiedmann, C. Brüne, A. Roth, H. Buhmann, L. W. Molenkamp, X. L. Qi and S. C. Zhang, *Science*, 2007, **318**, 766.
10. Nanoparticles and quantum dots, in *Handbook of Chalcogen Chemistry*, ed. F. A. Devillanova and W. W. du Mont, RSC Publishing, Cambridge, 2012.
11. S. L. Gu and Z. Z. Ye, in *Research Progress of Semiconductor Materials*, ed. Z. G. Wang and Y. L. Zhen, High Education Press, Beijing, 2012, vol.1, p. 469. (in Chinese).
12. O. Madelung, *Semiconductors: Data Handbook*, Springer, Berlin, 3rd edn, 2004.
13. S. Adachi, *Properties of Group-IV, III-V and II–VI Semiconductors*, John Wiley & Sons, Chichester, 2005.
14. J. B. Chaudhari, N. G. Deshpande, Y. G. Gudage, A. Ghosh, V. B. Huse and R. Sharma, *Appl. Surf. Sci.*, 2008, **254**, 6810.
15. V. M. Burlakov, A. P. Litvinchuk, V. N. Pyrkov, G. G. Tarasov and N. I. Vitrikhovskii, *Phys. Stat. Sol. (b)*, 1985, **128**, 389.
16. Z. T. Deng, H. Yan and Y. Liu, *J. Am. Chem. Soc.*, 2009, **131**, 17744.
17. H. Okuyama, Y. Kishita and A. Ishibashi, *Phys. Rev. B*, 1998, **57**, 2257.
18. F. E. H. Hassan, S. J. Hashemifar and H. Akbarzadeh, *Phys. Rev. B*, 2006, **73**, 195202.
19. O. V. Parasyuk, L. V. Piskach, Y. E. Romanyuk, I. D. Olekseyuk, V. I. Zaremba and V. I. Pekhnyo, *J. Alloys Compd.*, 2005, **397**, 85.

20. S. Schorr, G. Wagner, M. Tovar and D. Sheptyakov, *Mater. Res. Soc. Symp. Proc.*, 2007, **1012**, Y03–05.
21. A. F. Wells, *Structural Inorganic Chemistry*, Clarendon Press, Oxford, 5th edn, 1984.
22. G. Grebe, G. Roussos and H. J. Schulz, *J. Phys. C: Solid State Phys.*, 1976, **9**, 4511.
23. N. Amin, K. Sopian and M. Konagai, *Sol. Energy Mater. Sol. Cells*, 2007, **91**, 1202.
24. A. F. Wells, *Structural Inorganic Chemistry*, Oxford University Press, New York, 5th edn, 1984.
25. R. Castro-Rodrígueza, J. Mendez-Gamboab, I. Perez-Quintanab and R. Medina-Ezquivel, *Appl. Surf. Sci.*, 2011, **257**, 9480.
26. C. Fouassier, *Luminescence in Encyclopedia of Inorganic Chemistry*, John Wiley & Sons, New York, 1994.
27. L. I. Berger, *Semiconductor materials*, CRC Press, Cincinnati, 1996, p. 202.
28. J. Zhao, Y. P. Zeng, Q. M. Yang, Y. Y. Li, L.j. Cui and C. Liu, *J. Cryst. Growth*, 2011, **329**, 1.
29. Q. M. Yang, J. Zhao, M. Guan, C. Liu, L. J. Cui, D. J. Han and Y. P. Zeng, *Appl. Surf. Sci.*, 2011, **257**, 9038.
30. C. Ma, Y. Ding, D. Moore, X. Wang and Z. L. Wang, *J. Am. Chem. Soc.*, 2004, **126**, 708.
31. M. Califano, A. Zunger and A. Franceschetti, *Appl. Phys. Lett.*, 2004, **84**, 2409.
32. R. D. Schaller, M. A. Petruska and V. I. Klimov, *Appl. Phys. Lett.*, 2005, **87**, 253102.
33. E. Hendry, M. Koeberg, F. Wang, H. Zhang, C. D. M. Donegá, D. Vanmaekelbergh and M. Bonn, *Phys. Rev. Lett.*, 2006, **96**, 057408.
34. *Solar Markets: Overall Growth and Size by Country*, 2012, http://solarcellcentral.com/.
35. R. Scheer and H. W. Schock,*Chalcogenide Photovoltaics: Physics, Technologies, and Thin Film Devices*, Wiley-VCH, Weinheim, 2011.
36. P. Capper, ed., *Properties of Narrow-Gap Cadmium-Based Compounds*, INSPEC, London, 1994.
37. A. M. Hao, C. X. Gao, M. Li, C. Y. He, X.W Huang, D. M. Zhang, C. L. Yu, H. W. Liu, Y. Z. Ma, Y. J. Tian and G. T. Zou, *J. Phys.: Condens. Matter*, 2007, **19**, 425222.
38. N. N. Greenwood and A. Earnshaw, *Chemistry of the Elements*, Pergamon, Oxford, 1984, p. 1406.
39. A. M. Hao, C. X. Gao, M. Li, C. Y. He, X.W Huang, D. M. Zhang, C. L. Yu, R. Guan and G. T. Zou, *Chinese Phys.*, 2007, **16**, 2087.
40. M. V. Yakushev, A. A. Babenko and Y. G. Sidorov, *Inorganic Materials*, 2009, **45**, 13.
41. M. Cardona, R. K. Kremer, R. Lauck and G. Siegle, *Phys. Rev. B*, 2009, **80**, 195204.
42. S. Mnasri, S. A. Nasrallahl, N. Sfinal, N. Bouarissa and M. Said, *Semicond. Sci. Technol.*, 2009, **24**, 095008.

43. Z. W. Lu, D. Singh and H. Krakauer, *Phys. Rev. B*, 1989, **39**, 10154.
44. X. L. Wang, S. X. Dou and C. Zhang, *NPG Asia Mater.*, 2010, **2**, 31.
45. A. M. Shuvaev, G. V. Astakhov, A. Pimenov, C. Brüne, H. Buhmann and L. W. Molenkamp, *Phys. Rev. Lett.*, 2011, **106**, 107404.
46. X. Chen, Y. Wang, T. Cui, Y. M. Ma, G. T. Zou and T. Iitaka, *J. Chem. Phys.*, 2008, **128**, 194713.
47. K. Jóźwikowski, M. Kopytko, J. Piotrowski, A. Jóźwikowska, Z. Orman and A. Rogalski, *Solid-State Electron.*, 2011, **63**, 8.
48. R. S. Patrick, A. B. Chen, A. Sher and M. A. Berding, *J. Vac. Sci. Technol.*, 1988, **A6**, 2643.
49. Y. V. Korostelin, V. J. Kozlovskij, A. S. Nasibov and P. V. Shapkin, *J. Cryst. Growth*, 1996, **159**, 181.
50. J. F. Wang, A. Omino and M. Isshiki, *Mater. Sci. Eng.*, 2001, **83**, 185.
51. S. H. Song, J. F. Wang, G. M. Lalev, L. He and M. Isshiki, *J. Cryst. Growth*, 2003, **252**, 102.
52. Y. M. Tairov and V. F. Tsvetkov, *J. Cryst. Growth*, 1978, **43**, 209.
53. F. P. Doty, J. F. Butler, J. F. Schetzina and K. A. Bowers, *J. Vac. ScI. Technol. B*, 1992, **10**, 1418.
54. T. E. Schlesingera, J. E. Toneyb, H. Yoonc, E. Y. Leed, B. A. Brunettd, L. Franksd and R. B. Jamesd, *Mater. Sci. Eng., R*, 2001, **32**, 103.
55. M. Schiebera, T. E. Schlesingerb, R. B. Jamesc, H. Hermond, H. Yoone and M. Goorsky, *J. Cryst. Growth*, 2002, **237–239**, 2082.
56. C. Szeles, D. Bale, J. Grosholz, Jr., G. L. Smith, M. Blostein and J. Eger, *Proc. of SPIE*, 2006, **6319**, 191.
57. C. Szeles, S. E. Cameron, S. A. Soldner, J. O. Ndap and M. D. Reed, *J. Electron. Mater.*, 2004, **33**, 742.
58. N. Zhang, A. Yeckel, A. Burger, Y. L. Cui, K. G. Lynn and J. J. Derby, *J. Cryst. Growth*, 2011, **325**, 10.
59. M. D. Reed, Cs. Szeles and S. E. Cameron, *J. Cryst. Growth*, 2006, **289**, 494.
60. P. Norton, *Opto-Electron. Rev.*, 2002, **10**, 159.
61. P. Capper, C. Maxey, C. Butler, M. Grist and J. Price, *J. Mater. Sci. - Mater. Electron.*, 2004, **15**, 721.
62. P. Capper, C. Maxey, C. Butler, M. Grist and J. Price, *J. Cryst. Growth*, 2005, **275**, 259.
63. E. Saucedo, P. Rudolph and E. Dieguez, *J. Cryst. Growth*, 2008, **310**, 2067.
64. C. H. Su, S. L. Lehoczky, B. Raghothamachar and M. Dudley, *Mater. Sci. Eng., B*, 2008, **147**, 35.
65. C. H. Su and S. LLehoczky, *J. Cryst. Growth*, 2011, **319**, 4.
66. M. Ohmori, Y. Iwase and R. Ohno, *Mater. Sci. Eng. B*, 1999, **16**, 283.
67. R. Triboulet, *Prog. Cryst. Growth Char. Mater.*, 1994, **28**, 85.
68. Y. Wang, K. Kudo, Y. Inatomi, R. B. Ji and T. Motegi, *J. Cryst. Growth*, 2005, **284**, 406.
69. H. Shiraki, M. Funaki, Y. Ando, A. Tachibana, S. Kominami and R. Ohno, *IEEE Trans. Nucl. Sci.*, 2009, **56**, 1717.

70. H. Chen, S. A. Awadalla, R. Redden, G. Bindley, A. E. Bolotnikov, G. S. Camarda, G. Carini and R. B. James, *IEEE Trans. Nucl. Sci.*, 2007, **54**, 811.
71. H. Chen, S. A. Awadalla, K. Iniewski, P. H. Lu, F. Harris, J. Mackenzie, T. Hasanen, W. Chen, R. Redden, G. Bindley, Irfan Kuvvetli, Carl Budtz-Jørgensen, A. E. Bolotnikov, G. S. Camarda, Y. Cui, A. Hossain and R. B. James, *J. Appl. Phys.*, 2008, **103**, 014903.
72. S. B. Mirov, V. V. Fedorov, D. V. Martyshkin, I. S. Moskalev, M. S. Mirov and V. P. Gapontsev, *Opt. Mater. Express*, 2011, **1**, 898.
73. H. H. Woodbury and R. S. Lewandowski, *J. Cryst. Growth*, 1971, **10**, 6.
74. K. Byrappa, in *Hydrothermal Growth of Crystal*, ed. K. Byrappa, Pergamon, Oxford, 1991.
75. R. Triboulet, *Cryst. Res. Technol.*, 2003, **38**, 215.
76. A. Lopez-Otero, *Thin Solid Films*, 1978, **49**, 1.
77. H. M. Manasevit and W. I. Simpson, *J. Electrochem. Soc.*, 1971, **118**, 644.
78. L. L. Chang, R. Ludeke, in *Epitaxial Growth*, Part A, J. W. Matthews, ed. Academic Press, New York, 1975, p. 37.
79. E. Veuhoff, W. Pletschen, P. Balk and H. Luth, *J. Cryst. Growth*, 1981, **55**, 30.
80. T. Suntola, *Mater. Sci. Rep.*, 1989, **4**, 261.
81. P. Lilley, P. L. Jones and C. N. W. Litting, *J. Mater. Sci.*, 1970, **5**, 891.
82. T. Matsumoto, T. Morita and T. Ishida, *J. Cryst. Growth*, 1987, **53**, 225.
83. P. Besomi and B. W. Wessels, *J. Cryst. Growth*, 1981, **55**, 477.
84. T. Kyotani, M. Isshiki and K. Masumoto, *J. Electrochem. Soc.*, 1989, **136**, 2376.
85. N. Stucheli and E. Bucher, *J. Electron. Mater.*, 1989, **18**, 105.
86. M. Nishio, Y. Nakamura and H. Ogawa, *Jpn. J. Appl. Phys.*, 1983, **22**, 1101.
87. N. Lovergine, R. Cingolani, A. M. Mancini and M. Ferrara, *J. Cryst. Growth*, 1992, **118**, 304.
88. O.De. Melo, E. Sánchez, S.De. Roux and F. Rábago-Bernal, *Mater. Chem. Phys.*, 1999, **59**, 120.
89. M. Kasuga, H. Futami and Y. Iba, *J. Cryst. Growth*, 1991, **115**, 711.
90. S. D. Sordo, L. Abbene, E. Caroli, A. M. Mancini, A. Zappettini and P. Ubertini, *Sensors*, 2009, **9**, 3491.
91. H. M. Manasevit, *Appl. Phys. Lett.*, 1968, **12**, 1530.
92. S. Fujita, M. Isemura, T. Sakamoto and N. Yoshimura, *J. Cryst. Growth*, 1988, **86**, 263.
93. H. Mitsuhashi, I. Mitsuishi and H. Kukimoto, *J. Cryst. Growth*, 1986, **77**, 219.
94. P. J. Wright, P. J. Parbrook, B. Cockayne, A. C. Jones, E. D. Orrell, K. P. O. Donnell and B. Henderson, *J. Cryst. Growth*, 1989, **94**, 441.
95. S. Hirata, M. Isemura, Sz. Fujita and Sg. Fujita, *J. Cryst. Growth*, 1990, **104**, 521.
96. S. Nishimura, N. Iwasa, M. Senoh and T. Mukai, *Jpn. J. Appl. Phys.*, 1993, **32**, L425.

97. K. P. Giapis, K. F. Jensen, J. E. Potts and S. J. Pachuta, *Appl. Phys. Lett.*, 1989, **55**, 463.
98. S. J. Pachuta, K. F. Jensen and S. P. Giapis, *J. Cryst. Growth*, 1991, **107**, 390.
99. M. Danek, J. S. Huh, L. Foley and K. F. Jenson, *J. Cryst. Growth*, 1994, **145**, 530.
100. W. Kuhn, A. Naumov, H. Stanzl, S. Bauer, K. Wolf, H. P. Wagner, W. Gebhardt, U. W. Pohl, A. Krost, W. Richter, U. Dümichen and K. H. Thiele, *J. Cryst. Growth*, 1992, **123**, 605.
101. J. K. Menno, J. W. Kerri and F. H. Robert, *J. Phys. Chem. B*, 1997, **101**, 4882.
102. H. P. Wagner, W. Kuhn and W. Gebhardt, *J. Cryst. Growth*, 1990, **101**, 199.
103. N. R. Taskar, B. A. Khan, D. R. Dorman and K. Shahzad, *Appl. Phys. Lett.*, 1993, **62**, 270.
104. Y. Fujita, T. Terada and T. Suzuki, *Jpn. J. Appl. Phys.*, 1995, **34**, L1034.
105. J. Wang, T. Miki, A. Omino, K. S. Park and M. Isshiki, *J. Cryst. Growth*, 2000, **221**, 393.
106. M. Niraula, K. Yasuda, A. Watanabe, Y. Kai, H. Ichihashi, W. Yamada, H. Oka, T. Yoneyama, H. Nakashima, T. Nakanishi, K. Matsumoto, D. Katoh and Y. Agata, *IEEE Trans. Nucl. Sci.*, 2009, **56**, 836.
107. M. Tahashi, S. Uzura, H. Goto and T. Ido, *J. Cryst. Growth*, 2008, **310**, 3440.
108. D. Greiffenberg, R. Sorgenfrei, K. H. Bachem and M. Fiederle, *IEEE Trans. Nucl. Sci.*, 2007, **54**, 773.
109. M. Niraula, K. Yasuda, H. Ohnishi, H. Takahashi, K. Eguchi, K. Noda and Y. Agata, *J. Electron. Mater.*, 2006, **35**, 1257.
110. M. Niraula, K. Yasuda, H. Ohnishi, K. Eguchi, H. Takahashi, K. Noda and Y. Agata, *J. Cryst. Growth*, 2005, **284**, 15.
111. Z. Z. Zhang, D. Z. Shen, J. Y. Zhang, C. X. Shan, Y. M. Lu, Y. C. Liu, B. H. Li, D. X. Zhao, B. Yao and X. W. Fan, *Thin Solid Films*, 2006, **513**, 114.
112. Y. Kume, Q. X. Guo, T. Tanaka, M. Nishio, H. Ogawa and W. Z. Shen, *J. Cryst. Growth*, 2007, **298**, 441.
113. M. Traversa, N. Lovergine, P. Prete, L. Tapfer and A. M. Mancini, *Cryst. Res. Technol.*, 2005, **40**, 1003.
114. A. Toda, T. Margalith, D. Imanishi, K. Yanashima and A. Ishibashi, *Electron. Lett.*, 1995, **31**, 1921.
115. A. Cho, *J. Vac. Sci. Tech.*, 1971, **8**, S31.
116. C. T. Foxon, *J. Cryst. Growth*, 2003, **251**, 130.
117. R. F. C. Farrow, G. R. Jones, G. M. Williams and I. M. Young, *Appl. Phys. Lett.*, 1981, **39**, 954.
118. M. Reddy, J. M. Peterson, D. D. Lofgreen, J. A. Franklin, T. Vang, E. P. G. Smith, J. G. A. Wehner, I. Kasai, J. W. Bangs and S. M. Johnson, *J. Electron. Mater.*, 2008, **37**, 1274.
119. R. Singh, S. Velicu, J. Crocco, Y. Chang, J. Zhao, L. A. Almeida, J. Markunas, A. Kaleczyc and J. H. Dinan, *J. Electron. Mater.*, 2005, **34**, 885.

120. T. J. Delyon, R. D. Rajavel, B. Z. Nosho, S. Terterian, M. L. Beliciu, P. R. Patterson, D. T. Chang, M. F. Boag-O'brien, B. T. Holden, R. N. Jacobs and J. D. Benson, *J. Electron. Mater.*, 2010, **39**, 1058.
121. M. Reddy, J. M. Peterson, T. Vang, J. A. Franklin, M. F. Vilela, K. Olsson, E. A. Patten, W. A. Radford, J. W. Bangs, L. Melkonian, E. P. G. Smith, D. D. Lofgreen and S. M. Johnson, *J. Electron. Mater.*, 2011, **40**, 1706.
122. R. Sporken, S. Sivananthan, K. K. Mahavadi, G. Monfroy, M. Boukerche and J. P. Faurie, *Appl. Phys. Lett.*, 1989, **55**, 1879.
123. R. Najjar, R. André, H. Boukari, H. Mariette and S. Tatarenko, *Surf. Sci.*, 2008, **602**, 744.
124. R. Sorgenfrei, D. Greiffenberg, K. H. Bachem, L. Kirste, A. Zwerger and M. Fiederle, *J. Cryst. Growth*, 2008, **310**, 2062.
125. J. Zhao, Y. P. Zeng, Q. M. Yang, Y. Y. Li, L. J. Cui and C. Liu, *J. Cryst. Growth*, 2011, **329**, 1.
126. J. Fan, L. Ouyang, X. Liu, D. Ding, J. K. Furdyna, D. J. Smith and Y. H. Zhang, *J. Cryst. Growth*, 2011, **323**, 127.
127. S. P. Suprun, V. N. Sherstyakova and E. V. Fedosenko, *Semiconductors*, 2009, **43**, 1570.
128. E. Veuhoff, W. Pletschen, P. Balk and H. Luth, *J. Cryst. Growth*, 1981, **55**, 30.
129. M. B. Panish and S. Sumski, *J. Appl. Phys.*, 1984, **55**, 3571.
130. Y. P. Chen, G. Brill and N. K. Dhar, *J. Cryst. Growth*, 2003, **252**, 270.
131. H. Kato, M. Sano, K. Miyamoto and T. Yao, *J. Cryst. Growth*, 2002, **237–239**, 538.
132. Y. Kume, Q. X. Guo, Y. Fukuhara, T. Tanaka, M. Nishio and H. Ogawa, *J. Cryst. Growth*, 2007, **298**, 445.
133. M. Imaizumi, M. Adachi, Y. Fujii, Y. Hayashi, T. Soga, T. Jimbo and M. Umeno, *J. Cryst. Growth*, 2000, **221**, 688.
134. W. Xie, D. C. Grillo, M. Kobayashi, R. L. Gunshor, G. C. Hua, N. Otsuka, H. Jeon, J. Ding and A. V. Nurmikko, *Appl. Phys. Lett.*, 1992, **60**, 463.
135. S. Guha, A. Madhukar and K. C. Rajkumar, *Appl. Phys. Lett.*, 1990, **57**, 2110.
136. J. F. Wang, K. Kikuchi, B. H. Koo, Y. Ishikawa, W. Uchida and M. Isshiki, *J. Cryst. Growth*, 1998, **187**, 373.
137. J. Humenberger, G. Linnet and K. Lischka, *Thin Solid Films*, 1984, **121**, 75.
138. F. Sasaki, T. Mishina and Y. Masumoto, *J. Cryst. Growth*, 1992, **117**, 768.
139. B. J. Kim, J. F. Wang, Y. Ishikawa, S. Sato and M. Isshiki, *Phys. Stat. Sol. (a)*, 2002, **191**, 161.
140. A. Rogalski and J. Piotrowski, *Prog. Quantum Electron.*, 1988, **12**, 87.
141. Q. Jianga, J. T. Mullinsb, J. Tomana, T. P. Hasea, B. J. Cantwellb, G. Lloydc, A. Basub and A. W. Brinkman, *J. Cryst. Growth*, 2008, **310**, 1652.
142. I. R. B. Ribeiro, J. Suela1, J. E. Oliveira1, S. O. Ferreira and P. Motisuke, *J. Phys. D: Appl. Phys.*, 2007, **40**, 4610.
143. G. M. Lalev, J. Wang, S. Abe, K. Masumoto and M. Isshiki, *J. Crystal Growth*, 2003, **256**, 20.

144. G. M. Lalev, J. F. Wang, J. W. Lim, S. Abe, K. Masumoto and M. Isshiki, *Appl. Surf. Sci.*, 2005, **242**, 295.
145. M. A. Herman and J. T. Sadowski, *Cryst. Res. Technol.*, 1999, **34**, 153.
146. M. Ahonen, M. Pessa and T. Suntola, *Thin Solid Films*, 1980, **65**, 301.
147. M. Ritala and M. Leskelä, *Nanotechnology*, 1999, **10**, 19.
148. B. H. Flowers, T. L. Wade, J. W. Garvey, M. Lay, U. Happek and J. L. Stickney, *J. Electroanal. Chem.*, 2002, **524**, 273.
149. M. K. Mathe, S. M. Cox, V. Venkatasamy, U. Happek and J. L. Stickney, *J. Electrochem. Soc.*, 2005, **152**, C751.
150. M. Cavallini, M. Facchini, C. Albonetti, F. Biscarini, M. Innocenti, F. Loglio, E. Salvietti, G. Pezzatini and M. L. Foresti, *J. Phys. Chem. C*, 2007, **111**, 1061.
151. H. Nakamura and M. Aoki, *Jpn. J. Appl. Phys.*, 1981, **20**, 11.
152. C. Werkhoven, B. J. Fitzpatrik, S. P. Herko, R. N. Bhargave and P. J. Dean, *Appl. Phys. Lett.*, 1981, **38**, 540.
153. H. Nakamura, S. Kojima, M. Wasgiyama and M. Aoki, *Jpn. J. Appl. Phys.*, 1984, **23**, L617.
154. V. M. Skobeeva, V. V. Serdyuk, L. N. Semenyuk and N. V. Malishin, *J. Appl. Spectrosc.*, 1986, **44**, 164.
155. L. F. Lou and W. H. Frye, *J. Appl. Phys.*, 1984, **56**, 2253.
156. S. Neretina, R. A. Hughes, N. V. Sochinskii, M. Weber, K. G. Lynn, J. Wojcik, G. N. Pearson, J. S. Preston and P. Mascher, *J. Vac. Sci. Technol. A*, 2006, **24**, 606.
157. T. Ganguli, S. Porwal, T. Sharma, A. Ingale, S. Kumar, P. Tiwari, A. K. Balamurugan, S. Rajagopalan, A. K. Tyagi, K. S. Chandrasekaran, B. M. Arora and K. C. Rustagi, *Thin Solid Films*, 2007, **515**, 7834.
158. M. Liu, B. Y. Man, X. C. Lin, X. Y. Li and C. S. Chen, *Appl. Surf. Sci.*, 2007, **253**, 9291.
159. K. L. Narayanan, M. Yamaguchi, J. A. Davila-Pintle, R. Lozada-Morales, O. Portillo-Moreno and O. Zelaya-Angel, *Vacuum*, 2007, **81**, 1430.
160. H. Moualkia, S. Hariech, M. S. Aida1, N. Attaf1 and E. L. Laifa, *J. Phys. D: Appl. Phys.*, 2009, **42**, 135404.
161. N. K. Allouche, T. Ben Nasr, N. Turki Kamoun and C. Guasch, *Mater. Chem. Phys.*, 2010, **123**, 620.
162. T. B. Nasr, N. Kamoun, M. Kanzari and R. Bennaceur, *Thin Solid Films*, 2006, **500**, 4.
163. W. Y. Li, X. Cai, Q. L. Chen and Z. B. Zhou, *Mater. Lett.*, 2005, **59**, 1.
164. S. A. Quiñones, S. M. Ammu, A. Escobedo, M. Rodriguez, J. Cruz-Campa, J. McClure and D. Zubia, *J Mater Sci: Mater Electron*, 2007, **18**, 1085.
165. J. Luschitz, K. Lakus-Wollny, A. Klein and W. Jaegermann, *Thin Solid Films*, 2007, **515**, 5814.
166. P. Prathap, Y. P.V Subbaiah, K. T. Ramakrishna Reddy and R. W. Miles, *J. Phys. D: Appl. Phys.*, 2007, **40**, 5275.
167. Q. Jiang, B. J. Cantwell, J. T. Mullins, A. Basu and A. W. Brinkman, *J. Cryst. Growth*, 2008, **310**, 1664.

168. J. L. Cruz-Campa and D. Zubia, *Sol. Energy Mater. Sol. Cells*, 2009, **93**, 15.
169. S. Larramendi, F. C. Zawislak, M. Behar, E. Pedrero, M. Hernández Vélez and O. e Melo, *J. Cryst. Growth*, 2010, **312**, 892.
170. E. M. Larramendi, K. Gutiérrez, Z-B, C. Arens, U. Woggon, D. Schikora and K. Lischka, *J. Appl. Phys.*, 2010, **107**, 103510.
171. J. D. Major and K Durose, *Sol. Energy Mater. Sol. Cells*, 2011, **95**, 3165.
172. V. Kumar, K. L. A. Khan, G. Singh, T. P. Sharma and M. Hussain, *Appl. Surf. Sci.*, 2007, **253**, 3543.
173. M. Öztas, M. Bedir, S. Ocak and R. G. Yildirim, *J Mater Sci: Mater Electron*, 2007, **18**, 505.
174. M. D. Archbold, D. P. Halliday, K. Durose, T. P. A. Hase, D. S. Boyle, S. Mazzamuto, N. Romeo and A. Bosio, *Thin Solid Films*, 2007, **515**, 2954.
175. J. Schaffner, E. Feldmeier, A. Swirschuk, H.-J. Schimper, A. Klein and W. Jaegermann, *Thin Solid Films*, 2011, **519**, 7556.
176. F. A. Kroger and H. J. Vink, In *Solid State Physics*, ed. F. Seitz, D. Turnbull, Academic Press, New York, 1956, 310.
177. R. F. Brebrick, *Phys. Chem. Solids*, 1958, **4**, 190.
178. U. V. Desnica, *Prog. Crystal Growth and Charact.*, 1998, **36**, 291.
179. W. Faschinger, *J. Cryst. Growth*, 1996, **159**, 221.
180. W. Faschinger, *J. Cryst. Growth*, 1999, **197**, 557.
181. H. Cheng, J. M. DePuydt, J. E. Potts and T. L. Smith, *Appl. Phys. Lett.*, 1988, **52**, 147.
182. T. Yasuda, I. Mitsuishi and H. Kukimoto, *Appl. Phys. Lett.*, 1988, **52**, 57.
183. R. M. Park, M. B. Troffer, C. M. Rouleau, J. M. DePuydt and M. A. Haase, *Appl. Phys. Lett.*, 1990, **57**, 2127.
184. S. Zhang, H. Kinto, T. Yatabe and S. Iida, *J. Cryst. Growth*, 1988, **86**, 372.
185. S. Iida, T. Yatabe and H. Kinto, *Jpn. J. Appl. Phys.*, 1989, **28**, L535.
186. K. Ohkawa, T. Karasawa and T. Mitsuyu, *J. Crystal Growth*, 1991, **111**, 797.
187. J. Qiu, J. M. DePuydt, H. Cheng and M. A. Haase, *Appl. Phys. Lett.*, 1991, **59**, 2992.
188. T. Ohtsuka, J. Kawamata, Z. Zhu and T. Yao, *Appl. Phys. Lett.*, 1994, **65**, 466.
189. M. A. Haase, J. Qui, J. DePuydt and H. Cheng, *Appl. Phys. Lett.*, 1991, **59**, 1272.
190. P. J. Sebastian, *Appl. Phys. Lett.*, 1993, **62**, 2956.
191. Y. Kashiwaba, I. Kanno and T. Ikeda, *Jpn. J. Appl. Phys.*, 1992, **31**, 1170.
192. H. K. Xie, C. J. Tian, W. Li, L. H. Feng, J. Q. Zhang, L. L. Wu, Y. P. Cai, Z. Lei and Y. J. Yang, *Appl. Surf. Sci.*, 2010, **257**, 1623.
193. R. L. Gunshor and A. V. Nurmikko, ed., *II–VI Blue Green Light Emitters: Device Physics and Epitaxial Growth*, Academic Press, Chestnut Hill, 1997.

194. K. D. Mynbaev and V. I. Ivanov-Omskiĭ, *Semiconductors*, 2006, **40**, 3.
195. B. Ghosh, *Appl. Surf. Sci.*, 2008, **254**, 4908.
196. B. Ghosh, *Microelectron. Eng.*, 2009, **86**, 2187.
197. H. Morkoc, S. Strite, G. B. Gao, M. E. Lin, B. Sverdlov and M. Burns, *J. Appl. Phys.*, 1994, **76**, 1363.
198. Y. Fan, J. Han, L. He, J. Saraie, R. L. Gunshor, G. C. Hua and N. Otsuka, *Appl. Phys. Lett.*, 1992, **61**, 3160.
199. A. Ishibashi and Y. Mori, *J. Crystal Growth*, 1994, **138**, 677.
200. S. Itoh, N. Nakayama, S. Matsumoto, M. Nagai, K. Nakano, M. Ozawa, H. Okuyama, S. Tomiya, T. Ohata, M. Ikeda, A. Ishibashi and Y. Mori, *Jpn. J. Appl. Phys.*, 1994, **33**, L938.
201. Z. Fang, X. C. Wang, H. C. Wu and C. Z. Zhao, *Int. J. Photoenergy*, 2011, **2011**, 297–350.
202. J. Y. Tsao, M. E. Coltrin, M. H. Crawford and J. A. Simmons, *Proc. IEEE*, 2010, **98**, 1162.
203. N. Zheludev, *Nature Photon.*, 2007, **1**, 189.
204. R. Braunstein, *Phys. Rev.*, 1955, **99**, 1892.
205. Nick Holonyak and S. F. Bevacqua, *Appl. Phys. Lett.*, 1962, **1**, 82.
206. R. N. Hall, G. E. Fenner, J. D. Kingsley, T. J. Soltys and R. O. Carlson, *Phys. Rev. Lett.*, 1962, **9**, 366.
207. M. G. Craford, Roger W. Shaw, A. H. Herzog and W. O. Groves, *J. Appl. Phys.*, 1972, **43**, 4075.
208. T. P. Pearsall, B. I. Miller, R. J. Capik and K. J. Bachmann, *Appl. Phys. Lett.*, 1976, **28**, 499.
209. D. R. Frankl, *Phys. Rev.*, 1958, **111**, 1540.
210. B. L. Crowder, F. F. Morehead and P. R. Wagner, *Appl. Phys. Lett.*, 1966, **8**, 148.
211. J. Nishizawa, K. Itoh, Y. Okuno and F. Sakurai, *J. Appl. Phys.*, 1985, **57**, 2210.
212. T. Niina, T. Minato and K. Yoneda, *Jpn. J. Appl. Phys.*, 1982, **21**, L387.
213. J. Nishizawa, K. Itoh, Y. Okuno and F. Sakurai, *J. Appl. Phys.*, 1985, **57**, 2210.
214. H. Cheng, J. M. DePuydt, J. E. Potts and T. L. Smith, *Appl. Phys. Lett.*, 1988, **52**, 147.
215. R. M. Park, H. A. Mar and N. M. Salansky, *J. Appl. Phys.*, 1985, **58**, 1047.
216. J. M. DePuydt, M. A. Haase, H. Cheng and J. E. Potts, *Appl. Phys. Lett.*, 1989, **55**, 1103.
217. K. Ohkawa, T. Karasawa and T. Mitsuyu, *Jpn. J. Appl. Phys.*, 1991, **30**, L152.
218. R. Park, *J. Vac. Sci. Technol. A*, 1991, **10**, 701.
219. M. A. Haase, H. Cheng, J. M. DePuydt and J. E. Potts, *J. Appl. Phys.*, 1990, **67**, 448.
220. J. Ren, K. A. Bowers, B. Sneed, D. L. Dreifus, J. W. Cook, Jr., J. F. Schetzina and R. M. Kolbas, *Appl. Phys. Lett.*, 1990, **57**, 1901.
221. H. Roemer, *Proc. IEEE*, 1963, **51**, 1782.
222. A. R. Adams, *Electron. Lett.*, 1986, **22**, 249.

223. E. Yablonovitch and E. O. Kane, *J. Lightwave Technol.*, 1986, **LT-4**, 504.
224. N. Nakayama, S. Itoh, T. Ohata, K. Nakano, H. Okuyama, M. Ozawa, A. Ishibashi, M. Ikeda and Y. Mori, *Electron. Lett.*, 1993, **29**, 1488.
225. N. Nakayama, S. Itoh, H. Okuyama, M. Ozawa, T. Ohata, K. Nakano, M. Ikeda, A. Ishibashi and Y. Mori, *Electron. Lett.*, 1993, **29**, 2194.
226. A. Waag, F. Fischer, H.-J. Lugauer, Th. Litz, T. Gerhard, J. Niirnberger, U. Lunz, U. Zehnder, W. Ossau, G. Landwehr, B. Roos and H. Richter, *Mater. Sci. Eng., B*, 1996, **43**, 65.
227. C. Onodera, T. Shoji, Y. Hiratate and T. Taguchi, *Jpn. J. Appl. Phys.*, 2008, **47**, 4602.
228. A. Ishibashi, *J. Cryst. Growth*, 1996, **159**, 555.
229. J. M. Phillips, M. E. Coltrin, M. H. Crawford, A. J. Fischer, M. R. Krames, R. Mueller-Mach, G. O. Mueller, Y. Ohno, L. E. S. Rohwer, J. A. Simmons and J. Y. Tsao, *Laser & Photon. Rev.*, 2007, **1**, 307.
230. K. Katayama and T. Nakamura, *J. Appl. Phys.*, 2004, **95**, 3576.
231. M. Adachi, K. Ando, T. Abe, N. Inoue, A. Urata, S. Tsutsumi, Y. Hashimoto, H. Kasada, K. Katayama and T. Nakamura, *Phys. Stat. Sol. (b)*, 2006, **243**, 943.
232. Ichirou Nomura, Yuki Nakai, Koichi Hayami, Takumi Saitoh and Katsumi Kishino, *phys. stat. sol. (b)*, 2006, **243**, 924.
233. M. Adachi, Y. Hashimoto, K. Kanzaki, S. Ohashi, Y. Morita, T. Abe, H. Kasada, K. Ando and M. Tajima, *J. Mater. Sci.: Mater. Electron.*, 2008, **19**, S299.
234. K. Ando, Y. Hashimoto, K. Kanzaki, S. Ohashi, Y. Morita, T. Abe, H. Kasada and M. Adachi, *J. Korean Phys. Soc.*, 2008, **53**, 2857.
235. J. S. Wang, W. J. Chen, C. S. Yang, Y. H. Tsai, H. H. Wang, R. H. Chen, J. L. Shen and C. D. Tsai, *Appl. Phys. Lett.*, 2011, **98**, 021908.
236. S. Guha, J. M. DePuydt, M. A. Haase, J. Qiu and H. Cheng, *Appl. Phys. Lett.*, 1993, **63**, 3107.
237. T. Nakamura, S. Fujiwara, H. Mori and K. Katayama, *Jpn. J. Appl. Phys.*, 2004, **43**, 1287.
238. T. Nakamura, *Electrical Engineering in Japan*, 2006, **154**, 42.
239. S. Nakamura, G. Fasol and S. J. Pearton, *The Blue Laser Diode The Complete Story,* Springer, Berlin, 2nd edn, 2000.
240. B. J. Park, H. S. Seo, J. T. Ahn, J. H. Song, W. J. Chung and D. Y. Jeon, *J. Mater. Res.*, 2011, **26**, 2394.
241. G. Boutaud, W. M. Cranton, D. C. Koutsogeorgis, R. M. Ranson, C. Tsakonas and C. B. Thomas, *Mater. Sci. Eng., B*, 2009, **165**, 202.
242. *http://www.ceelite.com/technology/technology-components.asp*, 2012.
243. A. Daud, M. Kitagawa, S. Tanaka and H. Kobayashi, *Rep. Faculty of Engineering, Tottori University (Japan)*, 1994, **25**, 153.
244. R. Bhargava (ed), *Properties of Wide Bandgap II–VI Semiconductors*, Institution of Engineering and Technology, Stevenage, 1997.
245. K. Nakanishi, I. Suemune, H. Masato, Y. Kuroda and M. Yamanishi, *Jpn. J. Appl. Phys.*, 1990, **29**, L2420.

246. H. Wang, K. S. Wong, I. K. Sou and G. K. L. Wong, *Appl. Phys. Lett.*, 1995, **66**, 3140.
247. S. Colak, B. J. Fitzpatrick and R. N. Bhargava, *J. Cryst. Growth*, 1985, **72**, 1.
248. D. Hervé, R. Accomo, E. Molva, L. Vanzetti, J. J. Paggel, L. Sorba and A. Franciosi, *Appl. Phys. Lett.*, 1995, **67**, 2144.
249. A. L. Gurskii, V. V. Gruzinskii, A. N. Gavrilenko, I. I. Kulak, A. I. Mitskovets, G. P. Yablonskii, M. Scholl, J. Söllner and M. Heuken, *J. Appl. Phys.*, 1995, **77**, 5394.
250. P. I. Kuznetsov, G. G. Yakushcheva, V. A. Jitov, L. Yu. Zakharov, B. S. Shchamkhalova, V. I. Kozlovsky, D. A. Sannikov, Ya. K. Skasyrsky and M. D. Tiberi, *Phys. Stat. Sol. (C)*, 2006, **3**, 771.
251. 3M claims record efficacy for green LED, *LEDs Magazine*, 23 Nov. 2009.
252. M. A. Haase, J. Xie, T. A. Ballen, J. Zhang, B. Hao, Z. H. Yang, T. J. Miller, X. Sun, T. L. Smith and C. A. Leatherdale, *Appl. Phys. Lett.*, 2010, **96**, 231116.
253. D. A. Davies, A. Vecht, J. Silver, P. J. Marsh and J. A. Rose, *J. Electrochem. Soc.*, 2000, **147**, 765.
254. G. F. Brown and J. Q. Wu, *Laser & Photon. Rev.*, 2009, **3**, 394.
255. G. Conibeer, *materialstoday*, 2007, **10**, 42.
256. J. W. Garland, T. Biegala, M. Carmody, C. Gilmore and S. Sivananthan, *J. Appl. Phys.*, 2011, **109**, 102423.
257. T. L. Chu and S. S. Chu, *Solid-State Electron.*, 1995, **38**, 533.
258. *Publications, Presentations, and News Database: Cadmium Telluride*, National Renewable Energy Laboratory.
259. K. Zweibel, J. Mason and V. Fthenakis, *Sci. Am.*, Jan. 2008.
260. Solar Power Lightens Up with Thin-Film Technology, *Sci. Am.*, Apr. 2008.
261. A. Luque and S. Hegedus (eds), *Handbook of Photovoltaic Science and Engineering*, John Wiley & Sons, Chichester, 2nd edn, 2011.
262. D. A. Jenny and R. H. Bube, *Phys. Rev.*, 1954, **96**, 1190.
263. D. A. Cusano, *Solid-State Electron.*, 1963, **6**, 217.
264. B. Goldstein, *Phys. Rev.*, 1958, **109**, 601.
265. D. A. Cusano, *Solid-State Electron.*, 1963, **6**, 217.
266. D. Bonnet and H. Rabenhorst, *Proc. 9th IEEE Photovoltaic Specialist Conference*, 1972, 129.
267. Y. S. Tyan and E. A. Perez-Albuerne, *Proc. 16th IEEE Photovoltaic Specialist Conference*, 1982, 794.
268. K. Zweibel, *Thin Films: Past Present and Future*, National Renewable Energy Laboratory, 1995.
269. R. Noufi and K. Zweibel, High-Efficiency CdTe and CIGS Thin-Film Solar Cells: Highlights and Challenges, IEEE 4th World Conference on Photovoltaic Energy Conversion (WCPEC-4), 2006, **1**, 317.
270. M. Afzaal and P. O'Brien, *J. Mater. Chem.*, 2006, **16**, 1597.
271. T. L. Chu, S. S. Chu, J. Britt, C. Ferekides, C. Wang, C. Q. Wu and H. S. Ullal, *IEEE Electron. Device Lett.*, 1992, **13**, 303.

272. J. Britt and C. Ferekides, *Appl. Phys. Lett.*, 1993, **62**, 2851.
273. X. Wu, *Proc. 17th Eur. Photovoltaic Sol. Energy Conference,* Munich, 2001, 995.
274. X. Z. Wu, *Solar Energy*, 2004, **77**, 803.
275. X. Mathew, J. P. Enriquez, A. Romeo and A. N. Tiwari, *Solar Energy*, 2004, **77**, 831.
276. E. Hughes, Flexible CdTe thin-film solar cells reach efficiency of 12.4%, *PV-tech.org*, 21 August 2009.
277. A. Rogalski, *Rep. Prog. Phys.*, 2005, **68**, 2267.
278. M. B. Reine, *Fundamental properties of mercury cadmium telluride Encyclopedia of Modern Optics*, Academic Press, London, 2004.
279. W. D. Lawson, S. Nielson, E. H. Putley and A. S. Young, *J. Phys. Chem. Solids*, 1959, **9**, 325.
280. A. Rogalski, *Infrared detectors*, CRC Press, Boca Raton, 2nd edn, 2011, 366.
281. J. P. Zanatta, G. Badano, P. Ballet, C. Largeron, J. Baylet, O. Gravrand, J. Rothman, P. Castelein, J. P. Chamonal, A. Million, G. Destefanis, S. Mibord, E. Brochier and P. Costa, *J. Electron. Mater.*, 2006, **35**, 1231.
282. W. A. Radford, E. A. Patten, D. F. King, G. K. Pierce, J. Vodicka, P. Goetz, G. Venzor, E. P. Smith, R. Graham, S. M. Johnson, J. Roth, B. Nosho and J. Jensen, Third generation FPA development status at Raytheon Vision Systems, *Proc. of SPIE*, 2005, **5783**, 331.
283. R. Breiter, T. Ihle, J. C. Wendler, H. Lutz, S. Rutzinger, T. Schallenberg, K. C. Hofmann and J. Ziegler, *Opt. Eng.*, 2011, **50**, 061010.
284. A. Singh, V. Srivastav and R. Pal, *Opt. Laser Technol.*, 2011, **43**, 1358.
285. A. Rogalski, *Infrared Phys. Technol.*, 2002, **43**, 187.
286. V. P. Ponomarenko, *Phys. Usp.*, 2003, **46**, 629.
287. C. R. Becker, X. C. Zhang and K. Ortner, *Thin Solid Films*, 2002, **412**, 129.
288. K. C. Mandal, S. H. Kang, M. Choi, J. A. Wei, L. L. Zheng, H. Zhang, G. E. Jellision, M. Groza and A. Burger, *J. Electron. Mater.*, 2007, **36**, 1013.
289. C. Szeles, *phys. stat. sol. (b)*, 2004, **241**, 783.
290. T. Takahashi and S. Watanabe, *IEEE Trans. Nucl. Sci.*, 2001, **48**, 950.
291. O. Limousin, *Nucl. Instrum. Methods Phys. Res., A*, 2003, **504**, 24.
292. M. Fiederle, V. Babentsov, J. Franc, A. Fauler and J. P. Konrath, *Cryst. Res. Technol.*, 2003, **38**, 588.
293. R. Leckie and H. Lenos, *Chip Scale Rev.*, 2009, **13**.
294. M. Prokesch, D. S. Bale and C. Szeles, *IEEE Trans. Nucl. Sci.*, 2010, **57**, 2397.
295. J. Liptac, R. Parker, V. Tang, Y. Peysson and J. Decker, *Rev. Sci. Instrum.*, 2006, **77**, 103504.
296. J. Hong, B. Allen, J. Grindlay, N. Chammas, S. Barthelemy, R. Baker, N. Gehrels, K. E. Nelson, S. Labov, J. Collins, W. R. Cook, R. McLean and F. Harrison, *Nucl. Instrum. Methods Phys. Res., A*, 2009, **605**, 364.
297. M. Djurašević, I. Vukanac, A. Kandić, L. Nađđerđ, Z. Milošević and M. Radenković, *J. Environ. Radioact.*, 2007, **95**, 75.

CHAPTER 11.5

Nanoparticles and Quantum Dots

LIHUI YUWEN AND LIANHUI WANG*

Institute of Advanced Materials (IAM), School of Materials Science and Engineering, Nanjing University of Posts and Telecommunications, 9 Wenyuan Road, Nanjing 210046, P. R. China, Email: iamlhwang@njupt.edu.cn
*Email: wlhui@fudan.edu.cn

11.5.1 Introduction

Since the famous talk given by Feynman decades ago, nanoscience and nanotechnology have achieved tremendous success and continue to impact multidisciplinary fields including material science, chemistry, optoelectronics, biomedicine, energy, environment, and so on.[1] Large numbers of nanomaterials with different compositions, structures, and functionalities have been prepared. Among them, nanoparticles are the most normal building blocks for constructing various complex materials and provide typical models for fundamental physics and chemistry study.[2–3]

Nanoparticles are ultra-small particles with dimensions in the nanometre range, often in the range of 1–100 nm.[4] As an intermediate state between molecule and bulk material, nanoparticles with hundreds to thousands of atoms show distinct variation in fundamental properties compared with their bulk state, such as size-dependent optical, electronic, catalytic, magnetic, and mechanical properties.[5] Among different kinds of nanoparticles with various compositions, metal chalcogenide semiconductor nanoparticles represent a series of important functional nanomaterials with great application potential in optoelectronics and life science. When the size of semiconductor nanoparticles is smaller than their bulk exciton Bohr radius (usually in the range 1–10 nm),

Handbook of Chalcogen Chemistry: New Perspectives in Sulfur, Selenium and Tellurium, 2nd Edition, Volume 2
Edited by Francesco Antonio Devillanova and Wolf-Walther du Mont

Published by the Royal Society of Chemistry, www.rsc.org

the carriers in nanocrystals are confined in all three dimensions and these semiconductor nanoparticles are called quantum dots (QDs).[6–7] Quantum confinement effect facilitates the tuning of the bandgap energy of QDs to emit different colours just using one semiconductor material, as shown in Figure 11.5.1. Due to their unique optoelectronic properties, QDs have found many potential applications such as light-emitting diodes (LEDs), solar cells, and biomedicine.

QDs can be prepared with different semiconductor materials, including Group II–VI, III–V, IV–VI, and I_2–VI. Metal chalcogenide semiconductors, such as CdS, CdSe, CdTe, ZnSe, PbS, PbSe, provide good candidates for QDs preparation (Table 11.5.1). By choosing proper compositions and control over the size, different kinds of QDs with different emission colours can be synthesized conveniently.

Since QDs have been involved in various research areas and are still under intensive research, plenty of reports on them have appeared in recent years. In this chapter we will focus on the typical metal chalcogenide semiconductor QDs, mainly the Group II–VI QDs which have been studied systematically and thoroughly. First, we will give a brief overview of the research history of QDs and their basic properties, and then the development of state-of-the-art synthesis methods are discussed in detail, especially the relatively "greener" aqueous solution methods. Moreover, research on exploiting QDs as functional building blocks in optical and electronic devices has progressed rapidly, and this will also be mentioned in the latter part of this chapter.

11.5.1.1 Brief History of Quantum Dots Research

In 1977, Katzschmann *et al.* observed that when the size of Group II–VI semiconductor nanoparticles doped in glass is smaller than 5 nm, their colour changes.[10] By using small-angle X-ray diffraction analysis, they found the energy gap of the semiconductor nanoparticles changed with size; however, they considered that this was caused by the shrinkage of the crystal lattice

Figure 11.5.1 CdTe QDs of different sizes synthesized in aqueous solution (taken under UV irradiation at 365 nm).

Table 11.5.1 Bandgaps and structure parameters of typical II–VI, IV–VI, I_2–VI semiconductor materials

Group	*Name*	*Bandgap (E_g, eV)*	*Crystal structure*	*Lattice constant (Å)*
II–VI[a]	ZnS	3.8	Wurzite	a = 3.811, c = 6.234
		3.68	Zinc blende	5.406
	ZnSe	2.67	Zinc blende	5.667
	ZnTe	2.25	Zinc blende	6.104
	CdS	2.42	Zinc blende	5.832
		2.53	Wurzite	a = 4.135, c = 6.749
	CdSe	1.71	Wurzite	a = 4.299, c = 7.010
	CdTe	1.45	Zinc blende	6.482
IV–VI[b]	PbS	0.42	Rock salt	5.936
	PbSe	0.28	Cubic	6.124
	PbTe	0.31	Cubic	6.462
I_2–VI[b]	Ag_2S	1.0	Monoclinic	a = 4.23, b = 6.91, c = 7.87
	Ag_2Se	0.15	Orthorhombic	a = 4.333, b = 7.062, c = 7.764
	Ag_2Te	0.67	Monoclinic	a = 8.15, b = 4.47, c = 8.09

Data from ref.[8];
Data from ref.[9]

rather than the quantum confinement effect. In 1980, Ekimov *et al.* also observed the absorption peak of CuCl nanocrystals dispersed in silicate glasses shifts to the short-wave region and the size-dependence optical phenomenon was interpreted as a "quantum size effect".[11] Using quantum mechanics to interpret the size-dependent photophysical properties observed from the semiconductor nanocrystals opens the door to understanding the fundamental physics of zero-dimensional semiconductor nanomaterials. After that, Efros *et al.* pioneered the theory study of electronic structures of QDs in 1982.[12] They adopted effective mass approximation (EMA) and assumed spherical geometry of QDs, parabolic conductance and valence bands, infinite spherical symmetry potential wells, and neglected the Coulomb interactions between electrons and holes. The theory provides a standard for evaluating the quantum confinement effect, Bohr radius, and quantizes the energy level originating from quantum confinement, which builds the theoretical basis for further study.[12] In 1983, Brus *et al.* reported a colloidal chemistry method to synthesize aqueous dispersed CdS nanocrystals which show continuous bathochromic shift of the absorption peak during growth.[13] By analysing the size-dependent optical properties of the CdS nanocrystals using quantum mechanics, Brus established a model involving electron–hole interaction and further developed the EMA method.[14–16]

For a long period, many technical terms were used to describe semiconductor nanocrystals with quantum confinement effect, such as semiconductor clusters, artificial atoms, Q-particles, quantum crystals, quantum spheres, *etc.*[7,17–21] Compared with two different quantized systems, "quantum wells" and "quantum wires", Reed defined the three-dimensional quantized system as "quantum dots", where the carriers have zero-dimensional freedom.[6–7,22]

11.5.2 Basic Properties of Quantum Dots

Because of their unique electronic structure, QDs have different optical, electrical, mechanical, magnetic, and thermal properties compared with their bulk form.[7,16,23]

11.5.2.1 Quantum Confinement Effect

When the size of semiconductor nanocrystals is smaller than their bulk exciton Bohr radius, the movement of the electrons and holes are confined in three-dimensional space, and the continuous energy band splits into independent energy levels. When the size of semiconductor nanocrystals decreases further, the kinetic energy of the carriers increases and their optical absorption peak blue-shifts, which is the quantum confinement effect.[22,24–25]

Efros *et al.* first gave a theoretical treatment of the quantum confinement of semiconductor nanocrystals. According to the relationship between particle size and the Bohr radius, the confinement can be divided into three types.[12,26] As a fundamental property of low-dimensional semiconductor nanomaterials, the quantum confinement effect can be described as follows:[24]

$$a_B = \frac{h^2 \varepsilon}{4\pi^2 e^2}\left[\frac{1}{m_e^*} + \frac{1}{m_h^*}\right] \tag{1}$$

where a_B is the Bohr radius, ε is the dielectric constant of vacuum, and m_e* and m_h* are effective mass of electron and hole, respectively.

Based on this progress, Brus, Kayanuma, and Wang developed quantitative relationship between energy gap of QDs and their size (Equation 2), involving the interaction between electrons and holes:[15–16,27]

$$\Delta E = \frac{h^2}{8r^2}\left[\frac{1}{m_e^*} + \frac{1}{m_h^*}\right] - \frac{1.786e^2}{\varepsilon r} - 0.248E_{Ry} \tag{2}$$

where ΔE is the energy shift for the QDs in the exciton energy, r is the radius of the QDs, ε is dielectric constant of vacuum, m_e* and m_h* are effective mass of electron and hole, respectively, and E_{Ry} is the effective Rydberg energy. According to Equation 2, when the size of the semiconductor nanocrystals decreases, the quantum confinement effect causes exciton energy to increase. As shown in Figure 11.5.2, when atoms aggregate to form clusters, the discrete HOMO and LUMO energies change from single value to a series of energy levels, which finally form quasi-continuous energy bands and finally approach the energy bandgap of the bulk materials.

11.5.2.2 Optical Properties

Compared with traditional organic fluorescent dyes, QDs have several advantages: narrow emission peak and broad absorption peak, large Stokes shift, tunable emission colour, high photostability, high photoluminescence quantum yield (PLQY).[28–29]

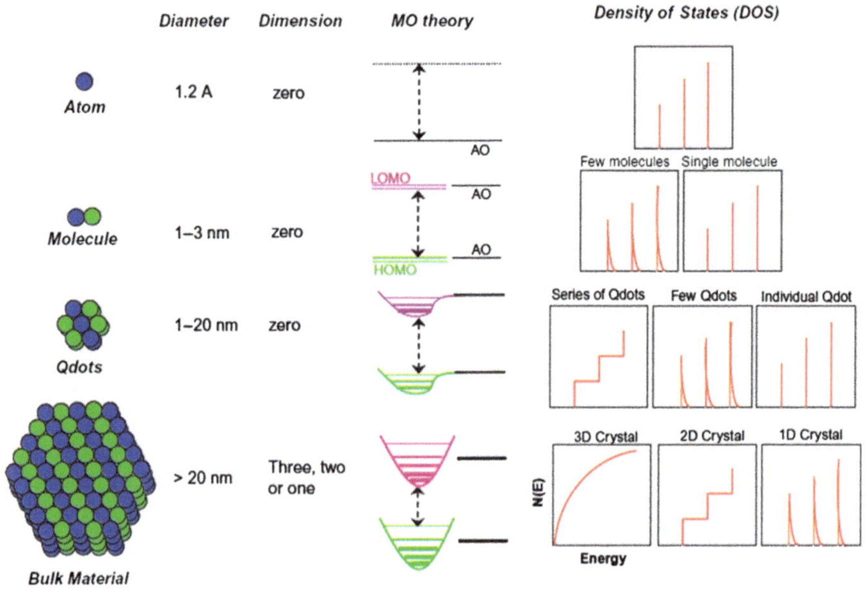

Figure 11.5.2 Schematic illustration of the influence of the density of states (DOS) with changes in the number of atoms in materials. AO, atomic orbital, MO, molecular orbital.
(Reprinted with permission from ref. 8. Copyright 2008, John Wiley & Sons, Inc.)

As shown in Figure 11.5.3, the absorption peaks of ordinary organic dyes are in a limited range, whereas QDs can be excited at a wavelength shorter than absorption peaks. So it is possible for QDs of different colours to be excited simultaneously with a single-wavelength light source, which is very useful for multiplex detection in bioanalysis (Figure 11.5.4). The FWHM (full width at half maximum) of QD's emission peak is usually smaller than 30 nm, while for organic dyes the FWHM is usually much larger and often causes self-absorption issues.

Due to the quantum confinement effect, decreasing the size of QDs will increase the carrier energy and shift the absorption and emission peak to shorter wave regions. For example, II–VI semiconductors such as CdSe ($E_g = 1.71$ eV) or CdTe ($E_g = 1.45$ eV) can be used to prepare QDs with tunable emission covering the whole visible range; II–VI QDs with larger bandgaps can emit UV light; Emission of PbS and PbSe QDs can easily be tuned in the infrared range. The facile tunability of the emission colours of QDs is hardly possible for organic dyes.

Photobleaching is often observed for organic dyes under intensive light excitation, however, QDs show excellent photostability due to their inorganic composition, which is especially useful for biophysical applications. As shown in Figure 11.5.5, under the same irradiation intensity, after being continuously irradiated for 180 seconds, fluorescent signal (green light) from the organic dye Alexa488 labelled on the surface of cells disappeared, while the yellow light signal from QDs was still bright.

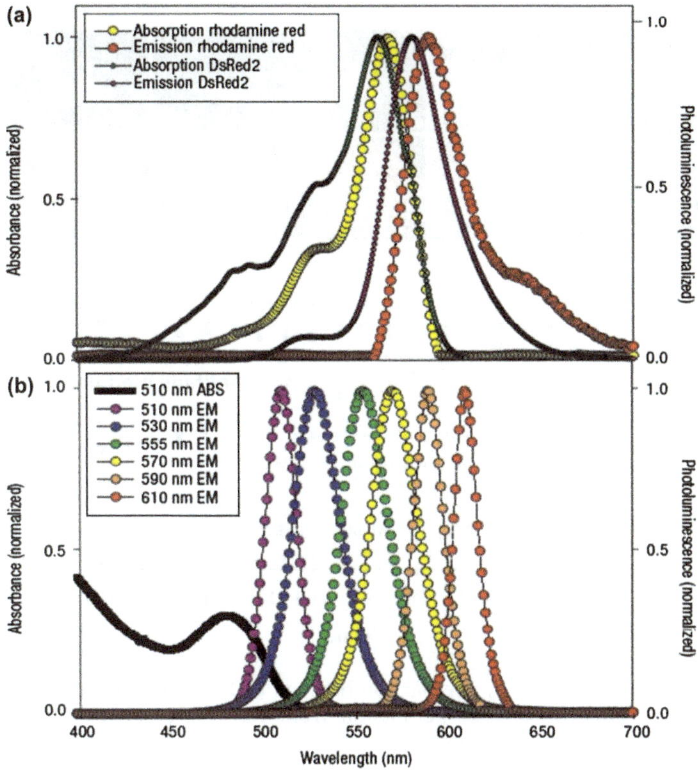

Figure 11.5.3 The absorption and emission peaks of CdSe QDs with different emission colours and organic dyes (rhodamine red and DsRed2). (Reprinted with permission from ref. 28. Copyright 2005, Nature Publishing Group.)

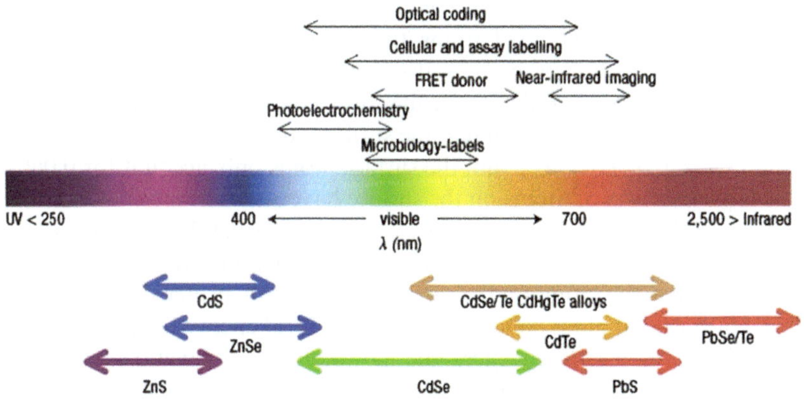

Figure 11.5.4 The range of emission spectrum of typical II–VI and IV–VI QDs and their use for bioanalysis.
(Reprinted with permission from ref. 28. Copyright 2005, Nature Publishing Group.)

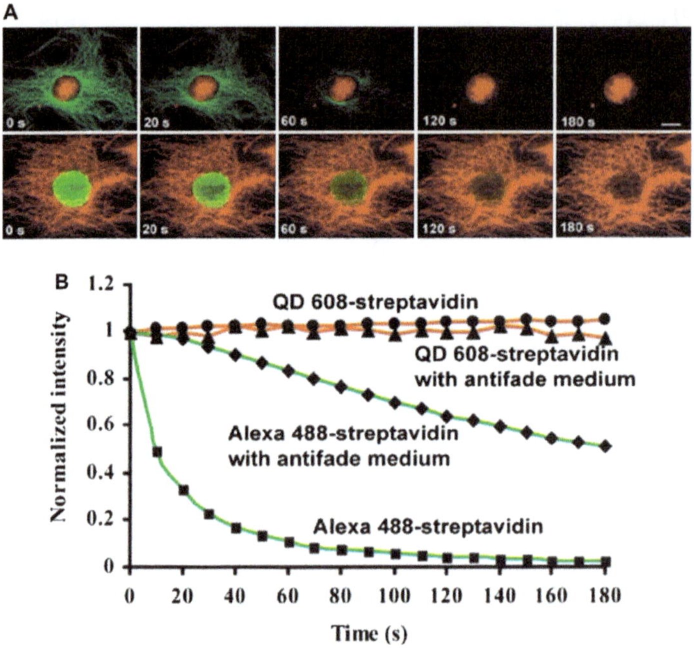

Figure 11.5.5 Photostability comparison between QDs and organic dye Alexa 488. (Reprinted with permission from ref. 30. Copyright 2003, Nature Publishing Group.)

11.5.2.3 Surface State

Surface state is another important property of QDs. As shown in Figure 11.5.6, the typical size of QDs is in the nanoscale region (1–10 nm); decreasing their size will significantly increases their specific surface area. Assuming a cube-shaped particle formed from one type of atoms, the average distance between two atoms is 0.2 nm, and the crystal structure is also cubic, then if the particle size is 2 nm, the proportion of atoms on the nanocube surface is about 54% in all; when the size of the particle increases to 20 nm, the surface atom ratio drops to about 6%; if the nanocube is as large as 2 μm, the surface atoms only account for 0.06% in all. The large numbers of atoms on the surface have a significant influence on the properties of QDs.[31] From the point of view of chemistry, QDs formed by atoms without any surfactant molecules are 'pure'; however, according to semiconductor physics, the atoms on the surface of QDs with insufficient coordinating number can form dangling bonds and introduce local energy levels which act as carrier traps.[7] As a consequence, the non-radiative relaxation of excitons increases and the PLQY of QDs reduces. In order to obtain highly luminescent QDs, the surface should be coated by other semiconductor materials with larger bandgap to confine the excitons in the nanocrystals. Using surface passivation, various core–shell structure QDs with high quantum yield have been prepared, and this has become a standard method to obtain high-quality luminescent QDs.[7]

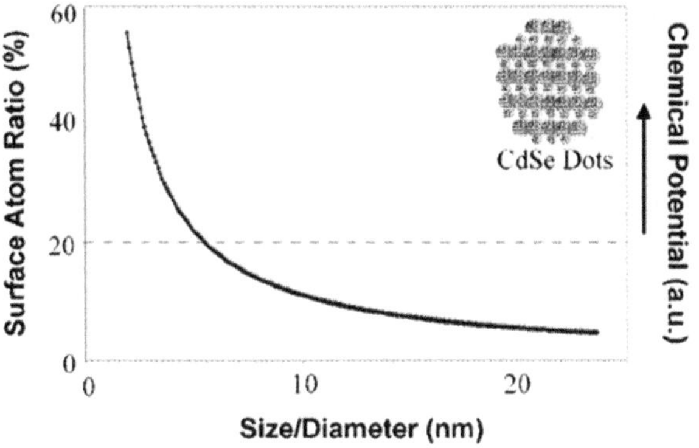

Figure 11.5.6 Dependence of surface atom ratio on the size change of CdSe QDs. (Reprinted with permission from ref. 31. Copyright 2002, American Chemical Society.)

11.5.3 Synthesis Methods

As the basis of application, material preparation techniques play an important role in the development of the QD-related research. QDs can be prepared by either a top-down or a bottom-up approach.[7] However, using modern nanofabrication techniques developed in the microchip industry to prepare large-scale QDs is hard to realize.[22] On the other hand, the chemical approach, especially colloidal chemistry developed in the 20th century, is a highly controllable, scalable, and cost-effective method to synthesize various kinds of QDs with different compositions, structures, and morphologies.

According to the solvents used in the synthesis of QDs by the colloidal chemistry route, synthetic methods can be roughly divided into two types: organic-phase synthesis and aqueous-phase synthesis. Also, combined with many special techniques, such as photochemical, microwave, biomineralization, *etc.*, many derived methods have been developed.

11.5.3.1 Organic-Phase Synthesis

In 1988, Brus, Steigerwald *et al.* explored highly reactive silylselenium and silyltellurium reagents as organic Se and Te sources to synthesize CdSe nanocrystals in inverse micellar solutions.[32] Steigerwald synthesized larger CdSe nanoparticles by thermolysis of molecular precursor compounds $Cd(SePh)_2$ and $[Cd(SePh)_2]_2[Et_2PCH_2CH_2PEt_2]$ in 4-ethylpyridine solution.[33]

In 1993, Bawendi *et al.* developed the versatile and effective high-temperature organic-phase method to prepare high-quality CdE (E = S, Se, Te) QDs, which is a milestone in QDs synthesis.[34] In this method, highly reactive $Cd(CH_3)_2$ is used as cadmium source, TOP and TOPO are used as solvents, and the complex of Se and TOP is used as selenium source. The reaction system is heated to about 300 °C under the protection of argon, and the cadmium sources or chalcogenide sources are injected rapidly into the reaction system, so this method is also called "hot injection". Under high temperature, the precursors quickly form lots of crystal nuclei. CdE QDs with different sizes from 1.5 to 12 nm and small size distributions can be obtained.[34–35] However, this synthesis route requires highly inflammable, explosive, and toxic precursors.

Peng developed a relatively "greener" organic-phase synthesis by using CdO and $Cd(Ac)_2$ as safer and cheaper cadmium sources and choosing fatty acids and fatty amines as ligands, and octadecene as a high boiling-point solvent. The PLQY of the nearly monodisperse high-quality CdSe QDs prepared by this method is as high as 85%; the FWHM is only 23 nm.[31,36–39] Through careful choice of precursors, the synthesis of QDs in organic-phase solution becomes safer, easier, and cheaper. The reproducibility of the synthesis is much higher than ever, which also makes it a versatile approach for other nanocrystals' synthesis.[40–41]

Since the nucleation and growth for QDs synthesized by the high-temperature organic method are well separated, the size distribution of the QDs is very narrow. As shown in Figure 11.5.7, according to the La Mer model for monodisperse colloid growth, immediately after the injection of the precursors, the reaction temperature drops and the concentration of precursors exceeds the nucleation threshold (critical concentration) significantly, which leads to a 'nucleation burst' and forms plenty of nuclei. When the precursor concentration is lower than the nucleation threshold during the growth process, the synthesis enters the growth stage. Until the precursors are nearly depleted, further growth of the QDs depends on Ostwald ripening. In this stage, larger QDs keep growing by consuming the smaller ones, which causes broadening of the size distribution.[35]

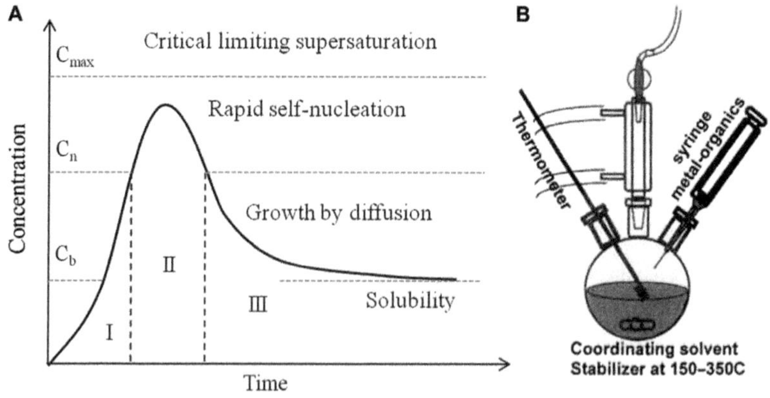

Figure 11.5.7 The LaMer model for monodispersed nanoparticle formation (C_b: solubility of the bulk materials; C_n: critical concentration for nucleation; C_{max}: maximum concentration for nucleation; I: stage before nanoparticle nucleation; II: rapid nucleation stage; III: growth and Ostwald Ripening stage).
(Reprinted with permission from ref. 35. Copyright 1950, American Chemical Society.)

The growth kinetics of II–VI QDs was studied in hot organic media.[42] The results showed that extra injection of precursors can "focus" the size distribution of the QDs, which revealed that the precursor concentration has a remarkable influence on the growth dynamics. When the precursor solution is injected into the high-temperature reaction system, nucleation occurs immediately and the precursors are gradually depleted until the monomer concentration is lower than the critical threshold. The nucleation stage is very transient and hard to characterize, while in the growth process, the relationship between solubility of the nanocrystals, and their radius can be expressed by a Thomson–Gibbs equation (Equation 3), assuming the growth process is diffusion-limited. S_r and S_b refer to solubility of the nanoparticles and their bulk materials, respectively, σ is the surface energy of the nanoparticles, V_m is the molar volume of the bulk materials, R is the gas constant, and T is the temperature.[43]

$$S_r = S_b \exp(2\sigma V_m / rRT) \tag{3}$$

$$\frac{dr}{dt} = K\left(\frac{1}{r} + \frac{1}{\delta}\right)\left(\frac{1}{r^*} - \frac{1}{r}\right) \tag{4}$$

If $2\sigma V_m / rRT \ll 1$, the growth of nanocrystals is under diffusion-limited control and the growth rate is shown in Equation 4, where K is a constant proportional to diffusion constant, δ is the thickness of the diffusion layer, r^* is the critical size under present precursor concentration, and r is the real average size of the nanocrystals. If the average size of nanocrystals is equal to the critical size r^*, the growth rate is zero, which means the nanocrystals will stop growing; as shown in Figure 11.5.8, if the average size of nanocrystals is larger than the critical size r^*, the growth rate is positive, which means the

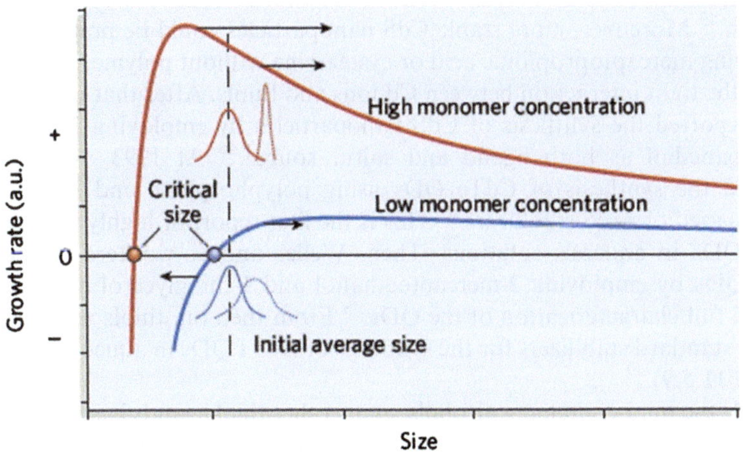

Figure 11.5.8 The kinetic influence of precursor concentration on the growth rate of colloidal nanoparticles.
(Reprinted with permission from ref. 44. Copyright 2005, Nature Publishing Group.)

nanocrystals will grow bigger; if precursor solution is introduced to the reaction system by a second injection, the smaller nanocrystals will grow faster than the larger ones, which will narrow or focus the size distribution.[44] When the precursor is nearly depleted, the critical size gradually increases and only nanocrystals larger than the critical size can keep on growing, accompanying the dissolving of the smaller ones, which broadens or defocuses the nanocrystal size distribution.

11.5.3.2 Aqueous-Phase Synthesis

11.5.3.2.1 Traditional Aqueous Route

From the synthetic view of point, the aqueous solution method has some advantages due to the nature of the solvent. First, the reaction can be carried out under relative mild conditions, which simplifies the equipment and manipulation of the experimental procedure; second, there are plenty of water-soluble reagents can be used as the precursors for QDs; third, the as-prepared QDs can be applied directly without further surface modification required for those prepared in organic phase in biological environment or other applications with water as solvent such as devices fabricated by layer-by-layer processes.

So far, thiols are the best ligands to control the nucleation and growth of II–VI semiconductor nanocrystals. However, the exploration of suitable stabilizers for high-quality QDs synthesis lasted for a long period. In 1988, Nosaka reported that by introducing mercaptoethanol as additive and using $CdCl_2$, Na_2S, sodium hexametaphosphate (HMP) or polymeric viologen as cadmium source, sulfur source, and stabilizer, respectively, homogeneous CdS nanoparticles of different sizes can be obtained reproducibly in aqueous solution.[45] Moreover, more stable CdS nanoparticles could be prepared solely employing mercaptopropionic acid or cysteamine without polymeric stabilizers due to the tight interaction between Cd ions and thiols. After that work, Hayes *et al.* reported the synthesis of CdS nanoparticles by employing 3-mercapto-1,2-propanediol as both ligand and sulfur source.[46] At 1993, Nozik *et al.* reported the synthesis of CdTe QDs using polyphosphate and 3-mercapto-1,2-propanediol as co-stabilizers.[47] This is the first report of highly luminescent II–VI QDs in aqueous solutions. Then, Weller and co-workers synthesized CdTe QDs by employing 2-mercaptoethanol and 1-thioglycerol as stabilizers and did full characterization of the QDs.[48] From then on, thiols were actually used as standard stabilizers for the synthesis of II–VI QDs in aqueous solutions (Figure 11.5.9).[49]

Beside the thiol-containing alcohols, many other thiol ligands has been utilized for II–VI QDs synthesis in aqueous solution (Figure 11.5.10), including thioglycolic acid (TGA),[50] 3-mercaptopropionic acid (3-MPA),[51] 2-mercaptopropionic acid (2-MPA),[52] mercaptosuccinic acid (MSA),[53–54] dihydrolipoic acid (DHLA),[55] 6-mercaptohexanoic acid (MHA),[56] 11-mercaptoundecanoic acid (MUA),[49,56] 2-mercaptoethylamine(MEM),[57–58] glutathione(GSH),[59–60]

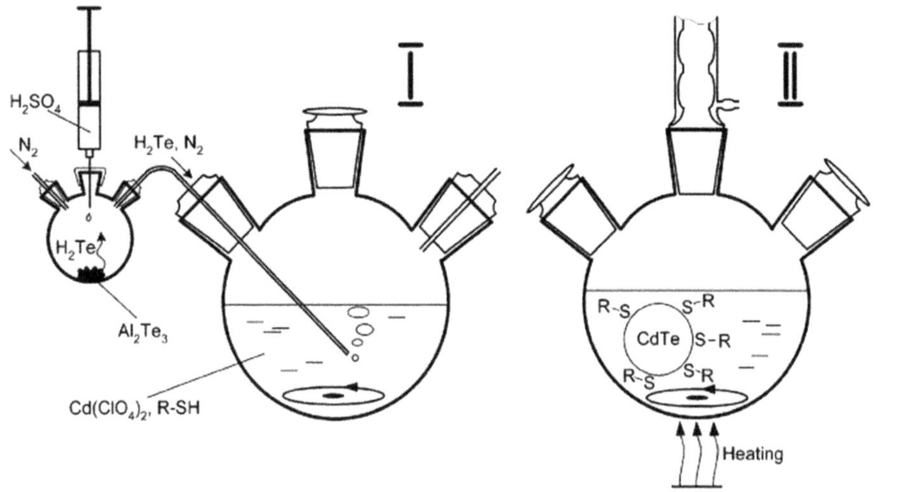

Figure 11.5.9 Schematic presentation of the synthesis of thiol-capped CdTe QDs. (I) Formation of CdTe precursors by introducing H_2Te gas into the reactor containing thiols and cadmium ions. (II) Nucleation and growth of CdTe QDs promoted by reflux.
(Reprinted with permission from ref. 50. Copyright 2002, American Chemical Society.)

cysteamine,[61] l-cysteine,[62] 5-mercaptomethyltetrazole,[63] and thiolated polyethylene glycol(PEG-SH).[64] TGA and MPA are the most popular stabilizers for aqueous II–VI QD synthesis, and the as-prepared CdTe QDs usually have high PLQY in the range of 40–60%.[65] Long alkyl chain stabilizers such as MHA and MUA are suitable for growing small CdTe QDs (0.6–2.5 nm in diameter) with relatively low PLQY(1–3%).[56] The chelating ligand–DHLA molecules containing two thiol groups can render QDs with better chemical and thermal stability and biocompatibility.[55,66] QDs capped by mercaptoamines are stable in acidic solution and the amino groups can be bioconjugated with aldehyde-modified biomolecules.[62] Thiol-containing oligopeptides such as GSH have also been reported to be effective ligands of CdTe QDs with PLQY ~45%.[59–60] GHS-coated CdTe QDs are biocompatible and can be further used for biolabelling. PEG-SH is a novel amphiphilic ligand which has been used to synthesize CdTe QDs in both hydrophobic organic solvents and polar solvents without using additives in the reaction and can render QDs with amphiphilicity, which may have potential applications in biological transport study.[64]

The ligands not only influence the optical property and the surface functionality of QDs, but also play an important role in the growth mechanism. Early study showed that when cysteine and TGA are used as stabilizers at the same time, CdTe grows in a one-dimensional manner.[68–69] In an in-depth study, Zhang *et al.* showed that using ligands with analogous structures to TGA (HS-CH(R)-COOH, R = H, CH_3, NH_2, *etc.*), such as L-cysteine (LCS), thiolactic acid (TLA), mercaptosuccinic acid (MSA), one-dimensional (1D)

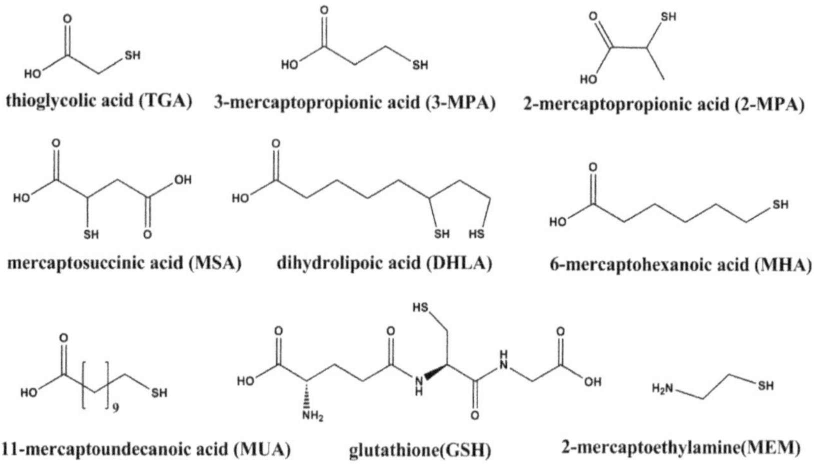

Figure 11.5.10 Typical chemical structures of thiol-containing ligands employed in the aqueous synthesis of QDs.

CdTe nanostructures could be prepared (shown in Figure 11.5.11).[70–71] Jin and co-workers studied the growth mechanism of CdTe 1D structures using LCS as ligands. They claimed that oriented attachment and Ostwald ripening dominate the anisotropic growth of the CdTe nanostructures which is dependent on the initial precursor concentration.[72]

Besides the study of growth mechanism of the thiol-capped method, improving the PLQY of the QDs is also of great importance for further application. QDs synthesized by the "classic" aqueous method often suffer from low PLQY due to surface traps, as for CdS and CdSe QDs the photoluminescence (PL) is usually dominated by defect-related emission rather than band-edge emission.[52,73–75] Although thiols have little influence on the optical properties of CdTe QDs compared to CdSe, the PLQY of CdTe QDs is relative low without proper surface modification. One effective method is using hydrolysis of thiol ligands to release sulfur atoms to the surface of CdTe QDs to form a CdS shell which can improve the PLQY of CdTe QDs up to 60%.[65,76–77] Also, photochemical decomposition of thiols is used to enhance the PL of CdTe QDs up to 80–98%.[78–79] However, these techniques are only effective for TGA-CdTe and the photochemical degradation is time consuming. Peng *et al.* reported CdTe/CdS core–shell QDs with PLQY as high as 58% by refluxing the TGA-capped CdTe QDs with thioacetamide as a sulfur source.[80] Zhong *et al.* synthesized CdTe/CdS QDs using a one-pot approach and the PLQY was improved from 40% to 80%.[81] Introducing inorganic material with larger bandgap to the surface of QDs not only enhances the PLQY but also improves the photostability and biocompatibility.[55,80,82–83]

Although synthesis of CdTe QDs is very successful, their emission spectra cannot cover the ultraviolet to blue and near-infrared to infrared regions. To extend the range of emission spectrum of QDs, many other II–VI QDs with

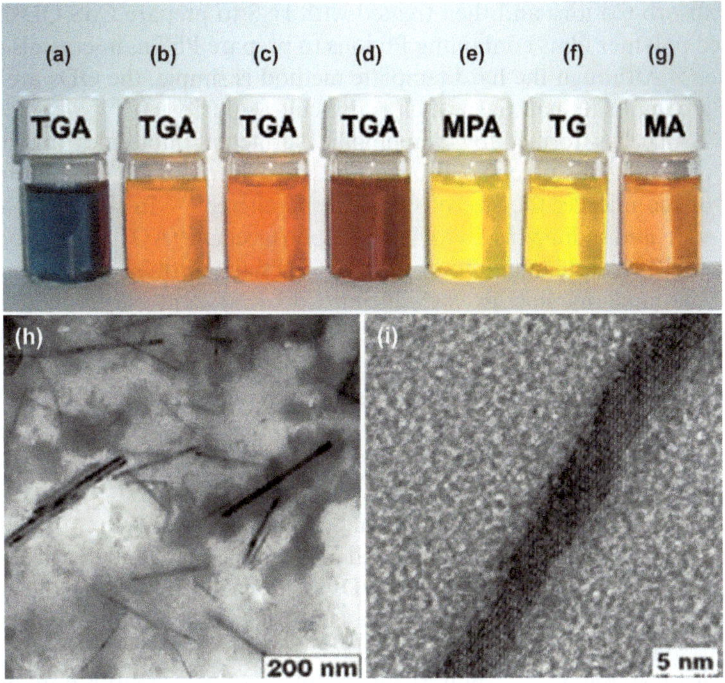

Figure 11.5.11 Photographs of aqueous solutions of CdTe nanocrystals formed by storing solutions of their precursors at different concentrations (a–d) and CdTe nanocrystals formed in the presence of (e) MPA,(f) TG, and (g) MA. (h) TEM and (i) HRTEM images of TGA-stabilized CdTe nanocrystals.
(Reprinted with permission from ref. 70. Copyright 2006, John Wiley & Sons, Inc.)

wide-band and narrow-band semiconductors or in the alloying and doping forms, have been prepared by the classic aqueous solution method. Without changing the size of the QDs, by tuning the compositions of the alloyed QDs the emitting colours can be adjusted continuously.[84] The alloyed QDs of $Zn_{1-x}Cd_xSe$,[85–86] $Zn_{1-x}Cd_xSe/ZnS$,[87] $Cd_xZn_{1-x}Te$,[88] $ZnSe_{1-x}Te_x$,[89] $CdSe_xTe_{1-x}$,[90] CdSeTe/ZnS,[91] CdSeTe,[92] CdHgTe/CdS,[93] CdHgTe.[94] Introducing transition metal(Cu or Mn) ions into ZnSe host can generate new emission centres and avoid the use of cadmium, such as ZnSe:Mn,[95–98] ZnSe:Cu.[95,99]

11.5.3.2.2 Template Synthesis

One of the earliest methods for QD preparation is template synthesis, by doping glass with semiconductors.[10–11,100] After that, solid porous materials were used as hard templates to synthesis semiconductor nanocrystals. Wang *et al.* used zeolite as a template to absorb Cd or Pb ions, then treated with H_2S at 400 °C to prepare CdS and PbS nanocrystals.[101] Nafion membrane was also used to absorb Cd ions and then treated with H_2S to prepare CdS QDs. Yang *et al.* used polymer films containing Pb ions to prepare PbS nanocrystals on the substrate.[102] Although the hard template method is simple, the QDs are in the form of a composite material, which is difficult to purify.

Reverse micelles formed by amphiphilic surfactant molecules are natural containers of aqueous precursors for *in situ* synthesis of QDs.[103] By controlling the size of the micelle, the size of QDs can be altered.[104] In the early studies, Fendler *et al.* used sodium bis-2-ethylhexylsulphosuccinate (AOT)/H_2O/isooctane reverse micelles as a template to synthesize CdS nanocrystals.[105] Whitten *et al.* reported a similar method by introducing $Cd(ClO_4)_2$ and Na_2S solution into AOT/heptane/H_2O reverse micelle solution to prepare CdS QDs.[106] Brus and co-workers synthesized small CdSe, CdTe nanocrystals using AOT/heptane/H_2O reverse micelle system.[32] The first synthesis of band-edge emitting II–VI QDs in reverse micelles was reported by Pileni *et al.*[107] They prepared CdTe QDs with different sizes by tuning the size of the water droplets confined in the reverse micelles.

Reverse micelles formed by amphiphilic block copolymers are also used to synthesize II–VI QDs. Eisenberg *et al.* used polystyrene-*b*-poly(acrylic acid) block polymer as stabilizers to synthesize CdS QDs in the water droplet.[108] Other similar block polymers were also used for synthesis of CdS QDs, including poly(acrylic acid)-poly(styrene)-poly(ethyleneoxide),[109–110] hydroxylated poly-(styrene-b-butadiene-b-styrene),[111] polystyrene-b-poly(acrylic acid)-b-poly-(methyl methacrylate),[112] and polystyrene-b-poly(4-vinylpyridine).[113]

Although micelle-based colloidal template method is facile and aggregation of QDs can be avoided during synthesis, the lack of ligand protection generates plenty of defects on the surface of QDs, which usually causes weak and defect-related PL emission.[114]

11.5.3.2.3 Hydrothermal Method

The reaction temperature of traditional aqueous-phase synthesis is usually near the boiling point of water, which is much lower than the organic-phase method (200–320 °C). The relatively low temperature leads to low growth rate and more surface defects.

By employing a hydrothermal method, the reaction can be carried out under high temperature (as high as 220 °C) and high pressure, which can significantly accelerate the growth rate and form high-quality QDs with fine crystal structure and improved PLQY. Zhang *et al.* first reported the synthesis of CdTe QDs at about 180 °C under hydrothermal conditions. For red-emitting CdTe QDs, the reaction time was reduced to less than 2 hours.[115] The PLQY of the as-prepared CdTe QDs was improved to as high as 30%. Guo *et al.* performed a detailed study of the influence of reaction conditions on the PL properties.[116] By using an optimized TGA/Cd molar ratio the PLQY of 50% can be reached. The hydrothermal method can be used to synthesize CdTeS alloyed near-infrared emitting (680–800 nm) QDs with high PLQY (as high as 68%).[117] The alloyed structure formed in the one-pot approach originated from hydrolysis of thiol-containing stabilizers accelerated by high temperature (180 °C). Moreover, distinct deviation of the X-ray diffraction peaks from pure CdTe phase and the tendency toward CdS phase have also been observed in other studies, thus the sulfur doping into the original crystal structure may be considered as a common features of hydrothermal method and has direct relationship to their superior PL properties.[116–119] The hydrothermal method is not only used for core-only QDs synthesis but also applicable for core–shell structure synthesis, for instance, CdTe/CdS,[119] CdTe/CdSe,[120] CdTe/CdS/ZnS.[121]

11.5.3.2.4 Microwave Irradiation Assisted Synthesis

Compared to classical heating techniques, dielectric heating by microwave irradiation (MI) can provide heating rates up to orders of magnitude greater and excellent heating homogeneity, which have been successfully used to accelerate the reaction rate, improve the yield of product, and increase the reaction selectivity for organic chemistry.[122–124] On the other hand, to obtain high-quality QDs with narrow size distribution and high PLQY needs precise control of experimental conditions to optimize both the nucleation and growth process, which is a weakness of conventional aqueous synthesis method heating *via* oil bath compared to the organometallic method. Combining the aqueous method and MI heating can greatly improve the synthetic conditions: the rapid, homogeneous, controllable heating and versatile choice of precursors can both be fulfilled at the same time.

In 2000, Zhu *et al.* first reported the synthesis of II–VI CdSe nanocrystals by MI in aqueous solutions. [125] Qian *et al.* reported the synthesis of CdTe QDs with high PLQY (40%–60%) and wide emission spectrum (505–733 nm) with MI.[126] Later, ZnSe(S) alloyed QDs with PLQY of 17% and biocompatible GSH-capped CdTe QDs were also prepared utilizing the same method.[60,127] In our group, He *et al.* developed a method called "program process of microwave irradiation" (PPMI) to synthesize CdTe QDs employing a CEM discover system.[128] The CdTe QDs can be synthesized within 30 min and the PLQY was in the range of 30–68%. Using optimum reaction conditions the as-prepared CdTe QDs with both high PLQY (82%) and low FWHM (27 nm) can be obtained, which represented the best results ever reported in aqueous solution.[79] After illumination under ambient environment the highest PLQY value is near identical (98%), which indicates that the MI-assisted method is a unrivalled approach for QDs synthesis in aqueous synthesis. Based on the successful synthesis of core QDs, our group developed the MI-assisted method and successfully synthesized high-quality CdTe/CdS, CdTe/CdS/ZnS core–shell structure QDs (Figure 11.5.12).[82,129] Other II–VI QDs synthesized by the MI method have been reported recently including $CdSe_xTe_{1-x}$,[130] CdSe,[131] CdSe/ZnS,[132–134] CdSe/CdS/ZnS,[135] ZnSe,[136] and ZnSe: Mn/ZnS.[137]

11.5.3.2.5 Sonochemical Method

Ultrasound is an unconventional heating method for chemistry. During ultrasonication, bubbles formed in solution collapse and form a hot spot with

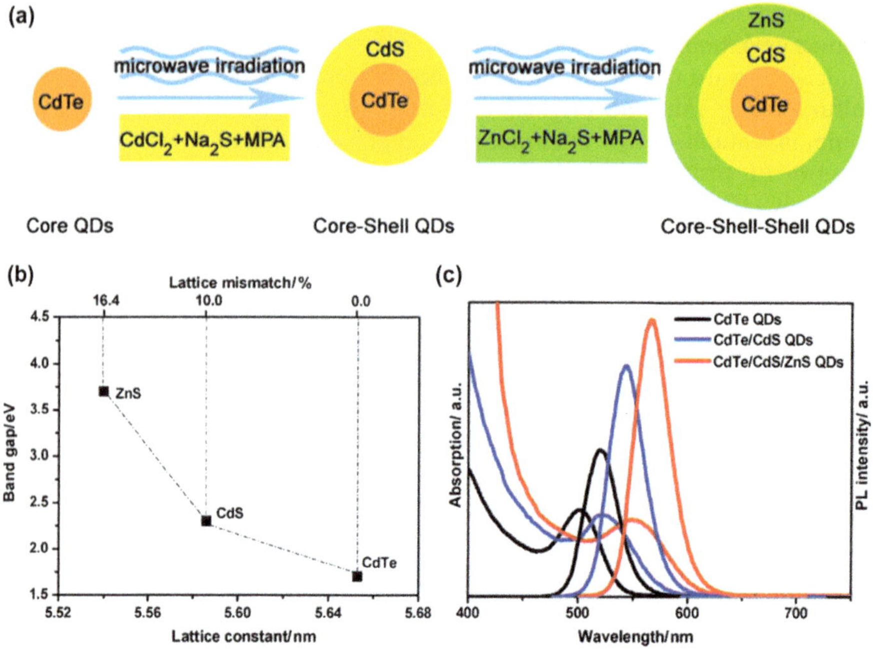

Figure 11.5.12 (a) Schematic illustration of microwave-assisted synthesis of water-dispersed CdTe/CdS/ZnS core–shell–shell (CSS) QDs (b) Interrelationship between bandgap energy, lattice constant, and lattice mismatch of bulk CdTe, CdS, and ZnS. Here, lattice mismatch refers to the mismatch compared to CdTe.(c) Representative absorption and photoluminescence (PL) spectra of CdTe core QDs, CdTe/CdS core–shell QDs, and corresponding CdTe/CdS/ZnS CSS QDs. (Reprinted with permission from ref. 82. Copyright 2008, John Wiley & Sons, Inc.)

temperature nearby 5000 K, pressure of 500 atmospheres, heating and cooling rates greater than 10^{10} K s^{-1}, which provides extreme conditions allowing access to unusual reactions.[138–139]

Gedanken and co-workers synthesized ZnS and ZnSe nanocrystals by sonochemical method in aqueous solution.[140–141] Li *et al.* synthesized CdSe nanocrystals in aqueous solution by a sonochemical method.[142] Yang *et al.* successfully grew CdS shells on 1-thioglycerol-capped CdTe QDs using a sonochemical method at 80 °C.[143]

11.5.3.2.6 Radiation Synthesis

As a powerful technique for synthesis, radiolysis can be used to synthesize QDs. Meisel *et al.* reported the radiolytic synthesis of CdS nanocrystals by γ-irradiation of an aqueous solution of thiols and cadmium salts.[46] Mostafavi *et al.* found that the dose and the dose rate influenced the size of CdS and ZnS clusters significantly.[144,145]

11.5.3.2.7 Bio-Assisted Strategy

Usually, room temperature conditions at ambient pressure and without the use of highly reactive precursors is not practicable for QD synthesis using common chemical approaches, but it is feasible using bio-assisted strategy with the help of biological systems.[146–147] During the long period of evolution, living organisms have gained the ability of biomineralization to prepare exquisite nanomaterials during metabolism, such as the well-known magnetotactic bacteria.[148]

In 1989, CdS QDs were reported to be synthesized by incubation cadmium salts in two yeasts: *Candida glabrata* and *Schizosaccharomyces pombe*.[149] Under the stimulation of toxic cadmium ions, metal-chelating peptides containing cysteine residues ($(\gamma$-Glu-Cys$)_n$-Gly, $n = 2$–3) are produced by yeast cells, which form Cd-peptides complexes that change to CdS QDs intracellularly. The successful bio-assisted synthesis of CdS QDs in microbes provides a unique synthetic technique compared to classical chemical routes.[150] By a freeze–thaw effect, CdS QDs with a diameter of about 1.8 nm were formed in the *S. pombe* cells without breaking the cells.[151] In order to get better control over the morphology and property of CdS QDs grown in cells, *Escherichia coli* was used.[152–153] The *E. coli* was genetically modified to produce phytochelatin (PC) to mimic the synthesis mechanism in *S. pombe*. CdS QDs synthesized by this method were monodisperse due to the well-controlled growth. Beside the intracellular synthesis, extracellular enzymatic reduction synthesis of CdS QDs was performed by *Fusarium oxysporum*, a kind of fungus.[154] CdS QDs were prepared on the surface of immobilized *Rhodobacter sphaeroides*.[155] The size of CdS QDs, in the range of 2.3 to 36.8 nm, can be controlled by simply tuning the culture time.

Other metal chalcogenide nanocrystals can also be prepared by this method. Living yeasts were used as biosynthesis platform by utilizing their intracellular metabolism of Na_2SeO_3 and detoxification of Cd ions at the same time. Pang *et al.* demonstrated the synthesis of fluorescent CdSe QDs, as shown in Figure 11.5.13.[156] The size of CdSe could be tuned from 2.7 to 6.3 nm by altering the incubation time from 10 h to 40 h and the emission wavelength shifted from 520 nm to 670 nm, respectively. By tuning their size, multiple colour CdTe QDs were synthesized in *E. coli* and yeast cells, which were used as biolabels and showed good biocompatibility due to the protein capping on the surface.[157–158] ZnS nanocrystals with diameter in the range 2–5 nm were prepared within natural biofilms of aerotolerant sulfate-reducing bacteria of the family Desulfobacteriaceae.[159] PbS nanocrystals with absorption peak near 330 nm were synthesized in yeast, *Torulopsis* species.[160] Since the band gap of bulk PbS is about 0.4 eV, the large blue-shift of the absorption of PbS nanocrystals is an obvious sign of a strong quantum confinement effect and the size of the nanocrystals characterized by TEM is only as small as 2–3 nm.

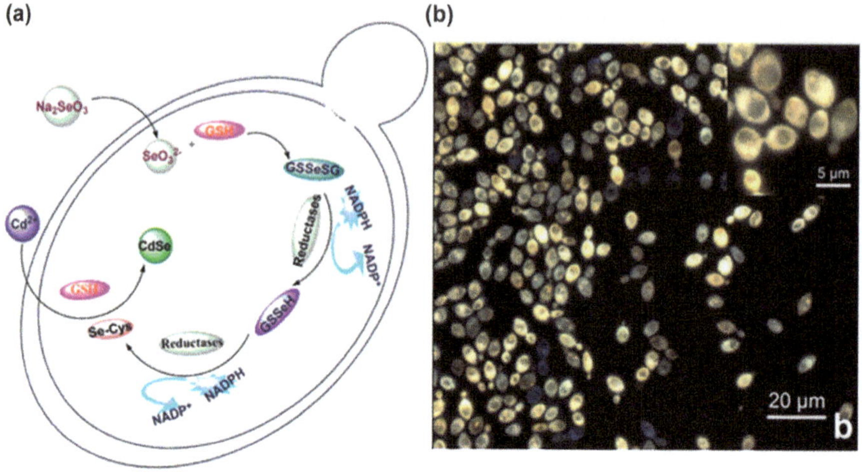

Figure 11.5.13 (a) Route for biosynthesis of fluorescent CdSe QDs and (b) images for subcellular location of intracellular fluorescence.
(Reprinted with permission from ref. 156. Copyright 2009, John Wiley & Sons, Inc.)

Various biomolecules which have a decisive influence on microbial synthesis have also been used to synthesize QDs *in vitro* independently, including peptides, proteins, and nucleotides.[161–163] As model compounds, thiol-containing phytochelatins (PCs) and glutathione (GSH) were used to produce CdS QDs.[162] PC-capped CdS QDs showed better control over the size distribution and superior stability to GSH. A spatially addressable combinatorial library of peptides was used to investigate the influence of PC structure on the growth of CdS QDs.[164] Results indicated that the size of CdS clusters increased with more spacer bonds between the cysteine residues. As shown in Figure 11.5.14, Chen *et al.* grew the PC-capped CdS QDs with enzyme-mediated synthesis.[165] By using immobilized PC synthase, the GSH molecules can be transformed to PCs with different cysteine residues in the reaction process, which further act as strong capping agents for CdS growth. Benson *et al.* reported the synthesis of CdSe QDs using polyhistidine fusion proteins as stabilizers.[166] The biofunctional peptide was demonstrated for programmed synthesis of CdSe/ZnS core–shell QDs.[167] Peptide containing both CdS-specific and ZnS-specific peptides was utilized as ligand to enable the fine tuning of the QDs structure and is also useful for the preparation of other core–shell QDs.

Mann *et al.* used cage-like apoferritin as template to synthesize the CdS–ferritin nanocomposite.[161] The 8 nm inner space of the apoferritin provides an excellent nanoreactor for the metal chalcogenide nucleation and growth, while the outer shell is suitable to conjugate various biomolecules and renders QDs with circularly polarized optical property (shown in Figure 11.5.15).[168–171] Baneyx *et al.* designed purified fusion proteins

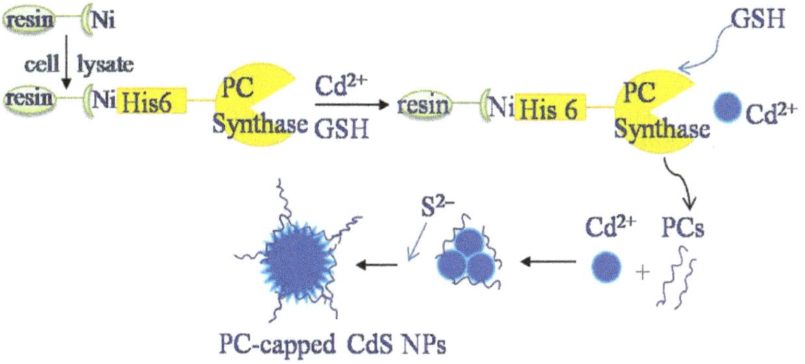

Figure 11.5.14 Schematic representation of the strategy for the PC template synthesis of CdS nanocrystals using purified PCS.
(Reprinted with permission from ref. 165. Copyright 2010, American Institute of Physics.)

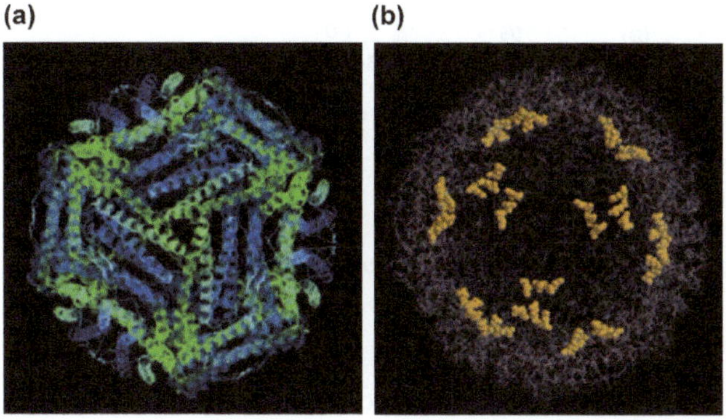

Figure 11.5.15 (a) Illustration of apoferritin from a threefold channel (ribbon model) and (b) a cross-sectional view. The glutamate residues on the interior surface are depicted as a yellow space-filling model.
(Reprinted with permission from ref. 171. Copyright 2010, John Wiley & Sons, Inc.)

containing ZnS binding peptide and adaptors for immunoglobin G (IgG) to prepare immune-ZnS QDs.[172] The combining of stabilization and conjugation functionalities into one fusion protein can facilitate the one-pot synthesis of biofunctional QDs probes.

In addition to peptides, nucleotides and nucleic acids also have some promising features for QD synthesis, such as diverse groups to coordinate with metal ions, well-defined molecular structures, and specific interaction with other biomolecules. Green *et al.* investigated adenosine 5′-triphoshate (ATP) , thymidine 5′-triphosphate(TTP), guanosine 5′-triphosphate (GTP), and cytidine

5′-triphosphate(CTP) as capping agents to synthesize CdS QDs.[163,173] Compared to ATP and CTP, GTP and TTP have better control over the growth of CdS QDs. Kelley *et al.* reported the use of RNA as ligands and templates to prepare CdS QDs.[174] The results revealed that the oxygen atoms from the phosphate groups and the nitrogen atoms on the purine and pyrimidine bases act as coordinating sites. Moreover, the structure of the RNA molecules has significant influence on the formation of CdS QDs. DNA molecules were used as template to synthesize conductive CdS nanowires under UV irradiation.[175] The nanowire has continuous CdS phase and is suitable for nanoelectronics applications. Due to their robust complementary pairing ability and programmability, modified DNA molecules were used as functional ligands to prepare luminescent and biofuctional CdTe QDs by a designed one-pot strategy.[176] By introducing phosphorothioates into the backbone of nucleotides, the modified DNA can act both as stabilizing agent and bimolecular adapter, as shown in Figure 11.5.16. The DNA-capped CdTe QDs have highly specific binding ability to various biomolecules and show good biocompatibility.

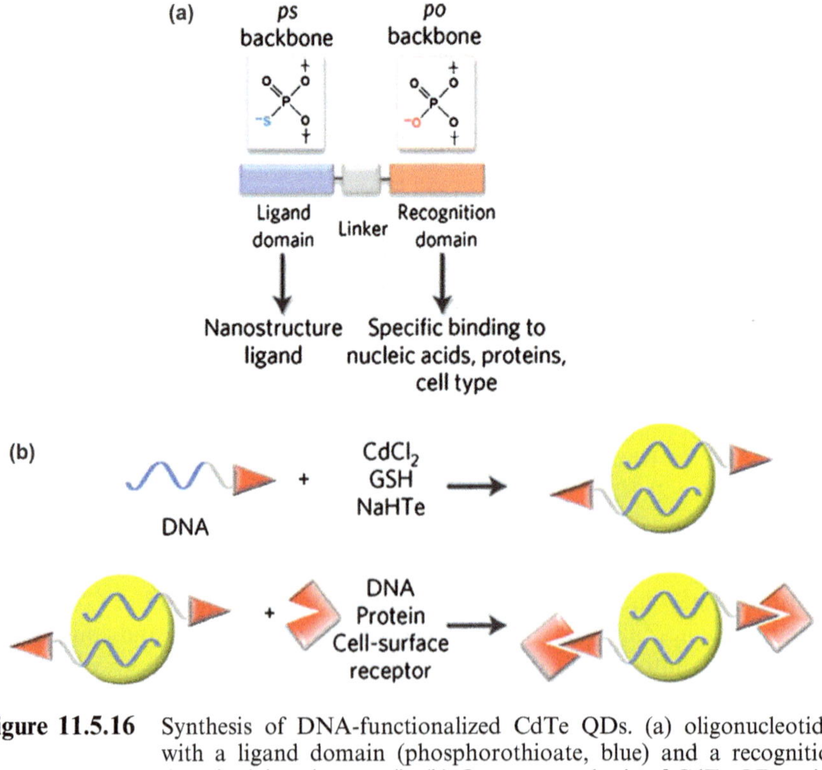

Figure 11.5.16 Synthesis of DNA-functionalized CdTe QDs. (a) oligonucleotides with a ligand domain (phosphorothioate, blue) and a recognition domain (phosphate, red). (b) One-pot synthesis of CdTe QDs using DNA and glutathione (GSH) as ligands.
(Reprinted with permission from ref. 176. Copyright 2009, Nature Publishing Group.)

11.5.4 Applications

In recent years, the synthetic techniques of colloidal QDs with narrow size distribution, tunable size, structure, and composition have gained great success. QDs with different emission colours and high PLQY can be obtained in ordinary laboratories, which has greatly advanced the exploration of QDs as basic building blocks to construct functional nanomaterials. Due to their unique electronic structures and excellent optoelectronic properties, QDs have great potentials for LEDs, solar cells, and biomedicine.[177–178]

11.5.4.1 QD LEDs

In 1994, Alivisatos and co-workers prepared the first QD LED device.[179] The structure of inorganic–organic hybrid electroluminescent device is ITO/CdSe/PPV/Mg. Yellow or red light emissions arise from the recombination of the carriers in CdSe QDs and p-paraphenylene vinylene (PPV) films; the operating voltage is 4 V with luminosity about 100 cd/m^2. Due to the imbalance of the carriers, the external quantum efficiency (EQE) is lower than 0.01%.

Since this pioneering work, various QD LEDs with different active layers and structure have been made. As shown in Figure 11.5.17, the typical structure of QD LEDs is composed of a QD emitting layer, an electron transport layer (ETL), a hole transport layer (HTL), and a transparent conductive oxide (TCO) electrode.

Bulovic *et al.* used a single CdSe/ZnS QDs layer sandwiched between two organic films to enhance the performance of the QD LEDs.[180] In this device N,N′-diphenyl-N,N′-bis(3-methylphenyl)-(1,1′-biphenyl)-4,4′-diamine (TPD) was used as HTL and tris-(8-hydroxyquinoline)aluminium (Alq_3) was used as ETL, which can improve the carrier injection efficiency. At 125 mA cm^{-2}, the brightness of device is 2000 cd m^{-2} with EQE 0.5%. In 2007, Sun *et al.* reported

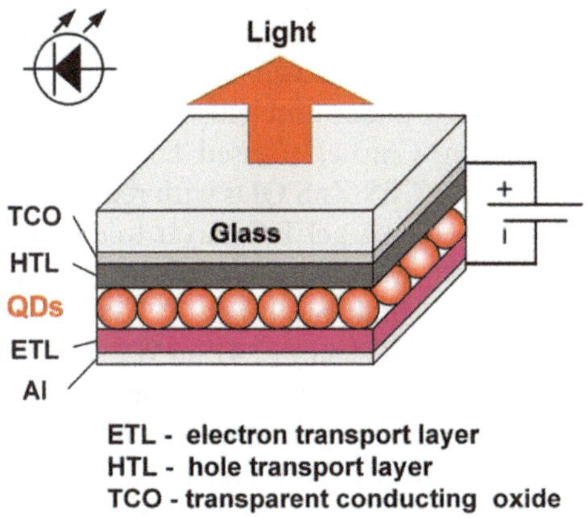

Figure 11.5.17 Illustration of typical device structure of QD LED. (Reprinted with permission from ref. 177. Copyright 2010, American Chemical Society.)

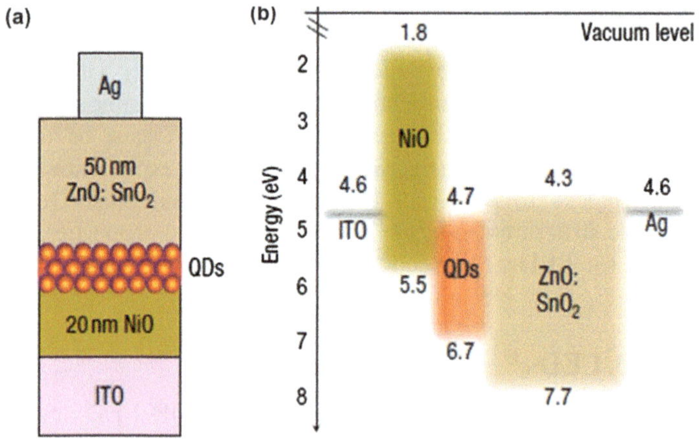

Figure 11.5.18 Device structure and energy levels of QD LED using inorganic oxide electron and HTLs.
(Reprinted with permission from ref. 181. Copyright 2008, Nature Publishing Group.)

high-performance QD LEDs with saturated red, yellow, and green emission. In their QD LEDs, a PEDOT:PSS layer was introduced between the ITO electrode and the HTL. The maximum brightness of the red-emission LEDs is over 9000 cd m^{-2}.

The organic compounds used in QD LEDs may partially decompose after working for a period, which causes stability issues. In order to solve the problem, Bulovic *et al.* utilized inorganic semiconductors as charge-transport layers instead of organic materials.[181] As shown in Figure 11.5.18, p-NiO and ZnO:SnO_2 were used as HTLs and ETLs, respectively, and the electroluminescent devices operated with peak brightness of 1950 cd m^{-2} and a maximum EQE of nearly 0.1%. Using inorganic charge-transport materials not only enhances the carrier transport process but also significantly improves the device stability when operated under ambient atmosphere conditions without extra packing and protection. Chio *et al.* used 1,7-diaminoheptane as bifunctional linker to crosslink CdSe/CdS/ZnS QDs with red emission colour as light-emitting layer and introduced a sol–gel TiO_2 layer to decrease the large energy barriers for electron injection into QDs. The device was fabricated with all-solution techniques and has a high brightness (12 380 cd m^{-2}), low turn-on voltage (1.9 V), and high power efficiency (2.41 lm W^{-1}).[182] Qian and colleagues reported high-performance QD LEDs with solution-processed multilayer structures.[183] To achieve this goal, they used water, chlorobenzene, toluene, and ethanol as orthogonal solvents for PEDOT:PSS, poly-TPD, CdSe-ZnS QDs, and ZnO nanoparticles, respectively, to avoid compromising the integrity of the underlying layers while depositing overlayers. The devices have maximum luminance and power efficiency values of 68 000 cd m^{-2} and 8.2 lm W^{-1} for green light emission, and the device drive voltage at 600 cd m^{-2} was typically 4 V.

11.5.4.2 QD Solar Cells

Solar energy is one of the most promising candidates for next-generation renewable energy.[184] If all of the Sun's light energy arriving at the Earth in an hour could be converted into electricity, it would meet the need of whole world for a year.[185] Although solar cells based on crystal silicon (first-generation solar cells) and semiconductor thin films (second-generation solar cells) have been intensively investigated, making low-cost and high-efficiency solar cells remains a great challenge.[186]

There are many advantages of using metal chalcogenide QDs (II–VI, IV–VI) as photoactive materials in solar cells. QDs usually have broad absorption spectrum and high extinction coefficients, which is important to help absorb more photons from sunlight; QDs can be synthesized in solution, which is inexpensive and easy to scale up; QD solar cells devices can be processed by solution techniques, such as spin-coating, ink-jet printing, or roll-to-roll printing, which can lower the cost of fabrication.[185,187]

In recent years, various types of QD solar cells have been studied and rapid progress has been made.

11.5.4.2.1 QD-Conjugated Polymer Hybrid Solar Cells

In this type of solar cell, both QDs and conjugated polymers (CP) can absorb the incident photons and form excitons. At the QD–CP interface, the excitons separate into holes and electrons and diffuse to the opposite electrode.

In 1996, Greenham *et al.* studied the photon absorption, exciton separation, and carrier transport properties of blends formed by CdSe QDs and conjugated polymer. Their results showed that when the organic ligands of the QDs are removed, the PL is quenched, which is consistent with the charge separation at the QD–CP interface. Huynh *et al.* utilized CdSe nanorods and poly(3-hexylthiophene) to fabricate hybrid solar cells with power conversion efficiency of 1.7% under Air Mass 1.5 Global (AM1.5G) solar conditions.[188] Although the CP–QD hybrid forms a donor–acceptor (D–A) heterostructure where efficient charge separation can occur, the control of morphology is a serious issue. If the polymer domains are as large as twice the exciton diffusion length (typically 5–20 nm), the separation of photogenerated excitons is limited. Gur and co-workers used hyperbranched CdSe nanocrystals and poly-(3-hexylthiophene) (P3HT) to form percolation networks to help the effective charge collection.[189] The energy conversion efficiency of the device under AM1.5G is as high as 2.2%. Ren *et al.* reported P3HT nanowire-CdS QDs bulkheterojuction solar cells (shown in Figure 11.5.19).[190] By controlling the phase separation of the donor and acceptor materials, the charge separation can be improved efficiently. The maximum power conversion efficiency of the device is as high as 4.1% under AM 1.5.

11.5.4.2.2 QD-Sensitized Solar Cells

Dye-sensitized solar cells (DSSC) have achieved great success and attract wide attention both from the scientific institutions and from the industrial world due to their high energy conversion efficiency (>10%) and low cost.[191] Since QDs

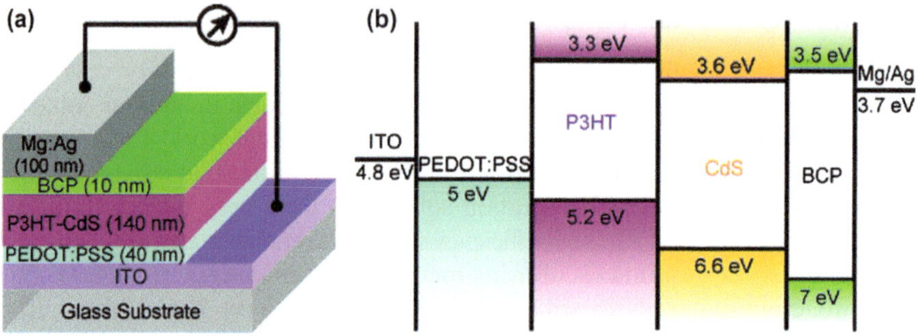

Figure 11.5.19 (a) CdS-P3HT hybrid solar cell device architecture and (b) the corresponding flat-band diagram.
(Reprinted with permission from ref. 190. Copyright 2011, American Chemical Society.)

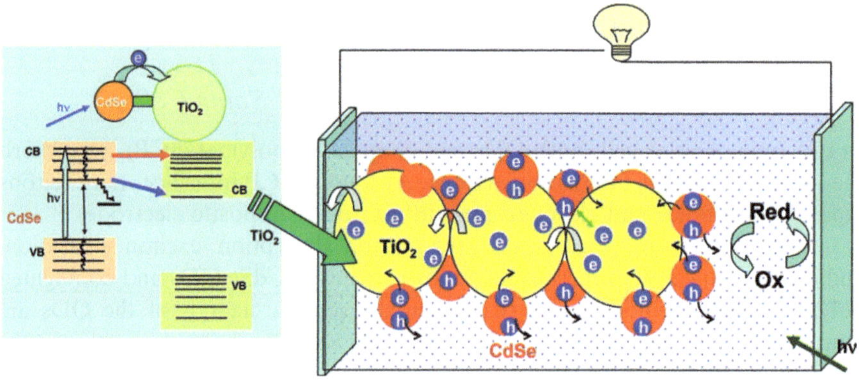

Figure 11.5.20 Operation of a typical QD-SSC.
(Reprinted with permission from ref. 192. Copyright 2008, American Chemical Society.)

have high extinction coefficients and can be prepared by low-cost colloidal synthesis, they are promising candidates as efficient and low-cost inorganic light harvesters. As shown in Figure 11.5.20, the incident light is first absorbed by QDs and forms excitons; The charge separation occurs at the QD–TiO_2 interface, where the electrons are collected by the TiO_2 electrode and charged QDs are reduced by the redox couples in the electrolyte, which diffuse to the cathode and are regenerated to the neutral state.[192]

Based on the understanding of photophysics of traditional DSSCs and the development of the synthesis of colloidal QDs, QD-sensitized solar cells (QD-SSCs) have been extensively studied during recent years and become an important type of QD solar cells.[193] Kamat *et al.* assembled CdSe QDs onto the surface of mesoscopic TiO_2 films by a linker with both thiol and carboxylic acid groups (shown in Figure 11.5.21).[194] The PL quenching was investigated and the carrier injection from the excited state of CdSe QDs into TiO_2 was confirmed, which can be used to generate photocurrent. Niitsoo *et al.* deposited

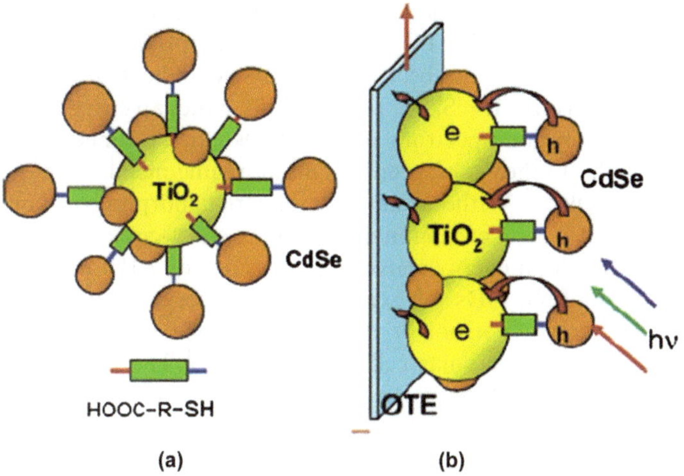

Figure 11.5.21 (a) Modification of TiO_2 electrodes with CdSe QDs linked by bifunctional mercaptan carboxylic acid molecules. (b) Light absorption and charge separation in a QD-sensitized TiO_2 electrode. (Reprinted with permission from ref. 194. Copyright 2006, American Chemical Society.)

CdS QDs onto nanoporous TiO_2 films and treated the electrode either by direct deposition of CdSe or selenization of the Cd-rich CdS layer to study the influence of deposition parameters on the performance of the QD-SSC.[195] The experimental results showed that the existence of Cd-rich CdS layer prior to CdSe deposition can improve the energy conversion efficiency to as high as 2.8%. Diguna *et al.* prepared high-efficiency CdSe-TiO_2 QD-SSCs by modifying the surface of CdSe QDs with a ZnS shell.[196] The energy conversion efficiency reaches as high as 2.7% under solar illumination of 100 mW cm^{-2}. Kongkanand and Peng used a TiO_2 nanotube array instead of the ordinary nanoporous TiO_2 photoelectrode.[197–198] The QD–TiO_2 nanotube structure enhances the charge transport in solar cells and significantly improves the energy conversion efficiency to higher than 4%.[198]

11.5.4.2.3 Inorganic QD Solar Cells

Since organic semiconductor materials can degrade during operation, QD–CP solar cells often suffer from a stability issue, which is also encountered in QD-SSCs due to the liquid electrolyte and sulfide redox couples normally used.[199–201]

Alivisatos *et al.* used CdTe and CdSe nanocrystals to construct donor–acceptor hybrid solar cells without CP.[199] The energy conversion efficiency approaches 3% under AM1.5G with open-circuit voltage about 0.45 V. After being tested for 13 000 hours under AM1.5G at ambient atmosphere the solar cells showed a 13.6% increase in energy conversion efficiency.

Since half of the Sun's energy reaching the Earth lies in the infrared range, in order to improve the efficiency of solar cells the bandgap energy of light harvesting materials should be less than 2 eV. However, the typical II–VI QDs (ME, M = Zn, Cd; E = S, Se, Te) have strong absorption only in the visible range, which limits their light absorption ability. As indicated in Table 11.5.1, lead chalcogenides (PbE, E = S, Se, Te) have bulk bandgap energy smaller than 0.5 eV, thus, absorption spectrum of PbE QDs can be tuned by size from infrared to visible. Luther *et al.* prepared all-inorganic metal/PbSe QDs/metal Schottky solar cells.[202] Without sintering or strict morphology control, the device produced an exceptionally large short-circuit photocurrent (>21 mA cm^{-2}) with an energy conversion efficiency of 2.1% under AM1.5G. Sargent reported a PbSe QDs solar cell with 3.6% energy conversion efficiency in the infrared range.[203] Choi *et al.* reported PbSe QD-based solar cells operated *via* the excitonic solar cell mechanism. [204] The energy conversion efficiency of the PbSe excitonic solar cells exceeds 3% under AM1.5G. Sargent *et al.* developed depleted bulk heterojunction structured solar cells, which showed superior device performance to planar heterojunction solar cells.[205] In this structure, all available photons can be effectively absorbed by the QD layer while ensuring the QD film is fully depleted, which increases the short-circuit current density, and under optimized parameters the energy conversion efficiency exceeds 5% under AM1.5G illumination.

Colloidal QDs used in solar cells are usually synthesized in organic solutions with the help of long alkyl chain ligands, such as oleic acid and oleic amine, which play an important role in the growth stage. However, the ligands capped on the surface of QDs form an insulating layer that hinders the charge separation and carrier transport. In order to improve the device performance, several methods have been used to exchange the ligands, including small molecular amines and thiols,[206–210] thermally reliable ligands,[211–212] and inorganic metal chalcogenide ligands.[213–215] Tang *et al.* developed atomic ligand passivation techniques to modify the surface of PbS QDs used in solar cells (shown in Figure 11.5.22).[216] The monovalent halide ions have strong affinity to the lead ions on the surface of the PbS QDs and form X-type bonds, which can reduce the defect density. The usually used short ligands are about 0.5 nm and the metal chalcogenide complex is approximately 0.7 nm ($Sn_2S_6^{4-}$), while the halide anion is only 0.1 nm, which can help minimize the interparticle spacing to promote carrier transport. After halide ion modification, PbS QD solar cells were fabricated under ambient atmosphere and the device showed up to 6% energy conversion efficiency under solar AM1.5G.

11.5.4.2.4 Next-Generation QD Solar Cells

As predicted by Nozik, the maximum thermodynamic energy conversion efficiency for single-junction solar cells can be improved from about 31% to 66% by utilizing extra energy produced from hot photogenerated carriers in QDs.[217]

Klimov *et al.* first observed the carrier multiplication(CM) phenomenon with very high efficiency in PbSe QDs in the ultrafast (picosecond) time scale, which has been well established in bulk semiconductors.[218] In a CM process, after absorbing one high-energy photon with energy at least twice the bandsgap,

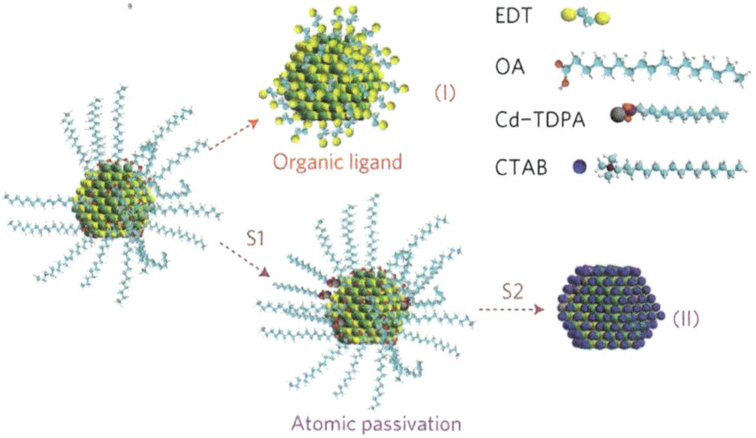

Figure 11.5.22 Organic and atomic passivation strategies for the surface treatment of PbS QDs. (S1) Cadmium–tetradecylphosphonic acid (Cd-TDPA) complex was first introduced to the PbS QDs surface to exchange the original deprotonated oleic acid (OA); (S2) Introducing Br^- to cap the Cd ions on the surface of PbS QDs by using cetyltrimethylammonium bromide (CTAB).
(Reprinted with permission from ref. 216. Copyright 2011, Nature Publishing Group.)

QDs generate multiple electron–hole pairs, which provides a possible way to effectively utilize the energy of the short wavelength light. Hence, great effort has been devoted to understand the basic physics of CM during recent years.[219–226] In 2011, Nozik *et al.* reported the first PbSe QD solar cells with peak external photocurrent quantum efficiency exceeding 100%, which is a great leap towards breaking the single-junction Shockley–Queisser limit and will significantly improve the energy conversion efficiency.[227]

11.5.5 Conclusion

After about three decades of development, many aspects of QDs research have progressed tremendously: fundamental physics, materials chemistry, biomedicine application, and optoelectronics devices. The first inaugural Kavli Prize in Nanoscience was awarded to Professor Brus in 2008 for his outstanding contributions to the development of semiconductor QDs, one of the most fundamental building blocks for nanoscience and nanotechnology.[228–229]

However, this research area is still far from mature and just in its infancy, because the understanding of basic physics and chemistry of QDs is still limited, and most applications of QDs remain on the proof-of-principle level.[230] From the viewpoint of materials science, although various metal chalcogenide QDs have been successfully synthesized, there is still much room to exploit novel QDs as building blocks to meet the rapidly growing need in multidisciplinary fields. First, QDs with excellent optical, electrical properties, and low toxicity are greatly in demand for practical biomedical or optoelectronics applications. To this end, numerous metal chalcogenide semiconductors provide unlimited possibility for material choice. Second, the growth mechanism of QDs has not been completely understood, which limits the control of the synthesis process. Moreover, further understanding of the basic physics of QDs will promote the research for practical uses in multiple areas, for example, high-performance next-generation QD solar cells.

In summary, due to their great potential for optoelectronics and biomedicine applications, QDs will continuously attract more researchers from interdisciplinary areas and great progress can be expected in the next decade.

Acknowledgment

This work was financially supported by the National Basic Research Program of China (No. 2012CB933301) and the Ministry of Education of China (No. IRT1148).

References

1. H. E. Schaefer, *Nanoscience: The Science of the Small in Physics, Engineering, Chemistry Biology and Medicine*. Springer, Heidelberg, 2010.
2. B. Bhushan, *Handbook of Nanotechnology*. Springer, Heidelberg, 2010.
3. U. Simon, G. Schmid, *Nanoparticles: From Theory to Application*. Wiley-VCH, Weinheim, 2004.
4. C. N. R. Rao, *The Chemistry of Nanomaterials: Synthesis, Properties and Applications*. Wiley-VCH, Weinheim, 2006, vol. 1.
5. E. L. Wolf, *Nanophysics and Nanotechnology*. Wiley Online Library, 2004.
6. M. A. Reed, R. T. Bate, K. Bradshaw, W. M. Duncan, W. R. Frensley, J. W. Lee and H. D. Shih, *J. Vac. Sci. Technol. B*, 1986, **4**(1), 358–360.
7. A. P. Alivisatos, *Science*, 1996, **271**(5251), 933–937.
8. A. Kitai, *Luminescent materials and applications*. Wiley, New York, 2008.
9. *SpringerMaterials http://www.springermaterials.com*.
10. R. Katzschmann, A. Rehfeld and R. Kranold, *Phys. Status Solidi A*, 1977, **40**(2), 161.
11. O. A. A. Ekimov and A. I., *JETP Lett*, 1984, **40**(8), 1136–1139.
12. A. L. E. Al.L Efros, *Sov. Phys. Semiconduct*, 1982, **16**, 772–775.
13. R. Rossetti, S. Nakahara and L. E. Brus, *J. Chem. Phys.*, 1983, **79**(2), 1086–1088.
14. L. E. Brus, *J. Chem. Phys.*, 1984, **80**(9), 4403–4409.
15. R. Rossetti, J. L. Ellison, J. M. Gibson and L. E. Brus, *J. Chem. Phys.*, 1984, **80**(9), 4464–4469.
16. L. Brus, *J. Phys. Chem.*, 1986, **90**(12), 2555–2560.
17. L. Brus, *Appl. Phys. A*, 1991, **53**(6), 465–474.
18. M. A. Kastner, *Physics Today*, 1993, **46**(1), 24–31.
19. R. C. Ashoori, *Nature*, 1996, **379**(6564), 413–419.

20. N. Gaponik, S. G. Hickey, D. Dorfs, A. L. Rogach and A. Eychmüller, *Small*, 2010, **6**(13), 1364–1378.
21. A. Eychmuller, *J. Phys. Chem. B*, 2000, **104**(28), 6514–6528.
22. M. Reed, *Sci. Am.*, 1993, 118–123.
23. A. M. Smith and S. Nie, *Acc. Chem. Res.*, 2010, **43**(2), 190–200.
24. Y. Wang and N. Herron, *J. Phys. Chem.*, 1991, **95**(2), 525–532.
25. A. P. Alivisatos, *J. Phys. Chem.*, 1996, **100**(31), 13226–13239.
26. A. L. Efros and M. Rosen, *Annu. Rev. Mater. Sci.*, 2000, **30**, 475–521.
27. Y. Kayanuma, *Phys. Rev. B*, 1988, **38**(14), 9797.
28. I. L. Medintz, H. T. Uyeda, E. R. Goldman and H. Mattoussi, *Nat. Mater.*, 2005, **4**(6), 435–446.
29. X. Michalet, F. F. Pinaud, L. A. Bentolila, J. M. Tsay, S. Doose, J. J. Li, G. Sundaresan, A. M. Wu, S. S. Gambhir and S. Weiss, *Science*, 2005, **307**(5709), 538–544.
30. X. Wu, H. Liu, J. Liu, K. N. Haley, J. A. Treadway, J. P. Larson, N. Ge, F. Peale and M. P. Bruchez, *Nat. Biotechnol.*, 2003, **21**(1), 41–46.
31. Z. A. Peng and X. G. Peng, *J. Am. Chem. Soc.*, 2002, **124**(13), 3343–3353.
32. M. L. Steigerwald, A. P. Alivisatos, J. M. Gibson, T. D. Harris, R. Kortan, A. J. Muller, A. M. Thayer, T. M. Duncan, D. C. Douglass and L. E. Brus, *J. Am. Chem. Soc.*, 1988, **110**(10), 3046–3050.
33. J. G. Brennan, T. Siegrist, P. J. Carroll, S. M. Stuczynski, L. E. Brus and M. L. Steigerwald, *J. Am. Chem. Soc.*, 1989, **111**(11), 4141–4143.
34. C. B. Murray, D. J. Norris and M. G. Bawendi, *J. Am. Chem. Soc.*, 1993, **115**(19), 8706–8715.
35. V. K. LaMer and R. H. Dinegar, *J. Am. Chem. Soc.*, 1950, **72**(11), 4847–4854.
36. Z. A. Peng and X. G. Peng, *J. Am. Chem. Soc.*, 2001, **123**(1), 183–184.
37. L. H. Qu, Z. A. Peng and X. G. Peng, *Nano. Lett.*, 2001, **1**(6), 333–337.
38. L. H. Qu and X. G. Peng, *J. Am. Chem. Soc.*, 2002, **124**(9), 2049–2055.
39. W. W. Yu and X. G. Peng, *Angew. Chem. Int. Ed.*, 2002, **41**(13), 2368–2371.
40. J. Park, J. Joo, S. G. Kwon, Y. Jang and T. Hyeon, *Angew. Chem. Int. Ed.*, 2007, **46**(25), 4630–4660.
41. X. Peng, *Nano Res.*, 2009, **2**(6), 425–447.
42. X. G. Peng, J. Wickham and A. P. Alivisatos, *J. Am. Chem. Soc.*, 1998, **120**(21), 5343–5344.
43. T. Sugimoto, *Adv. Colloid Interface Sci.*, 1987, **28**, 65–108.
44. Y. Yin and A. P. Alivisatos, *Nature*, 2005, **437**(7059), 664–670.
45. Y. Nosaka, K. Yamaguchi, H. Miyama and H. Hayashi, *Chem. Lett.*, 1988, **17**(4), 605–608.
46. D. Hayes, O. I. Micic, M. T. Nenadovic, V. Swayambunathan and D. Meisel, *J. Phys. Chem.*, 1989, **93**(11), 4603–4608.
47. T. Rajh, O. I. Micic and A. J. Nozik, *J. Phys. Chem.*, 1993, **97**(46), 11999–12003.
48. A. L. Rogach, L. Katsikas, A. Kornowski, D. S. Su, A. Eychmuller and H. Weller, *Ber. Bunsen. Phys. Chem.*, 1996, **100**(11), 1772–1778.

49. M. Green, H. Harwood, C. Barrowman, P. Rahman, A. Eggeman, F. Festry, P. Dobson and T. Ng, *J. Mater. Chem.*, 2007, **17**(19), 1989–1994.
50. N. Gaponik, D. V. Talapin, A. L. Rogach, K. Hoppe, E. V. Shevchenko, A. Kornowski, A. Eychmüller and H. Weller, *J. Phys. Chem. B*, 2002, **106**(29), 7177–7185.
51. H. Zhang and B. Yang, *Thin Solid Films*, 2002, **418**(2), 169–174.
52. H. Y. Acar, R. Kas, E. Yurtsever, C. Ozen and I. Lieberwirth, *J. Phys. Chem. C*, 2009, **113**(23), 10005–10012.
53. C. Wang, Q. Ma and X. G. Su, *J. Nanosci. Nanotechno.*, 2008, **8**(9), 4408–4414.
54. E. Ying, D. Li, S. Guo, S. Dong and J. Wang, *PloS one*, 2008, **3**(5).
55. Y. F. Liu, B. Xie, Z. G. Yin, S. M. Fang and J. B. Zhao, *Eur. J. Inorg. Chem.*, 2010, **10**, 1501–1506.
56. F. Aldeek, L. Balan, J. Lambert and R. Schneider, *Nanotechnology*, 2008, **19**(47), 9.
57. I. L. Radtchenko, G. B. Sukhorukov, N. Gaponik, A. Kornowski, A. L. Rogach and H. Mohwald, *Adv. Mater.*, 2001, **13**(22), 1684–1687.
58. V. K. Komarala, *et al*, *Nanotechnology*, 2006, **17**(16), 4117.
59. Y. G. Zheng, S. J. Gao and J. Y. Ying, *Adv. Mater.*, 2007, **19**(3), 376–380.
60. H. F. Qian, C. Q. Dong, J. F. Weng and J. C. Ren, *Small*, 2006, **2**(6), 747–751.
61. R. Osovsky, A. Shavel, N. Gaponik, L. Amirav, A. Eychmüller, H. Weller and E. Lifshitz, *J. Phys. Chem. B*, 2005, **109**(43), 20244–20250.
62. N. N. Mamedova, N. A. Kotov, A. L. Rogach and J. Studer, *Nano. Lett.*, 2001, **1**(6), 281–286.
63. V. Lesnyak, S. V. Voitekhovich, P. N. Gaponik, N. Gaponik and A. Eychmüller, *ACS Nano*, 2010, **4**(7), 4090–4096.
64. A. Dubavik, V. Lesnyak, W. Thiessen, N. Gaponik, T. Wolff and A. Eychmuller, *J. Phys. Chem. C*, 2009, **113**(12), 4748–4750.
65. A. L. Rogach, T. Franzl, T. A. Klar, J. Feldmann, N. Gaponik, V. Lesnyak, A. Shavel, A. Eychmuller, Y. P. Rakovich and J. F. Donegan, *J. Phys. Chem. C*, 2007, **111**(40), 14628–14637.
66. Z. Fang, L. Liu, L. L. Xu, X. G. Yin and X. H. Zhong, *Nanotechnology*, 2008, **19**(23), 235603–235609.
67. N. Gaponik and A. L. Rogach, *Phys. Chem. Chem. Phys.*, 2010, **12**(31), 8685–8693.
68. J. Li, X. Hong, D. Li, K. Zhao, L. Wang, H. Wang, Z. Du, J. Li, Y. Bai and T. Li, *Chem. Commun.*, 2004, **15**, 1740–1741.
69. B. Tang, J. Niu, C. Yu, L. Zhuo and J. Ge, *Chem. Commun.*, 2005, **33**, 4184–4186.
70. H. Zhang, D. Y. Wang and H. Mohwald, *Angew. Chem. Int. Ed.*, 2006, **45**(5), 748–751.
71. H. Zhang, D. Y. Wang, B. Yang and H. Mohwald, *J. Am. Chem. Soc.*, 2006, **128**(31), 10171–10180.

72. T. Wang, Z. G. Jin, Y. M. Shi, W. L. Li and J. X. Yang, *Cryst. Growth Des.*, 2009, **9**(12), 5077–5082.
73. A. L. Rogach, A. Kornowski, M. Y. Gao, A. Eychmuller and H. Weller, *J. Phys. Chem. B*, 1999, **103**(16), 3065–3069.
74. S. F. Wuister, C. D. Donega and A. Meijerink, *J. Phys. Chem. B*, 2004, **108**(45), 17393–17397.
75. Y. Wang, Z. Tang, M. A. Correa-Duarte, I. Pastoriza-Santos, M. Giersig, N. A. Kotov and L. M. Liz-Marzán, *J. Phys. Chem. B*, 2004, **108**(40), 15461–15469.
76. C. Li and N. Murase, *Chem. Lett.*, 2005, **34**(1), 92–93.
77. A. Shavel, N. Gaponik and A. Eychmüller, *J. Phys. Chem. B*, 2006, **110**(39), 19280–19284.
78. H. B. Bao, Y. J. Gong, Z. Li and M. Y. Gao, *Chem. Mater.*, 2004, **16**(20), 3853–3859.
79. Y. He, L. M. Sai, H. T. Lu, M. Hu, W. Y. Lai, Q. L. Fan, L. H. Wang and W. Huang, *Chem. Mater.*, 2007, **19**(3), 359–365.
80. H. Peng, L. J. Zhang, C. Soeller and J. Travas-Sejdic, *J. Lumin.*, 2007, **127**(2), 721–726.
81. Z. Y. Gu, L. Zou, Z. Fang, W. H. Zhu and X. H. Zhong, *Nanotechnology*, 2008, **19**(13), 135604.
82. Y. He, H. T. Lu, L. M. Sai, Y. Y. Su, M. Hu, C. H. Fan, W. Huang and L. H. Wang, *Adv. Mater.*, 2008, **20**(18), 3416–3421.
83. M. Green, P. Williamson, M. Samalova, J. Davis, S. Brovelli, P. Dobson and F. Cacialli, *J. Mater. Chem.*, 2009, **19**(44), 8341–8346.
84. X. Zhong, M. Han, Z. Dong, T. J. White and W. Knoll, *J. Am. Chem. Soc.*, 2003, **125**(28), 8589–8594.
85. Y. G. Zheng, Z. C. Yang and J. Y. Ying, *Adv. Mater.*, 2007, **19**(11), 1475–1479.
86. F.-C. Liu, T.-L. Cheng, C.-C. Shen, W.-L. Tseng and M. Y. Chiang, *Langmuir*, 2008, **24**(5), 2162–2167.
87. J. Cao, B. Xue, H. Li, D. W. Deng and Y. Q. Gu, *J. Colloid Interface Sci.*, 2010, **348**(2), 369–376.
88. W. W. Li, J. Liu, K. Sun, H. J. Dou and K. Tao, *J. Mater. Chem.*, 2010, **20**(11), 2133–2138.
89. V. Lesnyak, A. Dubavik, A. Plotnikov, N. Gaponik and A. Eychmuller, *Chem. Commun.*, 2010, **46**(6), 886–888.
90. N. Piven, A. S. Susha, M. Doeblinger and A. L. Rogach, *J. Phys. Chem. C*, 2008, **112**(39), 15253–15259.
91. G. X. Liang, L. L. Li, H. Y. Liu, J. R. Zhang, C. Burda and J. J. Zhu, *Chem. Commun.*, 2010, **46**(17), 2974–2976.
92. G. X. Liang, M. M. Gu, J. R. Zhang and J. J. Zhu, *Nanotechnology*, 2009, **20**, 41.
93. H. Qian, C. Dong, J. Peng, X. Qiu, Y. Xu and J. Ren, *J. Phys. Chem. C*, 2007, **111**(45), 16852–16857.
94. H. Sun, H. Zhang, J. Ju, J. Zhang, G. Qian, C. Wang, B. Yang and Z. Y. Wang, *Chem. Mater.*, 2008, **20**(21), 6764–6769.

95. J. Han, H. Zhang, Y. Tang, Y. Liu, X. Yao and B. Yang, *J. Phys. Chem. C*, 2009, **113**(18), 7503–7510.
96. C. Wang, X. Gao, Q. Ma and X. Su, *J. Mater. Chem.*, 2009, **19**(38), 7016–7022.
97. P. T. Shao, Q. H. Zhang, Y. G. Li and H. Z. Wang, *J. Mater. Chem.*, 2011, **21**(1), 151–156.
98. A. Aboulaich, M. Geszke, L. Balan, J. Ghanbaja, G. Medjahdi and R. L. Schneider, *Inorg. Chem.*, 2010, **49**(23), 10940–10948.
99. S. Jana, B. B. Srivastava, S. Acharya, P. K. Santra, N. R. Jana, D. D. Sarma and N. Pradhan, *Chem. Commun.*, 2010, **46**(16), 2853–2855.
100. A. Ekimov and A. Onushchenko, *JETP Lett*, 1981, **34**(6), 345–349.
101. Y. Wang and N. Herron, *J. Phys. Chem.*, 1987, **91**(2), 257–260.
102. J. Wang, W. Chen, A. Liu, G. Lu, G. Zhang, J. Zhang and B. Yang, *J. Am. Chem. Soc.*, 2002, **124**(45), 13358–13359.
103. D. Shchukin and G. Sukhorukov, *Adv. Mater.*, 2004, **16**(8), 671–682.
104. M. Pileni, *Nat. Mater.*, 2003, **2**(3), 145–150.
105. M. Meyer, C. Wallberg, K. Kurihara and J. H. Fendler, *J. Chem. Soc., Chem. Commun.*, 1984, **2**, 90–91.
106. T. Dannhauser, M. O'Neil, K. Johansson, D. Whitten and G. McLendon, *J. Phys. Chem.*, 1986, **90**(23), 6074–6076.
107. N. F. Dorothe Ingert, Laurent Levy, Pierre Gouzerh and Marie-Paule Pileni, *Adv. Mater.*, 1999, **11**(3), 220–223.
108. M. Moffitt, H. Vali and A. Eisenberg, *Chem. Mater.*, 1998, **10**(4), 1021–1028.
109. C. W. Wang and M. G. Moffitt, *Langmuir*, 2005, **21**(6), 2465–2473.
110. N. Duxin, F. Liu, H. Vali and A. Eisenberg, *J. Am. Chem. Soc.*, 2005, **127**(28), 10063–10069.
111. D. Wang, Y. Cao, X. Zhang, Z. Liu, X. Qian, X. Ai, F. Liu, D. Wang, Y. Bai, T. Li and X. Tang, *Chem. Mater.*, 1999, **11**(2), 392–398.
112. Y. Guo and M. G. Moffitt, *Macromolecules*, 2007, **40**(16), 5868–5878.
113. S. Anthony and J. Kim, *Chem. Commun.*, 2008.
114. C. W. Wang and M. G. Moffitt, *Langmuir*, 2004, **20**(26), 11784–11796.
115. H. Zhang, L. Wang, H. Xiong, L. Hu, B. Yang and W. Li, *Adv. Mater.*, 2003, **15**(20), 1712–1715.
116. J. Guo, W. L. Yang and C. C. Wang, *J. Phys. Chem. B*, 2005, **109**(37), 17467–17473.
117. W. Y. Mao, J. Guo, W. L. Yang, C. C. Wang, J. He and J. Y. Chen, *Nanotechnology*, 2007, **18**, 48.
118. W. H. Yang, W. W. Li, H. J. Dou and K. Sun, *Mater. Lett.*, 2008, **62**(17–18), 2564–2566.
119. D. Zhao, Z. He, W. Chan and M. Choi, *J. Phys. Chem. C*, 2008, **113**(4), 1293–1300.
120. J. Wang and H. Y. Han, *J. Colloid Interface Sci.*, 2010, **351**(1), 83–87.
121. J. N. Tian, R. J. Liu, Y. C. Zhao, Y. Peng, X. Hong, Q. Xu and S. L. Zhao, *Nanotechnology*, 2010, **21**, 30.
122. B. Hayes, *Microwave Synthesis: Chemistry at the Speed of Light*. CEM publishing Matthews, NC: 2002.

123. J. Tierney, P. Lidström, *Microwave Assisted Organic Synthesis*. Wiley-Blackwell, Oxford, 2005.
124. A. de la Hoz, A. Diaz-Ortiz and A. Moreno, *Chem. Soc. Rev.*, 2005, **34**(2), 164–178.
125. J. Zhu, O. Palchik, S. Chen and A. Gedanken, *J. Phys. Chem. B*, 2000, **104**(31), 7344–7347.
126. L. Li, H. F. Qian and J. C. Ren, *Chem. Commun.*, 2005, **4**, 528–530.
127. H. F. Qian, X. Qiu, L. Li and J. C. Ren, *J. Phys. Chem. B*, 2006, **110**(18), 9034–9040.
128. Y. He, H. T. Lu, L. M. Sai, W. Y. Lai, Q. L. Fan, L. H. Wang and W. Huang, *J. Phys. Chem. B*, 2006, **110**(27), 13352–13356.
129. Y. He, H. T. Lu, L. M. Sai, W. Y. Lai, Q. L. Fan, L. H. Wang and W. Huang, *J. Phys. Chem. B*, 2006, **110**(27), 13370–13374.
130. L. X. Song, J. L. Duan and J. H. Zhan, *Mater. Lett.*, 2010, **64**(16), 1843–1845.
131. M. Q. Chu, X. Y. Shen and G. J. Liu, *Nanotechnology*, 2006, **17**(2), 444–449.
132. J. Ziegler, A. Merkulov, M. Grabolle, U. Resch-Genger and T. Nann, *Langmuir*, 2007, **23**(14), 7751–7759.
133. W. Schumacher, A. Nagy, W. Waldman and P. Dutta, *J. Phys. Chem. C*, 2009, **113**(28), 12132–12139.
134. M. Roy, A. Herzing, S. De Paoli Lacerda and M. Becker, *Chem. Commun.*, 2008, **18**, 2106–2108.
135. H. Han, G. Francesco and M. Maye, *J. Phys. Chem. C*,8706–8715.
136. L. Huang and H. Han, *Mater. Lett.*, 2010, **64**(9), 1099–1101.
137. D. Zhu, X. Jiang, C. Zhao, X. Sun, J. Zhang and J. Zhu, *Chem. Commun.*, 2010, **46**(29), 5226–5228.
138. K. S. Suslick, *Science*, 1990, **247**(4949), 1439–1445.
139. J. H. Bang and K. S. Suslick, *Adv. Mater.*, 2010, **22**(10), 1039–1059.
140. N. A. Dhas, A. Zaban and A. Gedanken, *Chem. Mater.*, 1999, **11**(3), 806–813.
141. J. Zhu, Y. Koltypin and A. Gedanken, *Chem. Mater.*, 2000, **12**(1), 73–78.
142. J. P. Ge, Y. D. Li and G. Q. Yang, *Chem. Commun.*, 2002, **17**, 1826–1827.
143. H. Zhang Wang, Li Zhang, Sun and B. Yang, *J. Phys. Chem. C*, 2007, **111**(6), 2465–2469.
144. M. Mostafavi, Y. Liu, P. Pernot and J. Belloni, *Radiat. Phys. Chem.*, 2000, **59**(1), 49–59.
145. A. H. Souici, N. Keghouche, J. A. Delaire, H. Remita and M. Mostafavi, *Chem. Phys. Lett.*, 2006, **422**(1–3), 25–29.
146. R. N. Reese and D. R. Winge, *J. Biol. Chem.*, 1988, **263**(26), 12832–12835.
147. E. Dujardin and S. Mann, *Adv. Mater.*, 2002, **14**(11), 775–788.
148. M. B. Dickerson, K. H. Sandhage and R. R. Naik, *Chem. Rev.*, 2008, **108**(11), 4935–4978.
149. C. T. Dameron, R. N. Reese, R. K. Mehra, A. R. Kortan, P. J. Carroll, M. L. Steigerwald, L. E. Brus and D. R. Winge, *Nature*, 1989, **338**(6216), 596–597.
150. C. T. Dameron and D. R. Winge, *Inorg. Chem.*, 1990, **29**(7), 1343–1348.

151. P. Williams, E. Keshavarz-Moore and P. Dunnill, *Enzyme Microb. Technol.*, 1996, **19**(3), 208–213.
152. R. Y. Sweeney, C. B. Mao, X. X. Gao, J. L. Burt, A. M. Belcher, G. Georgiou and B. L. Iverson, *Chem. Biol.*, 2004, **11**(11), 1553–1559.
153. S. H. Kang, K. N. Bozhilov, N. V. Myung, A. Mulchandani and W. Chen, *Angew. Chem. Int. Ed.*, 2008, **47**(28), 5186–5189.
154. A. Ahmad, P. Mukherjee, D. Mandal, S. Senapati, M. I. Khan, R. Kumar and M. Sastry, *J. Am. Chem. Soc.*, 2002, **124**(41), 12108–12109.
155. H. J. Bai, Z. M. Zhang, Y. Guo and W. L. Jia, *Nanoscale Res. Lett.*, 2009, **4**(7), 717–723.
156. R. Cui, H. H. Liu, H. Y. Xie, Z. L. Zhang, Y. R. Yang, D. W. Pang, Z. X. Xie, B. B. Chen, B. Hu and P. Shen, *Adv. Funct. Mater.*, 2009, **19**(15), 2359–2364.
157. H. F. Bao, Z. S. Lu, X. Q. Cui, Y. Qiao, J. Guo, J. M. Anderson and C. M. Li, *Acta Biomater.*, 2010, **6**(9), 3534–3541.
158. H. Bao, N. Hao, Y. Yang and D. Zhao, *Nano Res.*, 2010, 1–9.
159. M. Labrenz, G. K. Druschel, T. Thomsen-Ebert, B. Gilbert, S. A. Welch, K. M. Kemner, G. A. Logan, R. E. Summons, G. D. Stasio, P. L. Bond, B. Lai, S. D. Kelly and J. F. Banfield, *Science*, 2000, **290**(5497), 1744–1747.
160. M. Kowshik, W. Vogel, J. Urban, S. K. Kulkarni and K. M. Paknikar, *Adv. Mater.*, 2002, **14**(11), 815–818.
161. K. K. W. Wong and S. Mann, *Adv. Mater.*, 1996, **8**(11), 928–932.
162. W. Bae and R. K. Mehra, *J. Inorg. Biochem.*, 1998, **69**(1–2), 33–43.
163. M. Green, D. Smith-Boyle, J. Harries and R. Taylor, *Chem. Commun.*, 2005, **38**, 4830–4832.
164. G. Spreitzer, J. M. Whitling, J. D. Madura and D. W. Wright, *Chem. Commun.*, 2000, **3**, 209–210.
165. F. Liu, S. H. Kang, Y. I. Lee, Y. H. Choa, A. Mulchandani, N. V. Myung and W. Chen, *Appl. Phys. Lett.*, 2010, **97**, 12.
166. B. P. Aryal and D. E. Benson, *Bioconjugate Chem.*, 2007, **18**(2), 585–589.
167. S. Singh, K. Bozhilov, A. Mulchandani, N. Myung and W. Chen, *Chem. Commun.*, 2010, **46**(9), 1473–1475.
168. I. Yamashita, J. Hayashi and M. Hara, *Chem. Lett.*, 2004, **33**(9), 1158–1159.
169. K. Iwahori, K. Yoshizawa, M. Muraoka and I. Yamashita, *Inorg. Chem.*, 2005, **44**(18), 6393–6400.
170. K. Iwahori, T. Morioka and I. Yamashita, *Phys. Status Solidi A*, 2006, **203**(11), 2658–2661.
171. M. Naito, K. Iwahori, A. Miura, M. Yamane and I. Yamashita, *Angew. Chem. Int. Ed.*, 2010, **49**(39), 7006–7009.
172. W. Zhou, D. T. Schwartz and F. Baneyx, *J. Am. Chem. Soc.*, 2010, **132**(13), 4731–4738.
173. M. Green, R. Taylor and G. Wakefield, *J. Mater. Chem.*, 2003, **13**(8), 1859–1861.

174. N. Ma, C. J. Dooley and S. O. Kelley, *J. Am. Chem. Soc.*, 2006, **128**(39), 12598–12599.
175. S. Kundu and H. Liang, *Adv. Mater.*, 2008, **20**(4), 826–831.
176. N. Ma, E. H. Sargent and S. O. Kelley, *Nat. Nanotechnol.*, 2009, **4**(2), 121–125.
177. D. V. Talapin, J.-S. Lee, M. V. Kovalenko and E. V. Shevchenko, *Chem. Rev.*, 2010, **110**(1), 389–458.
178. W. R. Algar, K. Susumu, J. B. Delehanty and I. L. Medintz, *Anal. Chem.*, 2011, **83**(23), 8826–8837.
179. V. L. Colvin, M. C. Schlamp and A. P. Alivisatos, *Nature*, 1994, **370**(6488), 354–357.
180. S. Coe, W.-K. Woo, M. Bawendi and V. Bulovic, *Nature*, 2002, **420**(6917), 800–803.
181. J. M. Caruge, J. E. Halpert, V. Wood, V. Bulovic and M. G. Bawendi, *Nat. Photonics*, 2008, **2**(4), 247–250.
182. K. S. Cho, E. K. Lee, W. J. Joo, E. Jang, T. H. Kim, S. J. Lee, S. J. Kwon, J. Y. Han, B. K. Kim, B. L. Choi and J. M. Kim, *Nat. Photonics*, 2009, **3**(6), 341–345.
183. L. Qian, Y. Zheng, J. G. Xue and P. H. Holloway, *Nat. Photonics*, 2011, **5**(9), 543–548.
184. N. S. Lewis, *Science*, 2007, **315**(5813), 798–801.
185. E. H. Sargent, *Nat. Photonics*, 2012, **6**(3), 133–135.
186. G. F. Brown and J. Wu, *Laser Photonics Rev.*, 2009, **3**(4), 394–405.
187. H. W. Hillhouse and M. C. Beard, *Curr. Opin. Colloid Interface Sci*, 2009, **14**(4), 245–259.
188. W. U. Huynh, J. J. Dittmer and A. P. Alivisatos, *Science*, 2002, **295**(5564), 2425–2427.
189. I. Gur, N. A. Fromer, C.-P. Chen, A. G. Kanaras and A. P. Alivisatos, *Nano. Lett.*, 2007, **7**(2), 409–414.
190. S. Ren, L.-Y. Chang, S.-K. Lim, J. Zhao, M. Smith, N. Zhao, V. Bulović, M. Bawendi and S. Gradečak, *Nano. Lett.*, 2011, **11**(9), 3998–4002.
191. M. K. Nazeeruddin, F. De Angelis, S. Fantacci, A. Selloni, G. Viscardi, P. Liska, S. Ito, B. Takeru and M. Grätzel, *J. Am. Chem. Soc.*, 2005, **127**(48), 16835–16847.
192. P. V. Kamat, *J. Phys. Chem. C*, 2008, **112**(48), 18737–18753.
193. P. V. Kamat, K. Tvrdy, D. R. Baker and J. G. Radich, *Chem. Rev.*, 2010, **110**(11), 6664–6688.
194. I. Robel, V. Subramanian, M. Kuno and P. V. Kamat, *J. Am. Chem. Soc.*, 2006, **128**(7), 2385–2393.
195. O. Niitsoo, S. K. Sarkar, C. Pejoux, S. Rühle, D. Cahen and G. Hodes, *J. Photochem. Photobiol. A*, 2006, **181**(2–3), 306–313.
196. L. J. Diguna, Q. Shen, J. Kobayashi and T. Toyoda, *Appl. Phys. Lett.*, 2007, **91**(2), 023116.
197. A. Kongkanand, K. Tvrdy, K. Takechi, M. Kuno and P. V. Kamat, *J. Am. Chem. Soc.*, 2008, **130**(12), 4007–4015.

198. W.-T. Sun, Y. Yu, H.-Y. Pan, X.-F. Gao, Q. Chen and L.-M. Peng, *J. Am. Chem. Soc.*, 2008, **130**(4), 1124–1125.
199. I. Gur, N. A. Fromer, M. L. Geier and A. P. Alivisatos, *Science*, 2005, **310**(5747), 462–465.
200. V. Chakrapani, D. Baker and P. V. Kamat, *J. Am. Chem. Soc.*, 2011, **133**(24), 9607–15.
201. L. Li, X. Yang, J. Gao, H. Tian, J. Zhao, A. Hagfeldt and L. Sun, *J. Am. Chem. Soc.*, 2011, **133**(22), 8458–60.
202. J. M. Luther, M. Law, M. C. Beard, Q. Song, M. O. Reese, R. J. Ellingson and A. J. Nozik, *Nano. Lett.*, 2008, **8**(10), 3488–3492.
203. G. I. Koleilat, L. Levina, H. Shukla, S. H. Myrskog, S. Hinds, A. G. Pattantyus-Abraham and E. H. Sargent, *ACS Nano*, 2008, **2**(5), 833–840.
204. J. J. Choi, Y. Lim, M. E. B. Santiago-Berrios, M. Oh, B. Hyun, L. Sung, A. C. Bartnik, A. Goedhart, G. G. Malliaras, H. D. Abruna, F. W. Wise and T. Hanrath, *Nano. Lett.*, 2009, **9**(11), 3749–3755.
205. D. A. R. Barkhouse, R. Debnath, I. J. Kramer, D. Zhitomirsky, A. G. Pattantyus-Abraham, L. Levina, L. Etgar, M. Grätzel and E. H. Sargent, *Adv. Mater.*, 2011, **23**(28), 3134–8.
206. E. J. D. Klem, H. Shukla, S. Hinds, D. D. MacNeil, L. Levina and E. H. Sargent, *Appl. Phys. Lett.*, 2008, **92**(21), 212105.
207. J. M. Luther, M. Law, Q. Song, C. L. Perkins, M. C. Beard and A. J. Nozik, *ACS Nano*, 2008, **2**(2), 271–280.
208. D. A. R. Barkhouse, A. G. Pattantyus-Abraham, L. Levina and E. H. Sargent, *ACS Nano*, 2008, **2**(11), 2356–2362.
209. N. Yaacobi-Gross, M. Soreni-Harari, M. Zimin, S. Kababya, A. Schmidt and N. Tessler, *Nat. Mater.*, 2011, **10**(12), 974–9.
210. K. S. Jeong, J. Tang, H. Liu, J. Kim, A. W. Schaefer, K. Kemp, L. Levina, X. H. Wang, S. Hoogland, R. Debnath, L. Brzozowski, E. H. Sargent and J. B. Asbury, *ACS Nano*, 2012, **6**(1), 89–99.
211. D. H. Webber and R. L. Brutchey, *J. Am. Chem. Soc.*, 2011, **134**(2), 1085–92.
212. A. W. Wills, M. S. Kang, A. Khare, W. L. Gladfelter and D. J. Norris, *ACS Nano*, 2010, **4**(8), 4523–4530.
213. M. V. Kovalenko, M. Scheele and D. V. Talapin, *Science*, 2009, **324**(5933), 1417–1420.
214. M. V. Kovalenko, M. I. Bodnarchuk, J. Zaumseil, J. S. Lee and D. V. Talapin, *J. Am. Chem. Soc.*, 2010, **132**(29), 10085–10092.
215. A. Nag, M. V. Kovalenko, J.-S. Lee, W. Liu, B. Spokoyny and D. V. Talapin, *J. Am. Chem. Soc.*, 2011, **133**(27), 10612–10620.
216. J. Tang, K. W. Kemp, S. Hoogland, K. S. Jeong, H. Liu, L. Levina, M. Furukawa, X. Wang, R. Debnath, D. Cha, K. W. Chou, A. Fischer, A. Amassian, J. B. Asbury and E. H. Sargent, *Nat. Mater.*, 2011, **10**(10), 765–771.
217. W. Shockley and H. J. Queisser, *J. Appl. Phys.*, 1961, **32**(3), 510–519.
218. R. D. Schaller and V. I. Klimov, *Phys. Rev. Lett.*, 2004, **92**(18), 186601.

219. R. J. Ellingson, M. C. Beard, J. C. Johnson, P. R. Yu, O. I. Micic, A. J. Nozik, A. Shabaev and A. L. Efros, *Nano. Lett.*, 2005, **5**(5), 865–871.
220. R. D. Schaller, M. Sykora, J. M. Pietryga and V. I. Klimov, *Nano. Lett.*, 2006, **6**(3), 424–429.
221. A. Shabaev, A. L. Efros and A. J. Nozik, *Nano. Lett.*, 2006, **6**(12), 2856–2863.
222. J. A. McGuire, M. Sykora, J. Joo, J. M. Pietryga and V. I. Klimov, *Nano. Lett.*, 2010, **10**(6), 2049–2057.
223. A. Pandey and P. Guyot-Sionnest, *Science*, 2008, **322**(5903), 929–932.
224. J. B. Sambur, T. Novet and B. A. Parkinson, *Science*, 2010, **330**(6000), 63–66.
225. W. A. Tisdale, K. J. Williams, B. A. Timp, D. J. Norris, E. S. Aydil and X.-Y. Zhu, *Science*, 2010, **328**(5985), 1543–1547.
226. W.-L. Chan, M. Ligges, A. Jailaubekov, L. Kaake, L. Miaja-Avila and X. Y. Zhu, *Science*, 2011, **334**(6062), 1541–1545.
227. O. E. Semonin, J. M. Luther, S. Choi, H. Y. Chen, J. B. Gao, A. J. Nozik and M. C. Beard, *Science*, 2011, **334**(6062), 1530–1533.
228. P. S. Weiss, *ACS Nano*, 2008, **2**(6), 1089–1096.
229. P. S. Weiss, *ACS Nano*, 2008, **2**(7), 1321–1321.
230. C. L. Choi and A. P. Alivisatos, *Annu. Rev. Phys. Chem.*, 2010, **61**(1), 369–389.

Miscellaneous Aspects

CHAPTER 12.1

Supramolecular Structures Based on Chalcogen–Halogen Secondary Bonds

WOLF-WALTHER DU MONT* AND
CRISTIAN GEORGE HRIB

Institut für Anorganische und Analytische Chemie der Technischen Universität Braunschweig, D-38023 Braunschweig, Germany
*Email: w.du-mont@tu-bs.de

12.1.1 General Aspects of Supramolecular Chemistry with Soft–Soft Interactions

From elemental sulfur to selenium and tellurium, intermolecular interactions (secondary bonds, soft–soft interactions) play an increasing role. According to N. W. Alcock,[1] the term 'secondary bond' describes interatomic distances longer than covalent single bonds but shorter than van der Waals interatomic distances.[1] In many cases secondary bonds can also be described as co-ordinative Lewis base–Lewis acid or charge-transfer (CT) (donor–acceptor) interactions. Such secondary bonds are formed by donation from the lone pair of a nucleophile into the σ* orbital of a covalent bond ('n → σ* attack') leading to unsymmetric or symmetric 3c–4e bond systems. Weak covalent bonds (implying low-energy σ* levels) are more easily attacked by $n \rightarrow \sigma^*$ overlap than strong bonds; this is why supramolecular arrays based on secondary interactions play a particular role in the chemistry of the heavier main group elements.

Handbook of Chalcogen Chemistry: New Perspectives in Sulfur, Selenium and Tellurium, 2nd Edition, Volume 2
Edited by Francesco Devillanova and Wolf-Walther du Mont

Published by the Royal Society of Chemistry, www.rsc.org

The role of secondary bond-like intermolecular interactions in inorganic compounds of selenium and tellurium is well documented[2–10] (see also previous chapters); the directed character of CT and halogen bond geometry has been analysed[11] (see also Chapter 12.4). The 'continuum of interactions'[7c] from van der Waals forces and secondary $n \rightarrow \sigma^*$ overlap via typical 3c–4e systems to 2c–2e covalent bonds was investigated by comparing a range of compounds with Se–I distances between 390 pm (about the sum of the van der Waals radii) and about 250 pm (single bonds).[12] Tellurium-centred secondary-bonding supramolecular synthons were covered by a recent review.[13] The present chapter reports a selection of supramolecular structures with chalcogen–halogen contacts, ordered under the aspects of functional groups, donor–acceptor relations, geometrical implications (such as T-structures or square-planar structures), and oxidation states.

12.1.2 Supramolecular Aspects of Structures from Some Classes of Compounds Discussed in Previous Chapters

12.1.2.1 Arrays of Binary Chalcogen Halides and Inorganic Halogenochalcogen Anions (see Chapter 8.1)

In the solid tetrahalides E_4X_{16} (E = Se: X = Cl, Br; E = Te: X = Cl, Br, I) the halogen atoms act as μ_2- and μ_3-bridging donors toward six-coordinated selenium and tellurium atoms (see Chapter 8.1).[7] The formally mixed-valence anions $Te_4I_{14}^{2-}$ and $Se_4Br_{14}^{2-}$ can be regarded as pairs of planar dimeric anions $E_2X_6^{2-}$ (E^{II}) linked by two I^+ or Br^+ cations through 3c–4e E–X–E bridges (Figure 12.1.1, left).[7,14] In these compounds Se and Te atoms can be regarded as electrophiles towards terminal halogen atoms and as nucleophiles towards the bridging halogens. The true mixed-valence compound $[(C_6H_5CH_2)(C_2H_5)_3N]_4[TeI_4][TeI_6]$[7] exhibits a supramolecular anion structure (Figure 12.1.1, right). It consists of layers of square-planar $[TeI_4]^{2-}$ anions and of slightly distorted octahedral $[TeI_6]^{2-}$ anions that can be regarded as assembly of columns (I···I contacts between the columns about 420 pm). Anion–anion I···I contacts are a feature of many iodotellurate structures; I_2 molecules can also be inserted between the anions.[15]

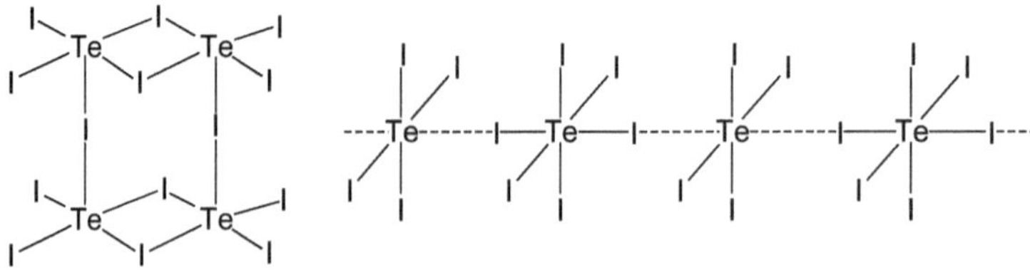

$Te_4I_{12}^{2-}$ $Te_2I_{10}^{4-}$

Figure 12.1.1 Tetranuclear and polymeric mixed-valence iodotellurates.

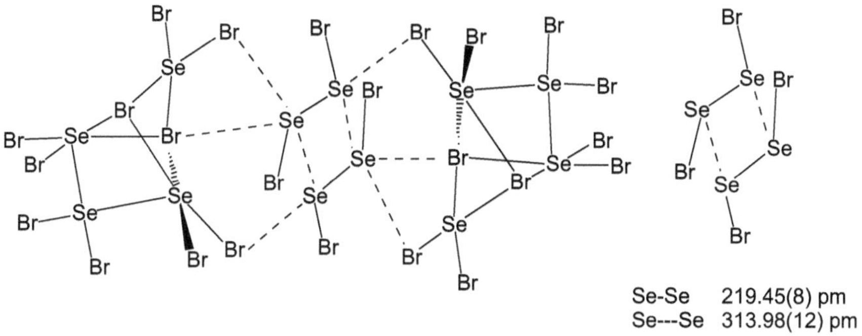

Figure 12.1.2 Left: the anion network of $(Me_4N)_4[(Se_4Br_{10})_2(Se_2Br_2)_2]$; right: the central $(Se_2Br_2)_2$ group from the anion.

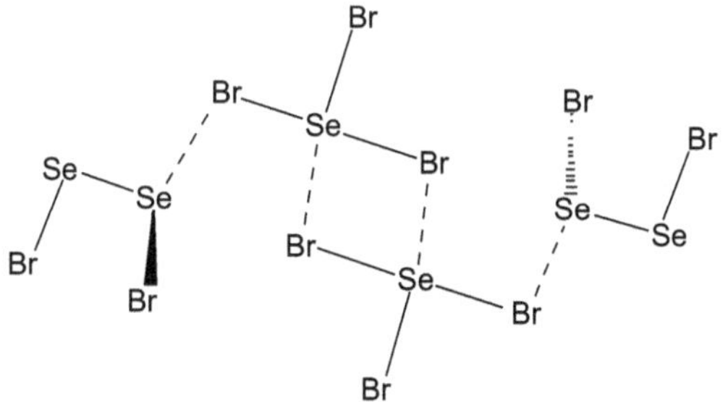

Figure 12.1.3 The anion network of $(PPh_4)_2[(Se_2Br_6)(Se_2Br_2)_2]$.

α-Se_2Br_2 (intermol. Se···Se 345 pm) and β-Se_2Br_2 (intermol. Se···Br 337 pm) display different types of their weak intermolecular contacts, leading to layers or three-dimensional (3D) nets, respectively.[16] In the tetramethylammonium salt $(Me_4N)_2[(Se_4Br_{10})(Se_2Br_2)_2]$ (Figure 12.1.2), 'α-Se_2Br_2-like' Se···Se contacts between the two Se_2Br_2 molecules lead to a cyclic Se_4Br_4 moiety that is connected to Br atoms of two neighbouring $Se_4Br_{10}{}^{2-}$ anions by Se···Br contacts.[17a] In the related salt $(PPh_4)_2[(Se_2Br_6)(Se_2Br_2)_2]$ the $Se_2Br_6{}^{2-}$ dianions are bridged through pairs of Se_2Br_2 molecules by 'β-Se_2Br_2-like' Se···Br contacts, leading to a chain structure (Figure 12.1.3).[17b]

A common feature of tellurium subhalides Te_3Cl_2, Te_2X (X = Br, I), and β-TeI is endless tellurium chains.[18] The unique compound $(Te_2)_2(I_2)$, made from GeTe, and Te with excess 67% HI, consists of alternating layers of square nets of Te_2 molecules (Te = Te 271.3 pm) and of layers of L-shaped I_4-dimers (I–I 286.6 pm).[19]

12.1.2.2 Supramolecular Chalcogen–Halogen Interactions Based on Cationic Species with Chalcogen–Chalcogen Bonds

Cationic species with S–S, Se–Se, or Te–Te bonds tend to exhibit significant interactions with halide or interhalide anions (Figure 12.1.4, see also Chapters

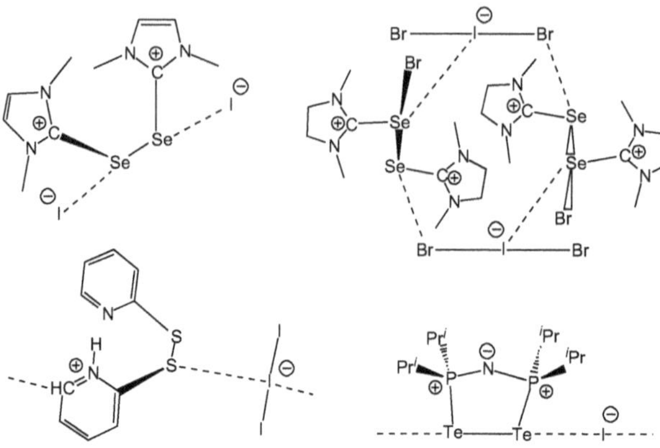

Figure 12.1.4 Examples of cation–anion interactions.

7.1–7.3 and 8.6.1). An increasing extent of cation–anion interactions correlates with chalcogen–chalcogen bond expansion.[20,21] The solid compound $[(C_5H_8N_2\text{-}Se)_2I_2$,[20a] prepared by reduction of $(C_5H_8N_2)\text{-}SeI_2$ with tellurium, exhibits cation–anion Se···I contacts of 341.0 and 333.0 pm (Figure 12.1.4, top left) in an almost linear I···Se–Se···I unit (<I–Se–Se 165.78° and 176.00°). Additional secondary contacts (Se···I 362.7 pm) lead to a a tetracoordination at the corresponding selenium atom and the formation of chains.

In the solid salt $[(C_5H_8N_2\text{-}Se)_2Br]^+$ IBr_2^- one bromine atom is attached to one atom of the Se–Se moiety, and the central iodine atom of the IBr_2^- anion exhibits a weak secondary contact to the other selenium atom (Figure 12.1.4, top right).[20a] In a related fashion, protonated bis(2-pyridyl)disulfide exhibits an S···I contact involving the central atom of the triiodide ions (Figure 12.1.4, bottom left), which are adjacent to a C–H function of the next neighbouring molecule.[21] Five-membered cyclic cations $[N(PR_2E)_2]^+$ (R = *i*Pr, E = Se, Te) [22] are bridged by their iodide counteranions leading to pairs of linear arrays (E-E···I···E-E···I···)$_x$ (Se–Se 248.4 pm, Te–Te 284.0 pm) that are connected by van der Waals-type E···E contacts (Figure 12.1.4, bottom right). The Se···I cation–anion distances are asymmetric (315 and 400 pm) whereas the Te···I···Te bridges are nearly symmetric (343 and 349 pm) (Figure 12.1.4, bottom right). The related mixed chalcogen derivative (E = Se, Te), the chloride $N(P\textit{i}Pr_2Te)_2Cl$, and the diphenylphosphorus derivative $N(PPh_2Te)_2I$ do not exhibit linear supramolecular arrays. The S, Te compound contains a Te–I single bond and van der Waals-like S···Te contacts in a linear array.[22]

A very unusual compound containing an Se–Se-bridged dication has been obtained by reacting *N*,*N′*-dimethylimidazolidine-2-selone with iodine in CCl_4 in a 1 : 1 molar ratio.[20b] The asymmetric unit contains both a molecular iodine adduct of the selenium donor molecule, and the dication $[(C_5H_{10}N_2\text{-}Se)_2]^{2+}$ balanced by two I_3^- ions (Figure 12.1.5). The crystal packing features chains of dications asymmetrically bridged by one I_3^- (Se···I 359.0 and 349.8 pm).

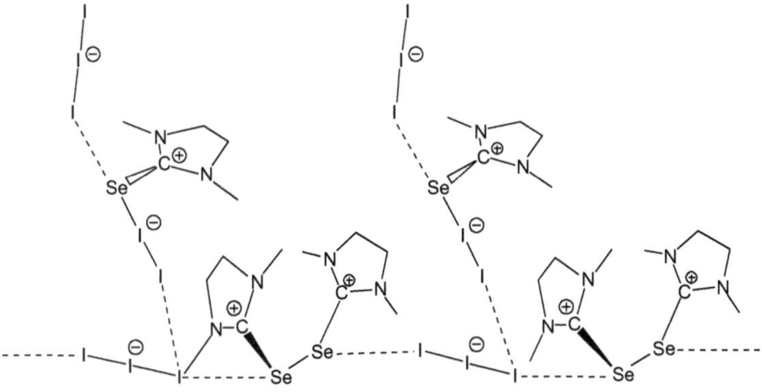

Figure 12.1.5 Se···I and I···I Interactions in solid $\{[(C_5H_{10}N_2\text{–}Se)_2]^{2+}$, $2I_3^-$, $C_5H_{10}N_2\text{–}Se\text{–}I_2\}$.

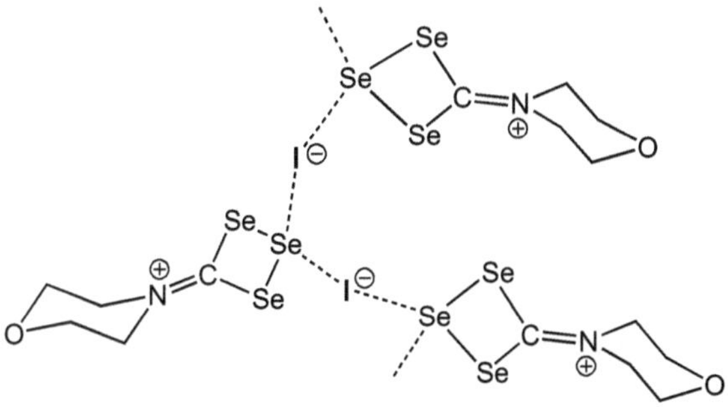

Figure 12.1.6 Chains from (*N*-morpholinocarbodiselenato)selenium(II) iodide.

The bridging I_3^- interacts at one end with the terminal iodine atom of the CT adduct unit by I···I long-range interactions of 419.3 pm. A further contact exists between the second molecule of I_3^- and the selenium atom of the adduct unit (386.7 pm), leading to an almost linear $I(I_3^-)\cdots[RSe\text{–}I]^+\cdots I^-$ system.[20b] Both the Se–Se (237.2 pm) and Se–I (268.3 pm) bonds are slightly longer than regular single bonds.[12]

Iodide-bridged cations containing the triselenetane ring system are present in $[O(C_2H_4)_2N{=}CSe_3I]$, which was isolated from the reaction of bis(morpholinoselenocarbonyl)triselenide with iodine.[23]

Solid (*N*-morpholinocarbodiselenoato)selenium(II)iodide (Figure 12.1.6) consists of chains of cationic moieties $[O(C_2H_4)_2N{=}CSe_3]^+$ that are μ_2-bridged through the central selenium atom by iodide anions. The cation may also be described as an Se^{2+} ion that is chelated by a $[O(C_2H_4)_2N{=}CSe_2]^-$ (morpholinodiselenocarbamato) ligand. Within the chain structure, there are two crystallographically different types of cations. Those with the longer Se···I contacts (318 pm) exhibit Se–Se bond lengths of 236 and 239 pm (longer than average Se–Se distances); the other type of cations allows closer Se···I contacts (304, 308 pm) that correlate well with a lengthening of the adjacent Se–Se bonds

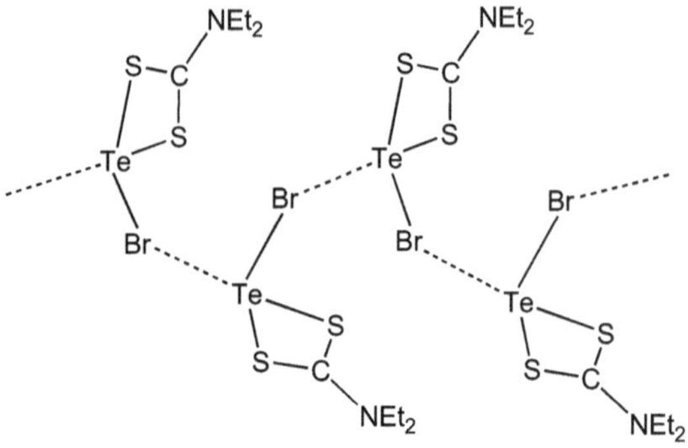

Figure 12.1.7 Chains from a dithiocarbamatotellurium(II) bromide.

(238, 244 pm). Each I^- acts as donor towards two neighbouring cations in a bent Se···I^-···Se sequence. A related chain-like dithiocarbamatoselenium iodide with μ_2-bridging iodide ions (Se–I 312.4 and 323.2 pm) has also been reported.[24] Dithiocarbamatotellurium halides exhibit related kinds of helical polymeric structures. From the chloride to the bromide (Figure 12.1.7) and iodide, the differences between the shorter ('primary') and the longer ('intermolecular') Te-X contacts decrease (Te–Cl: 268.6, 311.2; Te–Br: 286.7, 316.9; Te–I: 310.8, 327.9 pm).[24b,24c]

12.1.2.3 Supramolecular Structures of Adducts from Chalcogen Donors with Halogen Acceptors

12.1.2.3.1 Halogen Adducts of Diorganylsulfides and Selenides

With bromine, diorganosulfides tend to furnish CT adducts, whereas selenides can give equilibria between CT adducts and products of oxidative addition, *i.e.* molecular Se(IV) compounds (see Chapters 8.2, 10.3.3). A particular example of such equilibria is the bromination of selenanthrene reported by Nakanishi *et al.*[25]

With molecular iodine or iodine-like electrophiles I–X, diorganylsulfides and selenides react providing molecular CT complexes (see Chapter 8.2)[5,6,10,26,27] that may undergo intermolecular secondary interactions leading to supramolecular arrays. In the tetrahydroselenophene-diiodine adduct, the Se···I–I arrangement is essentially linear, and additional weak contacts between the 'terminal' iodine atom (having acquired donor properties) and the Se atom of a neighbouring molecule (Se···I 364 pm) lead to a polymeric chain structure. The 1,4-diselenane-diiodoacetylene adduct shows a related topology (Figure 12.1.8).[28]

The crystal structure of the complexes between iodine and 1,4-diselenane ([6]aneSe$_2$ · 2I$_2$, Se–I 282.9, I–I 287.0 pm, Se–I–I 180.0°)[6] shows discrete molecular units of the adducts hold together by Se···I contacts (388.9–402.2 pm) which contribute to the crystal packing together with several I···I long-range interactions. A slightly stronger Se···I [370.8 pm] secondary interaction is for the central feature of the crystal packing of [6]aneOSe · I$_2$.[6] The adducts of 1,4-diselenane with triiodomethane (Se···I 351.2 and 346.5 pm),

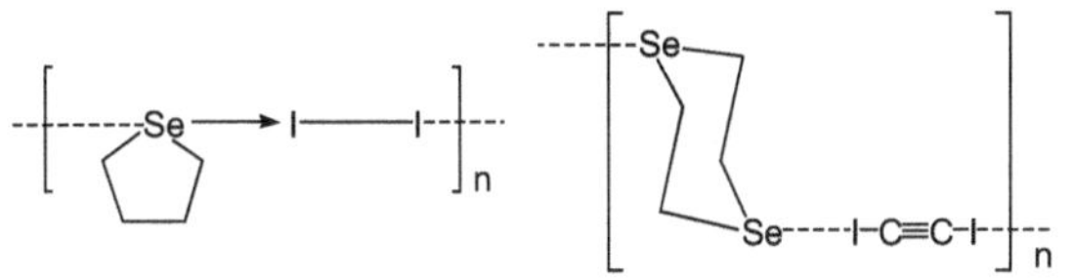

Figure 12.1.8 Topologies of chains from tetrahydroselenophene-I_2 (left) and 1,4 diselenenane-diiodoacetylene (right).

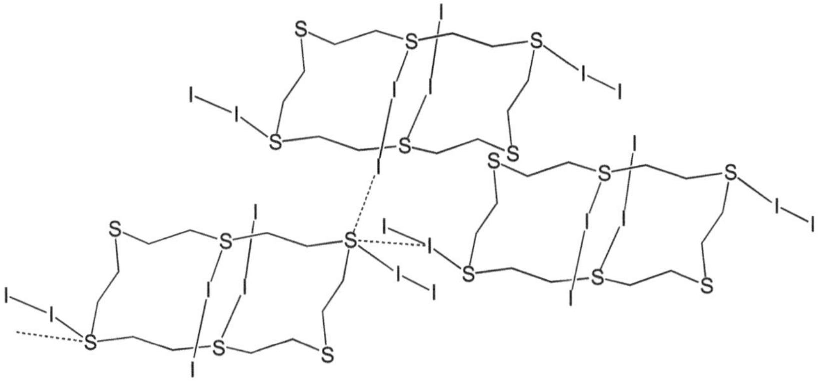

Figure 12.1.9 1 : 4 Adduct of [18]aneS_6 with diiodine; representative intermolecular S···I contacts are shown by dashed lines.

tetraiodoethylene (Se···I 341 pm) and diiodoacetylene (Se···I 333.6 pm) (Figure 12.1.5b) are cases of weak Se···I–C interactions between uncharged molecules.[28–30]

12.1.2.3.2 Thia-and Selena-Crown Ether Complexes

1,3,5-Triselenane ([6]aneSe_3) forms a molecular 1 : 1 adduct with iodine that exhibits weak intermolecular Se···I contacts (376.4–395.0 pm).[26] Thia-crown ethers tend to act as bifunctional bridging ligands toward molecular iodine, for instance in the 1 : 1 adducts of 1,3,5-trithiacyclohexane ([9]aneS_3) or of [18]aneS_6 with diiodine. [15]aneS_5 gives a 2 : 7 adduct with iodine: one I_2 molecule bridges two equivalent [15]aneS_5 ligands that contain three terminal S–I–I functions and one non-coordinated sulfur atom. In [18]aneS_6(I_2)$_4$ the molecules interact weakly through two independent S···I contacts of similar lengths (376 and 378 pm) building up a 3D network (Figure 12.1.9).

Interestingly, there is a tendency to obtain only certain stoichiometries of crown ether iodine complexes in the solid state, *i.e.* product formation is governed by packing behaviour and solubility of the particular complexes. In solution, 1 : 1 adducts are often the predominant species.[25,31–34]

The thia-crown ether [14]aneS_4 gives with IBr a 1 : 2 complex that exhibits intermolecular S···I and S···Br contacts; in [16]aneS_4(IBr)$_4$ all sulfur atoms are involved in coordination. Soft–soft interactions, established by I···I and I···Br

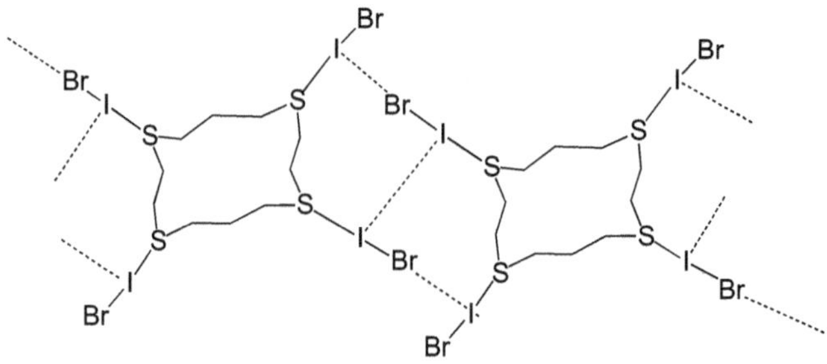

Figure 12.1.10 1 : 4 Adduct of 16]ane$S_4(IBr)_4$; partial view of the intermolecular I···I and I···Br contact pattern within a ribbon.

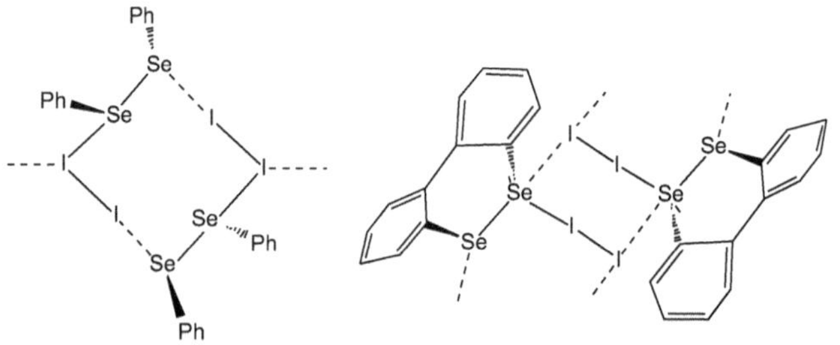

Figure 12.1.11 Intermolecular interactions in the iodine adducts of diphenyldiselenide (left) and of dibenzo[1,2]diselenine (right).

contacts (Figure 12.1.10), generate one-dimensional ribbons, which are linked by additional I···Br contacts forming a 3D grid.[35]

12.1.2.3.3 Reactions of Organic Diselenides with Iodine

In mixtures of organic diselenides with iodine the CT-like Se→I coordination competes with Se–Se bond cleavage.[36–39] Diphenyldiselenide gives a solid dimeric 1 : 1 adduct with molecular iodine.[36] One selenium atom of each diselenide acts as a donor towards molecular iodine (Se–I 299.2 pm, I–I 277.5 pm), whereas the other selenium atom behaves as a very weak acceptor (Se···I 358.8 pm) (Figure 12.1.11, left). In the crystal, the squares of $(Ph_2Se_2I_2)_2$ are connected by I···I contacts (412.7 pm).[36,37a] A different type of intermolecular association is found in the iodine adduct of dibenzo[1,2]diselenine (Figure 12.1.11, right: Se–I 290.95 pm, I–I 284.12 pm, intermolecular Se···I 375.7 pm).[38]

Addition of iodine to the bulkier diselenide $(2,4,6\text{-}i\text{Pr}_3C_6H_2Se)_2$ leads to a crystalline compound that can be regarded as an 'inclusion compound' of 0.5 equivalents of iodine within the solid diselenide.[37b] The iodine molecule is in contact (348.3 pm) with the selenium atoms of two different diselenide molecules. The Se–Se bond length (235.3 pm) of the iodine adduct is only slightly longer than that of the non-coordinated diselenide (233.9 pm).[39]

12.1.2.3.4 Coordination Compounds of Thio- and Selenocarbonyl Derivatives with Halogens

Molecular coordination compounds of thio- and selenocarbonyl derivatives with iodine are often regarded as CT complexes (see Chapter 8.2).[40,41] Starting from a series of different pentaatomic rings all containing a C=Se group, it has been shown that slight variations in the surroundings of the C=Se function or small changes in the reaction conditions can influence the nature of the final solid products, which may be molecular adducts [RR′C = Se→I–I],[41,42] ionic compounds $[(RR'C{=}Se)_2I]^+I_n^-$,[43] molecular compounds with three-coordinated selenium $[RR'CSeI_2]$,[37,42,44–46] or dicationic diselenides $[(RR'CSe)_2]^{2+}$. In some cases, ionic compounds with (2c–2e) Se–I bonds in cations $[RR'CSeI]^+$ can be formed, which exhibit anion–cation interactions.[46]

12.1.2.3.4.1 Intermolecular Association in Complexes [RR′C = E→I–I] (E = S, Se). In the 1 : 1 complex of 4,5-ethylenedithio-1,3-dithiol-2-thione with iodine, the C=S sulfur atom binds the iodine in the (usual) linear S–I–I fashion (S–I 275.5 pm, I–I 281.2 pm, S–I–I 175.4°). Packing as a polymeric chain of molecules is determined by the secondary contacts involving the sulfur atoms of the five-membered heterocycle, one of which is involved in a weak intermolecular S···I interaction, wheras the other exhibits a secondary S···S contact (Figure 12.1.12).[47] The reaction of the related 1,3-dithiol-2-thione-4,5-dicarboxylic ester ligand with IBr gives the molecular 2 : 3 adduct, which consists of centrosymmetric dimers formed by weak intermolecular S···I contacts (S···I 367 pm; Figure 12.1.13). The third equivalent of (disordered) IBr molecules connects the dimers, and additional C=O···HC contacts lead to a 3D network.[48a]

Attachment of a ferrocenyl substituent in the 4-position of ethylene-1,3-dithiol-2-thione leads to an ionic product, containing oxidized ferrocenium fragments and linear S–I–I moieties that display weak contacts to the V-shaped pentaiodide anions. The supramolecular structure is determined by a network of cation–cation, cation–anion, and anion–anion I···I interactions in the range 370–390 pm (Figure 12.1.14). This represents a supramolecular structure with

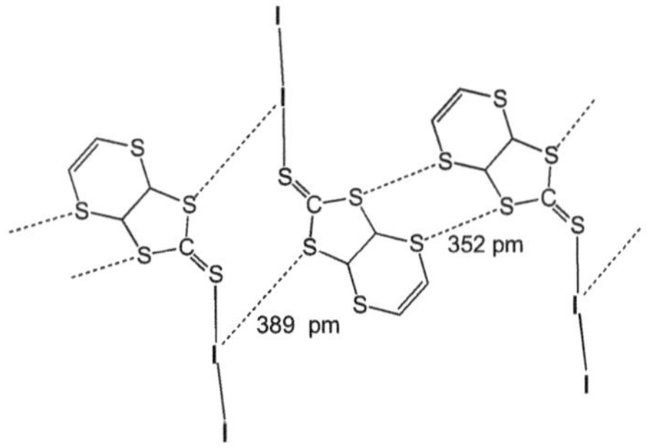

Figure 12.1.12 4,5-Ethylenedithio-1,3-dithiole-2-thione-I_2.

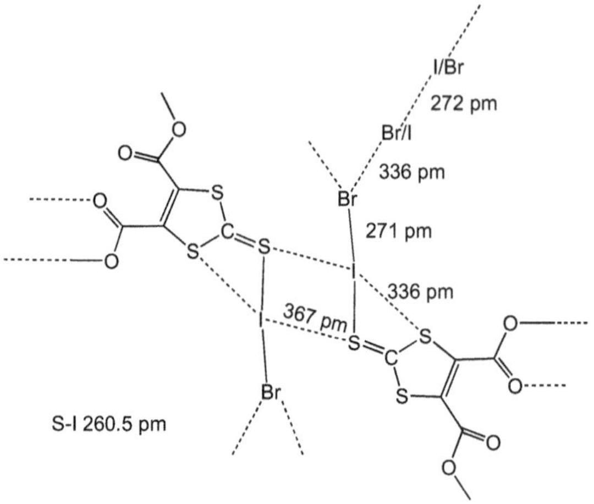

Figure 12.1.13 The 1,3-dithiole-2-thione-4,5-dicarboxylic ester 2:3 adduct with IBr.

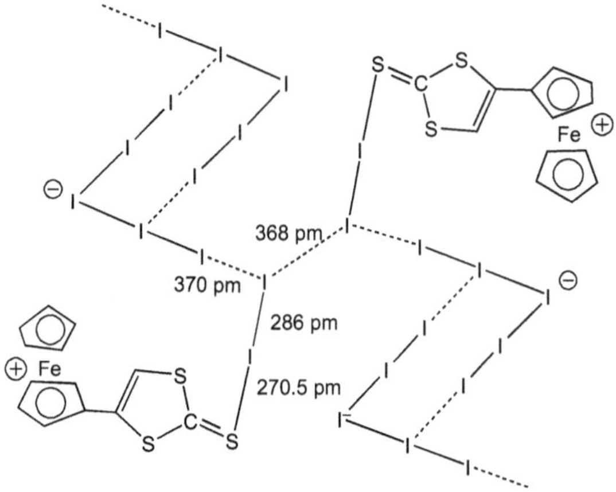

Figure 12.1.14 The polyiodide network determined by I···I cation/anion contacts in 4-ferrocenium-ethylene-1,3-dithiole-2-thione·I_2 adducts balanced by V-shaped pentaiodides.

paramagnetic centres (Fe^{III}). Magnetic measurements show a deviation from the Curie–Weiss law below 2.5 K, attributed to antiferromagnetic interactions.[48b]

The introduction of two bromomethyl functions in 4- and 5-positions of the ethylen-1,3-dithiole-2-thione diiodine adduct leads to a 3D network (Figure 12.1.15) defined by S···I, I···I, S···Br, I···Br, and CH···Br and contacts.[49]

12.1.2.3.4.2 Cations $[(RR'C{=}Se)_2I]^+$. In the case of *N*-methylbenzothiazole-2(*3H*)-selone (mbts), one equivalent of iodine provides a molecular adduct in dichloromethane solution;[50] crystallization, however, leads to an ionic compound of the type $[(RR'C{=}Se)_2I]^+I_3^-$.[43] This

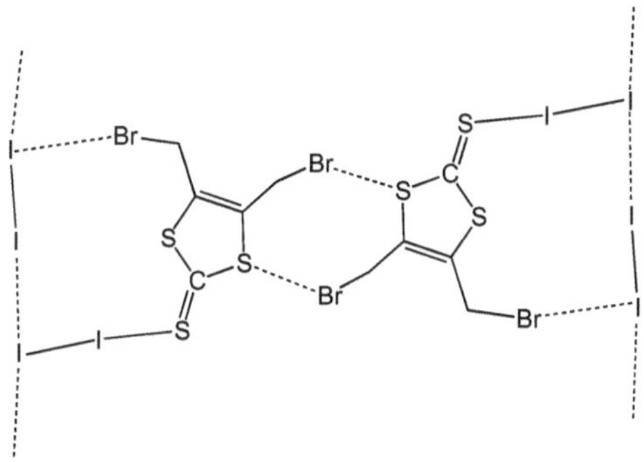

Figure 12.1.15 Secondary interactions in the 4,5-bis(bromomethyl)ethylene-1,3-dithiole-2-thione-bis(diiodine) complex.

compound was the first one reported in the literature featuring a two-coordinated iodine(I) complex with two donor molecules containing seleno-carbonyl groups. In the solid state two molecules of mbts linearly coordinate a central I^+ ion to give a slightly asymmetric Se–I–Se arrangement [Se–I 280.0, 271.9 pm, Se–I–Se 178.0°]. The unit cell contains two of these cations centrosymmetrically related and balanced by two I_3^- ions. Related I^+ cations with sulfur ligands are also well known.[43]

12.1.2.3.4.3 Bromothio- and Iodoselenocarbenium Salts. When the interaction between a selenium donor molecule and the linearly coordinated I-X (X = I, Br, Cl) acceptor is particularly strong, a remarkable fractional negative charge is induced on the terminal X atom which can behave as donor towards another molecule of acceptor I-X.[32,35,50] In the limiting case ionic $[RSe–I]^+ \cdots [X–I–X]^-$ systems are formed. This situation is well documented by the 1 : 2 adduct of mbts with IBr in which the first IBr molecule is so strongly bonded to the selenium atom that the Se–I distance (256.4 pm) can be formally regarded as a covalent single bond slightly perturbed by anion (donor)–cation (acceptor) interactions (see Chapter 3.3). Elongation of the IBr distance to of 312.9 pm and strong interaction of Br^- with a second IBr molecule leads to the formation of an asymmetric IBr_2^- anion [I–Br 280.3 and 264.5 pm]. The mbts donor molecule also gives a 1 : 2 molecular adduct with I_2. The structural features of the resulting bent RR′C-Se$\cdots I_2 \cdots I_2$ compound do not agree with an ion pair description as for mbtsI$^+$ BrIBr$^-$. In fact, in mbts · 2I_2 the interaction with the selenium atom [Se–I mean value 267.4 pm] lengthens the I–I bond to 303.0 pm (mean value). The further interaction of the terminal iodine atom of the Se–I_2 system with the second I_2 molecule [I$\cdots I_2$ mean value 319.2 pm] is not strong enough to form a triiodide ion. This system can be better described as RSe–$I^+ \cdots I^- \cdots I_2$, which resembles that the arrangement observed in 'V-shaped' pentaiodides in which an I^- ion binds two diiodine molecules $I_2 \cdots I^- \cdots I_2$.

The very nucleophilic C=S function of 1,3,4,5-tetramethylimidazolium-2-dithiocarboxylate gives a molecular 1 : 1 CT complex with iodine, but with bromine the formation of a 1 : 2 product is favoured. The solid compound is an

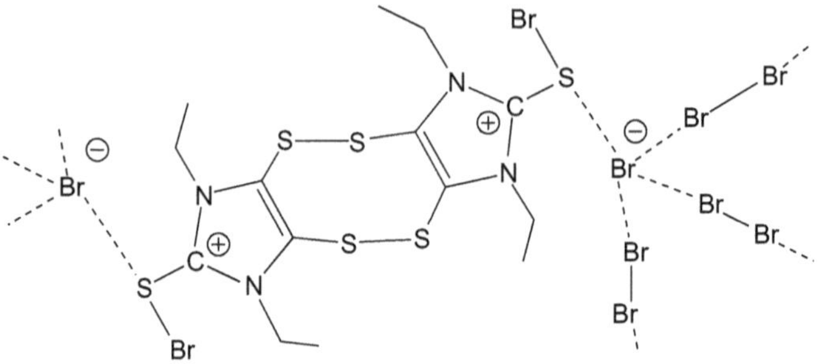

Figure 12.1.16 S···Br and Br···Br contacts leading to the 3D network of a bis-bromothiocarbenium polybromide.

ion pair from a cation containing a remarkably short S–Br single bond (214.6 pm) and a distorted tribromide anion exhibiting a significant cation–anion Br···Br contact (321.7 pm).[51] The reaction of [Pd(Et_2timdt)$_2$] with an excess of Br_2 yielded few crystals of 4,5,9,10-tetrathiocino-[1,2-*b*:5,6-*b'*]-1,3,6,8-tetraethyl-2,7-di(bromothio)diimidazolium dications bearing two terminal bromothiocarbenium functions, balanced by two Br^- counterions which linearly interact with the S-Br functions ($4p^2 \rightarrow \sigma^*_{S-Br}$) on the side of the sulfur atom. At the same time, each Br^- acts as a donor towards three Br_2 molecules (Figure 12.1.16), thus generating a 3D network via additional Br···Br contacts.[52]

12.1.2.3.5 Products from Reactions of Phosphane Sulfides and Selenides with Iodine

Depending on the conditions, Ph_3PS gives a molecular 1 : 1 adduct, a V-shaped 2 : 3 complex in which two of the 1 : 1 adducts are linked by a bridging I_2 molecule,[53] or a 1 : 3 complex that exists as a supramolecular network made up from stepped layers of polyiodide ribbons and domains of aryl groups.[54] Within this network, the ribbons may be regarded as another polymeric version of the I_5^- ion, involving iodide ions that are in contact as donors to three I_2 molecules within the ribbon and with the iodine atom of the terminal Ph_3PSI^+ unit. Stepwise packing of the ribbons leads to contacts of 410–417 pm between the iodine atoms of adjacent ribbons (Figure 12.1.17).[54]

Phosphane selenides are stronger soft donors than the sulfides.[55] The solid 1 : 1 adduct of iPr$_3$P=Se with I_2 is ionic (iPr$_3$PSe)$_2$I$^+$ I_3^-; reaction of tri-*t*-butylphosphane selenide with one equivalent of iodine, however, leads to a solid product that contains both a molecular 1 : 1 adduct tBu$_3$PSe–I–I *and* an ionic species (tBu$_3$PSe)$_2$I$^+$ I_3^- (Figure 12.1.18, left) in the same crystal.[56] Both kinds of species exhibit Se–I distances of about 276 pm that indicate bonds orders of 0.5. In other molecular adducts R_3PSe–I–I, more electron-releasing substituents (such as dialkylamino groups) lead to stronger Se–I bonds that correlate with decreasing bond orders within the PSe and within the I···I moieties. More electron-withdrawing substituents (such as R = Ph) have the opposite effect on the P–Se–I–I moieties.[56–58] The reaction of tri-*t*-butylphosphane selenide with 1.5 equivalents of iodine leads to a solid product that

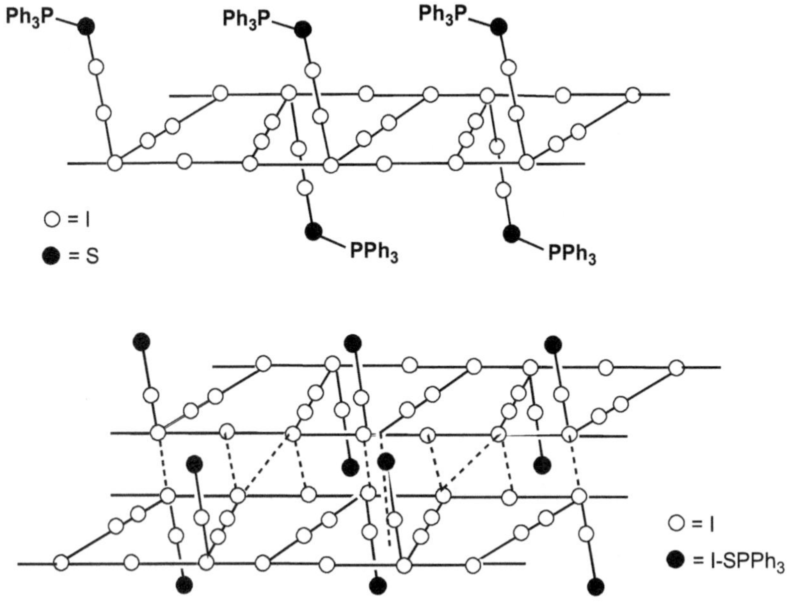

Figure 12.1.17 Top: ribbon structure of $Ph_3PS(I_2)_3$; bottom: stepped layers of $Ph_3PS(I_2)_3$ ribbons.

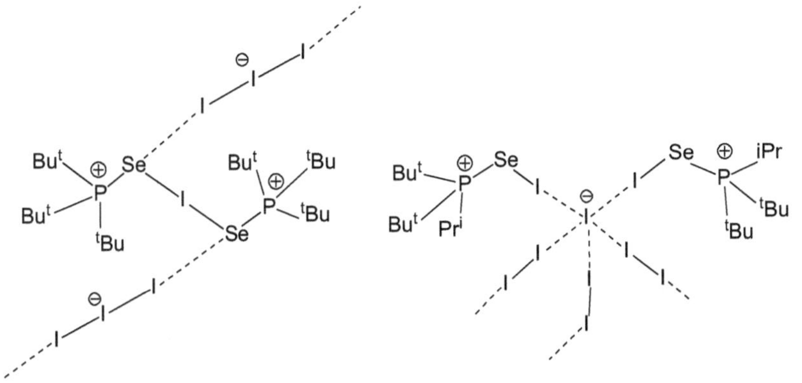

Figure 12.1.18 Left: Cation–anion chains of $(tBu_3PSe)_2I^+$ I_3^-; right: cation–anion contacts in $tBu_2iPrPSeI^+$ I_6^-.

contains pairs of $(tBu_3PSe)_2I^+$ cations intercalated between corrugated layers of the polymeric I_5^- anions. The $(tBu_3PSe)_2I^+$ cation pairs are well separated from the pentaiodide layers; they exhibit, however, pairs of Se···Se contacts within the cation pairs.[56]

With two or more equivalents of iodine, phosphane selenides can furnish iodoselenophosphonium salts $R_3PSeI^+I_n^-$ that exhibit various degrees of cation–anion I···I interactions. Reacting $(Me_2N)_3PSe$ and $(Morph)_3PSe$ (Morph = morpholino) with two equivalents of iodine provides the solid compounds $(Me_2N)_3PSeI_4$ and $(Morph)_3PSeI_5$ [57] (see Chapter 8.2); iPr_3PSe and *t*Bu_2*i*PrPSe with excess iodine provide *i*Pr_3PSeI_7 and *t*Bu_2*i*$PrPSeI_7$.[59] All these compounds are examples of $[R_3PSeI]^+$ cations with different extents of cation–anion interactions. The nucleophilicity of the iodide anion is suppressed by its coordination with increasing amounts of iodine; the decreasing nucleophilicity ($I^- > I^{-I}–I_2 > I^{-I}–2I_2 > \cdots$) correlates with increasing covalent Se–I bond orders, indicated by Se–I distances that are influenced by the different extents of ($n[I_x^-] - > \sigma^*[Se–I]$) overlap.

The slight steric differences between the alkyl groups of the cations lead to quite different extended structures of their polyiodide anions. In solid *t*Bu_2*i*$PrPSeI_7$ (Figure 12.1.18, right), one iodide anion bridges two of the *i*Pr_3PSeI^+ cations and is in further contact with three I_2 molecules; the other I^- anion is in contact with five I_2 molecules.[59]

12.1.3 Supramolecular Aspects of Structures from Halides of Divalent Chalcogens

12.1.3.1 Self-Association of Sulfenyl, Selenenyl and Tellurenyl Halides

The first structure determination of a sulfenyl iodide was that of thermally labile Ph_3CSI.

In the solid state the molecules pack in such a way that zigzag chains ···S–I···S–I··· (angular I···S–I, approximately linear S···I–S) are formed.[60] A similar Se(donor)···I(acceptor) chain topology is present in solid dureneselenenyl iodide solid DurSeI (2,3,5,6-Me_4C_6H = Dur; Figure 12.1.19, left).[61] Another compound with a related zigzag array is *p*-chlorophenylselenoacetylene which exhibits secondary Cl(donor, 8-Cl-2)-Se(acceptor,

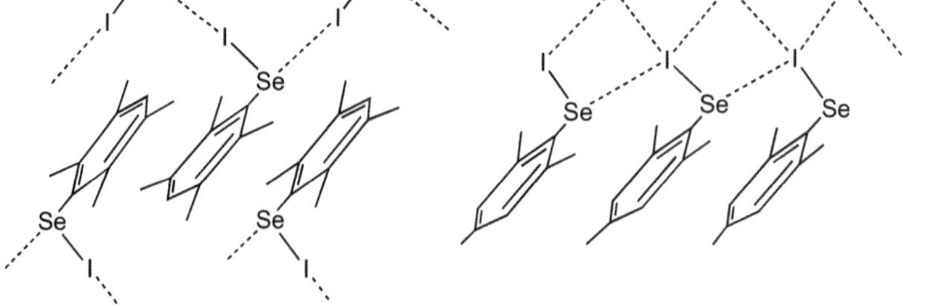

Figure 12.1.19 Intermolecular interactions in solid DurSeI [Dur = 2,3,5,6-$(CH_3)_4C_6H$-] (left) and in MesSeI [Mes = 2,4,6-$(CH_3)_3C_6H_2$-] (right).

10-Se-3) contacts (Cl···Se 370 pm) in the range of the sum of Cl and Se van der Waals radii (380 pm).[62]

In solid mesityleneselenenyl iodide (2,4,6-$Me_3C_6H_2SeI$), pairs of (Se–I···Se–I···) zigzag chains exhibit interchain I···I contacts, leading to double chains of molecules.[63] (Figure 12.1.19, right). Single chains with related patterns of (Te–Cl···Te–Cl···) contacts and aryl stacking exist in the deep blue solid 2,4,6-*t*$Bu_3C_6H_2TeCl$.[64]

Solid benzeneselenenyl bromide PhSeBr and -tellurenyl iodide PhEX (E = Se, X = Br and E = Te, X = I) are associated in a very particular way: Secondary E···E interactions connect four molecules leading to approximately planar E_4 units, whereby the participating chalcogen atoms can be classified as 2 + 2 coordinated (12-E_4).[65] The exocyclic bromine and iodine atoms exhibit inter-tetramer halogen–halogen contacts, leading to square X_4 units and eight-membered (E-E-X···X)$_2$ units in extended sheets.

In contrast to the above-mentioned zigzag (E–X···E–X)$_x$ chains of DurSeI, the corresponding tellurenyl iodide 2,3,5,6-Me_4C_6HTeI (Figure 12.1.20) consists of dimers with linear I–Te···Te–I moieties (Te–I 277.43 pm, Te···Te 314.58 pm), which in turn exhibit pairs of interdimer Te···I interactions within one-dimensional assemblies.[66]

A unique durenetellurenyliodide–diiodine adduct was isolated from DurTeI with molecular iodine. The monomeric units of DurTe(I)–I_2 aggregate pairwise through Te···I interactions (347 pm) leading to a rectangular six-atom cyclic dimer (Figure 12.1.21). Further pairs of interdimer Te···I contacts leads to a ribbon-like network of condensed six-membered and four-membered rings.

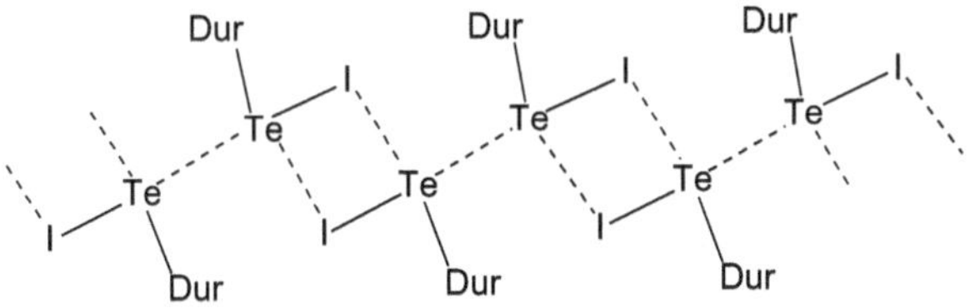

Figure 12.1.20 Chain structure of DurTeI.

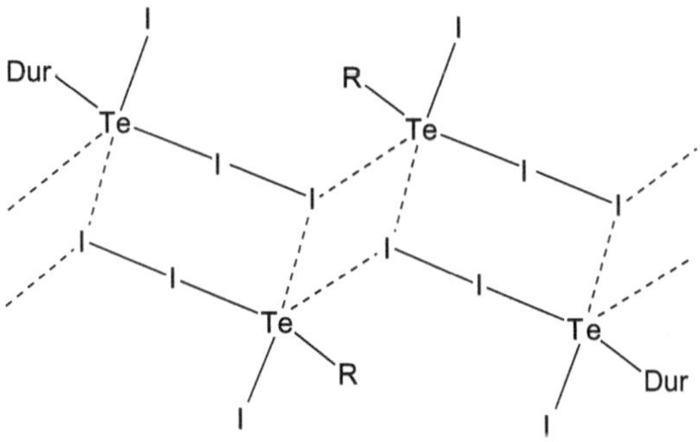

Figure 12.1.21 Ribbon structure of the DurTeI–I_2 adduct.

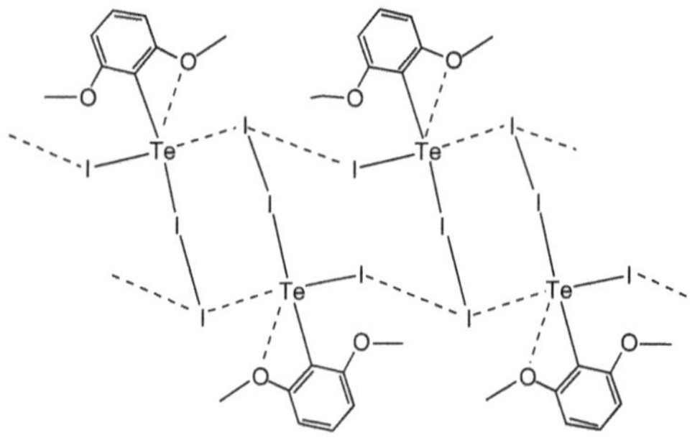

Figure 12.1.22 Ribbon structure of the 2,6-$(MeO)_2C_6H_3TeI–I_2$ adduct.

Each tellurium atom is in contact with one carbon atom and four iodine atoms.[66]

The formally related adduct of 2,6-$(MeO)_2C_6H_3TeI$ with molecular iodine also exhibits dimeric rectangle-like units (Figure 12.1.22).[67] In this case one MeO oxygen atom of each aryl substituent contributes to the coordination pattern of the tellurium atoms, which are surrounded by one carbon atom, one oxygen atom, and three iodine atoms. In this case, further pairs of interdimer I···I contacts lead to a ribbon-like network of condensed six-membered rectangular and chair-configurated eight-membered rings.[67]

12.1.3.2 Association of (10-E-3) Compounds (E = Se, Te) from Oxidative Halogen Addition to Chalcogenocarbonyl Functions $R_2C{=}E$ (E = Se, Te) and to Phosphine Chalcogenides $R_3P{=}E$

12.1.3.2.1 Chalcogenocarbonyl Dihalides R_2CEX_2

T-stacking via chalcogen–halogen contacts leading to (12-E_4) arrangements occurs in many T-shaped seleno- and tellurocarbonyl halogen adducts, as in the dichloromethane-solvated dimer of *N*-methylbenzothiazol-2-selone dibromide (Figure 12.1.23, left) and in the chain-like arrays of the solvent-free polymer of the same compound with linear Se-Br···Se bridges (Figure 12.1.23, right), *N,N′*-dimethylimidazoline-2-selone dibromide with angular Se-Br···Se bridges (Figure 12.1.24, left),[68–71] and 2,3-dihydro-1,3-diisopropyl-4,5-dimethylimidazol-2-tellurone diiodide (obtained from the reaction of 2,3-dihydro-1,3-diisopropyl-4,5-dimethylimidazol-2-ylidene with tellurium tetraiodide). The latter packs as a helical polymer via wide-angle Te–I···Te contacts[72] (Figure 12.1.24, right).

The first structurally known iodine derivatives of this kind were products from the iodination of 1,3-dimethyl-4-imidazoline-2-selone,[44] of 1,2-bis-(3-methyl-4-imidazoline-2-selone)ethane,[44] and of 1,3-diisopropyl-4,5-dimethyl-4-imidazoline-2-selone.[42] The I-Se–I fragments of all three compounds are asymmetric and roughly linear [I–Se–I ranging from 175.4 to 178.6 pm], and the CSeI2 moieties adopt T-shaped geometries (Se–C single bonds, Se–I bond orders close to 0.5), like those of internally chelated selenenyl

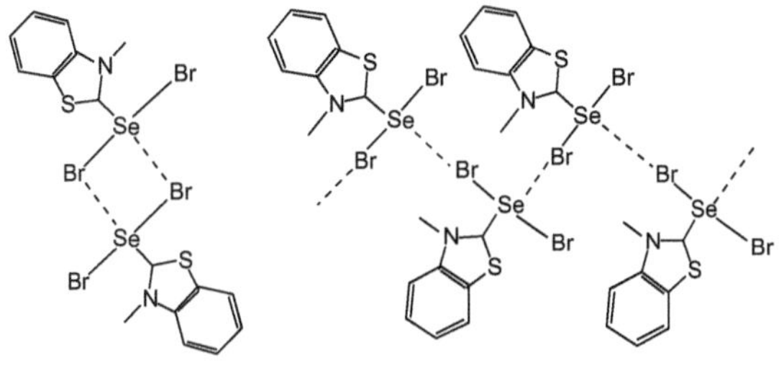

Figure 12.1.23 T-stacking of *N*-methylbenzothiazol-selone dibromide; left: the dimeric mbts dichloromethane solvate, right: polymeric solvent-free mbts.

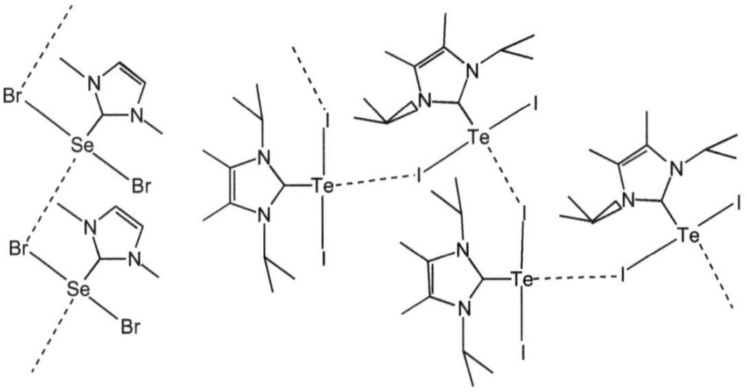

Figure 12.1.24 T-stacking of *N,N′*-dimethylimidazoline-2-selone dibromide (left), and 2,3-dihydro-1,3-diisopropyl-4,5-dimethylimidazol-2-tellurone diiodide (right).

iodides. An adduct of the composition (RR′CSeI_2.0.5 I_2) was obtained from 1,3-diisopropyl-4,5-dimethyl-4-imidazoline-2-selone with a slight excess of iodine. In this centrosymmetric compound, an iodine molecule (I–I 276.6 pm) is inserted (I···I 345.1 pm) between two T-shaped CSeI2 functions.[45] The Se–I bonds that participate in contacts with the inserted I_2 molecule are longer (285.1 pm) than the terminal ones (275.6).

From 1,1′-bis-(3-methyl-4-imidazoline-2-selone)methane with iodine bromide, a solid compound containing disordered molecules with T-shaped CSeI2 and Br–Se(C)–I functions in the same crystal was isolated; from 1,2-bis-(3-methyl-4-imidazoline-2-selone)ethane with iodine bromide, a solid compound containing a molecular species with two (ordered) T-shaped Br–Se(C)–I functions is formed.[73] Interestingly, the Se–I bonds (260.6, 262.9 pm) are shorter than Se–Br (289.7, 276.6 pm). The compound can therefore be regarded as derived from donor–acceptor interactions between Br^- anions and the Se atoms of $[Se–I]^+$ fragments.

Solid CBr_2Se_4 [1,2,3-triselenetane-4-(*Se,Se*-dibromoselone)] consists of a 3D network of planar molecules. The carbon atom is part of a T-shaped $CSeBr_2$ group and of a triselenetane ring. All four selenium atoms participate in intermolecular Se···Br or/and Se···Se interactions (Figure 12.1.25).[74]

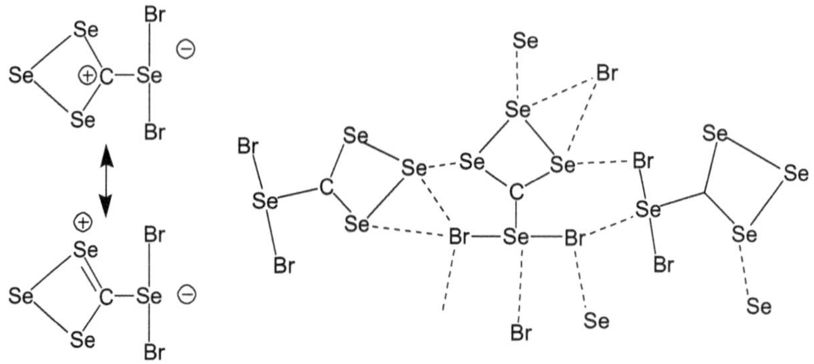

Figure 12.1.25 Left: resonance structures of CBr_2Se_4; right: intermolecular contacts in solid CBr_2Se_4.

12.1.3.2.2 Packing of Solid Phosphane Selenide Dibromides RR'_2PSeBr_2

Bromination of phosphane selenides leads to molecular phosphane selenide dibromides that contain T-shaped P-$SeBr_2$ functions. Structural investigation of a series of closely related trialkylphosphane selenide dibromides RR'_2PSeBr_2 (R,R′ = *i*Pr, *t*Bu) revealed remarkable differences in the packing of these T-shaped compounds.[75]

From $i$$Pr_3PSeBr_2$ two different crystalline phases were isolated. The intended preparation from $i$$Pr_3PSe$ with Br_2 led by recrystallization of the crude product from dichloromethane/pentane (vapour diffusion method) to monoclinic crystals (phase 1). Bromination of $i$$Pr_3PSe$ with an excess of bromine led to a mixture of products from which a few single crystals of phase 2 were collected.

In both phases the T-structures around Se are distorted by non-equidistant Se–Br bonds and by PSeBr angles that are larger than 90°. In the monomeric molecule $i$$Pr_3PSeBr_2$#1 the BrSeBr group differs from linearity by nearly 13° and from equidistance by about 12 pm (252.5 and 264.4 pm). One of the *i*-propyl groups adopts a conformation with a transoid H–C(1)–P–Se arrangement, this hydrogen atom is in contact with the selenium atom of an adjacent molecule. The orientation of the P–C(1) bond is approximately orthogonal to the BrSeBr vector. $i$$Pr_3PSeBr_2$#2 consists of two crystallographically different molecules that can be regarded as part of a centrosymmetric tetramer, when weak intermolecular Se···Br and Br···Br contacts are considered (Figure 12.1.26). The conformations of both $i$$Pr_3PSe$ groups of #2 are very similar to those in monomeric $i$$Pr_3PSeBr_2$#1. With respect to their BrSeBr functions, both molecules (terminal and bridging) of #2 are less distorted (Se–Br terminal molecule: 257.6 *vs* 260.2 pm; bridging molecule: 254.2 *vs* 260.6 pm) than that of $i$$Pr_3PSeBr_2$#1. The selenium atom of the terminal molecule of #2 exhibits a contact to one of the bromine atoms [Se(1)Br(3) 346.2 pm] of the bridging molecule; the other bromine atoms of two neighbouring bridging molecules exhibit a Br(4)···Br(4#) contact of 361.2 pm.

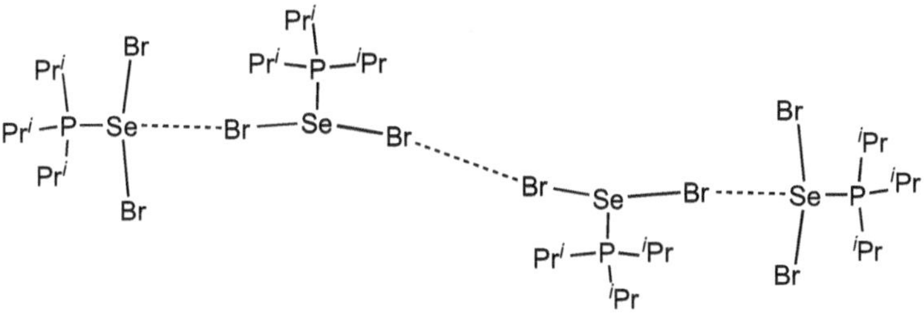

Figure 12.1.26 T-stacking of the iPr$_3$PSeBr$_2$ (phase #2) tetramers.

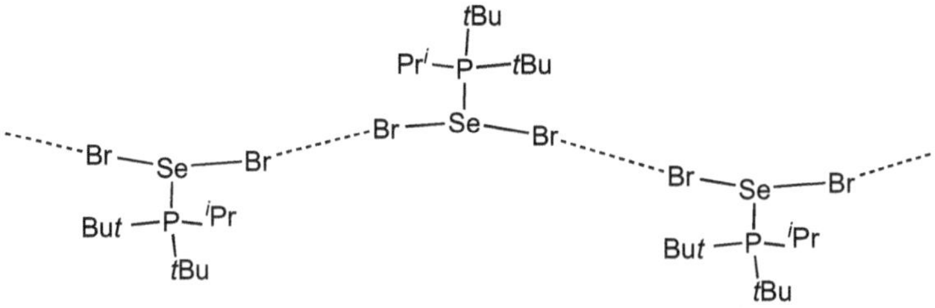

Figure 12.1.27 Array of molecules tBu$_2i$PrPSeBr$_2$ with Br···Br contacts.

In summary, the latter weak *soft* intermolecular interactions in phase 2 correlate with a decreasing distortion of the BrSeBr groups and with very slight (0.5 pm) expansion of the P–Se bonds. In tBu$_2$(iPr)PSeBr$_2$ intermolecular Br···Br contacts (358.90 pm) between T-shaped moieties lead to undulating chains (Figure 12.1.27) that are packed into layers. The conformation of the H–C–P–Se moiety in tBu$_2$(iPr)PSeBr$_2$ (involving the methine proton from the i-propyl group) is approximately synperiplanar ($\Theta = -43°$); the PSe bond is longer (230.39 pm) and the deviation of the BrSeBr group from linearity (BrSeBr 159°) larger than in any other known R$_3$PSeBr$_2$ structure.

Solid c-Hex$_3$PSeBr$_2$ (c-Hex = cyclohexyl)[76] exhibits a somewhat related arrangement of undulating chains; significant Br···Br contacts (361 pm) exist, however, only between pairs of one of the two crystallographically equivalent molecules, the other 50% of molecules do not exhibit soft–soft interactions. Quite different is the packing of iPr$_2t$BuPSeBr$_2$, which consists of centrosymmetric dimers arising from weak Se···Br interactions (Figure 12.1.28).[75] As in monomeric iPr$_3$PSeBr$_2$#1, intra- and intermolecular CH···Br contacts also play a significant part. Two T-shaped PSeBr$_2$ groups pack close together in such a way that the bridging Br atom of one molecule exhibits contacts (slightly less than van der Waals distances) to the Se atom (382 pm) and to the bridging Br atom (388 pm) of the neighbouring molecule. The Se–Br bond involved in these intermolecular interactions is shorter (253.6 pm) than the 'terminal' Se–Br bond (264.9 pm).[75]

In summary, steric strain from t-butyl groups correlates with increasing deviation from linearity of the Br–Se–Br groups and with increasing P–Se

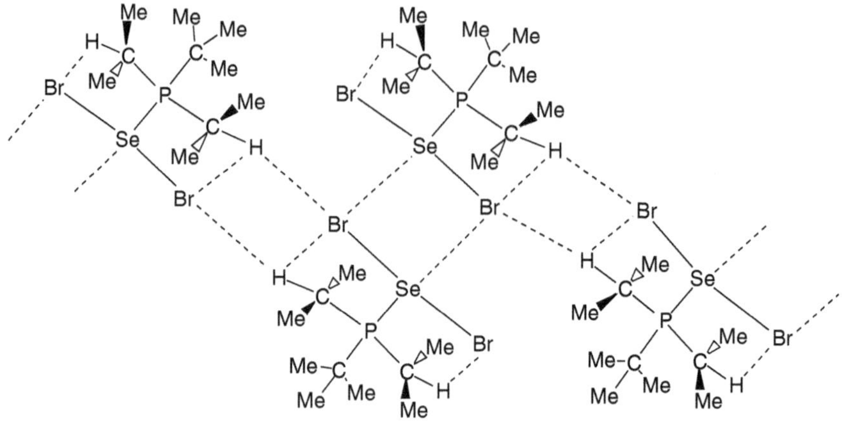

Figure 12.1.28 Se···Br, Br···Br, and CH···Br contacts in solid *t*Bu*i*Pr_2PSeBr_2.

distances. There is, however, no straightforward correlation of intermolecular interactions (Se···Br, Br···Br, Se···H, Br···H), steric effects, and non-equidistance of Se–Br bonds in the series of related trialkylphosphane selenide dibromides.

12.1.3.3 Supramolecular Structures of Complexes Between Divalent Tellurium Electrophiles and Chalcogen or Halogen Donors

Arenetellurenyl halides behave as dihalogen-like electrophiles: various ligands are known to coordinate at tellurium, generating in many cases T-shaped arrangements CTeXY with linear Y–Te–X units (Y = ligand with O, S, Se, or Te donor atom, X = Cl, Br, I).[9,77] Molecular tellurium dihalides are not stable at room temperature, but coordination with donor molecules Y has led to a number of square-planar complexes Y_2TeX_2.

12.1.3.3.1 Supramolecular Structures of Arenetellurenyl Halide Complexes with Chalcogen Donors

Solid T-shaped Y–Te(C)–X complexes such as tuTe(Ph)Cl (tu = thiourea) tend to exhibit dimeric structures through pairs of weak halide bridges; related weak cation–anion interactions occur in the 2 : 1 complex $[tu_2Te(Ph)]^+$ Cl^-. The weak contact is generally that *trans* to the Te-C bond ('*trans* effect' of the strong Te–C bond).[78,79]

The complex of dimesityltelluride with mesitylenetellurenyl iodide exhibits an interplay of Te···Te and Te···Arene interactions between linear Te–Te–I moieties leading to dimers that are linked through I···I contacts (Figure 12.1.29).[80]

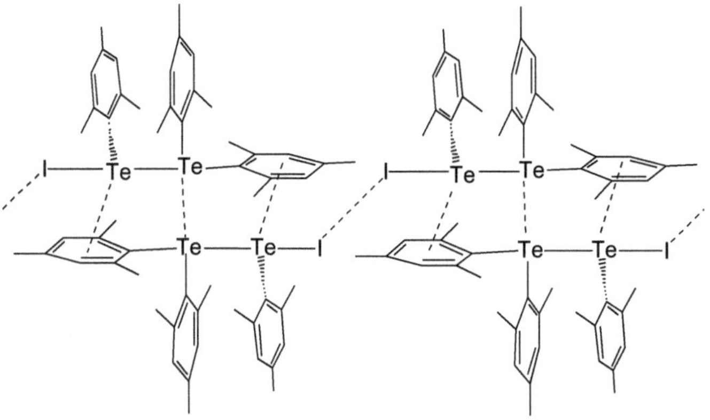

Figure 12.1.29 Supramolecular structure of Me_2Te–MesTeI.

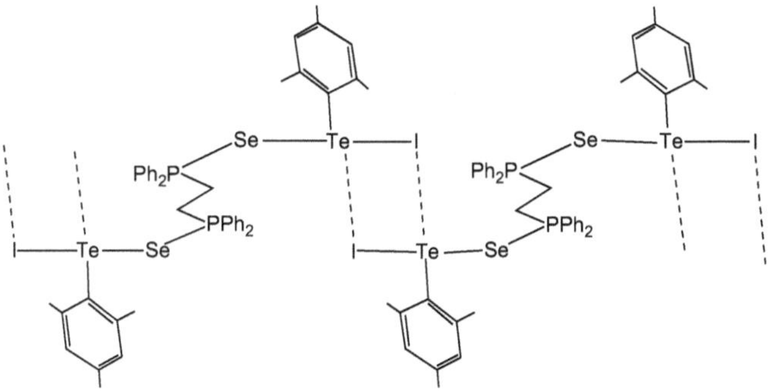

Figure 12.1.30 T-stacking in $dppeSe_2(MesTeI)_2$.

Supramolecular arrays involving T–T interactions can be generated by the use of bridging-bifunctional donors such as $Ph_2P(=Se)CH_2CH_2P(=Se)Ph_2$ ($dppeSe_2$). In solid $dppeSe_2[Te(I)Mes]_2$ the bidentate phosphane selenide ligand bridges two T-shaped SeTe(Mes)I moieties (Figure 12.1.30) to give centrosymmetric binuclear complexes.[81] The molecules are packed in a way that allows side-on intermolecular soft–soft interactions in pairs of approximately linear Se–Te–I units, thus linking the molecules into chain-like supramolecular arrays. The Te···I distances (399 pm; so-called 'T-T' interactions) between T-shaped CTe(Se)I moieties of neighbouring molecules are in the order of the sum of van der Waals radii of tellurium and iodine.

12.1.3.3.2 Supramolecular Structures of Tellurium Dihalide Complexes with Chalcogen Donors

The 1 : 1 complex of $dppeSe_2$ with TeI_2 crystallizes as a helical coordination polymer containing $dppeSe_2$ ligands between *trans*-configurated TeI_2 moieties,[81] whereas the 1 : 1 complex of dppm with TeI_2 crystallizes as a supramolecular

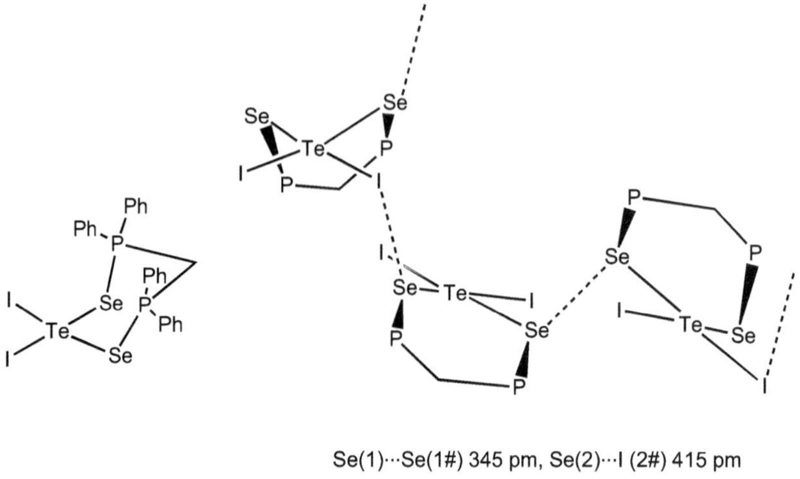

Figure 12.1.31 Left: molecular structure of $dppmSe_2TeI_2$; right: chain formation through intermolecular Se···Se and Se···I contacts.

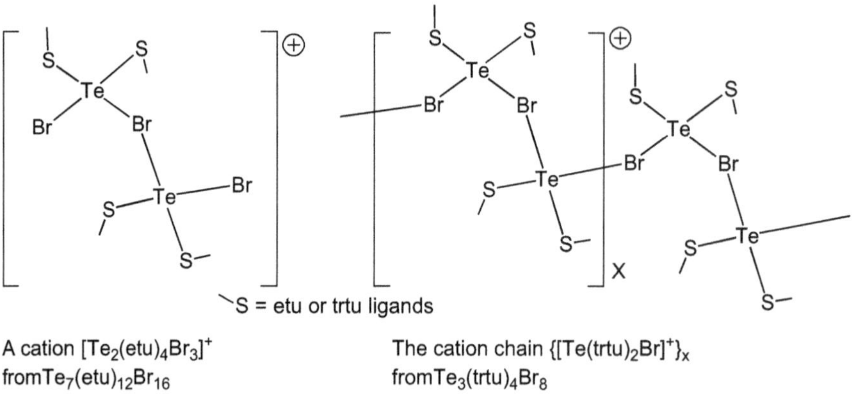

Figure 12.1.32 Topology of the cation $[Te_2(etu)_4Br_3]^+$ and of the cationic chain $\{[Te_2(trtu)_4Br]^+\}_x$.

chain of *cis*-chelate units that exhibit alternating intermolecular Se···Se and Se···I contacts (Figure 12.1.31).

In the compounds $Te_7(etu)_{12}Br_{16}$ [etu = ethylenebis(thiourea) and $Te_3(trtu)_4Br_8$ [trtu = trimethylenebis(thiourea)] bromide ions are abstracted by a stronger Lewis acid ($TeBr_4$) and the coordinative needs of Te(II) (preferring the 12-Te-4 configuration) in the cations $[Te_2(etu)_4Br_3]^+$)] (Figure 12.1.32, left) and coordination-polymeric $\{[Te_2(trtu)_4Br_2]^{2+}\}_x$ (Figure 12.1.32, right) are met by bromide bridging.[82]

In the salt $[Te(etu)_4]^{2+}$ $[TeCl_6]^{2-}$, cations and anions are packed in stacks to form infinite arrays in such a way that two opposite chlorine atom from octahedral $TeCl_6{}_-^2$ exhibit contacts (378 pm) with tellurium atoms of the cations.[83] Related cases of very unsymmetric halide bridges between tellurium atoms are the above-mentioned $[(C_6H_5CH_2)(C_2H_5)_3N]_4[TeI_4][TeI_6]$[7] (Figure 12.1.1) and triorganyltelluronium halogenotellurate salts (see Te^{IV} compounds, Figure 12.1.38).

12.1.3.3.3 Supramolecular Structures of Arenetellurenyliodide Complexes with Iodide Ions and Mixed-Valence Complexes

From *in situ* prepared deep blue 2,6-dimethylbenzenetellurenyl iodide ($2,6\text{-}Me_2C_6H_3TeI$) and pyridinium iodide, a dimeric 1:1 adduct $\{pyH^+[RTeI_2]^-\}_2$ and a polymeric 2:1 adduct $\{pyH^+[RTe(I)\text{-}I\text{–}Te(R)I]^-\}_n$ (R = $2,6\text{-}Me_2C_6H_3$) were prepared, depending on the amount of pyridinium iodide.[84] The anion of the 2:1 compound contains almost linear I(1)–Te–I(2) moieties [Te–I(1) 284.8 pm; Te–I(2) 311.7 pm] symmetrically bridged by angular μ-I(2) (148.56°). The terminal Te–I bonds of V-shaped I–Te–I–Te–I moieties exhibit pairwise secondary Te···I contacts, leading to a one-dimensional assembly (Figure 12.1.33).[84]

A mixed-valence compound of the composition $[(RTeI)(RTeI_3)]$ (R = $2,6\text{-}Me_2C_6H_3$) was obtained as black crystals on the addition of one equivalent of potassium iodide and a half equivalent of molecular iodine to *in situ* prepared $2,6\text{-}Me_2C_6H_3TeI$.[84] The solid compound contains cyclic dimers $[R(I)_2Te\text{–}I\text{–}Te(R)I]_2$ consisting of molecular units $R(I)_2Te\text{–}I\text{–}Te(R)I$ (Figure 12.1.34). Being linearly surrounded by the two tellurium atoms

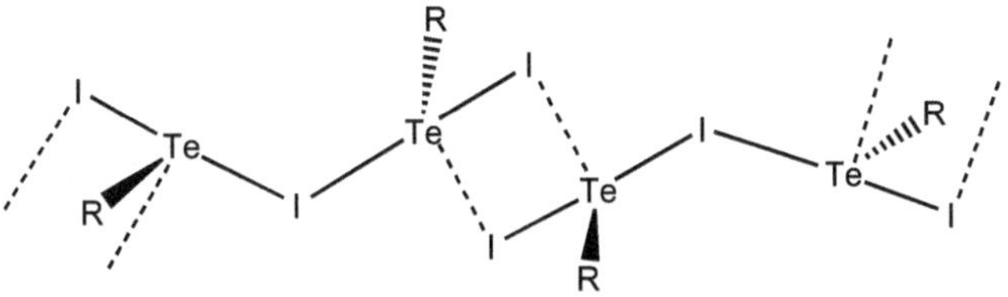

Figure 12.1.33 One-dimensional assembly of $\{pyH^+[RTe(I)\text{-}I\text{–}Te(R)I]^-\}_n$ (R = 2,6-$Me_2C_6H_3$).

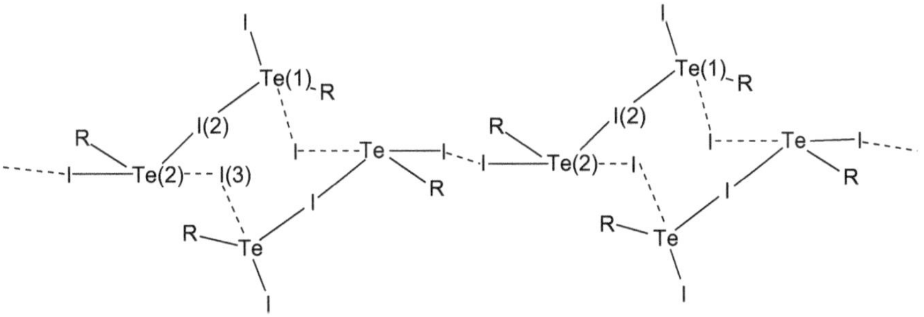

Figure 12.1.34 Polymeric assembly of $[(RTeI)(RTeI_3)]_n$ (R = $2,6\text{-}Me_2C_6H_3$).

[I(2)–Te(2) 291 pm, I(2)–Te(1) 302 pm], the central iodine atom I(2) of this unit apparently acts as an acceptor toward two RTeI molecules. Two of the cationic moieties $[R(I)Te]_2I^+$ are connected by two bridging iodide anions [I(3)···Te(2) 316 pm, I(3)···Te(1#) 327 pm], leading to the centrosymmetric dimer. Alternative descriptions would involve I(2) covalently bonded to Te(2) (as part of a $RTeI_3$ molecule), in turn acting as an acceptor towards Te(1) from the tellurenyl iodide RTeI, or I(2) could be regarded as part of a diiodotelluronium ion $RTeI_2^+$; Te(2), the centre of a diiodotellurate(II) anion $RTeI_2^-$, would then act as donor towards I(2). Iodine–iodine contacts (381.7) between the cyclic moities lead to a polymeric assembly.[84]

A related reaction of *in situ* prepared mesitylenetellurenyliodide (MesTeI) with potassium iodide furnished a compound of the composition $(MesTe)_3I_5$, formally a mixed-valence tellurium compound $(MesTeI)_2(MesTeI_3)$.[85] The chain-like polymeric network can be regarded as a central chain $[MesTeI_2^+, I^-, MesTeI]_n$ with a side chain MesTeI molecule coordinated to the μ_3-bridging iodide ion of the main chain. The iodide ion also acts as a donor towards the Te atom of the main chain RTeI unit and the Te atom of the $MesTeI_2^+$ cation (Figure 12.1.35). Since the Te atom of the main chain RTeI unit is also a donor towards an iodine atom of the nearest $MesTeI_2^+$ cation (slightly asymmetric linear Te–I–Te), both Te atoms of the main chain can be classified as (10–Te_4) (Ψ-trigonal-bipyramidal coordination sphere), whereas the side chain Te atom remains (10-Te_3, also Ψ-tbp), *i.e.* clearly Te(II). In short, the MesTeI units of the main chain are connected by linear ('I^+') and angular (I^-) atoms, leading to an array of Ψ-tbp $MesTeI_3$ units (Figure 12.1.35).[85]

When $[Co(bipy)_3]I_3$ was added to *in situ* prepared MesTeI, the compound $[Co(bipy)_3][(MesTeI_2)_2(\mu\text{-}I_2]I_3 \cdot 2\ CHCl_3$ was isolated as red crystals. The tellurium atoms of two diiodotellurate(II) anions are connected by a linearly bridging iodine molecule (I–I 291.4 pm, Te···I 322.5 pm). One iodine atom of each diiodotellurate(II) anion exhibits an I···Cl contact (385.6 pm) to a trichloromethane molecule. A pair of $CHCl_3$ molecules 'embraces' the single triiodide anion through I···Cl contacts (386.2 pm), generating a meandering chain made from $MesTeI_2^-$, I_2, $MesTeI_2^-$, $CHCl_3$, I_3^-, $CHCl_3$, $MesTeI_2^-$, and so on (Figure 12.1.36).[15]

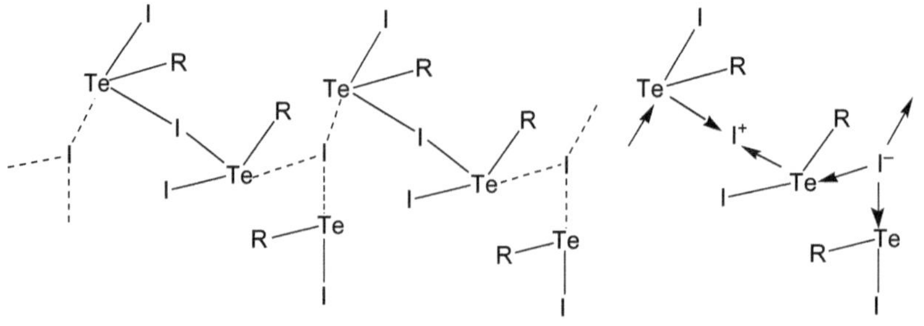

Figure 12.1.35 $Mes_3Te_3I_5$ ['$(MesTeI_3)(MesTeI_2)_2$']; left: chain topology; right: donor–acceptor relations within a formula unit.

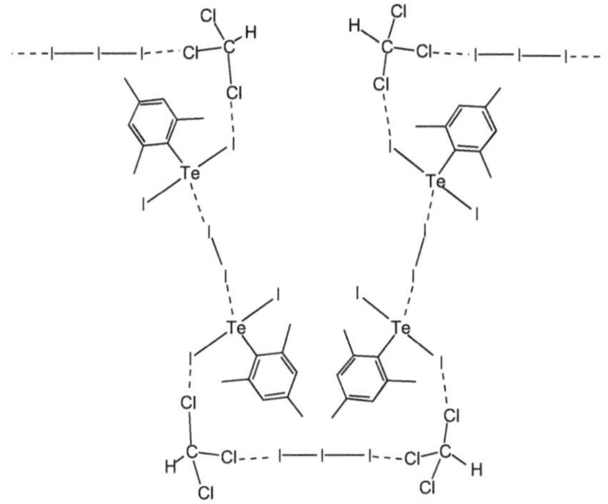

Figure 12.1.36 Anion network of $[Co(bipy)_3][(MesTeI_2)_2(\mu\text{-}I_2)]I_3 \cdot 2\ CHCl_3$.

12.1.4 Supramolecular Aspects of Structures from Tetravalent Chalcogen–Halogen Compounds

12.1.4.1 Cation–Anion Interactions Leading to Supramolecular Structures of Selenonium and Telluronium Salts

Most compounds R_3EX can be described as essentially ionic, exhibiting structures that involve directed n(halide)→σ*(E-C) contacts between pyramidal R_3E^+ cations and X^- anions.[3] Solid sulfonium and selenonium iodides exist as ion pairs with directed cation–anion contacts, so that the coordination environment of S and Se can be regarded as Ψ-trigonal-bipyramidal.[86]

In telluronium cations, however, the central Te atom tends to adopt higher coordination numbers, leading to chain-like polymers or cyclic aggregates.[87–90] In $Me_3TeCl \cdot H_2O$, tellurium undergoes two Te···Cl contacts approximately *trans* to Te–C and one further long Te···O contact (308.9 pm) as well as hydrogen bridging from H_2O to Cl leading to infinite chains. Solid Ph_3TeCl $^1/_2CHCl_3$ is a stair-like tetramer, whereas unsolvated Ph_3TeCl is based on dimeric units.[88] The solid fluorinated telluronium salt $(C_6F_5)_3TeCl$ (Figure 12.1.37, left) is an unsymmetrically chloride-bridged chain polymer.[90]

Even weakly coordinating anions such as $RTeX_2^-$, $RTeX_4^-$, and TeX_6^{2-} form directed anion–cation contacts with triorganyltelluronium ions, leading to dimers, oligomers, and extended chains. In Me_3Te^+ $MeTeI_4^-$ (=‘ß-Me_2TeI_2’),[91] the tellurium atom of the cation exhibits three strong Te–C bonds (201, 208, and 213 pm) and three Te···I secondary contacts (Te···I 384, 397, 400 pm), the tellurium atom of the anion participates in one (apical) TeC bond (215 pm), four (basal) TeI bonds (284, 289, 295, and 298 pm), and one secondary contact (Te···I 388 pm).[91] The structure of the solid mixed anion

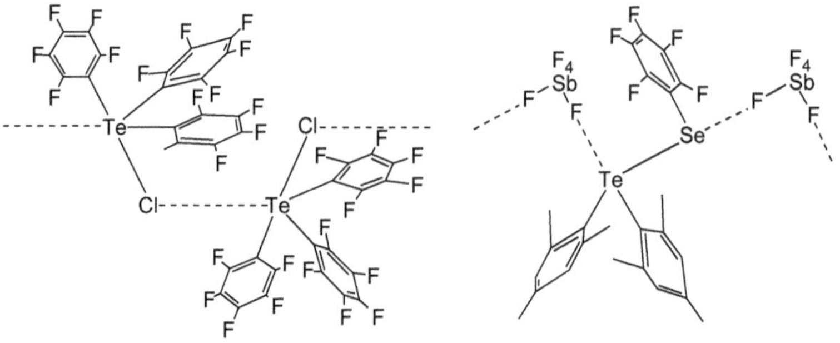

Figure 12.1.37 Left: anion bridging in solid tris(pentafluorophenyl)telluronium chloride; right: Anion bridging in (pentafluorophenylseleno)telluronium hexafluoroantimonate.

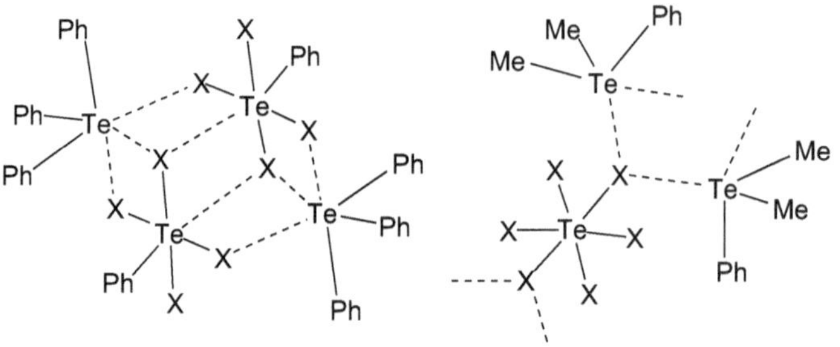

Figure 12.1.38 Left: centrosymmetric dimers of Ph_3Te^+ $PhTeX_4^{2-}$; right: asymmetric unit of $(PhMe_2Te^+)_2$ TeX_6^{2-}.

compound $(Ph_3Te^+)_4(MesTeI_2^-)_2(I^-)_2$ is closely related to that of tetrameric Ph_3TeX.[92]

Bridging hexahalogenotellurate(IV) dianions TeX_6^{2-} (X = Cl, I) act as tridentate donor towards two equidistant Ph_3Te^+ cations in ion triples $(Ph_3Te^+)_2$ TeX_6^{2-}.[93] In the solid compounds Ph_3Te^+ $PhTeX_4^-$ (X = Cl, Br, I) centrosymmetric dimers (Figure 12.1.38, left) involving cation–anion and anion–anion Te···X contacts are packed in layers. C-H···X contacts also play a role.[94] In $(PhMe_2Te^+)_2$ TeX_6^{2-} (X = Cl, Br), however, the dianions are linked by the telluronium cations in arrays of chains via secondary Te···X bonds (Figure 12.1.38, right).

Solid Ph_3Te^+ I_3^- exhibits a ribbon-like supramolecular structure, involving double chains of alternating anions and cations (Figure 12.1.39).[95] From each triiodide ion one terminal iodine atom (I3) is in contact with one telluronium ion (Te···I3′ 384 pm) whereas the other terminal iodine atom (I1) connects two different telluronium ions (Te···I1 377 pm and Te···I1′ 395 pm). The small angle Te–I1–I2 (71.8°), together with a van der Waals-like contact Te···I2 (401 pm) might indicate a kind of η^2-coordination of the I1–I2 unit with tellurium.

In telluronium ions with α-heteroatoms, such as Mes_2TeBr^+ SbF_6^-, $Mes_2TeTeMes^+$ SbF_6^-, or $Mes_2TeSeC_6F_5^+$ SbF_6^- (Figure 12.1.37, right), both

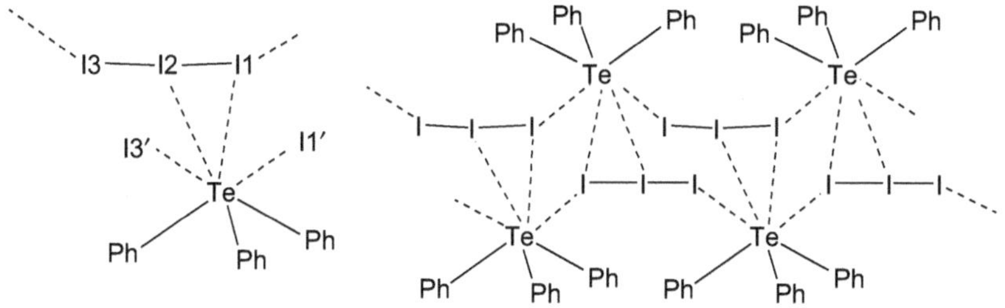

Figure 12.1.39 Ph_3Te^+ I_3^-. left: atom numbering of Ph_3TeI_3; right: ribbon structure.

the tricoordinated tellurium atom and the α-heteroatom exhibit electrophilic behaviour towards the hexafluoroantimonate couteranion, leading to rings or chains of ion pairs.[96]

12.1.4.2 Some New Supramolecular Aspects of Diorganylselenium and Tellurium Dihalide Structures

Compounds R_2EX_2 are in many cases dimers or coordination polymers;[3,6,89,97] the individual structures are influenced by electronic, steric, and π-stacking effects of the ligands at Se or Te. In the series of compounds Me_2SeX_2, the dichloride and the dibromide exhibit long-range Se···X contacts between discrete Ψ-tbp (10-Se_4) units (Cl 341–361 pm; Br 360 and 363 pm), whereas the diiodide is a coordination compound of Me_2Se with linear Se–I–I units (Se–I 276.8, I–I 291.6 pm) and intermolecular Se···I contacts (391 pm).

The comparison of *para*-substituted diaryltellurium dihalides (p-$Me_2NC_6H_4$)(p-YC_6H_4)TeX_2, and partially fluorinated compounds $R^F{}_2TeX_2$, allows one to recognize the interplay and the competition of π-stacking, Te···X, or X···X contacts (Figure 12.1.40).[98,99]

The exceptionally bright yellow colour of (p-$Me_2NC_6H_4$)$_2$$TeCl_2$,[100] and the metallic lustre of dark red or violet compounds R_2TeI_2, correlates well with halogen–halogen contacts in linear X–Te–X···X–Te–X··· arrays (Figure 12.1.40 top right),[101] different from the usual Te–X···Te bridging as in Ph_2TeX_2 (Figure 12.1.40, top left).[2,3] Extended chains with X···X contacts were not observed in mixed substituted compounds ArAr′$TeCl_2$. With Ar = p-EtOC_6H_4-, Ar′ = p-$Me_2NC_6H_4$- the packing of the monomeric molecules involves intermolecular Te···π(aryl) contacts (Figure 12.1.40, bottom left); with Ar = C_6H_5, Ar′ = p-$Me_2NC_6H_4$ a Te-X···Te bridge and a Te–Ph(π)···Te contact connect two molecules (Figure 12.1.40, bottom right). Interdimer slipped sandwich-like pairs of aminoaryls lead finally to centrosymmetric moieties of four molecules.[98]

π-Donating NMe_2 or OEt groups in the aryl substituents favour aryl stacking against halide bridging whereas electron-withdrawing fluorine atoms enhance the Lewis acidity of tellurium in $(R_F)_2TeX_2$, leading to chains with secondary Te···X contacts (X = F, Cl, Br).[99] In solid (4-$CF_3C_6F_4$)$_2$$TeF_2$ chains of doubly fluoride-bridged dimers are present. *Ortho*-fluorine atoms of the aryl

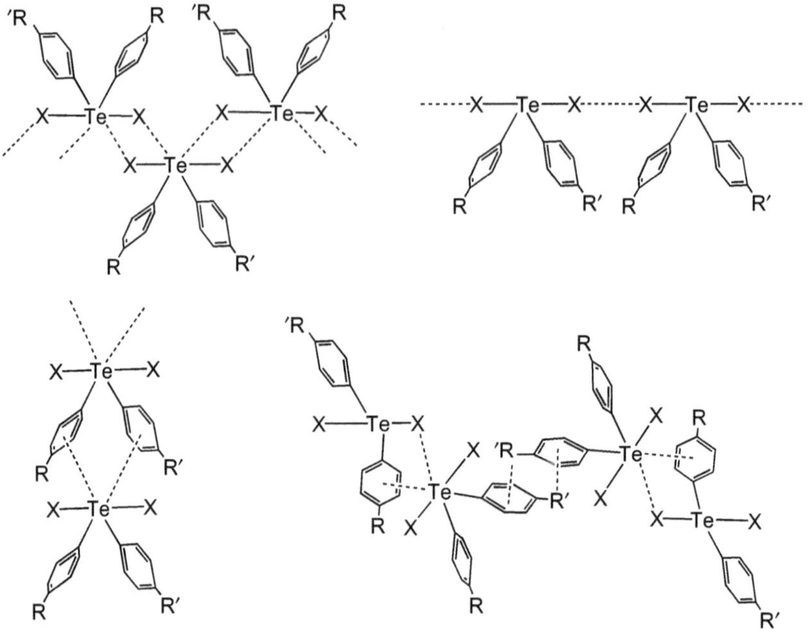

Figure 12.1.40 Four types of packing of compounds Ar_2TeX_2. Top left: Ph_2TeX_2; top right: $(p\text{-}Me_2NC_6H_4)_2TeX_2$ (X = Cl, Br, I); bottom left: $ArAr'TeCl_2$, Ar = p-$EtOC_6H_4$-, Ar' = p-$Me_2NC_6H_4$-; bottom right: $ArAr'TeCl_2$, : Ar = C_6H_5, Ar' = p-$Me_2NC_6H_4$-.

substituents contribute to tellurium coordination through *intra*molecular F···Te contacts in the range of 320–330 pm.[99]

In compounds $RR'TeI_2$ there is apparently no straightforward correlation between Te–I bond expansion and intermolecular Te···I and/or additional I···I interactions, since other contributions such as aryl stacking and CH···I contacts can also play a role. Te···I and I···I contacts can range from about 370 pm (weak secondary bonding) to 430 pm (about the van der Waals distance).[102–105]

A typical example is the cyclic dialkyltellurium diiodide $C_4H_7(CH_3)TeI_2$, which consists of asymmetric dimers (Te–I 283–304 pm, Te···I 410 pm) that are linked in a supramolecular network by additional I···I contacts (425 pm).[102] The incomplete Te–C cleavage of cyclic $C_4H_8TeI_2$ with an aniline-related dithiocarbamate furnished a few single crystals of solid {*trans*-$[(PhNH)_2CS]_2TeI_2 \cdot 4C_4H_8TeI_2$}. In this material, the $C_4H_8TeI_2$ molecules exhibit linear Te–I···I–Te contacts (366.3 and 391.1 pm), but the two different TeI_2-containing species are connected through Te···I (384.8 and 388.0 pm) interactions.[103]

From $Ph(Me_3SiCH_2)TeI_2$ two polymorphs were analysed. Polymorph **A** (triclinic) exhibits linear arrays of molecules with relatively strong I···I interactions (380 pm) and a 3D CH···I hydrogen bond network. Polymorph **B** (monoclinic) is composed of two alternating layers of molecules. In one type of layer, the molecules form a continuous quasi-2D puckered network through alternating Te···I and I···I interactions (similar to the sequence of T-stacking and Br···Br contacts in $i\text{Pr}_3\text{PSeBr}_2$, phase #2, see Figure 12.1.26), whereas the other layer consists of dimers with pairs of Te···I contacts.[104]

12.1.4.3 Supramolecular Association in Aryltellurium(IV) Halide Structures

A number of compounds of the types $RTeX_3$, $RTeX_4^-$, $RTeX_5^{2-}$ and related species were studied in classic papers on secondary bonds from N. W. Alcock *et al.*,[1] J. D. McCullough *et al.*,[6] and S. Husebye *et al.*[11,106] Substituent effects influencing the interplay of X···X, Te···X, and π stacking are a current field of research. In the following, a number of supramolecular structures of less usual compounds $RTeX_2TeR$, cations $RTeX_2^+$, and anions $RTeX_4^-$ will be discussed.

12.1.4.3.1 Supramolecular Ditelluride-1,1-Dihalide Structures

Bis(2,6-dimethoxyphenyl)ditelluride reacts with thionyl chloride and with molecular bromine or iodine yielding mixtures of ditelluride-1,1-dihalides $RTeTeX_2R$ [R = 2,6-$(MeO)_2C_6H_3$; X = Cl, Br, I],[107] in part accompanied by aryltellurium trihalides $RTeX_3$ (X = Cl, Br). The latter exhibit in the solid state halide-bridged chain structures (Te···Cl 338.7 pm, Te···Br 347.3 pm). In case of the iodide, the reaction has to be carried out at −10 °C in CH_2Cl_2, otherwise Te–Te bond cleavage and dismutations can take place, leading to RTeI or to $[R_4Te_2^{2+}][Te_4I_{14}^{2-}]$.[67] Solid $RTeTeCl_2R$ exists as a coordination polymer through secondary Te···Cl contacts (378.4 pm; Te1–Cl···Te2 121°) between one of the Cl atoms and the dicoordinated tellurium atom Te2. The bromine and iodine derivatives do not exhibit intermolecular Te···X distances that are under 400 pm; 'pseudodimeric association' was ascribed to centrosymmetric moieties $(RTeTeI_2R)_2$ with d(Te···I) = 401.6 pm.[107] The tetravalent tellurium atom of $RTeTeCl_2R$ [R = 2,6-$(MeO)_2C_6H_3$] does not participate in intermolecular Te···Cl contacts; it gains a coordination number of six through Te–C, Te–Te, two Te–Cl, and two intramolecular secondary Te···O bonds (Figure 12.1.41, left), whereas in solid $S_2N_2TeCl_2$ (Figure 12.1.41, right) three secondary Te···Cl contacts per tellurium atom lead to seven-coordination of the tellurium atoms in a supramolecular double-chain structure.[108]

12.1.4.3.2 Aryltelluriumtriiodides and Tetrahalogenotellurates(IV) as Building Blocks in Supramolecular Assemblies

Aryltelluriumtriiodides can add iodide, iodine, or both (= triiodide), furnishing oligonuclear anions. In one view of the solid pydidinium salt $(PyH)[DurTeI_3(I_3)]$ the '*equatorial*' iodine atom (I3) of a Ψ-tbp (10-Te-4) $DurTeI_3$ molecule appears to be coordinated by one of the terminal iodine (I4)

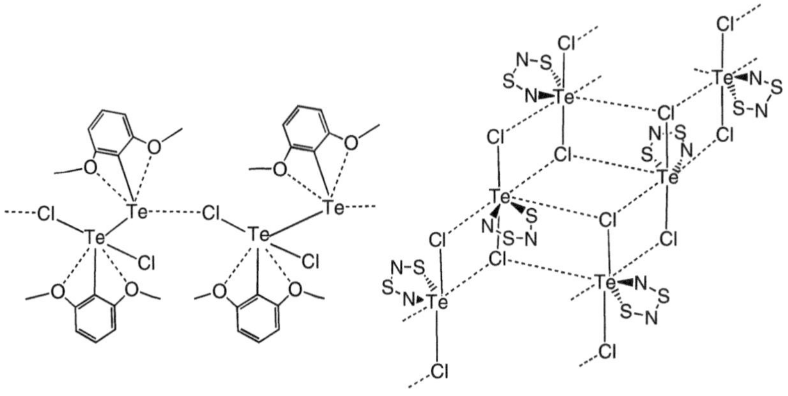

Figure 12.1.41 Left: Chain structure of $RTeTeCl_2R$ [R = 2,6-$(MeO)_2C_6H_3$-]; right: double-chain structure of solid $S_2N_2TeCl_2$.

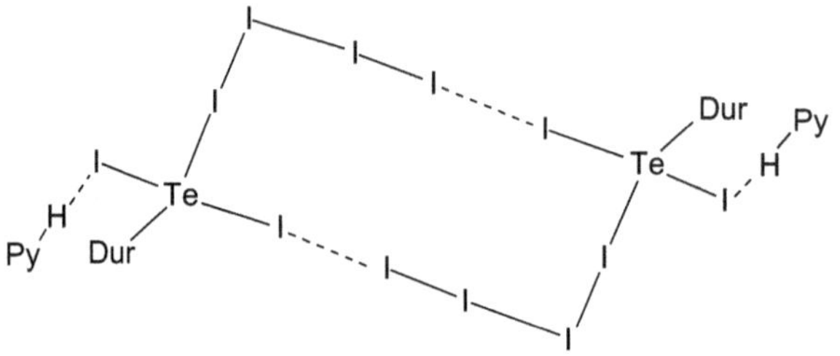

Figure 12.1.42 Dimers of $(PyH^+)[DurTeI_3(I_3^-)]$.

atoms of a very asymmetric I_3^- anion (I4–I5 306.3 pm, I5–I6 280.1 pm) (Fig. 12.1.42).[66] The distance I3–I4 (301.6 pm) is roughly the same as that of the adjacent bond in the triiodide. Alternatively, coordination of the divalent tellurium atom of a Ψ-tbp (10-Te-3) $DurTeI_2^-$ anion with an L-shaped I_4-unit could be imagined, leading to $[DurTeI_2\text{-}I_4]^-$ (10-Te-4, compare section 13.3.1, Figure 12.1.21: coordination of DurTeI with I_2). One of the apical iodine atoms of each $DurTeI_3(I_3)^-$ group exhibits a hydrogen bond to the pyridinium cation, the other one is in linear contact with the remaining terminal atom of the triiodide anion of a neighbouring $(PyH^+)[DurTeI_3(I_3^-)]$ moiety, leading to centrosymmetric dimers (Fig. 12.1.42).[66]

From related reactions of pyridinium iodide with *in situ* prepared MesTeI, the dimeric product $(PyH^+)_4[(MesTe^{II}I_2)_2(I_3)_2]$ and the polymeric compound $(PyH^+)_3\{[(MesTe^{IV}I_3)_2I](I_3)_2\}$ were isolated.[67] In the latter, two molecules of $MesTeI_3$ are connected through their equatorial iodine atoms with a bridging iodide anion, generating a V-shaped atom sequence Te–I(3)–I(4)–I(5)–Te (Figure 12.1.43). One of the apical iodine atoms of each $MesTeI_3$ molecule is weakly associated with the terminal iodine atom of a $PyH^+I_3^-$ unit, but of the other two apical Te–I functions, only one forms a further contact, being hydrogen-bonded with the third pyridinium cation of the asymmetric unit. The

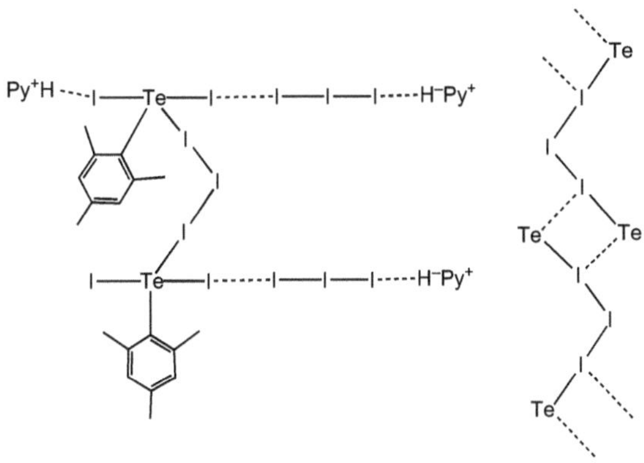

Figure 12.1.43 $(PyH^+)_3\{[(MesTe^{IV}I_3)_2I](I_3)_2\}$. Left: asymmetric unit; right: excerpt from the chain of V-shaped units.

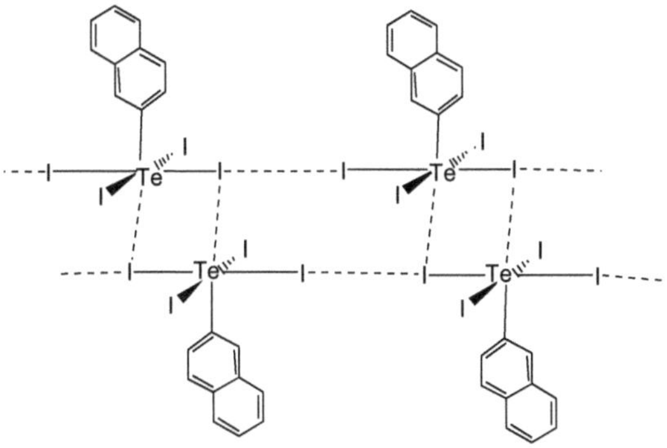

Figure 12.1.44 Ribbons of $[(\beta\text{-Naphtyl-TeI}_4^-)]_2$ units in solid $Et_4N(\beta\text{-Naphtyl-TeI}_4)$.

two equatorial Te–I functions from the V-shaped units undergo pairwise Te···I interactions (Figure 12.1.43, right), generating a chain network surrounded by the mesityl groups.[67]

The bridging of two $RTeI_3$ molecules by an iodide anion through I–I coordination (*not* I–Te coordination) is a very unusual case. Generally, $RTeX_3$ molecules add halides, generating $RTeX_4^-$ anions that exhibit square-pyramidal structures with the organic group R in apical position, and interanion secondary Te···X bonding is observed in most cases,[101,109–113] for instance in $Et_4N(\beta\text{-Naphtyl-TeI}_4)$ which consists of dimers, connected by Te···I contacts, which pack in linear arrays with I···I contacts (Figure 12.1.44).[113]

In this context, Schulz Lang *et al.* studied the role of cations in interanion attractive interactions in tetrahalogeno(phenyl)tellurate(IV) salts with help of a number of structure determinations.[109–112] From methanol solutions, the caesium salts $\{Cs[PhTeCl_4](CH_3OH)\}$ and $Cs[PhTeBr_4]$ were isolated.[109] In the former compound the $PhTeCl_4^-$ anions exist as parts of inversion symmetric Cl-bridged dimers (Te···Cl 367 pm). Pairs of caesium atoms are Cl^- and CH_3OH-bridged (Cs···O 311 pm; Cs···Cl 384 pm, see Figure 12.1.45). Altogether,

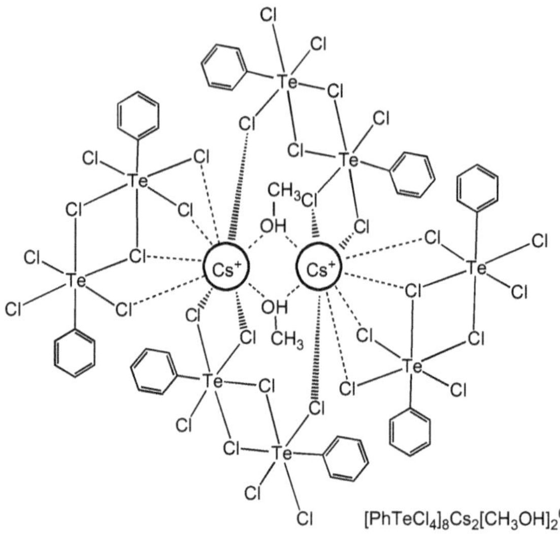

$[PhTeCl_4]_8Cs_2[CH_3OH]_2^{6-}$

Figure 12.1.45 Cation-anion and anion-anion interactions in $\{Cs[PhTeCl_4]CH_3OH\}$.

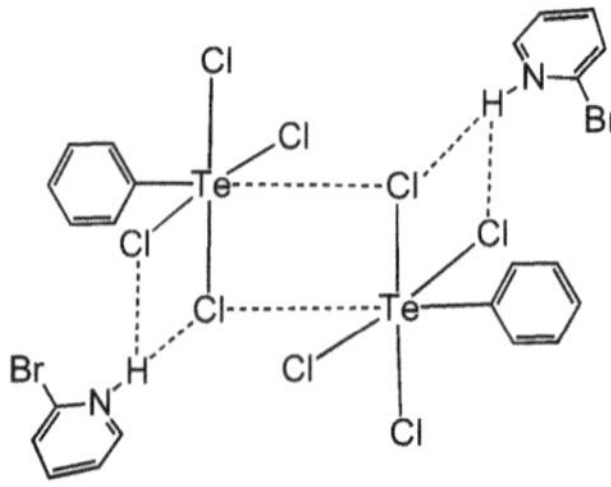

$\{[2\text{-}Br\text{-}C_5NH_5][PhTeCl_4]\}_2$

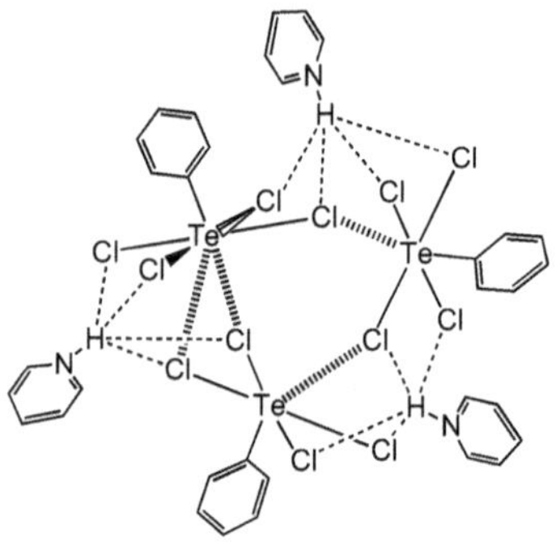

$\{[C_5NH_6][PhTeCl_4]\}_3$

Figure 12.1.46 Bromopyridinium (left) and pyridinium salts (right) of the $PhTeCl_4^-$ anion.

each caesium atom is in contact with eight chloride ions from five different $PhTeCl_4^-$ anions (Cs···Cl 349–384 pm) and with the oxygen atoms of two methanol molecules.

In $Cs[PhTeBr_4]$, which is solvent-free, 'chelating' bromine atoms of four single $PhTeBr_4^-$ anions surround each Cs^+ ion, achieving a quasi-pentagonal prism (seriously distorted dodecahedral, CN 8 at Cs). The pentagonal prisms are linked through faces and edges alternatively (Br–Cs–Br links).[109] Using 2-bromopydridinium cations, typical pairs $(PhTeCl_4^-)_2$ are formed; bifurcated hydrogen bonds exist between an acidic hydrogen atom of the cation and two Cl atoms (*cis* to each other), a terminal and a bridging one (Figure 12.1.46,

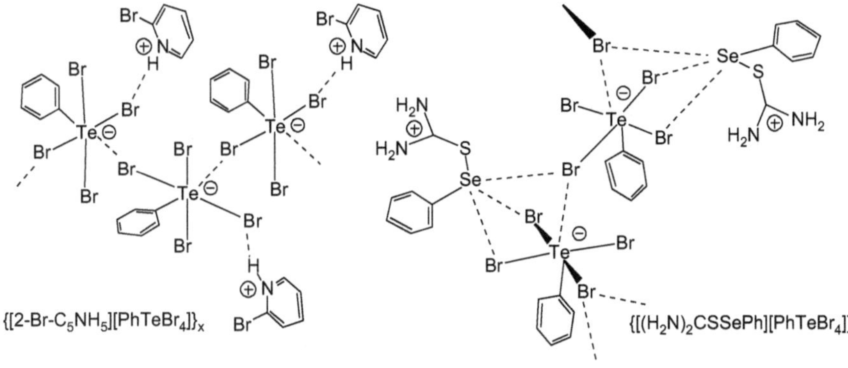

Figure 12.1.47 Left: N-H···Br contacts to $PhTeBr_4^-$ chains ; right: Se···Br contacts to $PhTeBr_4^-$ chains.

left).[110] With pyridinium cations instead of Cs^+, the $PhTeCl_4^-$ anions adopt a trimeric structure with two halogen atoms of each monomer forming secondary bonds to an adjacent tellurium atom. Four hydrogen bonds exist between the pyridinium proton and two pairs of chlorine atoms (Figure 12.1.46, right). In their 2-bromopyridium salts the $PhTeBr_4^-$ and $PhTeI_4^-$ anions aggregate as chain polymers.[111] In the iodide case, the iodine atom *trans* to that participating in the hydrogen bridging contact to the cation is terminal; in the bromide structure the bromine atom *trans* to that participating in the hydrogen bridging contact is part of Te-Br···Te bridge (Figure 12.1.47, left). In a related salt with the thiourea-derived diamino(phenylselenothio)carbenium cation,[112] the Se atom in β-position to the carbenium centre [Figure 12.1.47, right; compare Figure 12.1.4 (top) and 12.1.5] undergoes acceptor–donor interactions with two neighbouring Br atoms from one $PhTeBr_4^-$ anion (Se···Br 371 and 393 pm), and with one bridging Br atom from the next $PhTeBr_4^-$ anion (Se···Br 368 pm).[112]

12.1.4.4 Halide Bridging in Supramolecular Telluroxane Structures

Solid $Te_2Cl(OMe)_7$, formally an adduct of the two telluranes $TeCl(OMe)_3$ and $Te(OMe)_4$, exhibits two Te–O(Me)–Te bridges: one is very short and nearly symmetric (Te-O 226.8, 225.3), whereas the other is very asymmetric (Te–O 191.0, 307.8) (Figure 12.1.48).[114] The symmetrically bridging oxygen atom is at the apical position of the Ψ-tbp sawhorse structure-like $Te(OMe)_4$ unit from Te(2), but basal position with respect to the Ψ-octahedral square pyramid around Te(1). Further weak Te···O interactions lead to chains of the $Te_2Cl(OMe)_7$ moieties. Linear Te–Cl···Te contacts link these chains to each other. In CD_2Cl_2 solution at room temperature, rapid alkoxy group scrambling is indicated by ^{1}H- and ^{125}Te-NMR (Fig. 12.1.48).

The role of intramolecular n→σ* donor–acceptor interactions in compounds $TeCl_2(OR)_2$ and $TeCl_3OR$ (R = alkyl with β-donor atom) has been

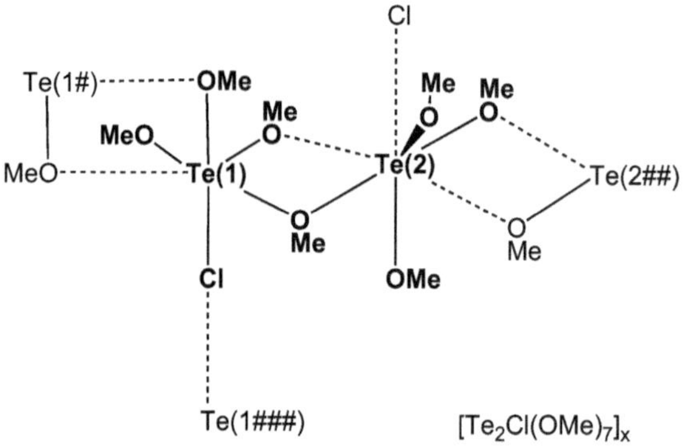

Figure 12.1.48 Ditellurium moiety and secondary bonding in $Te_2Cl(OMe)_7$.

elucidated experimentally and with help from *ab initio* calculations.[115] Solid $(MeOCH_2CH_2O)_2TeCl_2$ exhibits a network of intramolecular Te···O and intermolecular Te···O (267.4 pm) and Te···Cl (357.7 pm) contacts involving heptacoordinated tellurium.

The hexanuclear intramolecularly N-Te-coordinated cluster $(8\text{-}Me_2N\text{-}Napht\text{-}Te)_6O_8Cl_2$ (8-Me$_2$N-Napht = 8-dimethylaminonaphthyl) contains a central eight-membered cyclic moiety $(Te–O)_4$. The exocyclic Te–O bonds of two pairs of $8\text{-}Me_2N–Napht–TeO_3$ units (all Te coordination numbers 5) 'chelate' the two remaining 8-Me$_2$N–Napht–Te groups, each of which exhibits each a long contact to the terminal chloride ions (Te···Cl 307.4 pm). The octanuclear cluster $(8\text{-}Me_2N–Napht–Te)_6Te_2O_{12}Cl_2$ contains intracluster bridging chloride ions at the periphery of the cluster (Te···Cl 305.1 pm).[116]

The V-shaped units XTeOTeX of telluroxanes $[Ar_2Te(X)]_2O$ tend to exhibit intermolecular interactions through pairs of Te···X contacts. This leads to dimers for $[(4\text{-}MeO\text{-}C_6H_4)_2Te(I)]_2O$ and of $[Me_2(I)Te]_2O$[117] and to infinite chains for $[(4\text{-}MeO\text{-}C_6H_4)_2Te(Cl)]_2O$ and of $[4\text{-}Me_2N\text{-}C_6H_4Te(Cl)]_2O$.[118] In dimers of $[(4\text{-}MeO\text{-}C_6H_4)_2Te(I)]_2O$[117] the four tellurium atoms of dicationic moieties $[(4\text{-}MeO\text{-}C_6H_4)_2Te)_4O_2I_2]^{2+}$ with approximately square units $[Te–O–Te–I\text{-}]_2$ are bridged by two μ_4-iodide ions. In dimeric $[Me_2(I)Te]_2O$ (JOMC) and in polymeric $[4\text{-}Me_2N\text{-}C_6H_4Te(Cl)]_2O$ association occurs through pairs of 'V–V' interactions of I–Te–O–Te–I units[117,118] which are comparable to those of the above-mentioned V-shaped atom sequences I–Te–I–Te–I and Te–I–I–I–Te (see Figure 12.1.33 and 12.1.43).

The 'hydroxo-hydrate' $(Me_2Te)_2O(I)(OH)$, H_2O[117] consists of chains of the types $[Me_2TeO\text{-}Me_2TeO]_n$ and $\{[TeMe_2O(H)\text{-}TeMe_2(OH)]^{2+}\}_n$ (Figure 12.1.49). The cationic groups $[TeMe_2(OH)TeMe_2(OH)]^{2+}$ are associated via short secondary Te···OH interactions. μ_2-Iodide ions link pairs of Te atoms within chains $[Me_2TeO\text{-}Me_2TeO]_n$, and μ_4-iodide anions link tellurium atoms within the cationic chain and between the two types of chains. The lengths of Te···I interactions vary between 366.7 and 419.3 pm.

Base hydrolylsis of $C_4H_8TeI_2$ leads to $(C_4H_8Te)_3O_2I_2$, which consists of chains $I–Te(C_4H_8)–O–Te(C_4H_8)–O–Te(C_4H_8)I$. Besides the in-chain primary

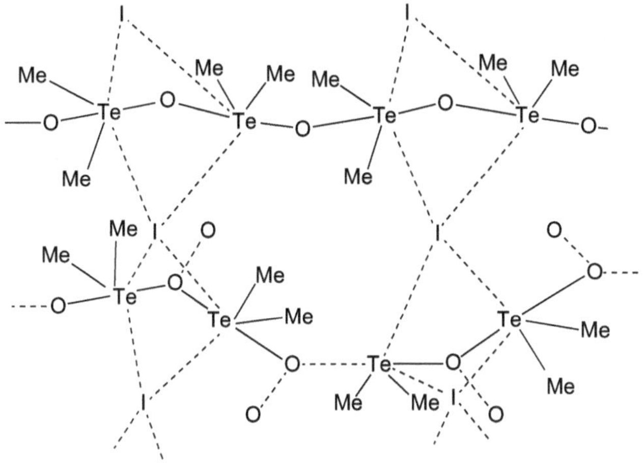

Figure 12.1.49 Polymer string of $(Me_2Te)_2O(I)(OH)$, H_2O.

Te–I bonds, a number of secondary Te···I contacts ranging from 363.6 to 403.1 pm lead to supramolecular association of the molecules.[117]

12.1.5 Base-Pair-Like Chalcogen–Halogen Arrays

12.1.5.1 Intermolecular Donor–Acceptor Interactions in Solid Iodophosphane Selenides

Iodine atoms adjacent to $\sigma^4\lambda^5$-phosphonium centres such as R_3PI^+ behave as soft electrophiles, phosphine selenide $R_3P{=}Se$ functions act as soft nucleophiles.[56,119] The combination of these two kinds of functional groups leads to novel kinds of base-pairing building blocks for supramolecular chemistry, when soft donor and acceptor functions are expressed in one type of molecule. The simplest combination of a $\sigma^4\lambda^5$–P–I acceptor function with a P=Se donor function within one molecule is expressed in iodophosphane selenides $R_nP({=}Se)I_{3-n}$.[56,119] A spatially extended version of the P=Se-donor/P–I-acceptor pair is represented by novel iminobisphosphane derivatives $R_2P({=}E)N{=}PR_2I$ (E = Se, Te) in which one four-coordinated phosphorus atom displays a P=Se or P=Te donor function and the other possesses a P–I bond adjacent to an ylidic P=N bond (iminophosphorane).[120]

The solid iodophosphane selenide [iPr$_2$P(=Se)I] consists of chains of molecules held together by secondary Se···I interactions (Figure 12.1.50, left).[119] Directed weak Se···H-C contacts interlink these chains. [(tBu$_2$P(=Se)I] crystallizes with two independent molecules. The compound can be described as a helical chain-like polymer when intermolecular P–Se···I–P contacts are taken into consideration. Bond angles and distances within the two molecules are very similar; the most significant deviation between molecules I and II is the slightly larger P–Se distance in molecule II (P1–Se1: 210.4(2) pm,

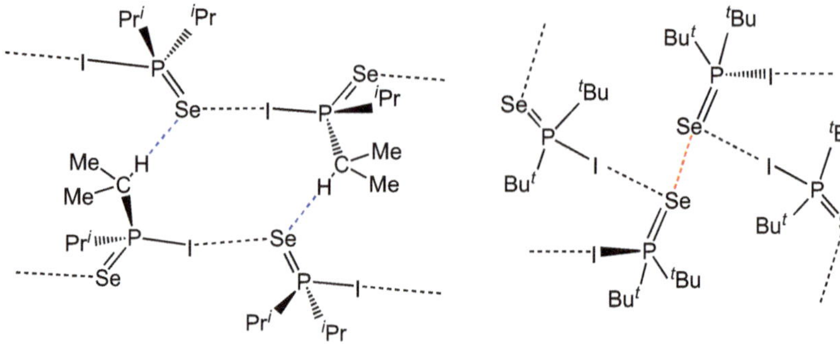

$^iPr_2P(=Se)I$ $^tBu_2P(=Se)I$

Figure 12.1.50 Left: PSe···I links and CH···Se crosslinks in *i*$Pr_2P(Se)I$; right: PSe···I links and Se···Se crosslinks in *t*$Bu_2P(Se)I$.

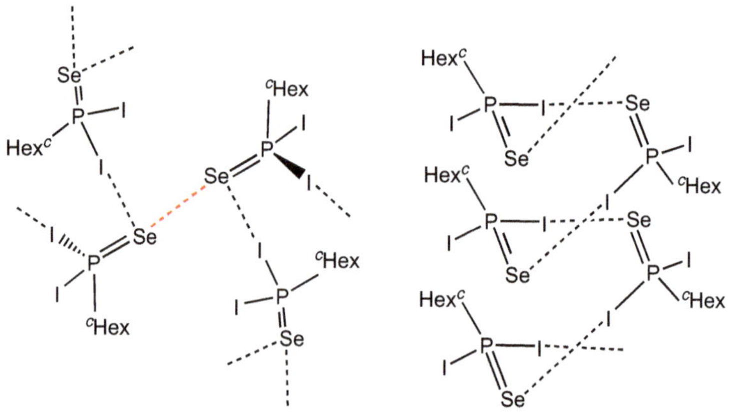

Figure 12.1.51 Left: PSe···I links and Se···Se crosslinks in *c*$Hex_2P(Se)I$; right: 'staircase' stacks in *c*$Hex_2P(Se)I$.

P2–Se2: 212.1(2) pm), which is involved in weak interchain Se···Se contacts [Se2···Se2′ 359.4(2) pm] (Figure 12.1.50, right).[56c]

The secondary Se···I contacts (Se1-I2′ 369 and Se2-I1 384.4 pm) are only slightly shorter than the sum of the van der Waals radii of Se and I. However, the 'L-shaped' geometry of the P=Se···I–P bridges, with the iodine atoms being approximately linearly coordinated, is consistent with the expectations for typical Se(donor)→I (acceptor) interactions [∠P1–I1–Se2 158.62(4)°, ∠P2–I2–Se1′ 167.68(4)°].

The related diiodophosphane selenide *c*-HexP(=Se)I_2 (Figure 12.1.51, left) contains a P=Se donor function; however, of the two electrophilic iodine atoms attached to the phosphorus atom, only one participates in soft–soft interactions. L-shaped P–Se···I–P donor–acceptor interactions (Se···I: 347.6 pm) form one motif of the supramolecular structure of *c*-HexP(=Se)I_2 and P–Se···Se–P contacts (361.2 pm) the other (Figure 12.1.45a). Staircase-like helices

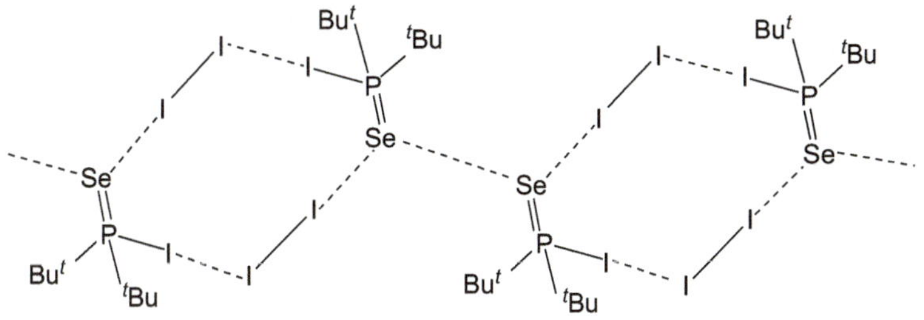

Figure 12.1.52 P–Se–I–I and I–I···I links in dimers $[t\mathrm{Bu_2P(Se{-}I{-}I)I}]_2$, and interdimer Se···Se contacts.

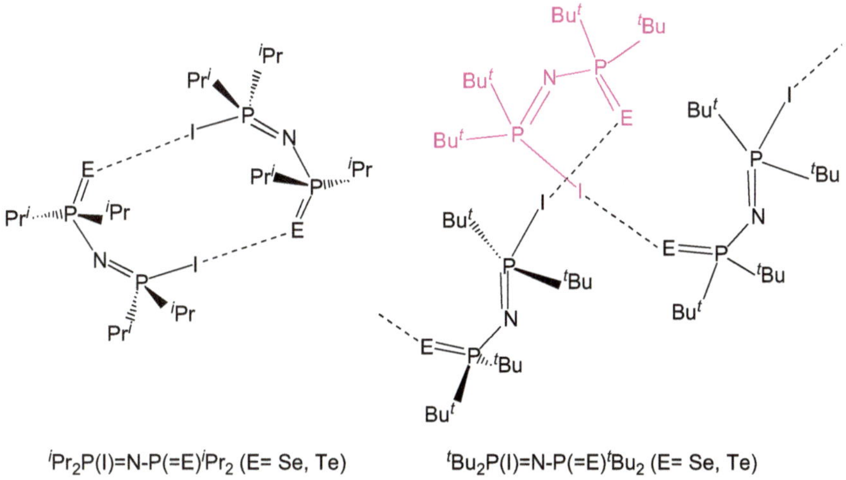

Figure 12.1.53 Schematic drawing of the coordination modes of iminobisphosphane-derived donor-acceptor systems $R_2P(=E)NPR_2I$ (E = Se, Te).

are formed by interconnection of two stacks of molecules through L-shaped P–Se···I–P bridges (Figure 12.1.51, right); layers are built up by connection of these helices through Se···Se contacts. The 'non-participating' terminal P–I groups approach in a herringbone pattern, creating domains of iodine atoms from adjacent layers.[56c]

Coordination of $[(t\mathrm{Bu_2P(=Se)I}]$ with molecular iodine furnishes a 1 : 1 adduct that crystallizes as a cyclic dimer (Figure 12.1.52).[56c] From each I_2 molecule within the structure, one atom acts as acceptor towards selenium, and the other is a very weak donor towards the iodine atom bonded to phosphorus. This compound fulfills the requirements of a 'soft–soft base pair'. As in the structure of the parent donor molecule $[(t\mathrm{Bu_2P(=Se)I}]$, selenium atoms of the dimer are involved in long secondary Se···Se contacts leading to chains from dimers.

As iminobisphosphane-derived P=E-donor–P–I-acceptor systems, the compounds $i\mathrm{Pr_2P(=E)N{=}P}i\mathrm{Pr_2I}$ (E = Se, Te) crystallize as centrosymmetric dimers (Se···I 395.2 pm, Te···I 400.0 pm), whereas the bulkier analogues $t\mathrm{Bu_2P(=E)N{=}P}t\mathrm{Bu_2I}$ exhibit chain strucutres with helical packing around a 2_1

screw axis (Se···I 379.7 pm, Te···I 379.5 pm) (Figure 12.1.53). The solid chloride $i\mathrm{Pr_2P({=}E)N{=}P}i\mathrm{Pr_2Cl}$, however, is a monomeric molecular species.[120]

12.1.6 Conclusion

The enhanced tendency of the heavier chalcogens (S ≪ Se < Te) and halogens (Cl ≪ Br < I) to expand their coordination spheres, either by donor–acceptor interactions with Lewis acids or bases, or by intermolecular secondary bonds, allows the crystallographic characterization of numerous supramolecular structures. Towards the heavier halogens, chalcogen atoms can act as donors (Lewis bases) or as acceptors (Lewis acids) that can be distinguished by the coordination geometry of their adducts, since in many cases the interactions are directed in accordance with the VSEPR rules and with the concept of n(donor) → σ*(acceptor) overlap. In a number of compounds, however, selenium, tellurium, bromine, and iodine atoms appear to play the roles of donor and acceptor at the same time, so that donor–acceptor assignments and formal oxidation numbers can become arbitrary.

The collection of structure determinations presented above reveals that in most cases the particular supramolecular networks were observed, not designed. An improved control of the interplay of different types of intermolecular interactions will be necessary in the future to develop 'supramolecular architecture' with heavier chalcogen–halogen interactions.

References

1. N. W. Alcock, *Advances in Inorganic Chemistry and Radiochemistry*, 1972, **15**, 1.
2. I. Haiduc and F. T. Edelmann, *Supramolecular Organometallic Chemistry*, Wiley-VCH, Weinheim, 1999; I. Haiduc, R. B. King and M. G. Newton, *Chem. Rev.*, 1994, **94**, 301.
3. S. Patai and Z. Rappoport, *The Chemistry of Organic Selenium and Tellurium Compounds, Vol. 1*, John Wiley and Sons, Chichester, 1986.
4. K. Akiba, *Chemistry of Hypervalent Compounds*, Wiley-VCH, New York, 1999.
5. H. A. Bent, *Chem. Rev.*, 1968, **68**, 590.
6. G. Y. Chao and J. D. Mc Cullough, *Acta Cryst.*, 1961, **14**, 940; H. Hope and J. D. Mc Cullough, *Acta Cryst.*, 1962, **15**, 806; H. Maddox and J. D. Mc Cullough, *Inorg. Chem.*, 1966, **5**, 522.
7. (a) B. Krebs and F.-P. Ahlers, *Adv. Inorg. Chem.*, 1990, **35**, 235; B. Krebs, *Nova Acta Leopoldina*, 1985, **59**, 131; (b) Unkonventionelle Wechselwirkungen in der Chemie metallischer Elemente, ed. B. Krebs, VCH, Weinheim, Germany, 1992; (c) P. Pyykkö, *Chem. Rev.* 1997, **97**, 597–636.
8. T. Klapötke and J. Passmore, *Can. Acc. Chem. Res.*, 1989, **22**, 234; J. Passmore, *Studies in Inorg. Chem.*, 1992, **14**, 373.

9. S. Husebye, *Phosphorus, Sulfur*, 1988, **38**, 271; S. Husebye and S. V. Lindemann, *Main Group Chemistry News*, 1995, **3**(4), 8; S. Husebye, *Phosphorus, Sulfur, Silicon, Rel. Elem.*, 1998, **136–138**, 377.
10. M. C. Aragoni, M. Arca, F. Demartin, F. A. Devillanova, A. Garau, F. Isaia, V. Lippolis and G. Verani, *Trends Inorg.Chem.*, 1999, **6**, 1–18; M. C. Aragoni, M. Arca, F. A. Devillanova, A. Garau, F. Isaia, V. Lippolis and G. Verani, *Coord. Chem. Rev.*, 1999, **184**, 271.
11. C. Ouvrard, J. Y. Le Questel, M. Berthelot and C. Laurence, *Acta Cryst. B*, 2003, **59**, 512.
12. W.-W. du Mont, A. Martens-von Salzen, F. Ruthe, E. Seppälä, G. Mugesh, F. A. Devillanova, V. Lippolis and N. Kuhn, *J. Organomet. Chem.*, 2001, **623**, 14–28.
13. A. F. Cozzolino, P. J. W. Elder and I. Vargas-Baca, *Coord. Chem. Rev.*, 2011, **255**, 1426–1438.
14. S. Hauge, K. Maroy and T. Odegard, *Acta Chem. Scand.*, 1988, **A42**, 51.
15. E. Faoro, G. Manzoni de Oliveira and E. Schulz Lang, *Polyhedron*, 2009, **28**, 63–68.
16. R. Kniep, L. Korte and D. Mootz, *Z. Naturforsch*, 1983, **38B**, 1.
17. (a) V. Janickis and M. Necas, *Z. Anorg. Allg. Chem.*, 2010, **636**, 818–823; (b) V. Janickis and M. Necas, *Z. Anorg. Allg. Chem.*, 2008, **634**, 905–910.
18. R. Kniep, D. Mootz and A. Rabenau, *Z. Anorg. Allg. Chem.*, 1976, **422**, 17.
19. R. Kniep and H. J. Beister, *Angew. Chem.*, 1985, **97**, 399; *Angew. Chem. Int. Ed. Engl.*, 1985, **97**, 399.
20. (a) F. Bigoli, F. Demartin, P. Deplano, F. A. Devillanova, F. Isaia, V. Lippolis, M. L. Mercuri, M. A. Pellinghelli and E. F. Trogu, *Inorg. Chem.*, 1996, **35**, 3194; (b) F. Demartin, F. A. Devillanova, F. Isaia, V. Lippolis and G. Verani, *Inorg. Chim. Acta*, 1997, **255**, 203.
21. C. D. Antoniadis, S. K. Hadjikakou, N. Hadjiliadis, M. Kubicki and I. S. Butler, *Eur. J. Inorg. Chem.*, 2004, **21**, 4324.
22. (a) J. Konu, T. Chivers and H. M. Tuonen, *Chem. Commun.*, 2006, 1634–36; (b) J. Konu, T. Chivers and H. M. Tuonen, *Inorg. Chem.*, 2006, **45**, 10678–10687; (c) S. D. Robertson, T. Chivers and H. M. Tuononen, *Inorg. Chem.*, 2008, **47**, 10634–10643; (d) T. Chivers, J. S. Ritch, S. D. Robertson, J. Konu and H. M. Tuononen, *Acc. Chem. Res.*, 2010, **43**, 1053–1062.
23. F. Bigoli, E. Leporati, M. A. Pellinghelli, G. Crisponi, P. Dcplano and E. F. Trogu, *J. Chem. Soc., Dalton Trans.*, 1983, 1763.
24. (a) S. Rajashree, R. K. Kumar, G. Aravamudan, M. R. Udupa, K. Sivakumar and H.-K. Fun, *Acta Cryst*, 1999, **C55**, 1320; (b) R. K. Kumar, G. Aravamudan, M. R. Udupa, M. Sehasayee and T. A. Hamor, *Polyhedron*, 1993, **12**, 2201; (c) R. K. Kumar, G. Aravamudan, M. R. Udupa and M. Sehasayee, *Polyhedron*, 1996, **52**, 3123.
25. (a) W. Nakanishi and S. Hayashi, *J. Organomet. Chem.*, 2000, **611**, 178; (b) W. Nakanishi, S. Hayashi, S. Yamaguchi and K. Tamao, *J. Chem. Soc., Chem. Commun.*, 2004, 140.

26. H. Hope and J. D. McCullough, *Acta Cryst.*, 1964, **17**, 712.
27. (a) M. Arca, F. Cristiani, F. A. Devillanova, A. Garau, F. Isaia, V. Lippolis and G. Verani, *Polyhedron*, 1997, **16**, 1983; (b) S. M. Godfrey, C. A. McAuliffe, R. G. Pritchard and S. Sarwar, *J. Chem. Soc., Dalton Trans.*, 1997, 3501; (c) J. D. McCullough, *Inorg. Chem.*, 1964, **3**, 1425.
28. T. Bjorvatten, *Acta Chem. Scand.*, 1963, **17**, 2292.
29. T. Dahl and O. Hassel, *Acta Chem. Scand.*, 1965, **19**, 2000.
30. (a) O. Holmesland and C. Romming, *Acta Chem. Scand.*, 1966, **20**, 2601; (b) S. Tomoda, Y. Takeuchi and Y. Nomura, *Synthesis*, 1985, 212.
31. M. Arca, F. Cristiani, F. A. Devillanova, A. Garau, F. Isaia, V. Lippolis, G. Verani and F. Demartin, *Polyhedron*, 1997, **16**, 1983.
32. A. J. Blake, F. A. Devillanova, A. Garau, L. M. Gilby, R. O. Gould, F. Isaia, V. Lippolis, S. Parsons, C. Radek and M. Schroder, *J. Chem. Soc., Dalton Trans.*, 1998, **12**, 2037.
33. A. J. Blake, F. Cristiani, F. A. Devillanova, A. Garau, L. M. Gilby, R. O. Gould, F. Isaia, V. Lippolis and S. Parsons, *J. Chem. Soc., Dalton Trans.*, 1997, **8**, 1337.
34. P. K. Baker, S. D. Harris, M. C. Durrant, D. L. Hughes and R. L. Richards, *Acta Cryst.*, 1995, **C51**, 697–700.
35. A. J. Blake, F. A. Devillanova, A. Garau, F. Isaia, V. Lippolis, S. Parsons and M. Schroder, *J. Chem. Soc., Dalton Trans.*, 1999, **4**, 525.
36. S. Kubiniok, W.-W. du Mont, S. Pohl and W. Saak, *Angew. Chem. Int. Ed.*, 1988, **27**, 431.
37. (a) W.-W. du Mont in *Reviews in Heteroatom Chemistry Vol. 1.* ed. S. Oae, MYU K. K. Tokyo, 1988, p. 138; (b) W.-W. du Mont, A. Martens, S. Pohl and W. Saak, *Inorg. Chem.*, 1990, **29**, 4848.
38. T. M. Klapötke, B. Klemm and M. Scherr, *Z. Anorg. Allg. Chem.*, 2010, **636**, 1955–1961.
39. (a) J. Jeske, A. Martens-von Salzen, W.-W. du Mont and P. G. Jones, *Acta Cryst.*, 1998, **C54**, 1873; (b) Z. Majeed, W. R. McWhinnie and P. R. Lowe, *Acta Cryst.*, 2000, **56**, 105.
40. M. C. Aragoni, M. Arca, F. A. Devillanova, A. Garau, F. Isaia, V. Lippolis and G. Verani, *Coord. Chem. Rev.*, 1999, **184**, 271.
41. F. Cristiani, F. Demartin, F. A. Devillanova, F. Isaia, G. Saba and G. Verani, *J. Chem. Soc., Dalton Trans.*, 1992, 3553.
42. N. Kuhn, T. Kratz and G. Henkel, *Chem. Ber.*, 1994, **127**, 849.
43. F. Demartin, P. Deplano, F. A. Devillanova, F. Isaia, V. Lippolis and G. Verani, *Inorg. Chem.*, 1993, **32**, 3694.
44. F. Bigoli, A. M. Pelinghelli, P. Deplano, F. A. Devillanova, V. Lippolis, M. L. Mercuri and E. F. Trogu, *Gazz. Chim. Ital.*, 1994, **124**, 445.
45. N. Kuhn, R. Fawzi, T. Kratz and M. Steimann, *Phosphorus, Sulfur and Silicon*, 1996, **112**, 225.
46. F. Cristiani, F. Demartin, F. A. Devillanova, F. Isaia, V. Lippolis and G. Verani, *Inorg. Chem.*, 1994, **33**, 6315.

47. F. Bigoli, P. Deplano, A. Ienco, C. Mealli, M. L. Mercuri, M. A. Pellinghelli, G. Pintus, G. Saba and E. F. Trogu, *Inorg. Chem.*, 1999, **38**, 4626.
48. (a) P. J. Skabara, N. Bricklebank, S. Rory Berridge, M. E. Long, S. J. Light, Coles and M. B. Hursthouse, *J. Chem. Soc., Dalton Trans.*, 2000, 3235; (b) D. W. Allen, R. Berridge, N. Bricklebank, S. D. Forder, F. Palacio, S. J. Coles and M. B. Hursthouse, *Inorg. Chem.*, 2003, **42**, 3975.
49. L. Lee, D. J. Crouch, S. P. Wright, R. Berrigde, P. J. Skabara, N. Bricklebank, S. J. Coles, M. E. Light and M. B. Hursthouse, *CrystEngComm.* 2004, **6**, 612.
50. F. Cristiani, F. A. Devillanova, A. Diaz and G. Verani, *J. Chem. Soc., Perkin Trans. 2*, 1984, **8**, 1383.
51. N. Kuhn, H. Bohnen and G. Henkel, *Z. Naturforsch.*, 1994, **B49**, 1473.
52. M. C. Aragoni, M. Arca, F. A. Devillanova, F. Isaia, V. Lippolois, A. Mancini, L. Pala, A. M. Z. Slawin and J. D. Woollins, *J. Chem. Soc, Chem. Commun.*, 2003, 2226.
53. (a) W.-W. Schweikert and E. A. Meyers, *J. Phys. Chem.*, 1968, **72**, 1561; J. W. Bransford and E. A. Meyers, *Cryst. Struct. Commun.*, 1978, **7**, 697; (b) D. C. Apperley, N. Bricklebank, S. L. Burns, D. E. Hibbs, M. B. Hursthouse and K. M. Abdul Malik, *J. Chem. Soc., Dalton Trans.*, 1998, **8**, 1289.
54. M. Arca, F. Demartin, F. A. Devillanova, A. Garau, F. Isaia, V. Lippolis and G. Verani, *J. Chem. Soc., Dalton Trans.*, 1999, **17**, 3069.
55. R. A. Zingaro and E. A. Meyers, *Inorg. Chem.*, 1962, **1**, 771.
56. (a) W.-W. du Mont, *Main Group Chemistry News*, 1994, **2**(3), 18; (b) F. Ruthe, Ph.D. thesis, Techn. Universität Braunschweig, 2004; (c) J. Jeske, W.-W. du Mont and P. G. Jones, *Chem. Eur. J.*, 1999, **5**, 385.
57. M. D. Rudd, S. V. Lindemann and S. Husebye, *Acta Chem. Scand.*, 1997, **51**, 689.
58. S. M. Godfrey, S. L. Jackson, C. A. McAuliffe and R. G. Pritchard, *J. Chem. Soc., Dalton Trans.*, 1997, **23**, 4499.
59. E. Seppälä, F. Ruthe, W.-W. du Mont and P. G. Jones, *J. Chem. Soc., Chem. Commun.*, 1999, **12**, 1471.
60. R. Minkwitz, U. Nass and H. Preut, *Z. Anorg. Allg. Chem.*, 1986, **538**, 143.
61. A. Martens-von Salzen, J. Jeske, W.-W. du Mont and P. J. Jones, *Phosphorus, Sulfur and Silicon*, 1998, **136–138**, 545–548.
62. S. Husebye, E. A. Meyers, R. A. Zingaro, A. L. Braga, J. V. Comasseto and N. Petragnani, *Acta Cryst.*, 1986, **42**, 1789.
63. J. Jeske, P. G. Jones, A. Martens-von Salzen and W.–W. du Mont, *Acta Cryst*, 2002, **E58**, 1–3.
64. J. Beckmann, S. Heitz and M. Hesse, *Inorg. Chem.*, 2007, **46**, 3275–3282.
65. (a) E. Schulz Lang, R. M. Fernandez Junior, E. T. Silveira, U. Abram and E. M. Vazquez-Lopez, *Z. Anorg. Allg. Chem.*, 1999, **625**, 1401; (b) P. D. Boyle, W. I. Cross, S. M. Godfrey, C. A. McAuliffe,

R. G. Pritchard, S. Sarwar and J. M. Sheffield, *Angew. Chem., Int. Ed.*, 2000, **39**, 1796; (c) N. A. Barnes, S. M. Godfrey, R. T. A. Halton, I. Mushtaq, R. G. Pritchard and S. Sarwar, *Dalton Trans.*, 2006, 1517–1523.
66. E. Faoro, G. Manzoni de Oliveira, E. Schulz Lang and C. Bicca Pereira, *J. Organomet. Chem.*, 2010, **695**, 1480–1486.
67. G. Manzoni de Oliveira, E. Faoro and E. Schulz Lang, *Inorg. Chem.*, 2009, **48**, 4607–4609.
68. M. C. Aragoni, M. Arca, F. Demartin, F. A. Devillanova, A. Garau, P. Grimaldi, F. Isaia, F. Lelj, V. Lippolis and G. Verani, *Eur. J. Inorg. Chem.*, 2004, 2363.
69. F. A. Devillanova, V. Lippolis, M. C. Aragoni, M. Arca, A. Garau, F. Isaia and G. Verani, *Chem. Eur. J.*, 2001, **7**, 3122.
70. P. D. Boyle, W. I. Cross, S. M. Godfrey, C. A. McAuliffe, R. G. Pritchard and S. Teat, *J. Chem. Soc., Dalton Trans.*, 1999, 2845.
71. F. A. Devillanova, P. Deplano, F. Isaia, V. Lippolis, M. L. Mercuri, S. Piludu, G. Verani and F. Demartin, *Polyhedron*, 1998, **17**, 305.
72. N. Kuhn, A. Abu-Rayyan, C. Piludu and M. Steimann, *Heteroatom Chem.*, 2005, **16**, 1339.
73. M. C. Aragoni, M. Arca, A. J. Blake, F. A. Devillanova, W.-W. du Mont, A. Garau, F. Isaia, V. Lippolis, G. Verani and C. Wilson, *Angew. Chem.*, 2001, **113**, 4359; *Angew. Chem. Int. Ed. Engl.*, 2001, **40**, 4229.
74. S. Larsen and L. Henriksen, *Acta Chem. Scand.*, 1984, **38**, 289.
75. C. G. Hrib, F. Ruthe, E. Seppälä, M. Bätcher, C. Druckenbrodt, C. Wismach, P. G. Jones, W.-W. du Mont, V. Lippolis, F. A. Devillanova and M. Bühl, *Eur. J. Inorg. Chem.*, 2006, 88–100.
76. S. M. Godfrey, S. L. Jackson, C. A. McAuliffe and R. G. Pritchard, *J. Chem. Soc., Dalton Trans.*, 1998, **24**, 4201.
77. O. Hauge and S. Vikane, *Acta Chem. Scand.*, 1988, **A42**, 87.
78. O. Foss, S. Husebye and K. Maroy, *Acta Chem. Scand.*, 1963, **17**, 1806.
79. O. Foss and S. Husebye, *Acta Chem. Scand.*, 1966, **20**, 132.
80. E. Faoro, G. Manzoni de Oliveira, E. Schulz Lang and C. Bicca Pereira, *J. Organomet. Chem.*, 2011, **696**, 2438–2444.
81. C. G. Hrib, P. G. Jones, W.-W. du Mont, V. Lippolis and F. A. Devillanova, *Eur. J. Inorg. Chem.*, 2006, 1294–1302.
82. O. Foss and K. Maroy, *Acta Chem. Scand.*, 1986, **A40**, 669.
83. H. K. Ault and S. Husebye, *Acta Chem. Scand.*, 1978, **A32**, 157.
84. E. Faoro, G. Manzoni de Oliveira and E. Schulz Lang, *J. Organomet. Chem.*, 2009, **694**, 1557–1561.
85. E. Faoro, G. Manzoni de Oliveira, E. Schulz Lang and C. Bicca Pereira, *J. Organomet. Chem.*, 2006, **691**, 5867–5872.
86. (a) A. F. Wells, *Structural Inorganic Chemistry* 5th. ed. Clarendon press, Oxford 1984, 739; (b) A. F. Wells, *Acta Cryst.*, 1966, **20**, 610.
87. (a) H. Fleischer and D. Schollmeyer, *Acta Cryst.*, 1999, **C55**, 1572; (b) M. J. Collins, J. A. Ripmeester and J. F. Sawyer, *J. Am. Chem. Soc.*, 1988, **110**, 8583.
88. R. F. Ziolo and M. Extine, *Inorg. Chem.*, 1980, **19**, 2964.

89. R. K. Chadha and J. E. Drake, *Can. J. Chem.*, 1986, **299**, 331.
90. T. M. Klapötke, B. Krumm, P. Mayer, K. Polborn and O. P. Ruscitti, *J. Fluorine Chem.*, 2001, **112**, 207.
91. F. Einstein, J. Trotter and C. Williston, *J. Chem. Soc. A*, **1967**, 2018.
92. (a) W.-W. du Mont, J. Jeske and P. G. Jones, *Phosphorus, Sulfur and Silicon*, 1998, **136–138**, 305; (b) C. Hrib, J. Jeske, P. G. Jones and W.-W. du Mont, *Dalton Trans.*, 2007, 3483–3485.
93. S. N. Närhi, R. Oilunkaniemi, R. S. Laitinen and M. Ahlgrén, *Acta Cryst*, 2004, **E60**, 0798.
94. S. Santos dos Santos, E. Schulz Lang and G. Manzoni de Oliveira, *J. Organomet..Chem.*, 2007, **692**, 3081–3088.
95. W-W. du Mont, J. Jeske and P. G. Jones, *Phosphorus, Sulfur and Silicon*, 2010, **185**, 1243–1249.
96. J. Jeske, F. Ruthe, W.-W. du Mont and P. G. Jones, *Angew. Chem.*, 1997, **109**, 2304: *Angew. Chem. Int. Ed. Engl.*, 1997, **36**, 2219; J. Jeske, W.-W. du Mont, F. Ruthe, P. G. Jones, L. M. Mercuri and P. Deplano, *Eur. J. Inorg. Chem.*, 2000, 1591.
97. S. M. Godfrey, C. A. McAuliffe, R. G. Pritchard and S. Sarwar, *J. Chem. Soc., Dalton Trans.*, 1997, **6**, 1031.
98. J. Beckmann, D. Dakternieks, A. Duthie, C. Mitchell and M. Schürmann, *Aust. J. Chem.*, 2005, **58**, 119.
99. T. M. Klapötke, B. Krumm, P. Mayer, K. Polborn and O. P. Ruscitti, *Inorg. Chem.*, 2001, **40**, 5169.
100. G. Morgan and H. Burgess, *J. Chem. Soc.*, 1929, 1103.
101. J. D. McCullough, C. Knobler and R. F. Ziolo, *Inorg. Chem.*, 1985, **24**, 1814, and references therein.
102. P. C. Srivastava, S. Baypai, Sm. Bajpai, R. Kumar, S. Srivastava and R. J. Butcher, *Struct. Chem.*, 2007, **18**, 223–240.
103. P. C. Srivastava, S. Baypai, R. Lath, R. Kumar, V. Singh, S. Dwivedi, S. R. J. Butcher, S. Hayashi and W. Nakanishi, *Polyhedron*, 2008, **27**, 835–848.
104. M. J. Poropudas, L. Vigo, R. Oilunkaniemi and R. S. Laitinen, *Heteroatom Chem.*, 2011, **22**, 348–357.
105. A. K. S. Chauhan, A. Anamika, R. C. Klumar, R. J. Srivastava, J. Butcher, Beckmann and A. Duthie, *J. Organomet. Chem.*, 2005, **690**, 1350–1355.
106. S. Husebye and S. V. Lindemann, *Main Group Chem. News*, 1995, **3**(4), 8.
107. E. Faoro, G. Manzoni de Oliveira, E. Schulz Lang and C. Bicca Pereira, *J. Organomet. Chem.*, 2011, **696**, 807–812.
108. J. Konu, M. Ahlgren, S. M. Aucott, T. Chivers, S. H. Dale, M. R. J. Elsegoog, K. E. Holmes, S. L. H. James, P. F. Kelly and R. S. Laitinen, *Inorg. Chem.*, 2005, **44**, 4992–5000.
109. E. Schulz Lang, G. M. de Oliveira, R. M. Fernandes Jr. and E. M. Vazquez-Lopez, *Z. Anorg. Allg. Chem.*, 2004, **630**, 717.
110. E. Schulz Lang, R. M. Fernandes Jr., C. Peppe, R. A. Burrow and E. M. Vazquez-Lopez, *Z. Anorg. Allg. Chem.*, 2003, **629**, 215.

111. E. Schulz Lang, G. M. de Oliveira, R. M. Fernandes Jr. and E. M. Vazquez-Lopez, *Inorg. Chem. Commun.*, 2003, **6**, 869.
112. E. Schulz Lang, G. A. Casagrande, G. Manzoni de Oliveira, G. N. Ledesma, S. S. Lemos, E. E. Castellano and U. Abram, *Eur. J. Inorg. Chem.*, 2006, 958.
113. E. Schulz Lang, G. Manzoni de Oliveira and G. Nanci Ledesma, *Z. Anorg. Allg. Chem.*, 2005, **631**, 1524–1527.
114. H. Fleischer and D. Schollmeyer, *Inorg. Chem.*, 2001, **40**, 324.
115. H. Fleischer, Y. Dienes and D. Schollmeyer, *Eur. J. Inorg. Chem.*, 2002, **8**, 2073.
116. J. Beckmann, J. Bolsinger and A. Duthie, *Chem Eur. J.*, 2011, **17**, 930–940.
117. J. Beckmann, J. Bolsinger and J. Spandl, *J. Organomet. Chem.*, 2008, **693**, 957–964.
118. J. Beckmann, J. Bolsinger and A. Duthie, *Aust. J. Chem.*, 2008, **61**, 172–182.
119. C. Daniliuc, C. G. Hrib, P. J. Jones and W.-W. du Mont, *Cryst. Growth Des.*, 2012, **12**, 185–188.
120. S. Ritch, S. D. Robertson, M. Risto and T. Chivers, *Inorg. Chem.*, 2010, **49**, 4681–4686.

CHAPTER 12.2

Synthesis and Stereochemistry of Optically Active Chalcogen Compounds

TOSHIO SHIMIZU

Department of Chemistry, Graduate School of Science and Engineering, Tokyo Metropolitan University, Minami-ohsawa, Hachioji, Tokyo 192-0397, Japan
Email: shimizu-toshio@tmu.ac.jp

12.2.1 Introduction

A number of optically active organosulfur compounds, such as sulfoxides, sulfonium salts, sulfonium ylides, sulfilimines, sulfoximines, and sulfuranes, have been synthesized, and their stereochemistry, properties, and reactivities have been widely studied with results included in numerous books.[1] Moreover, numerous papers dealing with asymmetric synthesis utilizing these optically active organosulfur compounds have also been published.[1–3] On the other hand, the number of optically active selenium or tellurium compounds isolated so far is limited, although chiral organoselenium and tellurium compounds are felt to be important key intermediates for asymmetric syntheses.[4–6] Studies on the isolation and stereochemistry of optically active selenium and tellurium compounds have been addressed in some reviews.[6–8] Optically active selenium and tellurium compounds, such as telluroxides, chalcogenic acids, and seleninate esters, were isolated for the first time during the last two decades.

Handbook of Chalcogen Chemistry: New Perspectives in Sulfur, Selenium and Tellurium, 2nd Edition, Volume 2
Edited by Francesco Antonio Devillanova and Wolf-Walther du Mont

Published by the Royal Society of Chemistry, www.rsc.org

In this chapter, recent development on the synthesis and stereochemistry of optically active chalcogen compounds over the last 15 years are described, focusing mainly on chiral selenium and tellurium compounds.

12.2.2 Chalcogen-Oxides

Optically active sulfoxides are well known and several synthetic methods have been developed, with numerous papers published on asymmetric synthesis utilizing optically active sulfoxides as chiral sources.[1,9]

Optically active selenoxides are known to be unstable towards racemization. An optically active selenoxide having a steroidal frame was obtained for the first time by Jones and co-workers in 1970.[10] Enantiomeric selenoxides were prepared by Davis *et al.* in 1983,[11] and an enantiomerically pure selenoxide was isolated for the first time by us in 1989.[12] Many optically active selenoxides, which are kinetically stabilized by bulky substituents, were synthesized over the last two decades, and their stereochemistry and stability toward racemization were studied.[6,7,13] During the last decade, some optically active selenoxides, which were thermodynamically stabilized by the intramolecular coordination of a Lewis base to the selenium atom, have been isolated. Optically active selenoxides **1**, **2** and **3** were obtained by optical resolution on chiral columns, and their stereochemistry and stability toward racemization under various conditions were clarified (Scheme 12.2.1).[14–16]

On the other hand, optically active telluroxides have not been isolated until recently, although it has been surmised that they are key intermediates in asymmetric synthesis.[5,6] In 1997, optically active telluroxides **4**, stabilized by bulky substituents toward racemization, were isolated for the first time by liquid chromatography on optically active columns (Scheme 12.2.2).[17,18] The stereochemistry was determined by comparing their chiroptical properties with those of chiral selenoxides with known absolute configurations. The stability of the chiral telluroxides toward racemization was found to be lower than that of the corresponding selenoxides, and the racemization mechanism that involved formation of the achiral hydrate by reaction of water was also clarified. Telluroxides **5** and **6**, which were thermodynamically stabilized by

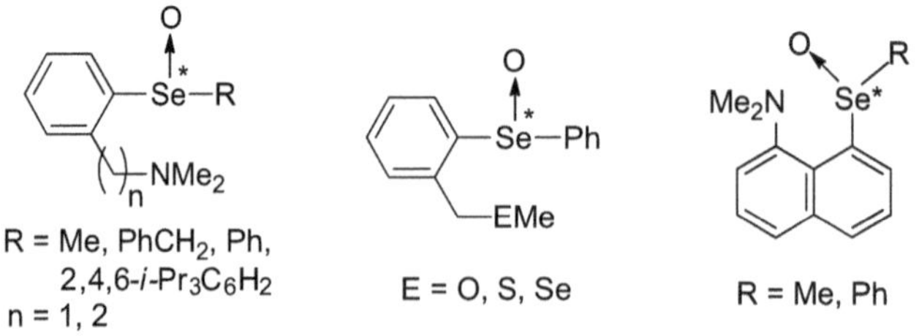

1 **2** **3**

Scheme 12.2.1

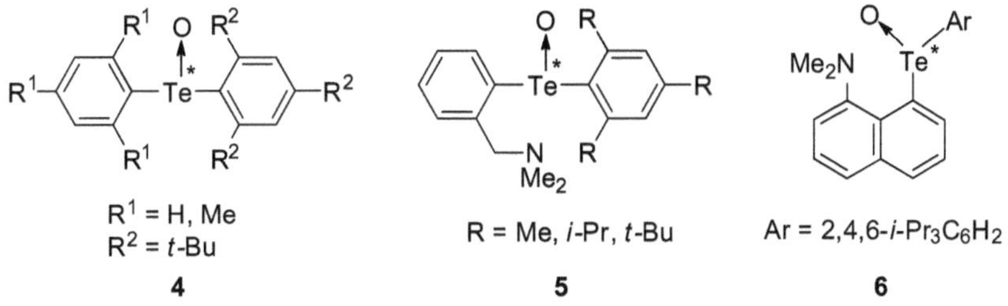

Scheme 12.2.2

nitrogen–tellurium interactions, were also optically resolved and their absolute configurations and stability were studied.[16,18]

12.2.3 Chalcogen-Onium Ylides

Chiral sulfonium ylides have been known for 40 years, and their stereochemistry and properties have been studied.[19] Optically active selenonium ylides were obtained for the first time by Sakaki and Oae in 1976, by reacting selenoxides with 1,3-cyclohexanedione under asymmetric conditions,[20] and also optically resolved by fractional recrystallization of the diastereomeric mixtures in early 1990s.[21] In 1995, optically active selenonium ylides **7** were obtained in over 99% diastereomeric excess (de) by nucleophilic substitution of optically active chloroselenurane or selenoxide with active methylene compounds with retention of configuration (Scheme 12.2.3).[22] The absolute configurations were determined by X-ray analysis of one isomer of the ylides.

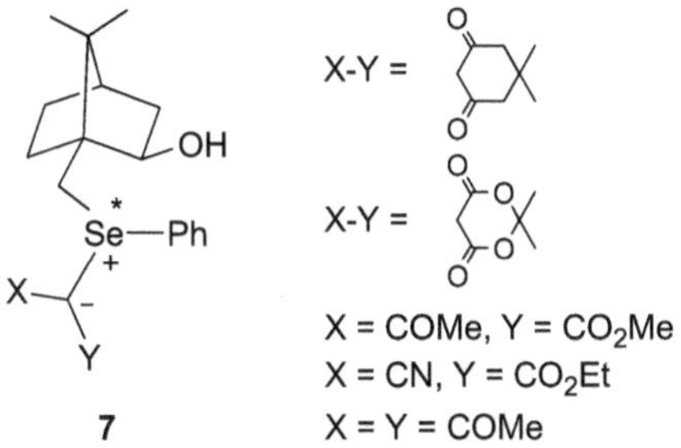

Scheme 12.2.3

Optically active telluronium ylides were not obtained for a long time. Optically active diastereomeric telluronium ylides **8** were obtained for the first time in 1995 by fractional recrystallization of the diastereomeric mixture (Scheme 12.2.4).[23] The absolute configurations of the chiral telluronium ylides were determined by comparing their specific rotations and circular dichroism spectra with those of the corresponding selenonium ylide with known absolute configuration. The telluronium ylides were found to be much more stable toward racemization than the sulfonium and selenonium ylides.

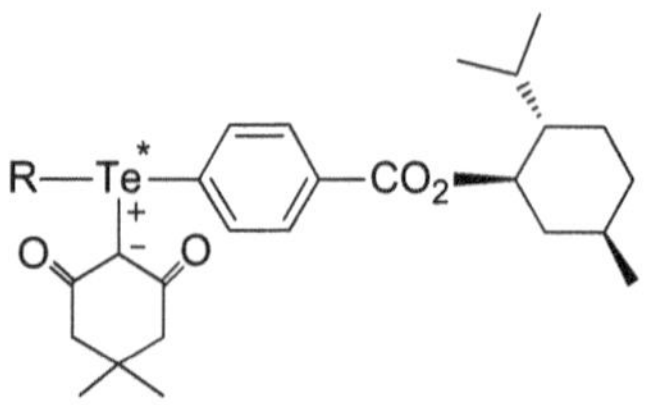

R = Me, 2,4,6-*i*-$Pr_3C_6H_2$

8

Scheme 12.2.4

The increasing stability toward racemization was supported by theoretical calculations carried out on the dimethyl chalcogen-onium ylides.[24] In fact, the activation energies for the pyramidal inversion of chalcogen-onium ylides, which is one of the possible pathways for the racemization, were estimated to be 27, 33, and 42 kcal mol^{-1}, for S, Se, and Te, respectively, in agreement with experimental results.

12.2.4 Chalcogen-Onium Imides

Some optically active sulfonium imides have been synthesized.[25] Optically active selenonium imides were reported for the first time by Krasnov and co-workers in 1981, although their optical purities were unknown.[26] Diastereomerically pure selenonium imide **9** was obtained for the first time in 1994 by fractional recrystallization of the diastereomeric mixture, and the absolute configuration was assigned based on the circular dichroism spectrum (Scheme 12.2.5).[27] Enantiomeric selenonium imide **10** was also obtained in 80% optical purity by stereoselective transformation of the corresponding optically active selenoxide with retention of stereochemistry, and a hypothesis was put forward as to the mechanism of the transformation.[28] Enantiomerically pure selenonium imides **11** were isolated by optical resolution using liquid chromatography on an optically active column in 1999.[29] The absolute configurations were determined by X-ray crystallographic analysis of one isomer, and the activation energies for the racemization of the optically active selenonium imides were found experimentally to be greater than those for sulfonium imides. In 2000, optically active selenonium imides **12** and **13** were synthesized by Uemura *et al.* by enantioselective imidation of prochiral selenides using optically active bis(oxazoline) or diastereoselective imidation of corresponding selenides in up to 36% enantiomeric excess (ee) and 76% de, respectively.[30] The enantioselective imidation was also utilized in the asymmetric synthesis of chiral allylic *N*-tosylamides. Recently, optically active selenonium imides **14**, which were stabilized by intramolecular coordination, were also isolated.[31]

On the other hand, telluronium imides **15** were isolated for the first time in 2002 by optical resolution of their racemic samples on an optically active column by medium-pressure column chromatography (Scheme 12.2.6).[32] The relationship between the absolute configurations and the chiroptical properties was clarified on the basis of their specific rotations and circular dichroism spectra. The racemization mechanism of the optically active telluronium imides, which involved the formation of corresponding telluroxides by hydrolysis of the telluronium imides, was proposed.

The pyramidal inversion energies of chalcogen-onium imides were estimated by MO calculations.[33] The values for model compounds, dimethyl chalcogen-onium imides, were evaluated to be 34, 44, and 54 kcal mol^{-1}, for S, Se, and Te, respectively, indicating that among the chalcogen-onium imides, telluronium imides are those that are most stable toward pyramidal inversion.

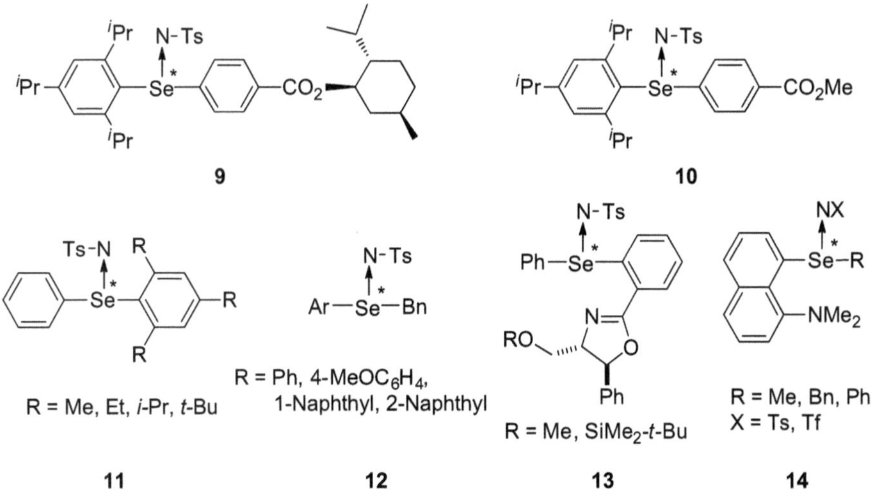

Scheme 12.2.5

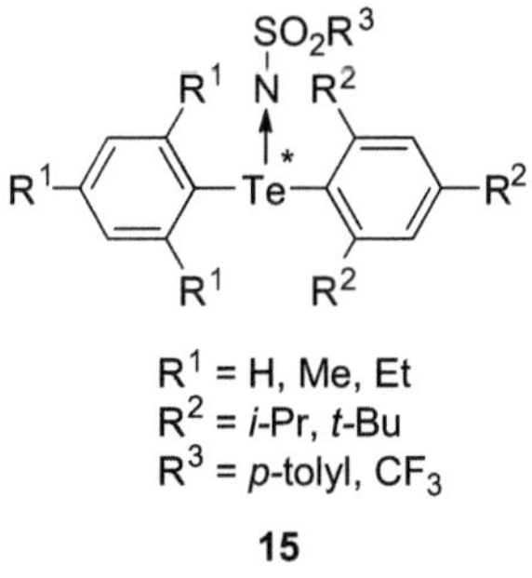

Scheme 12.2.6

12.2.5 Chalcogen-Onium Salts

Optically active sulfonium and selenonium salts are well known and the stereochemistry of the isomers has been studied.[1,4–6] Optically active cyclic diaryl(alkoxy)sulfonium salts **16**, **17**, and **18**, stabilized by intramolecular sulfur–oxygen interaction, were synthesized in 2000 by reacting optically active spirosulfuranes with trimethyloxonium tetrafluoroborate (Scheme 12.2.7).[34] The absolute configurations were assigned on the basis of the reaction mechanism. The sulfonium salts were hydrolysed in $KHCO_3$aq. to yield optically active sulfoxides in over 86% ee. Koizumi and co-workers synthesized diastereomeric selenonium salts **19** by reaction of chiral selenuranes with Grignard reagents (Scheme 12.2.8).[35]

Although two papers addressing the synthesis of optically active telluronium salts were published several decades ago,[36] their optical purities and absolute configurations remained unknown. In 1996, (*R*)- and (*S*)-diastereomerically pure ethylmethylphenyltelluronium salts **20** have been isolated by fractional recrystallization of the diastereomeric mixture,[37] and their absolute configurations been determined by X-ray crystallographic analysis (Scheme 12.2.9). Enantiomerically pure telluronium salts **21** have been prepared by anion-exchange reaction of the diastereomerically pure telluronium salts without loss of optical purities. The optically active telluronium salts have been found to show peculiar optical properties in terms of specific rotations and circular dichroism spectra in solution compared with those of the corresponding sulfonium and selenonium salts. In the first two papers reporting chiral

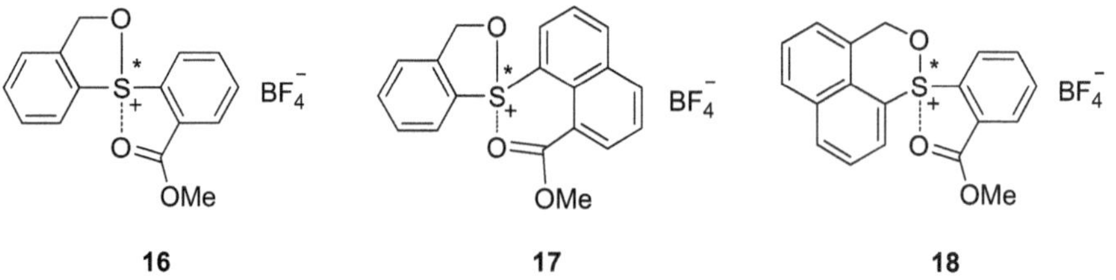

Scheme 12.2.7

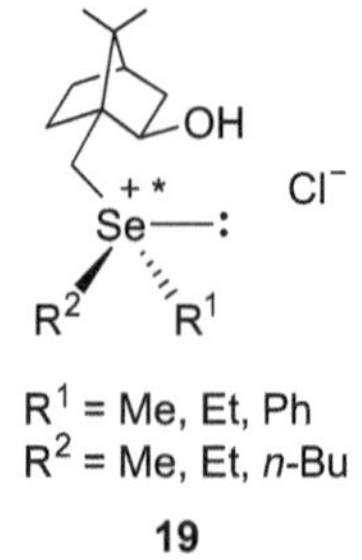

Scheme 12.2.8

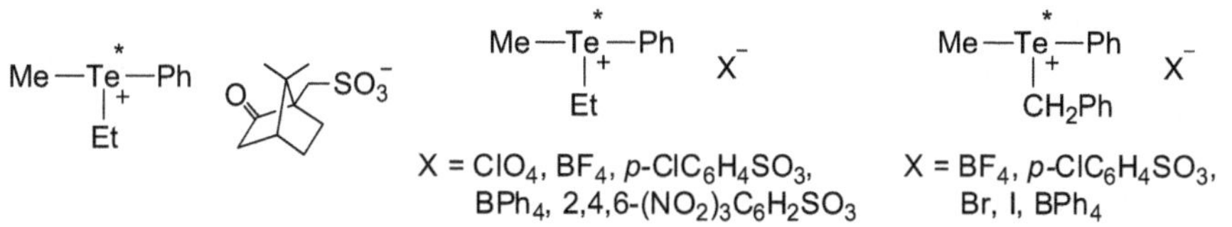

20 **21** **22**

Scheme 12.2.9

telluronium salts,[36] it was mentioned that mutarotation was observed in solutions; however, telluronium salts **20** and **21** are stable toward racemization even in solutions. The stereochemistry of diastereomerically pure ethylmethylphenyltelluronium salt **20** has been clarified by X-ray diffraction analysis together with corresponding selenonium salt.[38] Recently, optically active benzylmethylphenyltelluronium salts **22** were isolated by fractional recrystallization of the diastereomeric mixture followed by anion-exchange reactions.[39] The benzyltelluronium salts, possessing counter-anions with high nucleophilicity, racemize in solution and the racemization mechanism was proposed (Scheme 12.2.10).

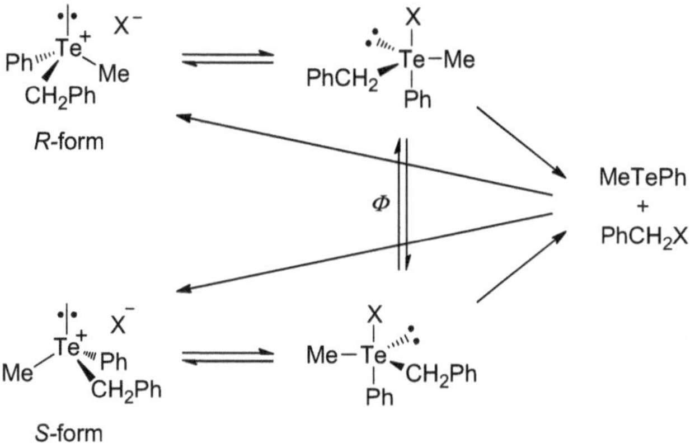

Scheme 12.2.10

12.2.6 Chalcogenic Acids

Chalcogenic acids, R-E(O)OH, are also tricoordinate and considered to have pyramidal structures. However, no studies on their optical activity have been reported since facile racemization of chalcogenic acids may occur *via* achiral chalcogenate anions, that are formed by the extrusion of a proton and/or by an intra- or intermolecular proton transfer reaction.

In 2001, optically active areneseleninic acids **23** were obtained for the first time as solution by optical resolution on a chiral column using liquid chromatography (Scheme 12.2.11).[40] Bulky substituents and electron-donating groups on the aromatic ring were effective in retarding the racemization. The absolute configurations of the optically active seleninic acids were assigned by comparing their circular dichroism spectra with that of optically active sulfinic acid **24**, the absolute configuration of which was determined by X-ray crystallographic analysis. Sulfinic acid **24** represents the first example of a chiral sulfinic acid showing a relationship between absolute configuration and optical properties.[41] Optically active methaneseleninic acid **25** was isolated as stable crystals for the first time by chiral crystallization.[42] The relationship between the absolute configuration and the circular dichroism spectra of the enantiomers was determined by X-ray crystallographic analysis. The optically active methaneseleninic acid was stable toward racemization in the solid state, although it racemized immediately in solution. Each enantiomer of methaneseleninic acid was also obtained in bulk by chiral crystallization in the presence of a seed crystal or from chiral solvents.[43] The absolute configuration of optically active methanesulfinic acid, which was also obtained by recrystallization as single crystals, was assigned by comparing its circular dichroism spectrum in the solid state with that of methaneseleninic acid.

Optically active arenetellurinic acid **26** was obtained for the first time by chromatographic resolution of the racemic sample on a chiral column in 2004 (Scheme 12.2.12).[44] It is stable toward racemization in solution, whereas the corresponding seleninic acids racemize in solution under similar conditions. Its absolute configuration was assigned by comparing the circular dichroism spectra with that of an optically active seleninic acid.

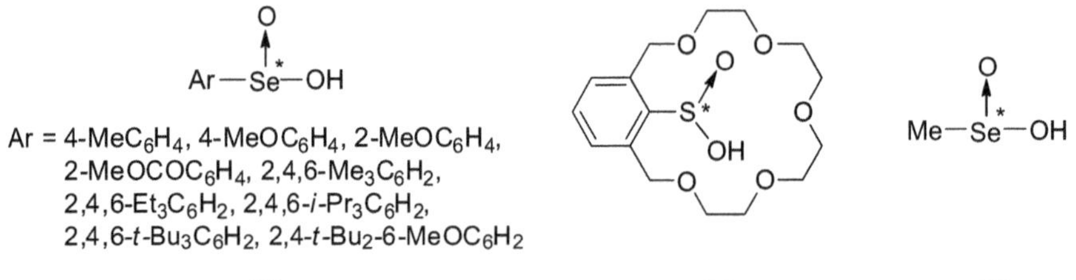

23 **24** **25**

Scheme 12.2.11

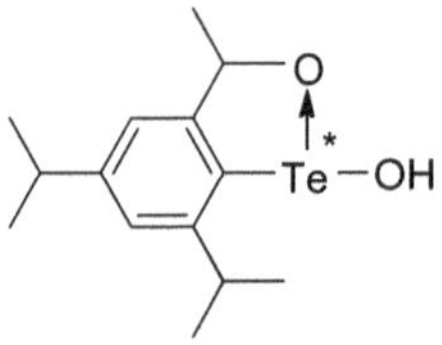

26

Scheme 12.2.12

Kinetic studies on the racemization, the oxygen exchange reaction using $H_2{}^{18}O$, and theoretical studies clarified the mechanism for racemization of optically active chalcogenic acids.[41,44] The mechanism of racemization of tellurinic acids is different from that of the corresponding seleninic acids. In fact, the mechanism for racemization of optically active tellurinic acids was found to involve an achiral tellurane that is formed by the addition of water to the tellurinic acid, whereas the racemization of the optically active sulfinic and seleninic acids involves a sulfinate or selenate anion that is formed by the extrusion of a proton (Scheme 12.2.13).

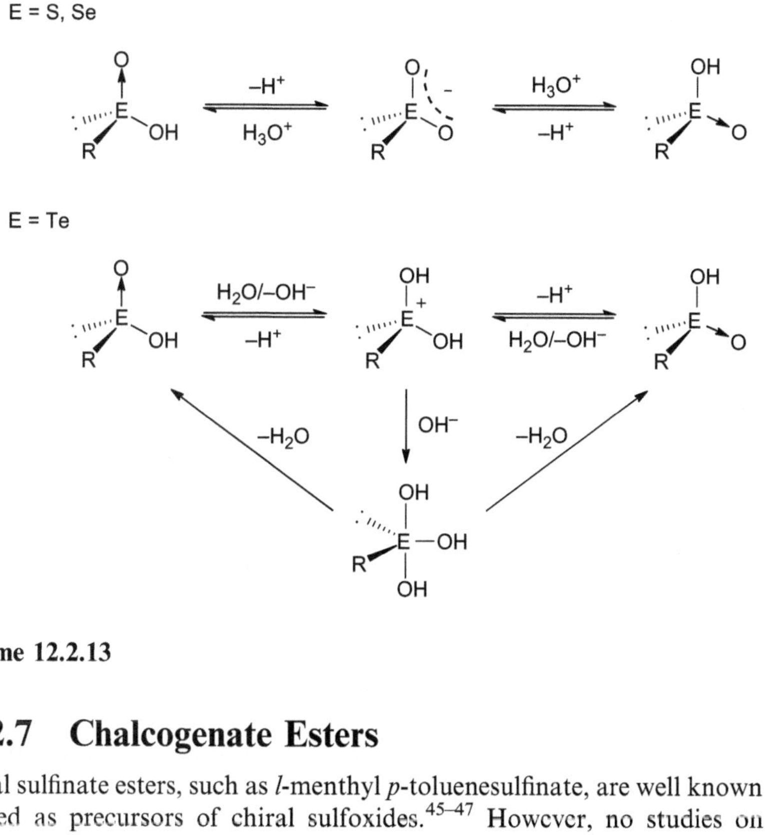

Scheme 12.2.13

12.2.7 Chalcogenate Esters

Chiral sulfinate esters, such as *l*-menthyl *p*-toluenesulfinate, are well known and utilized as precursors of chiral sulfoxides.[45–47] However, no studies on the corresponding chiral seleninate esters were reported until 2005. The reason for this lack is mainly due to the fact that acyclic seleninate esters are unstable toward hydrolysis. Optically active cyclic seleninate esters **27** were obtained for the first time in 2005 by chromatographic resolution on an optically active column (Scheme 12.2.14).[48] Their absolute configurations were determined by comparing their chiroptical properties with those of an analogous sulfinate ester, the absolute configuration of which was determined by X-ray analysis. The optically active seleninate esters were found to racemize in solution. Kinetic studies of racemization, the oxygen exchange reaction with $H_2{}^{18}O$, and theoretical studies have clarified that the racemization of the optically active

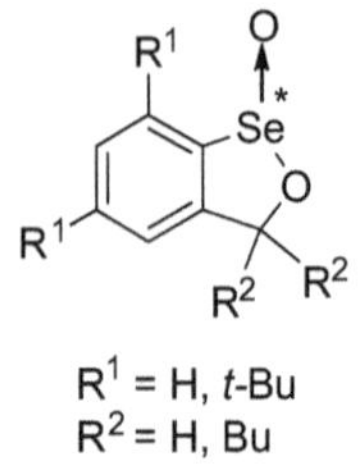

27

Scheme 12.2.14

seleninate esters in solution proceeds *via* an achiral hypervalent selenurane intermediate (hydrate), which is formed by the addition of water to the selenium atom. The optically active areneseleninate esters and arenesulfinate esters, which have bulky substituents at the *ortho* position of the benzene ring, react with Grignard reagents with retention of stereochemistry to give optically active selenoxides and sulfoxides, respectively.

12.2.8 Chalcogeninamides

Many reports on the isolation and properties of optically active sulfinamides have been published.[45] However, there have been no reports on the chiral selenium analogues, probably because of the instability of the seleninamides toward hydrolysis. In 2005, optically active seleninamides **28** were obtained for the first time by chromatographic resolution on an optically active column (Scheme 12.2.15).[49] Their absolute configurations were determined by comparing their chiroptical properties with those of analogous sulfinamides, the configuration of one of which is known whereas that of the other one was determined based on the reaction with Grignard reagent to form chiral sulfoxide with inversion of stereochemistry. The optically active seleninamides were found to racemize in solution. Kinetic studies of racemization and theoretical studies have clarified that also in this case, racemization in solution proceeds *via* an achiral hypervalent hydrate formed by reaction with water.

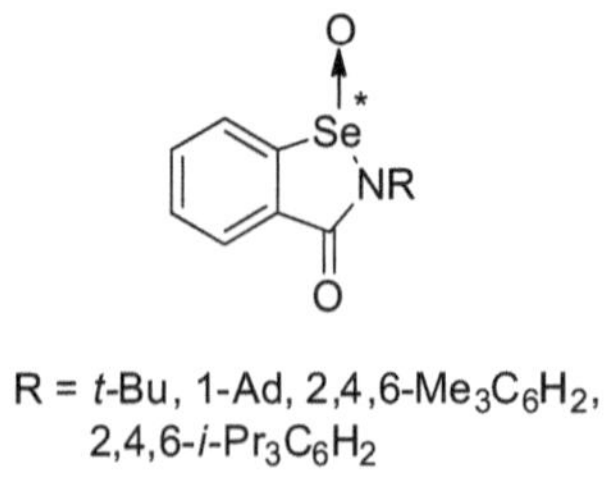

28

Scheme 12.2.15

12.2.9 Chalcogen-Oximines

Sulfoximines are an example of a tetracoordinated chiral compound, and their optical isomers have been isolated. Their stereochemistry was also studied.[50] Endocyclic sulfoximines are an example of chiral heteroaromatics. Endocyclic sulfoximines **29** were optically resolved on a chiral column for the first time by Allenmark and co-workers in 1995 (Scheme 12.2.16).[51] The stereoisomers were obtained by amination and subsequent cyclodehydration of optically active *o*-carboxyphenyl sulfoxide with hydrazoic acid or *O*-(mesitylenesulfonyl)hydroxylamine.

The corresponding optically active selenium compounds, selenoximines, have not been prepared. Optically active selenoximines **30** were isolated for the first time in 2004 by optical resolution using chromatography on a chiral column.[52] The absolute configurations were determined by comparing their specific rotations and circular dichroism spectra with those of the corresponding sulfur analogues with known absolute configurations. No racemization of the chiral selenoximines took place under several neutral, acidic, and basic conditions, and only decomposition was observed under aqueous acidic and basic conditions to give the corresponding selenone and *p*-toluenesulfonamide.

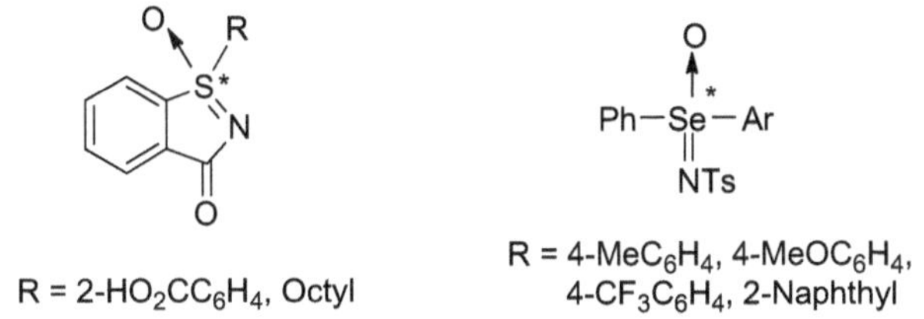

29 **30**

Scheme 12.2.16

12.2.10 Chalcogenuranes

Many optically active hypervalent chalcogen compounds, particularly sulfur compounds, have been synthesized and proposed as important key intermediates in various reactions of the chalcogen compounds.[53] Since the synthesis of spirosulfurane by Kapovits and Kálmán,[54] many optically active spirosulfuranes have been isolated in the last two decades. Spirosulfurane **31** was separated into enantiomers by kinetic resolution using a chiral host molecule and found to be optically stable by Drabowicz and Martin (Scheme 12.2.17).[55] Spirosulfurane **32** was separated into enantiomers by a chromatographic method by Allenmark and Claeson, and characterized by chiroptical methods.[56] Optically active compound **33** was obtained by dehydration of prochiral sulfoxide diol under asymmetric conditions.[57] The

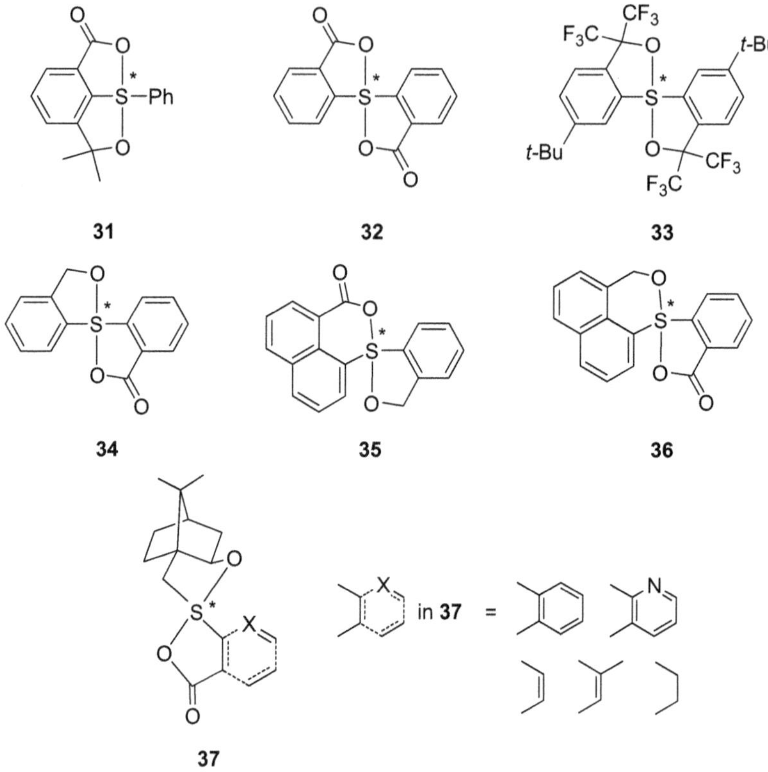

Scheme 12.2.17

dehydration of optically active diaryl sulfoxides also yielded optically active spirosulfuranes **34**, **35**, and **36**.[58] The molecular structures including absolute configurations were determined by the X-ray diffraction method. Optically active spirosulfuranes **37** were obtained as single diastereomers from chiral sulfides by Koizumi *et al.*, and were found to have a slightly distorted TBP structure by X-ray analysis.[59]

In the case of selenuranes, optically pure haloselenuranes **38** were synthesized by treatment of optically active selenoxide with HX (Scheme 12.2.18).[60] Chloroselenuranes **39** were also obtained by treatment of the corresponding allylic selenides with *tert*-butyl hypochlorite.[61] Optically active spiroselenurane was obtained by Lindgren in 1972, although the stereochemistry and the enantiopurity were not clear.[62] Optically pure spiroselenuranes **40** with 2-*exo*-hydroxy-10-bornyl group as the chiral ligand were obtained by treatment of selenides with *tert*-butylhypochlorite, followed by the addition of triethylamine.[63] Chiral spiroselenurane **41** was optically resolved by using an chiral column, and the crystal structure was determined by X-ray analysis.[64]

Selenurane oxides are also one of the hypervalent selenium compounds. The enantiomers of chiral selenurane oxide **42** were isolated for the first time in 2004

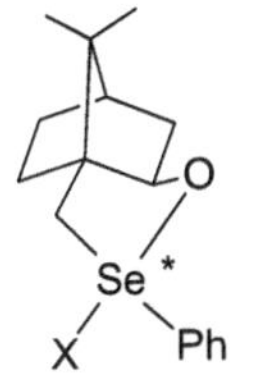

X = F, Cl, Br

38

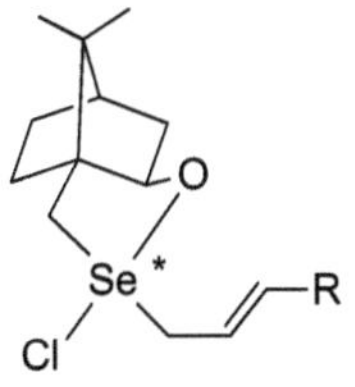

R = Ph, *i*-Pr

39

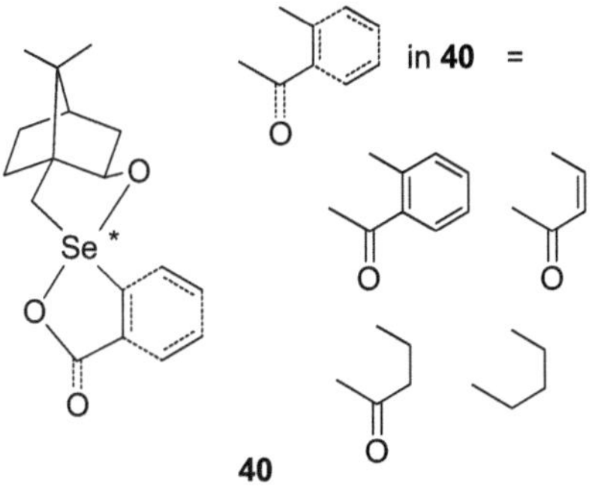

40

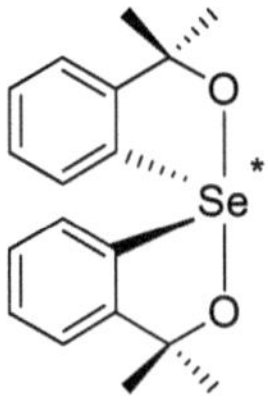

41

Scheme 12.2.18

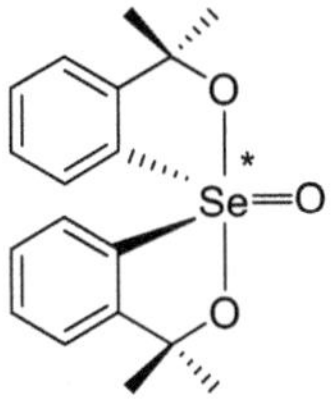

42

Scheme 12.2.19

by enantioselective liquid chromatography of the racemate or by spontaneous resolution that occurred during slow evaporation of its acetonitrile solution or slow crystallization from the same solvent (Scheme 12.2.19).[65] The absolute configurations of the enantiomers were determined by X-ray crystallographic analysis.

In the case of telluranes, treatment of telluride with *tert*-butylhypochlorite yielded optically active chlorotellurane **43** (Scheme 12.2.20).[66] Optically active fluoro-, bromo-, and iodo-telluranes were obtained from the chlorotellurane by treatment with AgF, NaBr, and NaI, respectively. Optically active spirotelluranes **44** were obtained by Koizumi *et al.*, and the stereochemistry was determined by X-ray crystallographic analysis.[63,66]

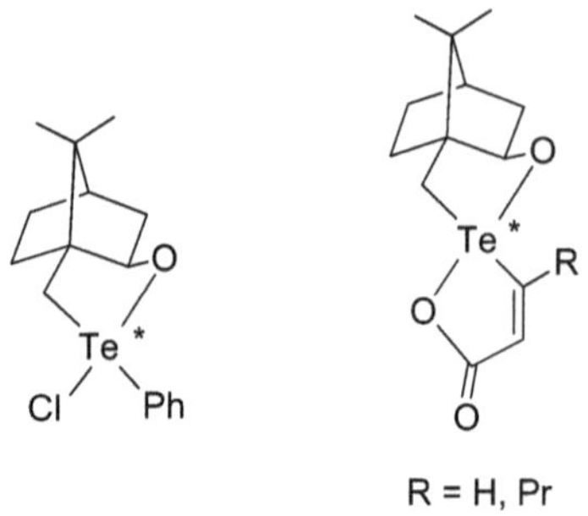

43 44

Scheme 12.2.20

12.2.11 Dichalcogenides

Acyclic dichalcogenides are well known to show skewed structures in the crystalline state or in the gas phase,[67] except for dichalcogenides with congested substituents, and many theoretical studies have indicated that such skewed geometries are global minima. Therefore, dichalcogenides are thought to exist in chiral *P*- and *M*-helical forms. However, the rotation barriers of chalcogen–chalcogen bonds for racemization are not sufficiently high to enable isolation of the optical isomers. The disulfide linked to the 2- and 2′-positions of a chiral 1,1′-binaphthyl group is considered to be a chiral disulfide.[68] However, this compound is configurationally restricted by atropisomerism of the binaphthyl group. Optically active simple dichalcogenides **45**, diphenyl disulfide, diphenyl diselenide, and diphenyl ditelluride, were isolated in bulk for the first time by chiral crystallization in 2001 (Scheme 12.2.21).[69] The relationship between the absolute configurations and the chiroptical properties in the solid state was clarified by X-ray crystallography of one isomer of the disulfide. Furthermore, the stereochemistry of chiral crystallization can be controlled by using an optically active crystal as seed crystal.

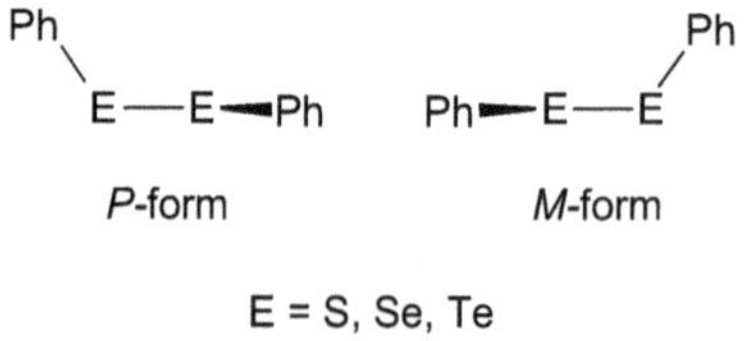

45

Scheme 12.2.21

12.2.12 Application for Asymmetric Synthesis

Recently, some applications of optically active selenium and tellurium compounds for asymmetric synthesis were reported. Enentioselective or diastereoselective selenoxide elimination of vinyl selenoxides yields optically active allenes. Important asymmetric synthesis utilizing chiral selemium and tellurium compounds is a route via [2,3]sigmatropic rearrangement. The [2,3]sigmatropic rearrangement of optically active allylic selenoxides, allylic selenonium imides, or allylic selenonium ylides affords optically active alcohols, amines, or alkylated products, respectively. In these reactions, optically active selenium compounds are transient key intermediates, and prepared *in situ*. These applications are summarized in some reviews.[5]

12.2.13 Conclusions

Recent studies on the synthesis and stereochemistry of chalcogen compounds focused on higher-periodic chalcogen compounds, such as selenium and tellurium compounds. In the chemistry of chiral sulfur compounds, asymmetric synthesis utilizing the chiral compounds is of interest. Many optically active selenium and tellurium compounds were obtained, and their stereochemistry and properties were clarified in the last two decades. Various applications of optically active chalcogen compounds to asymmetric synthesis have also been developed. In the next decade, further development leading to the application of chiral higher-periodic chalcogen compounds is expected.

References

1. For books, see: (a) P. Metzner and A. Tuillier, *Sulfur Reagents in Organic Synthesis*, Academic Press, London, 1993; (b) S. Patai, Z. Rappoport and C. J. M. Stirling, ed., *The Chemistry of Sulphones and Sulphoxides*, Wiley, New York, 1988; (c) M. Mikolajczyk and J. Drabowicz, in *Topics in Stereochemistry*, ed. E. L. Allinger, E. L. Eliel and S. H. Wilen, Wiley, New York, 1982, vol 13, p. 333; (d) C. J. M. Stirling and S. Patai, ed., *The Chemistry of the Sulphonium Group*, Wiley, New York, 1981.
2. For recent book, see: T. Toru and C. Bolm, ed., *Organosulfur Chemistry in Asymmetric Synthesis*, Wiley, Weinheim, 2008.
3. For recent reviews, see: (a) M. Mellah, A. Voituriez and E. Schulz, *Chem. Rev.*, 2007, **107**, 5133; (b) H. Pellissier, *Tetrahedron*, 2007, **63**, 1297.
4. For books, see: C. Paulmier, *Selenium Reagents and Intermediates in Organic Synthesis*, Pergamon Press, Oxford, 1986; S. Patai and Z. Rappoport, ed., *The Chemistry of Organic Selenium and Tellurium Compounds*, Wiley, New York, 1987, vol. 2.
5. For reviews, see: (a) C. Zhu and Y. Huang, *Cur. Org. Chem.*, 2006, **10**, 1905; (b) M. Tiecco, L. Testaferri, F. Marini, L. Bagnoli, C. Santi, A. Temperini, S. Sternativo and C. Tomassini, *Phosphorus, Sulfur, Silicon*, 2005, **180**, 729; (c) S. Uemura, *Phosphorus, Sulfur, Silicon*, 2001, **171**, 13; (d) Y. Nishibayashi and S. Uemura, *Top. Curr. Chem.*, 2000, **208**(201), 235; (e) Y. Nishibayashi and S. Uemura, *Rev. Heteroatom Chem.*, 1996, **14**, 83; (f) N. Komatsu and S. Uemura, *Adv. Detailed React. Mechanisms*, 1995,

4, 73; (g) S. Tomoda, Y. Usuki, K. Fujita and M. Iwaoka, *Rev. Heteroatom Chem.*, 1991, **4**, 249.

6. For review, see: T. Shimizu and N. Kamigata, *Org. Prep. Proced. Int.*, 1997, **29**, 603.
7. For reviews, see: T. Shimizu and N. Kamigata, *Rev. Heteroatom Chem.*, 1998, **18**, 11.
8. For reviews, see: (a) J. Drabowicz, J. Luczak, P. Lyzwa, P. Kielbasinski, M. Mikolajczyk, Y. Yamamoto, S. Matsukawa, K. Akiba, F. Wang, P. L. Polavarapu and M. W. Wieczorek, *Phosphorus, Sulfur, and Silicon*, 2005, **180**, 741; (b) N. Kamigata, *Phosphorus, Sulfur, Silicon*, 2005, **180**, 667.
9. For reviews, see: (a) M. C. Carreño, *Chem. Rev.*, 1995, **95**, 1717; (b) A. J. Walker, *Tetrahedron Asym.*, 1992, **3**, 961.
10. D. N. Jones, D. Mundy and R. D. Whitehouse, *J. Chem. Soc., Chem. Commun.*, 1970, 86.
11. F. A. Davis, J. M. Billmers and O. D. Stringer, *Tetrahedron Lett.*, 1983, **24**, 3191.
12. T. Shimizu, K. Kikuchi, Y. Ishikawa, I. Ikemoto, M. Kobayashi and N. Kamigata, *J. Chem. Soc., Perkin Trans.*, 1989, **1**, 597.
13. For review, see: N. Kamigata and T. Shimizu, *Rev. Heteroatom Chem.*, 1991, **4**, 226.
14. (a) T. Shimizu, M. Enomoto, H. Taka and N. Kamigata, *J. Org. Chem.*, 1999, **64**, 8242; (b) H. Taka, A. Matsumoto, T. Shimizu and N. Kamigata, *Chem. Lett.*, 2000, 726.
15. T. Soma, T. Shimizu, K. Hirabayashi and N. Kamigata, *Heteroatom Chem.*, 2007, **18**, 301.
16. H. Taka, A. Matsumoto, T. Shimizu and N. Kamigata, *Heteroatom Chem.*, 2001, **12**, 227.
17. T. Shimizu, Y. Yamazaki, H. Taka and N. Kamigata, *J. Am. Chem. Soc.*, 1997, **119**, 5966.
18. H. Taka, Y. Yamazaki, T. Shimizu and N. Kamigata, *J. Org. Chem.*, 2000, **65**, 2127.
19. For examples, see: (a) S. J. Campbell and D. Darwish, *Can. J. Chem.*, 1974, **52**, 2953; (b) M. Moriyama, S. Oae, T. Numata and N. Furukawa, *Chem. Ind.*, 1976, 163.
20. K. Sakaki and S. Oae, *Tetrahedron Lett.*, 1976, 3703.
21. (a) N. Kamigata, Y. Nakamura, H. Matsuyama and T. Shimizu, *Chem. Lett.*, 1991, 249; (b) N. Kamigata, Y. Nakamura, K. Kikuchi, I. Ikemoto, T. Shimizu and H. Matsuyama, *J. Chem. Soc., Perkin Trans. 1*, 1992, 1721.
22. T. Takahashi, N. Kurose, S. Kawanami, A. Nojiri, Y. Arai, T. Koizumi and M. Shiro, *Chem. Lett.*, 1995, 379.
23. N. Kamigata, A. Matsuhisa, H. Taka and T. Shimizu, *J. Chem. Soc., Perkin Trans.*, 1995, **1**, 821.
24. T. Shimizu, A. Matsuhisa, N. Kamigata and S. Ikuta, *J. Chem. Soc., Perkin Trans.*, 1995, **2**, 1805.

25. For examples, see: (a) J. Day and D. J. Cram, *J. Am. Chem. Soc.*, 1965, **87**, 4398; (b) M. Moriyama, T. Yoshimura, N. Furukawa, T. Numata and S. Oae, *Tetrahedron*, 1976, **32**, 3003.
26. V. P. Krasnov, V. I. Naddaka and V. I. Minkin, *Zh. Org. Khim.*, 1981, **17**, 445.
27. N. Kamigata, H. Taka, A. Matsuhisa, H. Matsuyama and T. Shimizu, *J. Chem. Soc., Perkin Trans.*, 1994, **1**, 2257.
28. T. Shimizu, N. Seki, H. Taka and N. Kamigata, *J. Org. Chem.*, 1996, **61**, 6013.
29. H. Taka, T. Shimizu, F. Iwasaki, M. Yasui and N. Kamigata, *J. Org. Chem.*, 1999, **64**, 7433.
30. Y. Miyake, M. Oda, A. Oyamada, H. Takada, K. Ohe and S. Uemura, *J. Organomet. Chem.*, 2000, **611**, 475.
31. T. Soma, N. Kamigata, K. Hirabayashi and T. Shimizu, *Bull. Chem. Soc. Jpn.*, 2007, **80**, 2389.
32. (a) T. Shimizu, Y. Machida and N. Kamigata, *Eur. J. Org. Chem.*, 2002, 2265; (b) T. Shimizu, Y. Machida and N. Kamigata, *Heteroatom Chem.*, 2003, **14**, 523.
33. T. Shimizu, N. Kamigata and S. Ikuta, *J. Chem. Soc., Perkin Trans.*, 1999, **2**, 1469.
34. D. Szabó, J. Varga, A. Csámpai and I. Kapovits, *Tetrahedron Asym.*, 2000, **11**, 1303.
35. J. Zhang, S. Saito and T. Koizumi, *Synth. Commun.*, 2001, **31**, 2441.
36. (a) T. M. Lowry and F. L. Gilbert, *J. Chem. Soc.*, 1929, 2867; (b) F. G. Holliman and F. G. Mann, *J. Chem. Soc.*, 1945, 37.
37. (a) T. Shimizu, T. Urakubo and N. Kamigata, *Chem. Lett.*, 1996, 297; (b) T. Shimizu, T. Urakubo and N. Kamigata, *J. Org. Chem.*, 1996, **61**, 8032.
38. T. Shimizu, T. Urakubo, P. Jin, M. Kondo, S. Kitagawa and N. Kamigata, *J. Organometal. Chem.*, 1997, **539**, 171.
39. (a) T. Shimizu, Y. Azami, R. Sakurai, K. Hirabayashi and N. Kamigata, *Chem. Lett.*, 2009, **38**, 1070; (b) T. Shimizu, R. Sakurai, Y. Azami, K. Hirabayashi and N. Kamigata, *Eur. J. Org. Chem.*, 2010, 6556.
40. (a) T. Shimizu, I. Watanabe and N. Kamigata, *Angew. Chem., Int. Ed.*, 2001, **40**, 2460; (b) T. Shimizu, Y. Nakashima, I. Watanabe, K. Hirabayashi and N. Kamigata, *J. Chem. Soc., Perkin Trans.*, 2002, **1**, 2151.
41. Y. Nakashima, T. Shimizu, K. Hirabayashi, M. Yasui, M. Nakazato, F. Iwasaki and N. Kamigata, *Tetrahedron Asym.*, 2004, **15**, 3791.
42. Y. Nakashima, T. Shimizu, K. Hirabayashi and N. Kamigata, *Tetrahedron Lett.*, 2004, **45**, 2301.
43. Y. Nakashima, T. Shimizu, K. Hirabayashi, M. Yasui, M. Nakazato, F. Iwasaki and N. Kamigata, *Bull. Chem. Soc. Jpn.*, 2005, **78**, 710.
44. Y. Nakashima, T. Shimizu, K. Hirabayashi and N. Kamigata, *Org. Lett.*, 2004, **6**, 2575.
45. For books, see: (a) S. Patai, ed., *The Chemistry of Sulphinic Acids, Esters and their Derivatives*, Wiley, Chichester, 1990; (b) D. H. Reid, *Organic*

Compounds of Sulfur, Selenium, and Tellurium, The Chemical Society, London, 1973, vol. 2.
46. I. Fernández and N. Khiar, *Chem. Rev.*, 2003, **103**, 3651.
47. K. K. Andersen, *Tetrahedron Lett.*, 1962, **3**, 93.
48. Y. Nakashima, T. Shimizu, K. Hirabayashi, F. Iwasaki, M. Yamasaki and N. Kamigata, *J. Org. Chem.*, 2005, **70**, 5020.
49. Y. Nakashima, T. Shimizu, K. Hirabayashi and N. Kamigata, *J. Org. Chem.*, 2005, **70**, 868.
50. For review, see: M. Reggelin and C. Zur, *Synthesis*, 2000, 1.
51. (a) S. Allenmark, C. Andersson and P. Widell, *Chirality*, 1995, **7**, 541; (b) S. Allenmark, S. Claeson and C. Löwendahl, *Tetrahedron Asym.*, 1996, **7**, 361.
52. T. Shimizu, K. Mitsuya, K. Hirabayashi and N. Kamigata, *Bull. Chem. Soc. Jpn.*, 2004, **77**, 375.
53. For reviews, see: (a) J. Drabowicz and G. Halaba, *Rev. Heteroatom Chem.*, 2000, **22**, 1; (b) J. Zhang and T. Koizumi, *Phosphorus, Sulfur, and Silicon*, 2000, **157**, 225.
54. I. Kapovits and A. Kálmán, *J. Chem. Soc., Chem. Commun.*, 1971, 649.
55. J. Drabowicz and J. C. Martin, *Tetrahedron Asym.*, 1993, **4**, 297.
56. (a) S. Allenmark and S. Claeson, *Tetrahedron Asym.*, 1993, **4**, 2329; (b) S. Allenmark and S. Claeson, *Enantiomer*, 1996, **1**, 423.
57. J. Drabowicz and J. C. Martin, *Pure and Appl. Chem.*, 1996, **68**, 951.
58. D. Szabó, S. Szendeffy, I. Kapovits, Á. Kucsman, M. Czugler, A. Kálmán and P. Nagy, *Tetrahedron Asym.*, 1997, **8**, 2411.
59. (a) J. Zhang, S. Saito and T. Koizumi, *J. Am. Chem. Soc.*, 1998, **120**, 1631; (b) J. Zhang, S. Saito and T. Koizumi, *J. Org. Chem.*, 1998, **63**, 9375.
60. T. Takahashi, N. Kurose, S. Kawanami, Y. Arai, T. Koizumi and M. Shiro, *J. Org. Chem.*, 1994, **59**, 3262.
61. N. Kurose, T. Takahashi and T. Koizumi, *J. Org. Chem.*, 1996, **61**, 2932.
62. B. Lindgren, *Acta. Chem. Scand.*, 1972, **26**, 2560.
63. J. Zhang, S. Takahashi, S. Saito and T. Koizumi, *Tetrahedron Asym.*, 1998, **9**, 3303.
64. J. Drabowicz, J. Luczak, M. Mikolajczyk, Y. Yamamoto, S. Matsukawa and K. Akiba, *Tetrahedron Asym.*, 2002, **13**, 2079.
65. J. Drabowicz, J. Luczak, M. Mikolajczyk, Y. Yamamoto, S. Matsukawa and K. Akiba, *Chirality*, 2004, **16**, 598.
66. T. Takahashi, J. Zhang, N. Kurose, S. Takahashi, T. Koizumi and M. Shiro, *Tetrahedron Asym.*, 1996, **7**, 2797.
67. For review, see: R. Steudel, *Angew. Chem., Int. Ed.*, 1975, **14**, 655.
68. (a) W. L. F. Armarego and E. E. Turner, *J. Chem. Soc.*, 1957, 13; (b) R. Kuroda and S. F. Mason, *Tetrahedron*, 1981, **37**, 1995.
69. T. Shimizu, H. Isono, M. Yasui, F. Iwasaki and N. Kamigata, *Org. Lett.*, 2001, **3**, 3639.

CHAPTER 12.3

Hypervalent Chalcogen Compounds

SATOKO HAYASHI AND WARO NAKANISHI*

Department of Material Science and Chemistry, Faculty of Systems Engineering, Wakayama University, 930 Sakaedani, Wakayama 640-8510, Japan
*Email: nakanisi@sys.wakayama-u.ac.jp

12.3.1 Introduction

Hypervalent chalcogen chemistry has been developed considerbly in the last decade. The main purpose of this chapter is to review the development of hypervalent chalcogen chemistry since the first version of this handbook,[1] published in 2006. The appearance of hypervalency in chalcogen atoms is closely related to the occurrence of the inter- and/or intramolecular interactions of the donor–acceptor type at the chalcogen atoms. Formal valency at chalcogen atoms will increase in this process. Singh and co-workers reviewed the role of intramolecular interactions in organoselenium chemistry.[2] The intramolecular interactions should occur much more easily if the steric conditions are suitably satisfied. 1,2-Positions in the benzene system, 1,8-positions in naphthalene system, 1,10-positions of phenanthrene system, 1,8,9-positions in anthracene and anthraquinone systems and the related systems are the typical candidates for the interactions. Woollins and co-workers reviewed the chemistry of 1,8-substituted naphthalene and related systems by Group 15 and 16 elements.[3] Hypervalency of chalcogens is demonstrated to play an import role also in biological, pharmaceutical,[4–6] and materials sciences.[7] The chemistry of such fields containing hypervalent

Handbook of Chalcogen Chemistry: New Perspectives in Sulfur, Selenium and Tellurium, 2nd Edition, Volume 2
Edited by Francesco Antonio Devillanova and Wolf-Walther du Mont

Published by the Royal Society of Chemistry, www.rsc.org

chalcogen compounds is discussed in more detail in chapters relating to these fields.

The classification of hypervalent chalcogen compounds is follows that in the first version, for convenience.

12.3.2 Hypervalent Chalcogen Compounds

12.3.2.1 10–E–3 (T-shaped)

Anionic T-shaped 10–E–3 compounds were prepared for $PyH^+[mesTeI_2]^-$ (**1**: mes = 2,4,6-$Me_3C_6H_2$), $(PyH)^+[mesTeIBr]^-$ (**2**), and $(PyH)^+[mesTeICl]^-$ (**3**) in the reaction of $[mesTe]_2$ with iodine followed by PyH^+X^- (X = I, Br, Cl).[8] The anionic fragments of **1** and **2** are assembled as dimers by reciprocal, secondary Te···X interactions, linked also to the pyridinium cations through μ-NH···X bonding. The Te atoms in $(PyH)^+[mesTeBr_2]^-$ (**4**), $(PyH)^+[mesTeClBr]^-$ (**5**), and $(PyH)^+[mesTeCl_2]^-$ (**6**) adopt a T-shaped configuration, which do not produce the dimeric forms.[9] Each anion in **1–4** does not interact with neighbouring anionic moieties, achieving also secondary NH···X bonding toward the pyridinium cation. Figure 12.3.1 shows the asymmetric unit of **4**,[9,10] which is also representative for the configuration of **5** and **6**.

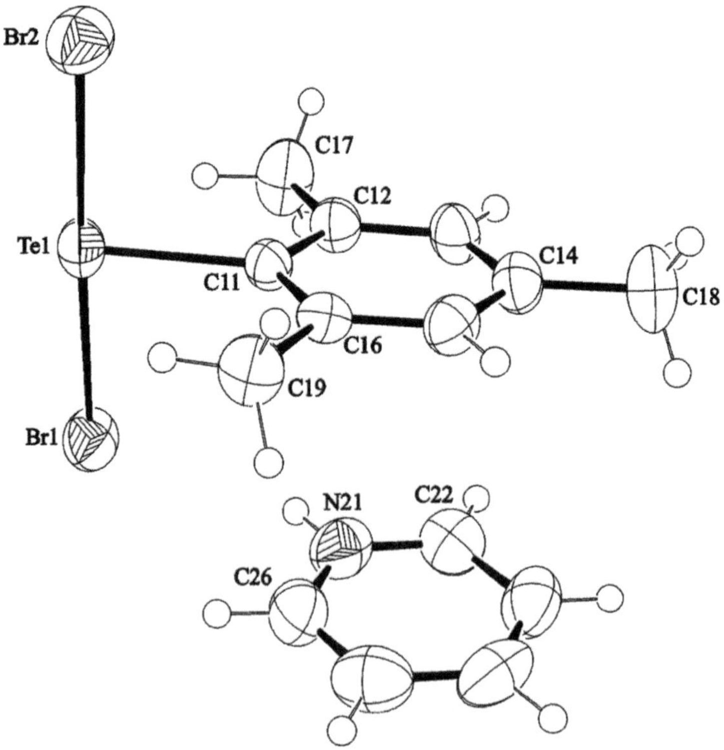

Figure 12.3.1 Molecular structure of **4** (asymmetric unit).

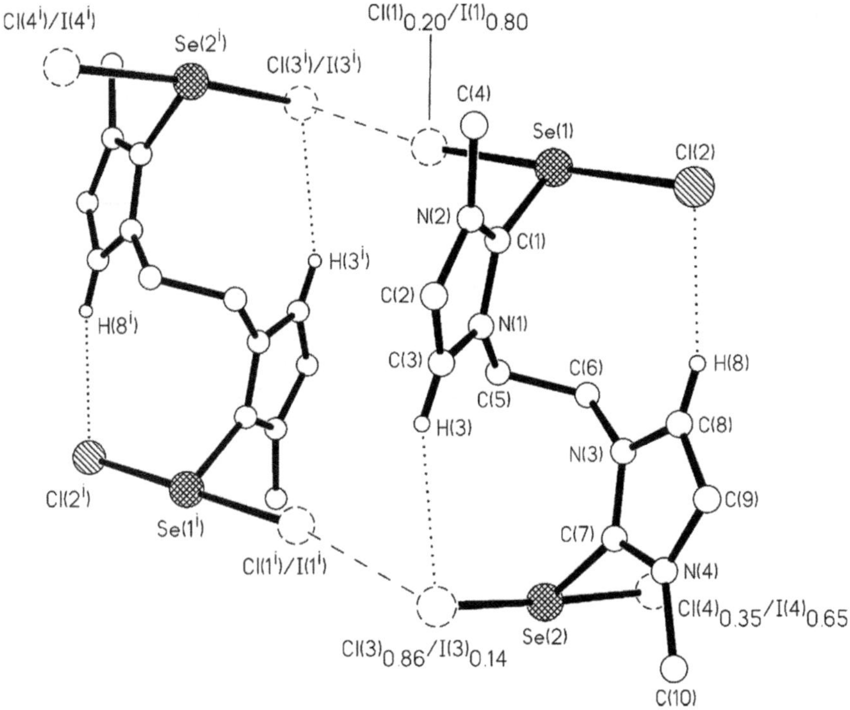

Figure 12.3.2 View of a dimeric unit of **8** · $(ICl)_2$. For clarity, only those hydrogen atoms involved in intramolecular hydrogen bonds (dotted lines) are shown.

Devillanova and co-workers investigated the reactivity of the imidazoline-2-selone derivatives, 1,1′-methylenebis(3-methyl-4-imidazoline-2-selone) (**7**) and 1,2-ethylenebis(3-methyl-4-imidazoline-2-selone) (**8**), towards IBr and ICl in the solid state.[11] They obtained T-shaped hypervalent chalcogen compounds featuring the extremely rare linear asymmetric I–E–X moieties (E = S, Se; X = Br, Cl). Figure 12.3.2 shows the structure of **8** · $(ICl)_2$. The 3c–4e nature of the bonding was established based on QTAIM (quantum theory of atoms-in-molecules) and ELF (electron localization function) analysis, employing the model compounds.

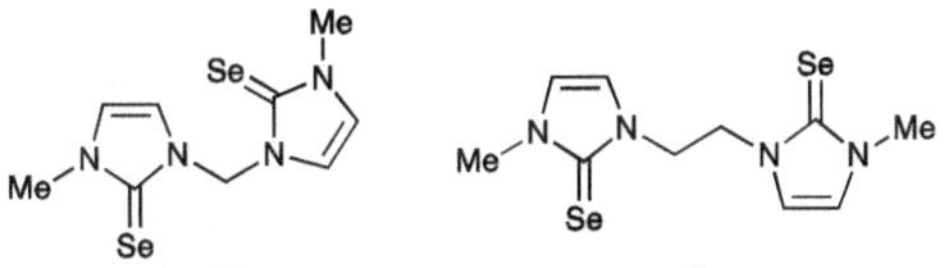

7 **8**

Stability, kinetics, and mechanisms have been investigated for the T-shaped hypervalent chalcogen (TY) and linear charge transfer (CT)-type adducts formed in the reactions between L = E donors [L: 1,3-dimethylimidazolyl framework; E: S (**9**), Se (**10**)] and halogens X_2 (X: Cl, Br, I), on the basis of the

Table 12.3.1 Stability of the CT adducts with respect to the TY isomers calculated for **9** (E = S) and **10** (E = Se).[a,b]

X_2	ΔE (**9** (E = S))	ΔE (**10** (E = Se))
Cl_2	11.5	21.3
Br_2	2.1	10.6
I_2	−3.7	3.5

[a]The ΔE value (= E (CT) − E(TY)) is given.
[b]In kcal mol^{-1}.

DFT calculations.[12] Table 12.3.1 shows the results. Hypervalent compounds are predicted to be always more stable than the CT in the gas phase, in agreement with the experimental results, except for the diiodine adducts of sulfur donors. Predicted reaction mechanism reveals that CT adducts are always the first products of the reactions and no transition states connecting the reactants directly to TY adducts have been found.

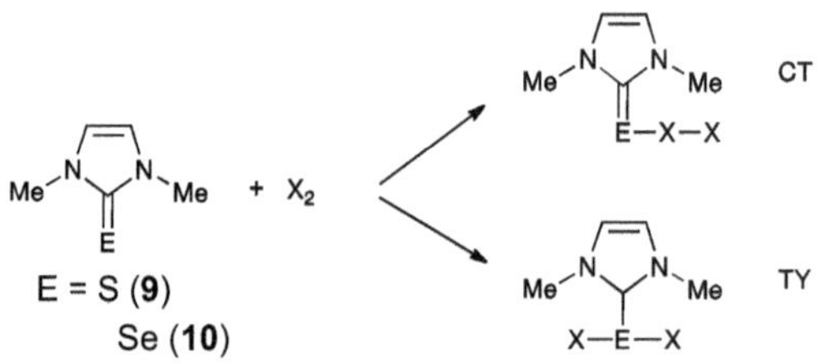

du Mont and co-workers obtained 10–Se–3 T-shaped $R_2R'P$–$SeBr_2$ (**12a–d**) in the reaction of bulky trialkylphosphane selenides, *t*-Bu_3PSe (**11a**), *i*-Pr_3PSe (**11b**), *t*-Bu_2(*i*-Pr)PSe (**11c**), and *t*-Bu(*i*-Pr)$_2$PSe (**11d**), with bromine.[13] The structures of **11b**, **12b–d**, and *t*-Bu_2(*i*-Pr)PBr_2 were determined by X-ray crystallography. Two different crystalline phases were isolated for **12b**. Figure 12.3.3 shows the structure of **12d**. ^{77}Se NMR shifts are shown not to correlate with increasing numbers of *t*-butyl groups in **11a–d**.

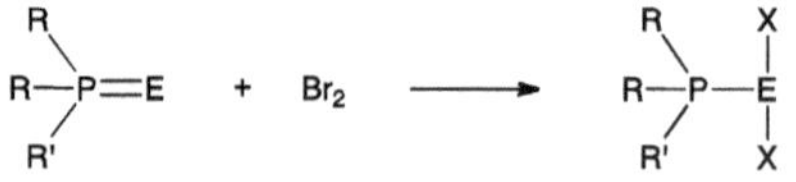

11a: R, R' = *t*-Bu
11b: R, R' = *i*-Pr
11c: R = *t*-Bu, R' = *i*-Pr
11d: R = *i*-Pr, R' = *t*-Bu

12a–d

T-shaped Y···E–X 3c–4e interactions occur through the attack of Y at E of polar E–X bonds. Such interactions are usually highly asymmetric.[14,15] Both side intramolecular interactions around E also lead to hypervalent X···E···X′ σ-type 3c–4e. There seems to be no border between the asymmetric T-shaped Y···E–X 3c–4e type interactions and the typical σ-type X···E···X 3c–4e forms, since the former come close to the latter successively as the Y···E interaction becomes stronger. The T-shaped 3c–4e interactions will also be stabilized by the cyclic 10π electron system in addition to the σ-type 3c–4e.

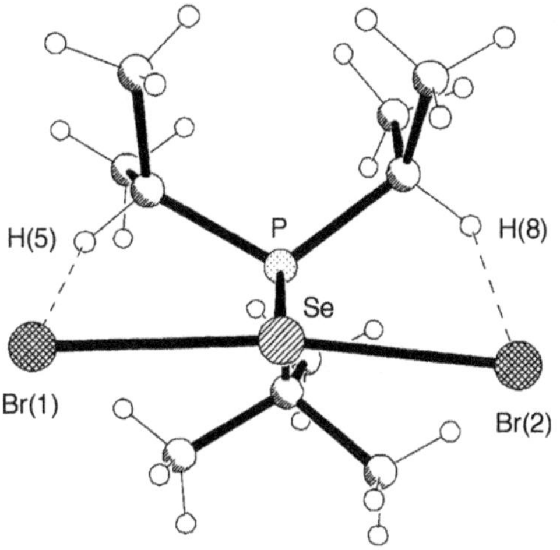

Figure 12.3.3 Molecular structure of **12d**: H(5)···Br(1) 2.99 Å, H(8)···Br(2) 2.67 Å.

Table 12.3.2 Substituent dependence of the energy differences.

	X	*Y*		$\Delta G^{\ddagger}_{298}$ (*kcal mol*$^{-1}$)	ΔE_{Total} (*kcal mol*$^{-1}$)	ΔE_{S-N} (*kcal mol*$^{-1}$)
14	H	H		16.6	15.7	56.5
15	Br	H	H-side	18.3	17.4	70.7
			Br-side	15.5	13.7	42.7

The hypervalent 3c–4e bond is the central feature of the structure and reactivity of hypervalent compounds.[16] Akiba and co-workers discussed the 3c–4e bond thoroughly.[17] The S···S···S bond of thiathiophthene (**13**), the so-called no-bond resonance compound, is realized as one of the examples of 3c–4e centred at the sulfur atom (10–S–3). A variety of reactions such as cycloaddition-elimination, bond-switching rearrangement, and bond-switching equilibration of sulfur containing heterocycles are shown to proceed through the presence of hypervalent 10–S–3 species.

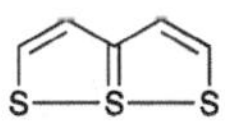

13

The nature of the hypervalent N···S···N bonds has been investigated thoroughly by Akiba and co-workers.[16] The N–S–N bond was investigated theoretically with DFT.[18] The total energy of the system was partitioned into atoms (E_{AA}) and interatomic regions (E_{AB}) by using the Bond-EDA technique. The energy for the N–S–N hypervalent bond in the global minimum (GM) of **14** (X = Y = H) is evaluated to be 60.6 kcal mol^{-1}, which corresponds to E_{AB} of the two S–N bonds. Table 12.3.2 summarizes the differences in energies between GM and the transition state (TS) for the compounds. The total energy of the system changes by 15.7 kcal mol^{-1} (ΔE_{Total}) where the hypervalent bond is broken upon rotation of the pyrimidine ring of a sulfurane compound. The results support previous experimental data.[17,19]

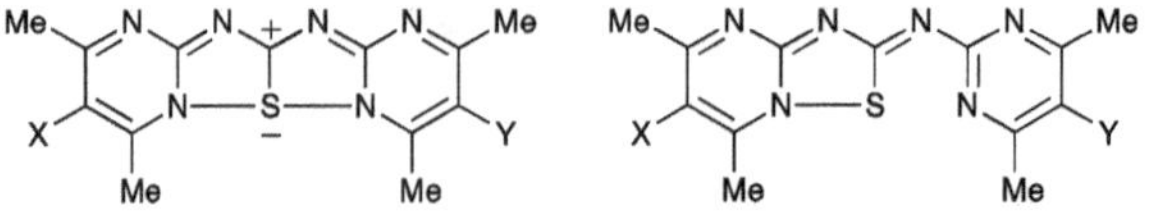

14 (X = Y = H): GM
15 (X = Br, Y = H): GM

14 (X = Y = H): TS
15 (X = Br, Y = H): TS

Reactions of the exocyclic amino groups of 2-aminobenzothiazoles with CS_2 gave series of molecules bearing thiourea, dithiocarbamate, dithiocarboimine, *S*-CH_3 and *O*-alkyl thiocarbamic ester, guanidine groups, and so on.[20] Figure 12.3.4 shows the X-ray structure of dibenzothiazolyltetraazathiapentalene **16**. The N3–S12 [2.051(2) Å] and N3A–S12 [1.868(2) Å] bonds are of different length, and longer than an N–S single bond (1.77 Å), but far from the sums of the van der Waals radii (S–N 3.35 Å). The C11–S12 bond length [1.777(4) Å] is slightly longer than a single C–S bond (1.75 Å). The T-shaped hypervalent sulfur atom in **16** was discussed as displaying 'single bond–no-bond resonance'.

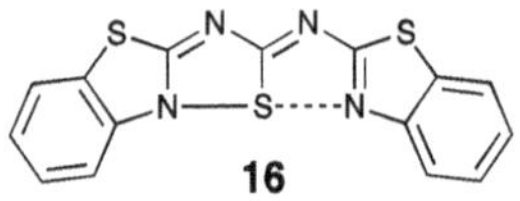

16

To clarify the nature of the NSN σ(3c–4e) bonding, together with the related NSeN and NTeN, the dynamic and static behaviours are investigated by applying AIM dual functional analysis on **17**–**19**.[21] The structures were optimized computationally, retaining C_{2v} symmetry. All bond-critical points

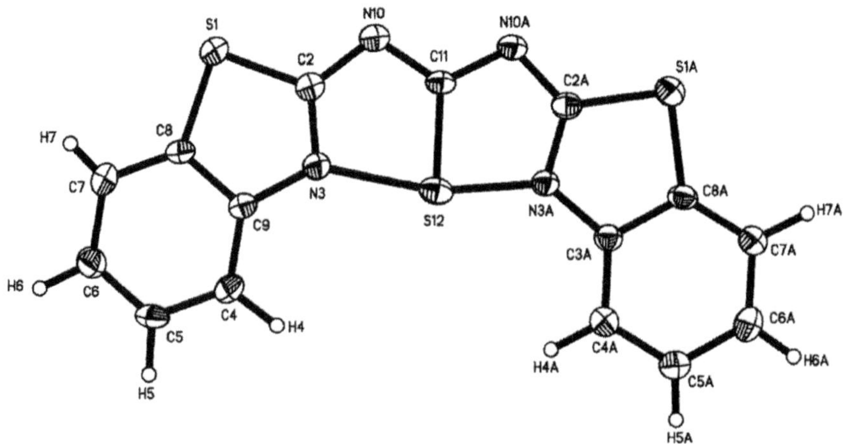

Figure 12.3.4 Molecular structure of **16**.

are detected as expected and exhibited on both sides of the N···E···N moiety, which supports the formation of NEN σ(3c–4e). It is confirmed that N···S···N is of the covalent nature close to Me_2S^+–Cl or Me_2Se^+–Br, whereas N···Se···N and N···Te···N have the (regular) CS nature close to the CT adducts of $Me_2S(–Cl)_2$ (TBP) and $Me_2Se···Br_2$ (MC), respectively. An experimental high-resolution charge density determination has been performed on NSN, which thoroughly supports the theoretical results. Details of the AIM analysis are discussed in Chapter 12.4.

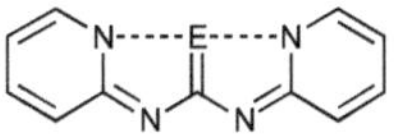

17 (E = S), **18** (E = Se), **19** (E = Te)

12.3.2.2 10–E–4

12.3.2.2.1 10–E–4 (X4: X = heteroatoms)

The molecular structure of a novel hypervalent tellurium (IV) compound, 2,2′-spirobi[4,7-diisopropyl-1,3,2-benzodithiatellurole] (**20**), has been determined by a single-crystal X-ray diffraction study.[22] The results represent the first crystal structural characterization of a hypervalent tellurium (IV) (tellurane) with only four sulfur ligands (Figure 12.3.5). The tellurane has a distorted pseudo-trigonal-bipyramidal geometry with the axial sulfur atoms bending toward the equatorial ones.

Srivastava and co-workers obtained $C_8H_8Te(S_2CNC_5H_{10})_2$ and $Te(S_2CNC_5H_{10})_2$ (**21**) or $C_8H_8TeI(S_2CNC_5H_{10})$ in the reaction of 1,3-dihydro-2λ^4-benzotellurole-2,2-diyl diiodide ($C_8H_8TeI_2$) with $NH_4S_2CNC_5H_{10}$ according to the reaction conditions.[23] In such a metathetical type of reaction, the formation of **21** is unprecedented and it corresponds to the first Te–C cleaved product. Figure 12.3.6 shows the structure of **21**. The planar structure is predicted around the TeS_4 moiety ($\Sigma\theta = 359.92°$). The planar structure is also predicted around the TeS_2I_2 moiety in $TeI_2(SCN_2Ph_2)_2$ with ∠STeS = ∠ITeI = 180.0°.

Kato and co-workers reported the synthesis of a series of selenium and tellurium bis(carbodithioates and carbothioates). Figure 12.3.7 shows the structures of $Se(SSCC_6H_4OMe\text{-}2)_2$ (**22**), $Te(SSCC_6H_4OMe\text{-}2)_2$ (**23**), and $Te(SSCC_6H_4Me\text{-}4)_2$ (**24**).[24] X-Ray structure analysis revealed that **22**–**24** have trapezoidal-planar configuration of ES_4 (E = Se, Te). The C=S···Te distances in **23** are comparable to the C=S···Se distances in **22**. QC calculations containing NBO analysis clarified the nature of the bondings.

12.3.2.2.2 10–E–4 (C2X2: X = heteroatoms)

Kawashima and co-workers prepared 1,2-dithietane **25** in 42% yield through treatment of **26** with *N*-bromosuccinimide and triethylamine in CCl_4 at 0 °C.[25] Figure 12.3.8 shows the crystal structure of **25**, which is the first example of a

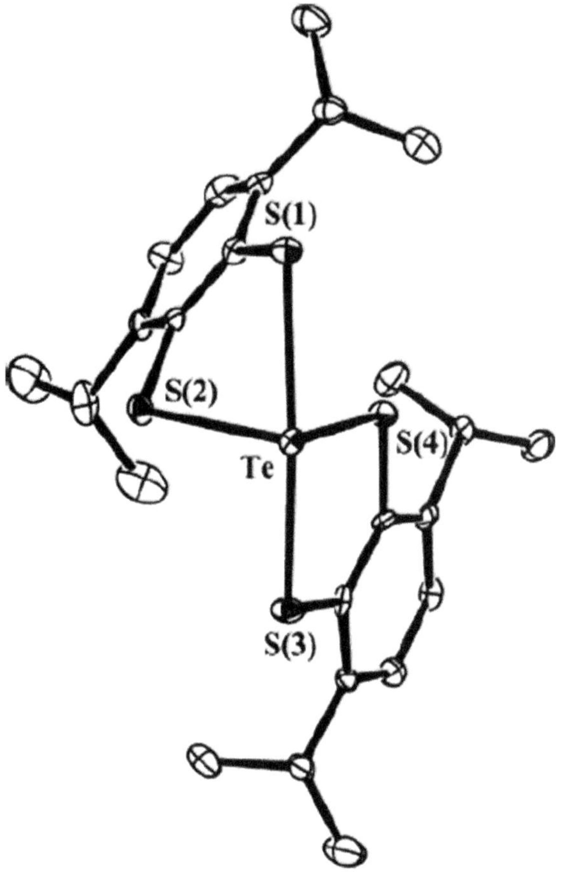

Figure 12.3.5 Molecular structure of **20** (50% ellipsoid).

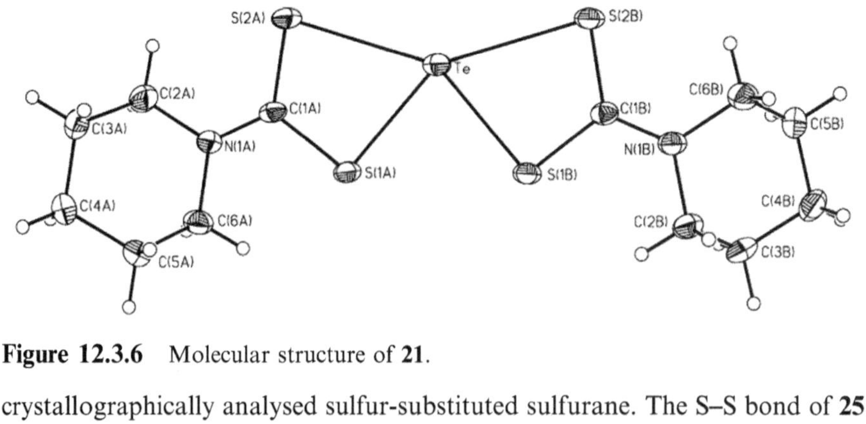

Figure 12.3.6 Molecular structure of **21**.

crystallographically analysed sulfur-substituted sulfurane. The S–S bond of **25** can be cleaved by hydride reduction without cleavage of the S–C bond. Reaction of **25** with $LiAlH_4$ gave **26** in 60% yield after quenching with aqueous NH_4Cl (Scheme 12.3.1). The formation of **26** shows reductive cleavage of the

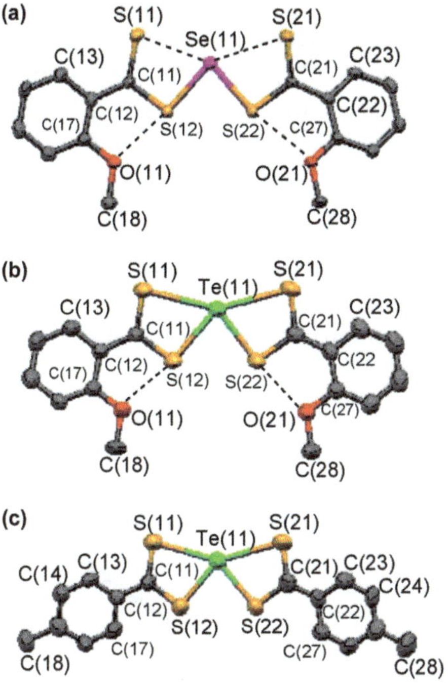

Figure 12.3.7 Molecular structures of **22** (a), **23** (b), and **24** (c) (50% ellipsoid).

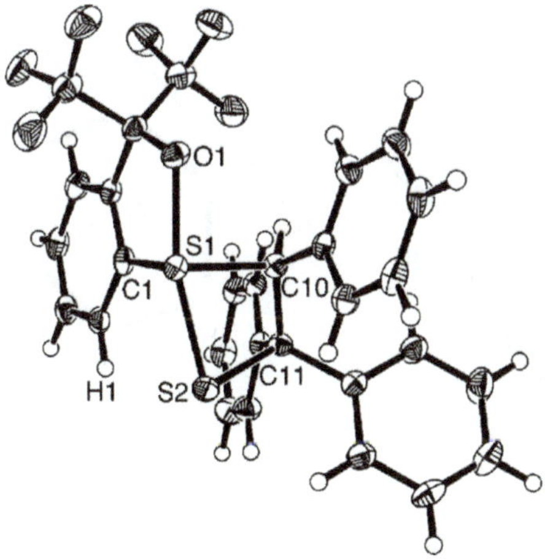

Figure 12.3.8 Molecular structure of **25** (50% ellipsoid).

S^{IV}–S^{II} bond accompanying the S^{IV}–O bond cleavage under the hydride reduction conditions. The oxidative formation of the S^{IV}–S^{II} bond is similar to the formation–cleavage processes of disulfide and thiol systems.

Woollins and co-workers prepared compounds **27** and **28** in the acenaphthene system, containing 10–Se–4 and 10–Te–4, respectively.[26] The

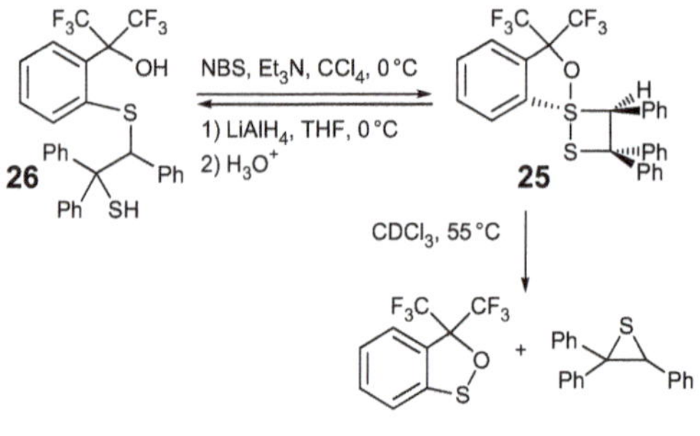

Scheme 12.3.1 Interconversion between **25** and **26** with the thermal decomposition of **25** in $CDCl_3$.

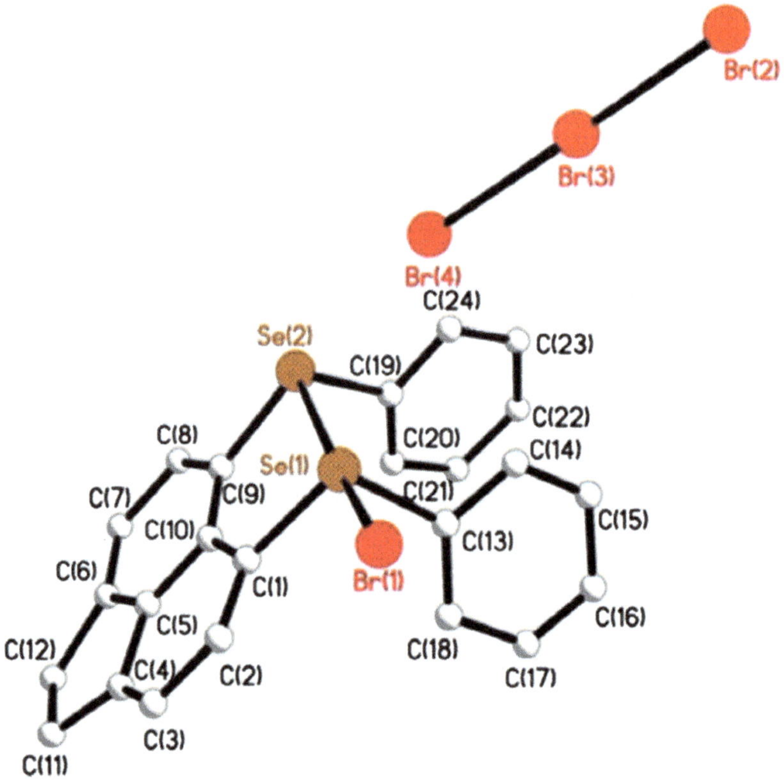

Figure 12.3.9 Molecular structure of bromoselanyl tribromide salt **27** (50% ellipsoid).

structures were confirmed by the X-ray crystatallographic analysis. Figures 12.3.9 and 12.3.10 show the structures of **27** and **28**, respectively. Weak interactions must also be operating at the peri-positions of the compounds, due to the suitable distances of the positions.

Compounds containing 10–Te–4 (C2X2) such as aryldiacyloxy-*n*-butyltellurane (**29**) was prepared, together with methyl 5-*t*-butyl-3-oxo-3*H*-benzo[*c*][1,2]oxatellurole-7-carboxylate **30** which has the

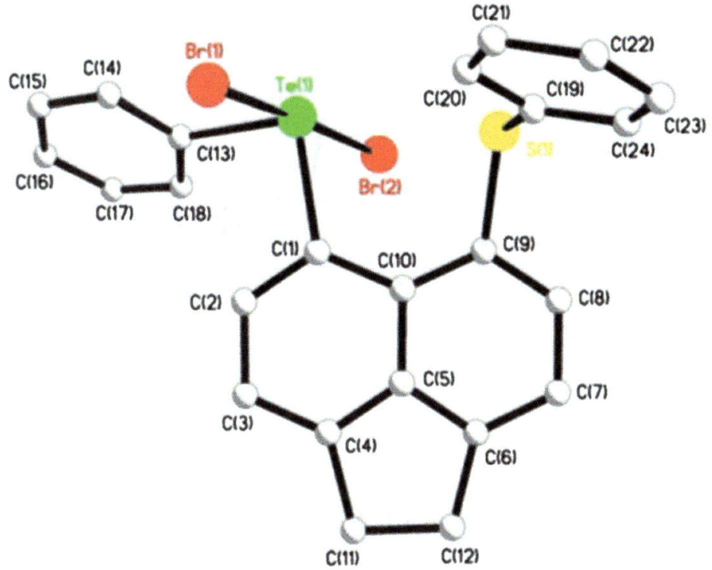

Figure 12.3.10 Molecular see-saw geometry adopted in the structure of tellurium insertion adduct **28**.

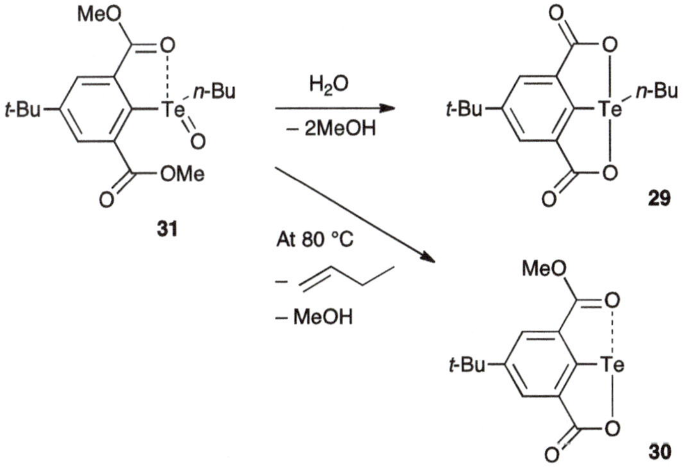

Scheme 12.3.2 Synthesis of **29** and **30** from **31**.

unsymmetric O···Te–O interactions (Scheme 12.3.2).[27] Reactions of Te compounds containing **31** were investigated and the structures containing the related compounds were also spectroscopically investigated. Figure 12.3.11 shows the structure of **29**, confirmed by X-ray analysis.

The enantiomers of 3,3,3′,3′-tetramethyl-1,1′-spirobi[3*H*,2,1]benzoxaselenole (**32**) have been separated on a chiral preparative chromatographic column.[28] The experimental vibrational circular dichroism (VCD) spectra have been

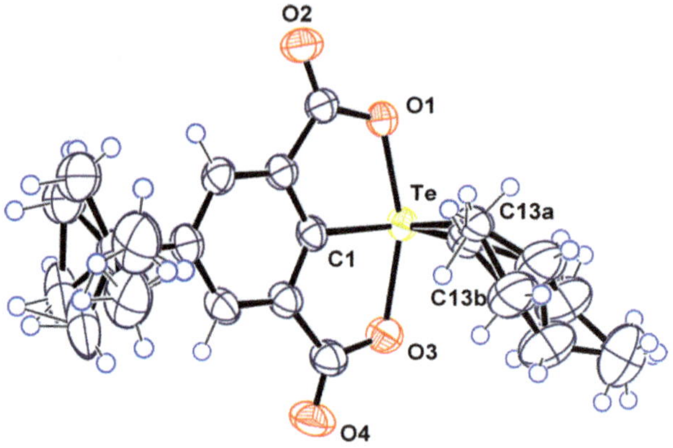

Figure 12.3.11 Molecular structure of **29** (50% ellipsoid): the *n*-Bu group being disordered.

obtained for both enantiomers in CH_2Cl_2. The absolute configuration of an enantiomer with positive specific rotation in CH_2Cl_2 at 589 nm is determined to be *R* through the calculation at the DFT (B3LYP) level. This conclusion has been verified by comparing results of experimental optical rotatory dispersion (ORD) and electronic circular dichroism (ECD) to predictions of the same properties using the B3LYP functional for **32**. In this connection, the stereochemistry of the hypervalent spirochalcogenuranes is reviewed, especially for the role of the central atom (S, Se, Te) for the chemical and configurational stability and the chiroptical and X-ray crystallographic techniques available for the determination of absolute configuration.[29]

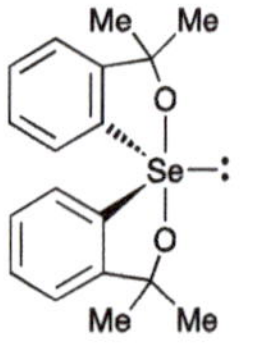

(−)$_{589}$-(*S*)-**32**

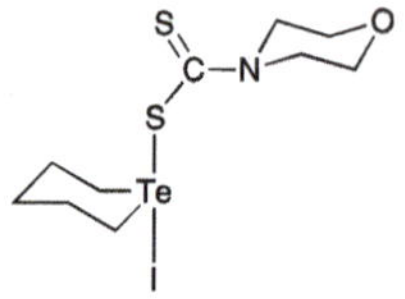

33

The structure of $C_5H_{10}TeI(S_2CNC_4H_8O)$ (**33**) determined by X-ray analysis was reported, which contains the telluracyclohexane ring.[23] Figure 12.3.12 shows the structure of **33**. The ring is present in chair conformation in which part of the ring is flattened but the part around the TeC_2SI moiety is considerably puckered. Hypervalent O–Te–O oligomers have also been investigated recently.[30]

12.3.2.3 12–E–6 and Higher

Several novel organoselenium and tellurium compounds were prepared and evaluated as mimetics of the selenoenzyme glutathione peroxidase, which protects cells from oxidative stress by reducing harmful peroxides with the thiol

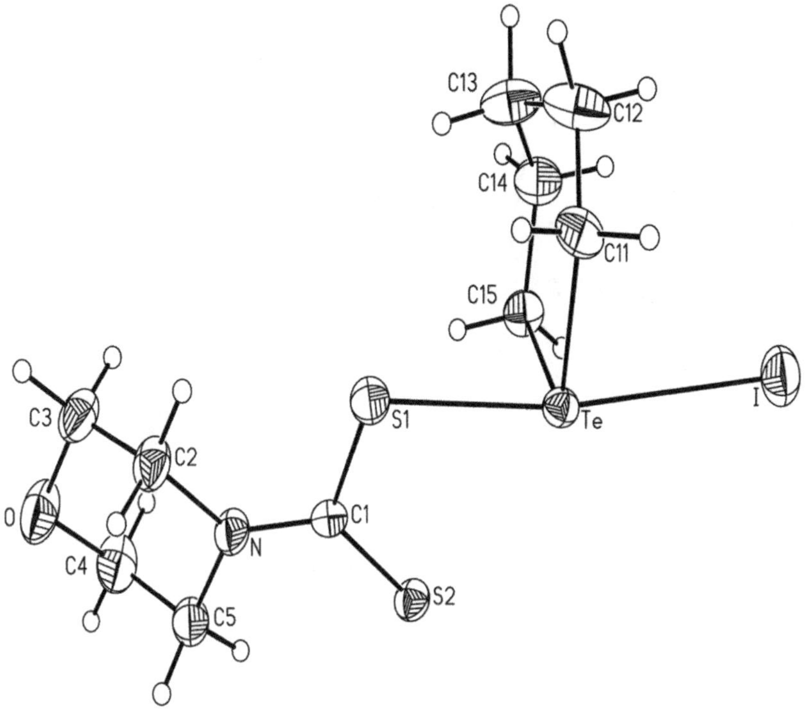

Figure 12.3.12 Molecular structure of **33** (50% ellipsoid).

glutathione. Among the compounds, spirodioxytellurane **34** produced the unusual and unexpected peroxide **35**, through oxidation with hydrogen peroxide, in which two hypervalent octahedral tellurium moieties are joined by ether and peroxide bridges.[31] Figure 12.3.13 shows the structure of **35**, which contains 12–Te–6.

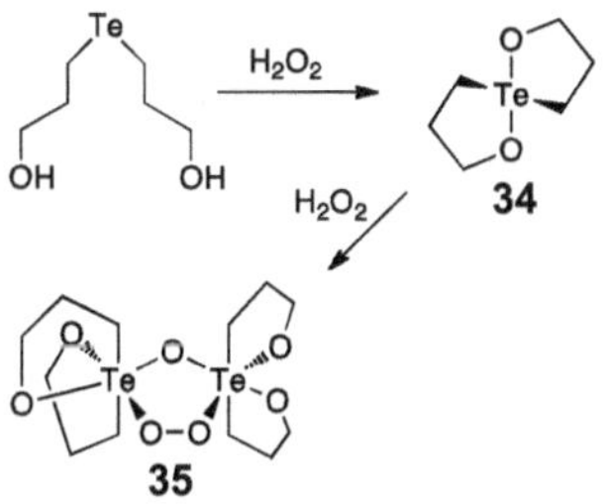

Liquid crystals that have ortho-fluorinated pentafluoro-λ^6-sulfanyl terminal groups are the most polar materials that are still compatible with current active matrix LCD technology. The synthesis of the first examples of this class of substance has been achieved by direct fluorination (Scheme 12.3.3).[32] The physico-chemical characterization, supported by DFT calculations, revealed some surprising features of these new compounds: the bulky SF_5 group is easily

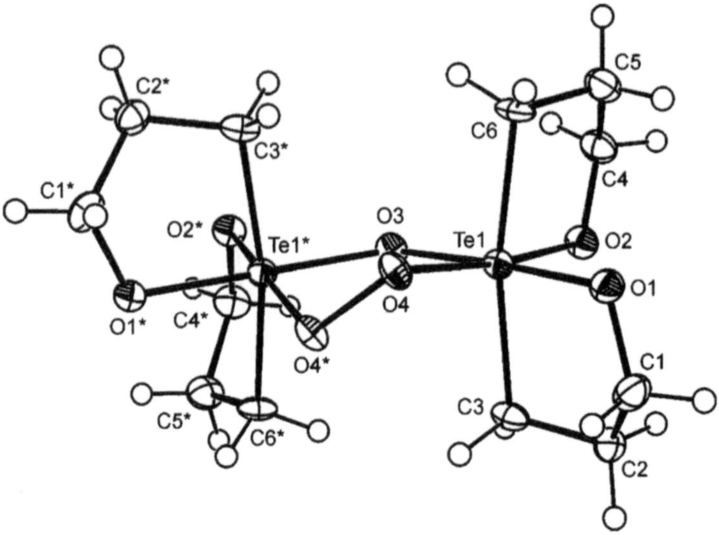

Figure 12.3.13 Molecular structure of peroxide **35** (50% ellipsoid).

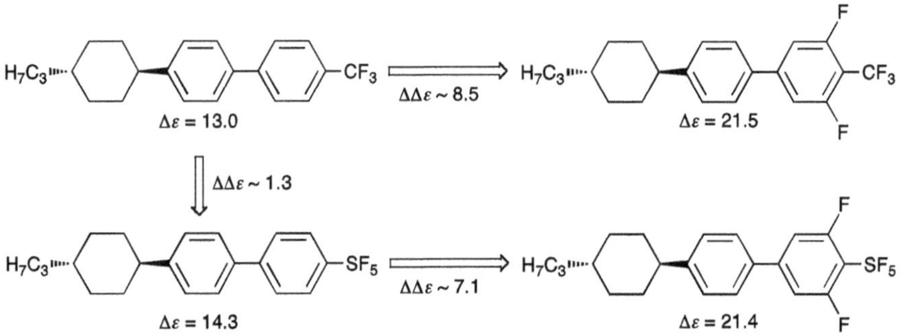

Scheme 12.3.3 Increase in the dielectric anisotropies (Δε) of liquid crystals induced by the replacement of CF_3 by the SF_5 group and by difluorination of the *ortho*-positions of the polar terminal group.

deformed and responds in a quite unusual way to the steric strain exerted by its *ortho*-substituents.

Selective direct fluorination provides a convenient access to new classes of liquid crystals and dyes carrying a trans-(trifluoromethyl)tetrafluorosulfuranyl group, a building block in organic chemistry with highly unusual and interesting properties.[33]

Srivastava and co-workers also obtained the oxidative addition product $CH_3TeI(S_2CNC_5H_{10})_2$ (**36**) in the reaction of $Te(S_2CNC_5H_{10})_2$ (**21**) with CH_3I.[23] Figure 12.3.14 shows the structure of **36**. Coordination to the tellurium atom in **36** is from four sulfur atoms, one methyl carbon, and an

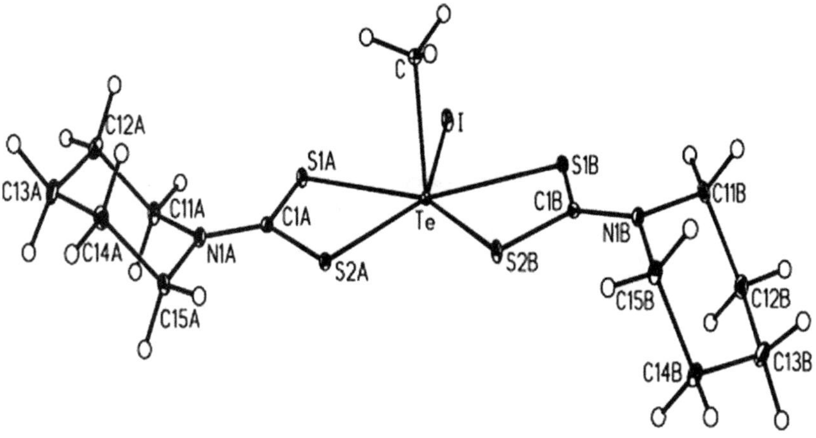

Figure 12.3.14 Molecular structure of **36**: Te–S = 2.671(7)–2.722(8) Å; Te–I = 2.973(4)Å; ∠S–Te–I = 77.08(17)–77.32(17)°; Te–C = 2.122(3) Å; ∠C–Te–S = 86.46(8)–88.62(8)°; ∠S–Te–S = 65.45(2)–73.66(2)° with ∠S(1A)–Te–S(1B) = 153.41(2)°.

iodine atom. The methyl group is perpendicular to the plane and the sterically active electron lone pair in the opposite side of the plane imparts an overall seven-coordinated pseudo-pentagonal bipyramidal environment about tellurium atom.

12.3.3 Weak Hypervalent Interactions

Weak hypervalent G···E–X 3c–4e type interactions[14,15,34] often control the fine structures of the compounds, which are closely related to T-shaped 3c–4e interactions.

Singh and co-workers prepared versatile types of chalcogen compounds, which are mainly related to the glutathione peroxidase-like activity and similar activities of the ebselen analogue. (The chemistry of ebselen and the related compounds is discussed in more detail in Chapter 10.2.) Most of the compounds contain weak E···O/E···N intramolecular interactions of the CS (closed-shell) type. The interactions lead to T-shaped 3c–4e interactions, which control the structures.

A chiral R*SeCl was obtained through reaction of R_2*Se_2 (**37**: R* = 2-$Me_2NCH(Me)C_6H_4$) with a Pd(II) chloride complex, together with the corresponding R*SeBr (**38**), in the process of preparing **37**.[35] On the other hand, the reaction of achiral ditelluride, R_2Te_2 (R = 2-$Me_2NCH_2C_6H_4$), with a Pd(II) chloride complex gave a novel dinuclear complex, $[PdCl(TeC_6H_4CH_2NMe_2\text{-}2)]_2$, and the dimeric tellurium complex of the di(μ-Cl) type $(RTeCl)_2$, **39**. Among them, the structures of **38** and **39** are shown in Figures 12.3.15 and 12.3.16, respectively. Lots of compounds containing the secondary interactions were obtained in the similar reactions.

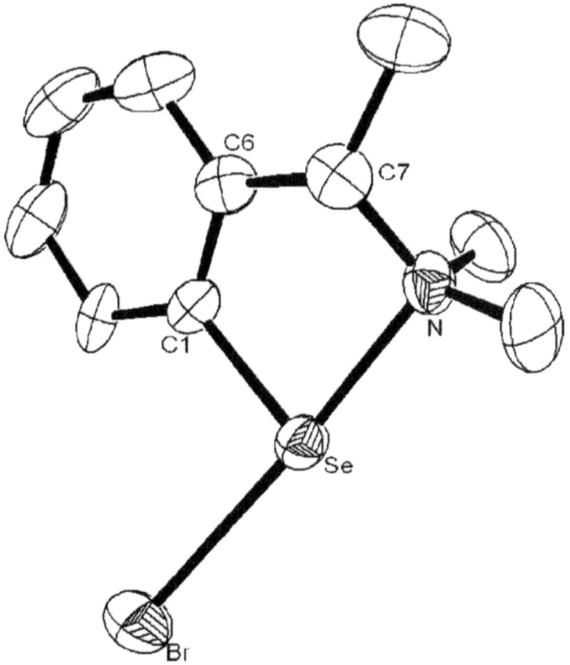

Figure 12.3.15 Molecular structure of **38**: Se–Br = 2.6114(12) Å; Se···N = 2.132(6) Å; ∠N–Se–Br = 176.06(15)°.

Bis[3-(4,5-dihydro-4,4-dimethyl-1,3-oxazol-2-yl)-4-(3,5-dimethylphenyl)-2-naphthyl] diselenide (**40**: RSeSeR) was prepared. Then RSeX [X = Cl (**41**), X = Br (**42**)] was obtained in the reaction of **40** with a stoichiometric amount of SO_2Cl_2 or Br_2.[36] Similar reaction afforded the novel selenenyl iodide RSeI (**43**), in which the selenium and iodine are covalently bonded to each other. Figure 12.3.17 shows the structure of **40**, where n(N)···σ*(Se–Se) 3c–4e interaction controls the structure (∠NSeSe = 173.88(17)°), together with the n(O)···σ*(Se–Se) 3c–4e interaction in other side (∠OSeSe = 152.28(2)°). The n(N)···σ*(Se–Se)···n(O) interaction of the 4c–6e type is not detected in **40**. Figure 12.3.18 shows the structure of **43**, of which the Se–I bond is stabilized by the n(N)···σ*(Se–I) 3c–4e interaction (∠NSeI = 175.26(14)°) (Table 12.3.3).

Diselenide, 2,2′-diselanediylbis(5-*t*-butylisophthalic acid) (**44**), undergoes intramolecular cyclization in methanol to form a cyclic selenenate ester, **45** (Scheme 12.3.4).[37] The cyclization reaction proceeds more rapidly in the presence of organic bases such as pyridine, adenine, and 4,4′-bipyridine, to form pyridinium **46**, adeninium **47** and 4,4′-bipyridinium **48**, respectively. Figure 12.3.19 shows the structure of adeninium **47**. Structural studies on the related compounds **49–51** shown below revealed that the Se···O(COO^-) secondary distances (2.170, 2.075, and 2.176 Å) were significantly shorter than the corresponding Se···O distances (2.465, 2.472, and 2.435 Å)

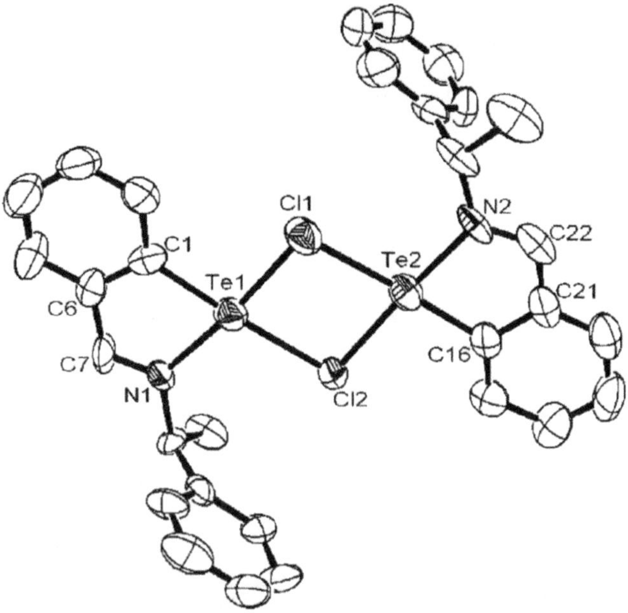

Figure 12.3.16 Molecular structure of **39**: Te1–Cl1 = 2.322(6) Å; Te1–Cl2 = 2.471(5) Å; Te1–N1 = 2.06(2) Å; Te1–C1 = 1.99(2) Å; Te2–Cl1 = 2.464(6) Å; Te2–Cl2 = 2.343(5) Å; Te2–N2 = 2.01(2) Å; Te2–C16 = 1.97(2) Å; ∠N1–Te1–C1 = 79.5(8)°; ∠Cl1–Te1–C1 = 94.8(6)°; ∠Cl1–Te1–N1 = 174.2(5)°; ∠Cl1–Te1–Cl2 = 85.3(2)°; ∠Cl2–Te1–N1 = 100.5(5)°; ∠Cl2–Te1–C1 = 172.6(6)°.

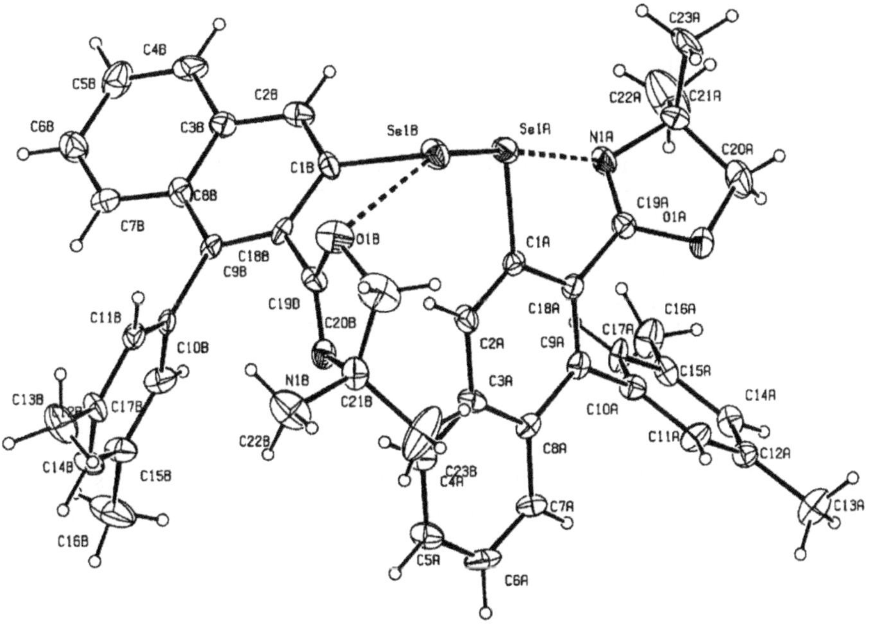

Figure 12.3.17 Molecular structure of **40**.

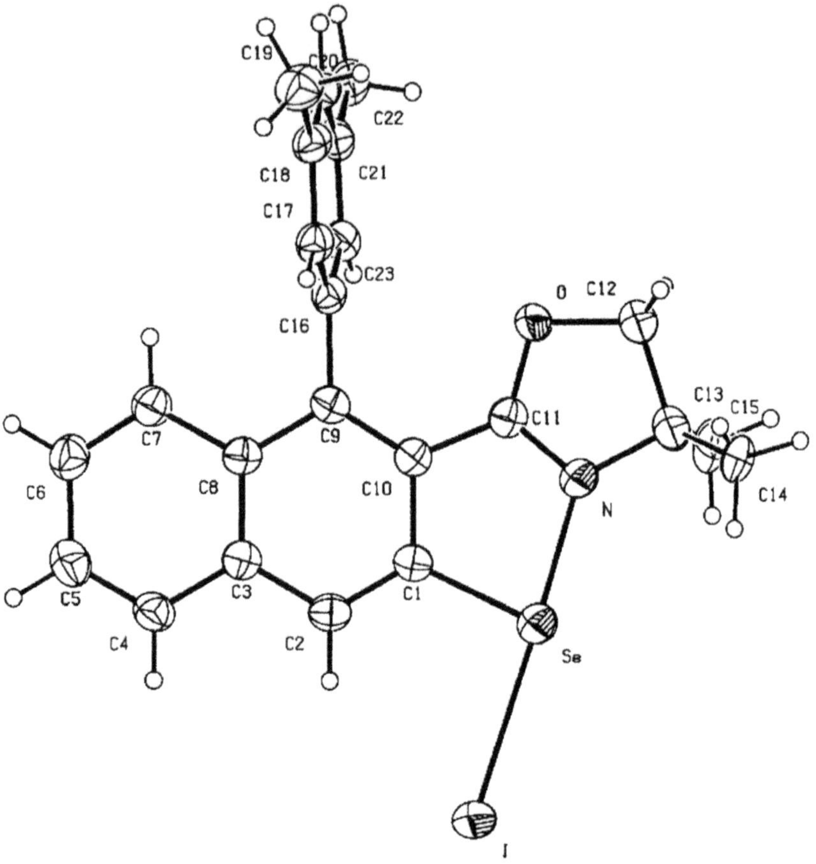

Figure 12.3.18 Molecular structure of **43**.

Table 12.3.3 Selected bond lengths and bond angles for **40**, **42**, and **43**.

Compound	*Se···N (Å)*	*Se–X (Å)*	*Se···O (Å)*	*∠NSeX*
40 (RSeSeR)	2.976(8)	2.3216(15)[a]	2.9815(5)	173.88(17)[b]
42 (RSeBr)	2.052(3)	2.6485(7)		175.15(10)
43 (RSeI)	2.059(5)	2.8297(8)		175.26(14)

[a]Se–Se distance.
[b]∠NSeSe.

observed for the selenenate esters stabilized by the neutral donors (CHO, COOH, and COOEt).[37] The existence of potential Se···O intramolecular nonbonding interactions was examined for a series of low-valent organoselenium compounds by IR, ^{1}H, and ^{77}Se NMR spectroscopy, X-ray crystallography, and/or computational studies.[38,39] The intramolecular interactions are summarized in Table 12.3.4.

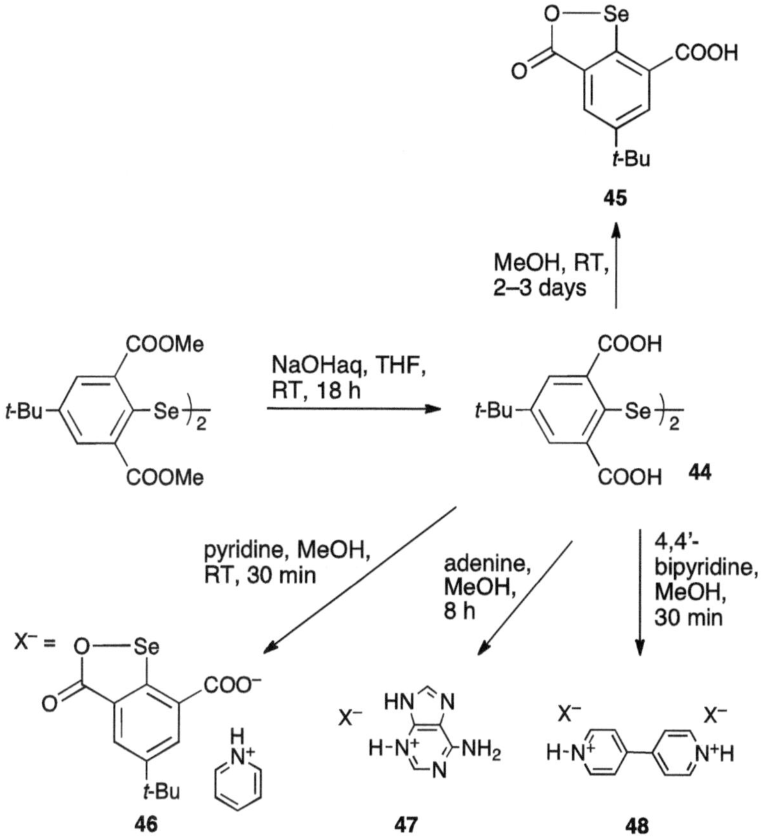

Scheme 12.3.4 Synthesis of diselenide **44** and its reactivity.

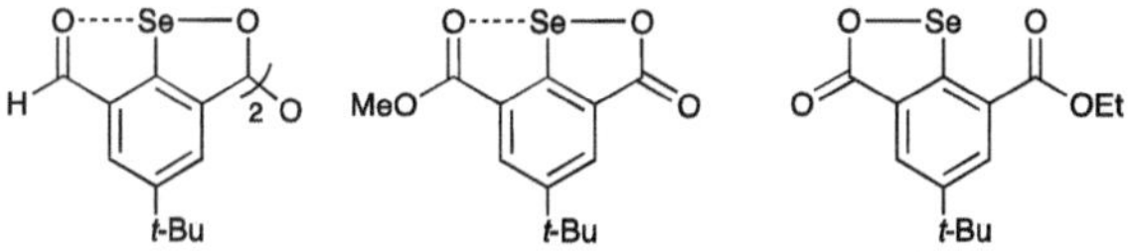

49 **50** **51**

Derivatives **53–55** of ebselen (**52**) and their S and Te derivatives **56–58** were prepared to examine the GPx-like catalytic activities.[40] The n(O)···σ*(E–N) 3c–4e type interaction is expected to operate in these compounds, as confirmed for **53** (r(O···Se) = 2.435(2) Å) and **54** (r(O···Se) = 2.553(3) Å) by the X-ray analysis.

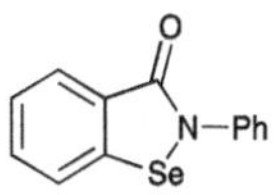

52

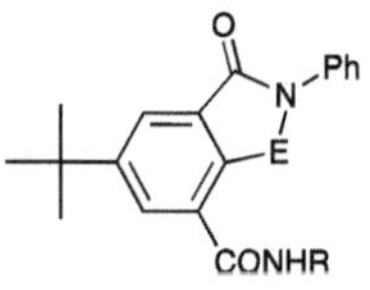

53: R = Ph, E = Se
54: R = Me, E = Se
55: R = *i*-Pr, E = Se
56: R = Ph, E = S
57: R = Me, E = S
58: R = Me, E = Te

The presence of intramolecular secondary Se···O interactions in seleninate and selenenate esters **59–62** and selenenic anhydride **63** has been confirmed

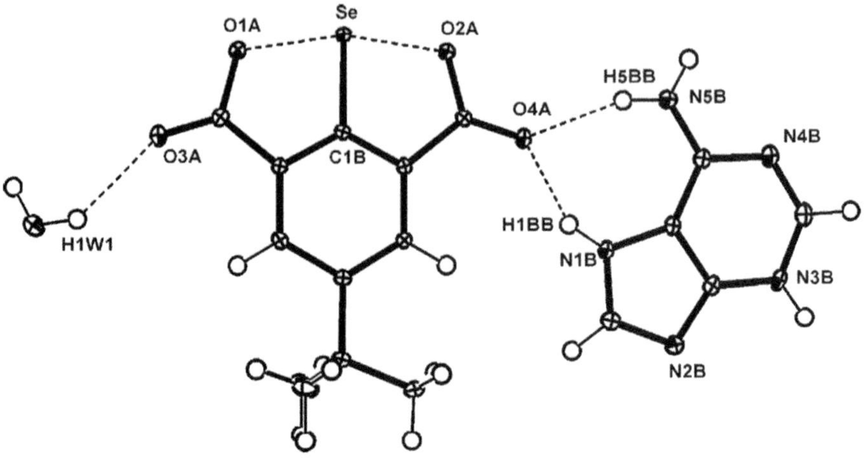

Figure 12.3.19 Molecular structure of **47** (50% ellipsoid).

Table 12.3.4 The Se–O and Se···O distances in **49**–**51**.[a]

Compound	*Se–O (Å)*	*Se···O (Å)*
49	1.855(5)	2.606(9)
	1.864(5)	2.466(6)
50	1.871(2)	2.472(2)
45	1.950(1)	2.281(1)
46	1.991(3)	2.170(3)
47	2.050(1)	2.071(1)
48	1.977(2)	2.176(2)
	2.061(2)	2.091(2)
51	1.883(2)	2.435(2)

[a]All the distances were taken from X-ray crystallographic data. See ref. 37.

by single-crystal X-ray diffraction studies as well as computational studies (Figure 12.3.20).[41] The Se–Br bond in **64** is also stabilized similarly (Figure 12.3.21). NBO and AIM calculations are also employed to detect the intramolecular Se···O interactions. Glutathione peroxidase-like (GPx) antioxidant activities of **59**–**62** and related heterocycles such as 7-nitro-1,2-benzisoselenol(2*H*)-3-one selenium oxide (**65**), 7-nitro-1,2-benzisoselenol(2*H*)-3-one (**66**), and **62** have been determined by the coupled reductase assay.

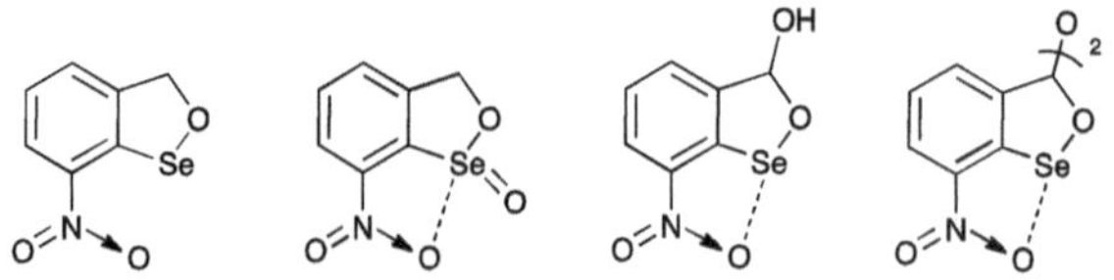

59 **60** **61** **62**

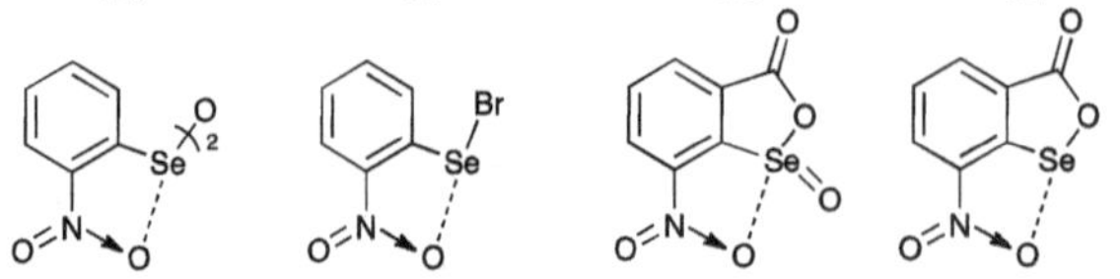

63 **64** **65** **66**

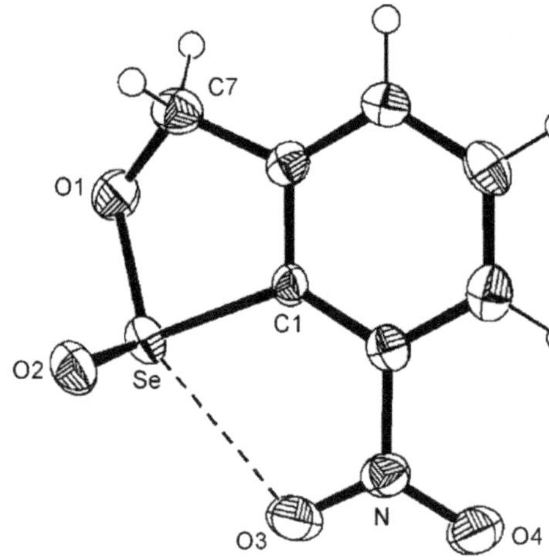

Figure 12.3.20 Molecular structure of **60** (50% ellipsoid).

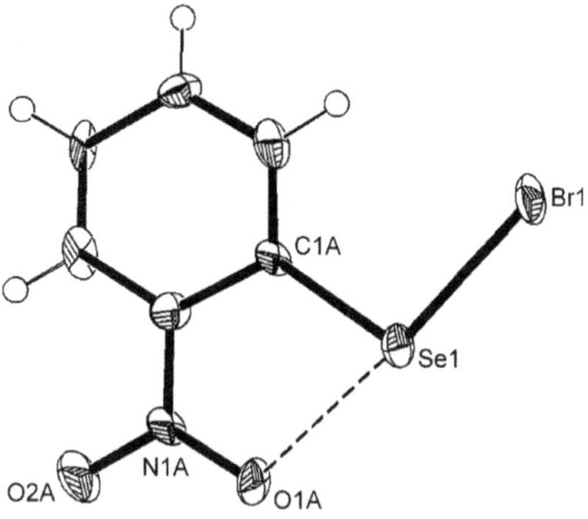

Figure 12.3.21 Molecular structure of **64** (50% ellipsoid).

Selenenium cations **67**–**69** are derived from *N*-[*n*-butyl-(3-nitrobenzylimino)]phenyl selenide (**70**) in relation to the high GPx-like activities (Scheme 12.3.5).[42] Oxidation of **68** with hydrogen peroxide afforded the ebselen derivative, 2-(4-bromophenyl)-7-nitro-1,2-benzisoselenazol-(2*H*)-3-one *Se*-oxide (**71**). The presence of non-bonding O···Se–N intramolecular interactions in **68** and **69** has been confirmed by single-crystal X-ray studies (see Figures 12.3.22 and 12.3.23, Table 12.3.5). Similar interactions are also predicted in **67** by the DFT calculations. The isolated selenoxide **71** shows the highest GPx-like activity among the series of compounds.

While $R_2SeHgCl_2$ (**72**: (R = 2-(4,4-dimethyl-2-oxazolinyl)phenyl)) is stable in solution, $R_2TeHgCl_2$ (**73**) undergoes slow dismutation in chlorinated solvents to give the fragments RTeCl (**74**) and RHgCl (**75**).[43] The intramolecular

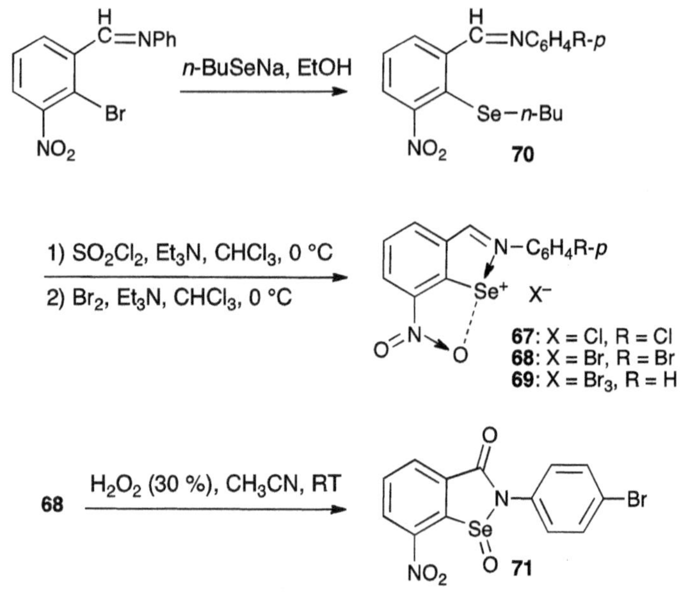

Scheme 12.3.5 Synthesis of selenenium cations **67–69** and its reactivity.

N···Te–Cl interaction is confirmed in **74**, which is very close to the N···Hg–Cl interaction in **75**, although the angles are very different: The former is around 180° (Figure 12.3.24), whereas the latter is 90°.

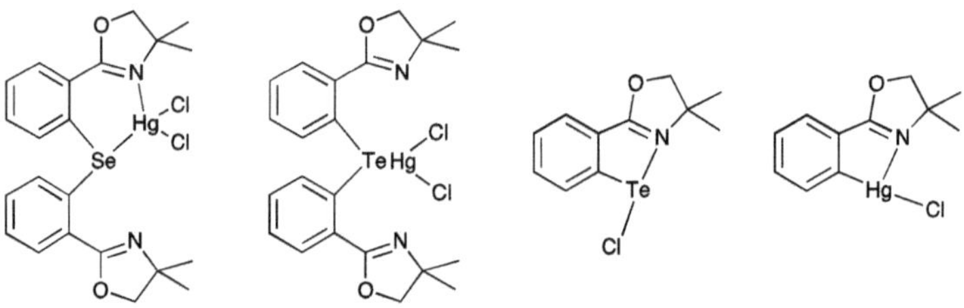

72 **73** **74** **75**

Di-cyclic 1-thia-6-oxa-6aλ^4-seleno-3-azapentalene having T-shaped O···Se–S 3c–4e (**76**) was obtained, of which *r*(Se, O), *r*(Se, S), and ∠OSeS were 2.1647(13) Å, 2.3097(6) Å, and 166.70(4)°, respectively.[44] Figure 12.3.25 shows the structure of **76**.

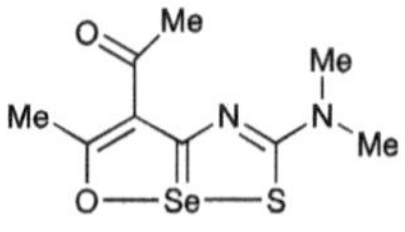

76

Similarly, a T-shaped compound with N···Se–Br 3c–4e in 2-$Me_2NCH_2C_6H_4SeBr$, was obtained of which *r*(Se, N), *r*(Se, Br), and ∠NSeBr were 2.132(6) Å, 2.6114(12) Å, and 176.06(15)°, respectively.[35]

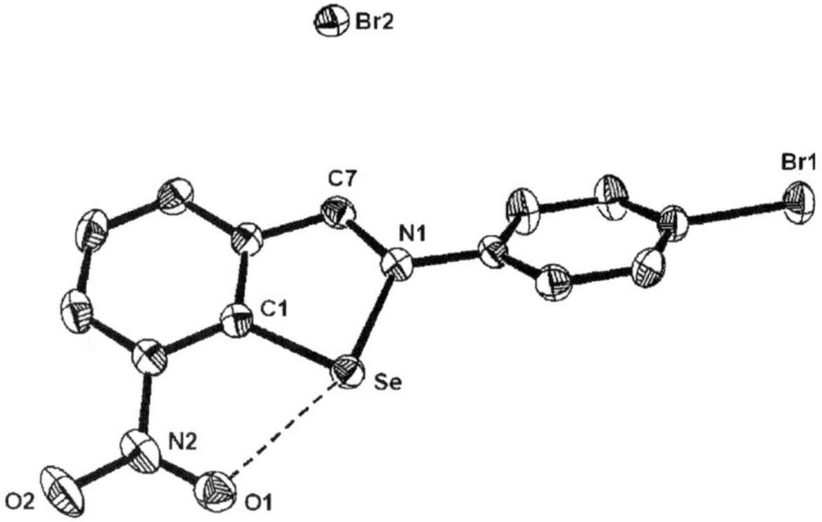

Figure 12.3.22 Molecular structure of **68** (50% ellipsoid).

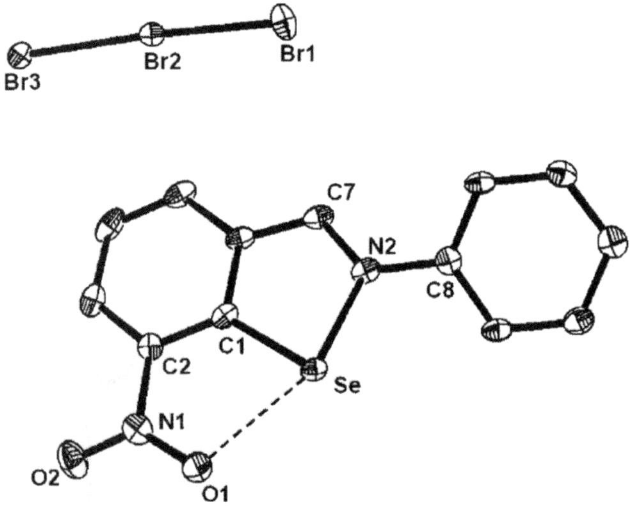

Figure 12.3.23 Molecular structure of **69** (50% ellipsoid).

Table 12.3.5 Calculated and observed Se···O distances, together with the second perturbation energies for the O···Se–N interactions in **67–69**.

Compound	$r_{Se\cdots O}$ [Å] (calcd.)	$r_{Se\cdots O}$ [Å] (obsd.)	$E_{Se\cdots O}$ (kcal mol^{-1})
67	2.690	–	19.21
68	2.532	2.544	17.3
69	2.445	2.522	21.3

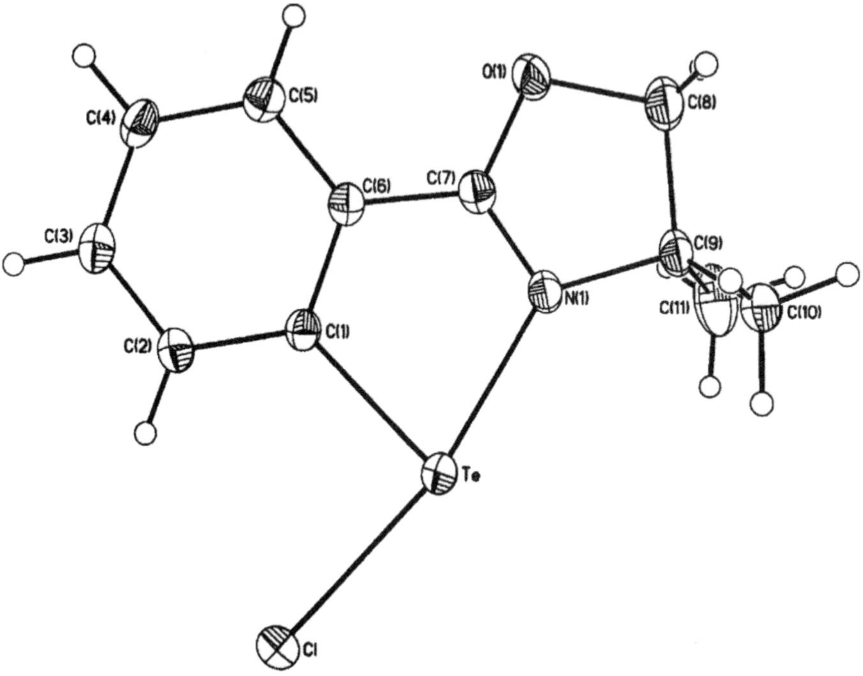

Figure 12.3.24 Molecular structure of **74.**

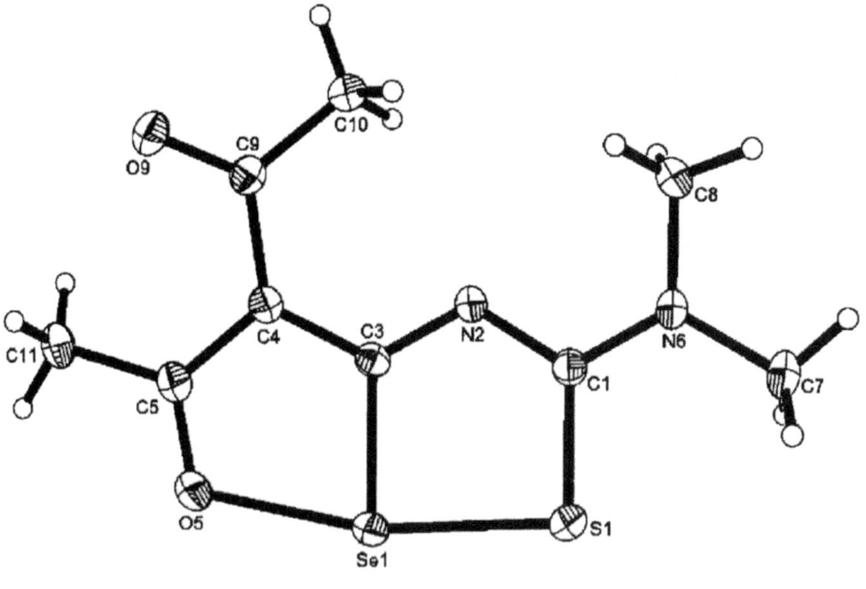

Figure 12.3.25 Molecular structure of **76** (35% ellipsoid).

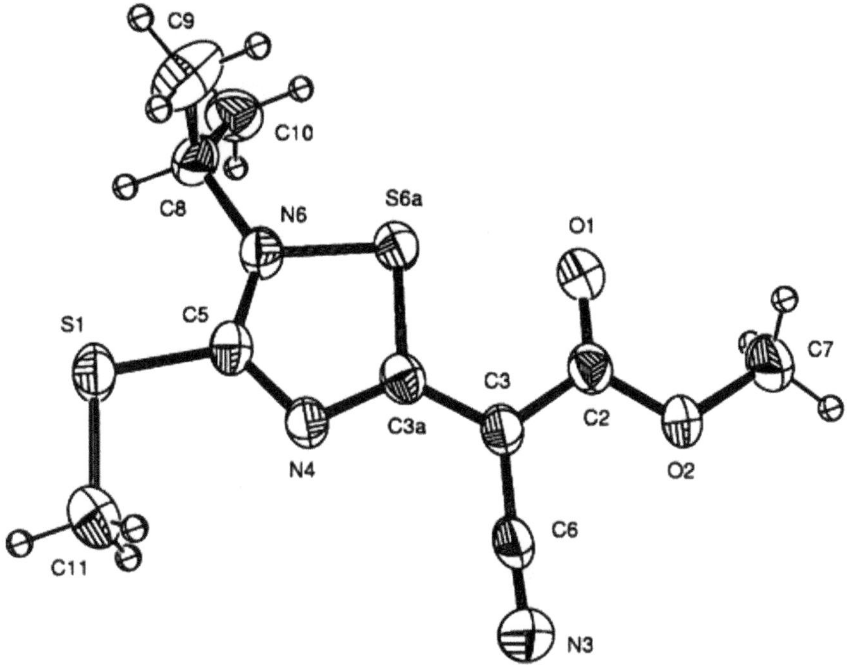

Figure 12.3.26 Molecular structure of **77a**.

The versatile chemical behaviour of 5-chloro-3-methylthio-1,2,4-thiadiazol-2-ium chlorides toward several nitrogen and carbon nucleophiles has been investigated.[45] π-Hypervalent sulfur compounds **77** were prepared. The structural assignments were discussed on the basis of IR and NMR spectroscopic data. A rather short O1–S6a distance of 2.395(5) Å was determined by X-ray analysis for **77a** (Figure 12.3.26).

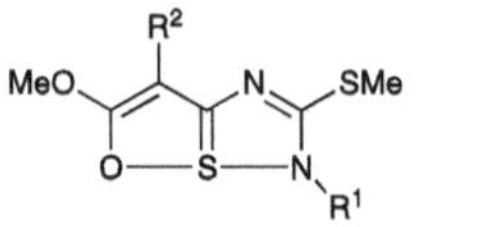

77a: R^1 = *i*-Pr, R^2 = CN
77b: R^1 = *i*-Pr, R^2 = CO_2Me
77c: R^1 = *t*-Bu, R^2 = CN

Naphthalene 1,8-positions serve as a good system to study non-bonded interactions.[46–50] Nakanishi and Hayashi investigated intramolecular homonuclear E···E interactions (E = O (**a**), S (**b**), Se (**c**), and Te (**d**)) at naphthalene 1,8-positions in 1,8-$(MeE)_2C_{10}H_6$ (**78a–78d**), 1-MeE-8-PhE$C_{10}H_6$ (**79a–79c**), and 1,8-$(PhE)_2C_{10}H_6$ (**80a–80d**). The delicate interactions are demonstrated to control the conformers of the compounds, by X-ray crystallographic analysis and quantum chemical calculations (Scheme 12.3.6).[51]

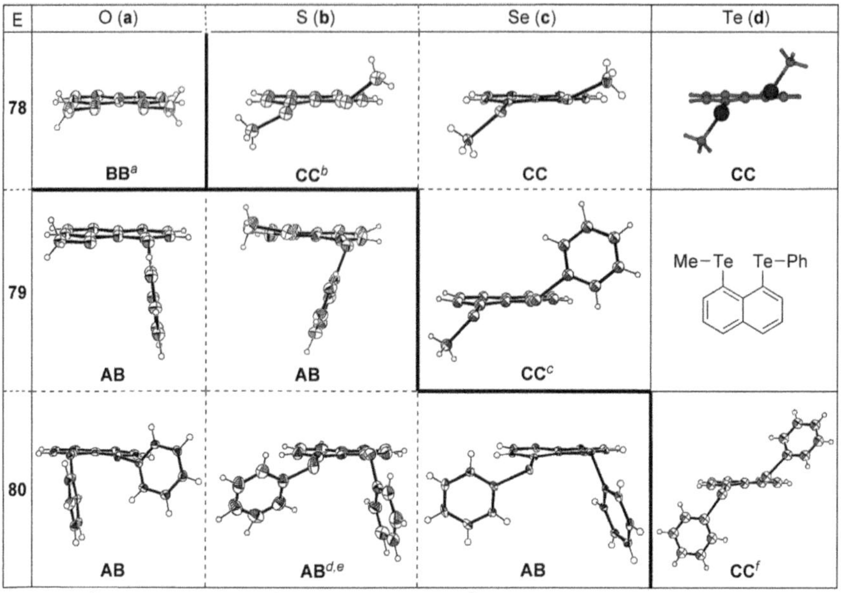

[a] Ref. 52. [b] Ref. 50g. [c] Ref. 46b. [d] Ref. 53. [e] Reexamined. [f] Ref. 50h.

Scheme 12.3.6 Observed structures in **78a–80d**.

The intramolecular Se···Se=O and Se···Se$^+$–OH interactions are similarly investigated at the 1,8-positions in **81–84**, where L and O stand for RSe and RSe(O) groups, respectively (Scheme 12.3.7). The stereochemistry and the mechanism of racemization in selenoxides are clarified by X-ray crystallographic analysis (Figure 12.3.27). A contour map of $\rho_b(\boldsymbol{r}_c)$ in the SeSeC$_9$ plane of **81** (LO) is depicted in Figure 12.3.28.[54] The intramolecular X···S=O (X = halogens) interactions at the 1,8-positions in **85–88** are investigated by the calculations, although G = H for **85**.

12.3.4 Extended Hypervalent Bonds; *m*c–*n*e ($m \geq 4$)

12.3.4.1 4c–6e

Hypervalent bonds higher than 3c–4e are called extended hypervalent bonds *m*c–*n*e ($m \geq 4$).[46,55,56] The first member of *m*c–*n*e ($m \geq 4$) is 4c–6e.[46] The driving force for the formation of 4c–6e is the CT interactions from the lone pair orbitals of E′ (n_p(E′)) at both sides of the central vacant orbital to the E–E bond (σ^*(E–E)). The CT interactions are the $n_p(E') \rightarrow \sigma^*(E\text{–}E) \leftarrow n_p(E')$ type, where E and E′ are S, Se, and Te or E′ could be atoms having lone pair orbitals, such as those of Groups 15 and 17.

The structure, stability and reactivity of E′···E–E···E′ ($E_2E'_2$) 4c–6e are examined employing naphthalene 1,8-positions in 1-(8-PhE′$C_{10}H_6$)EE($C_{10}H_6$E′Ph-8′)-1′

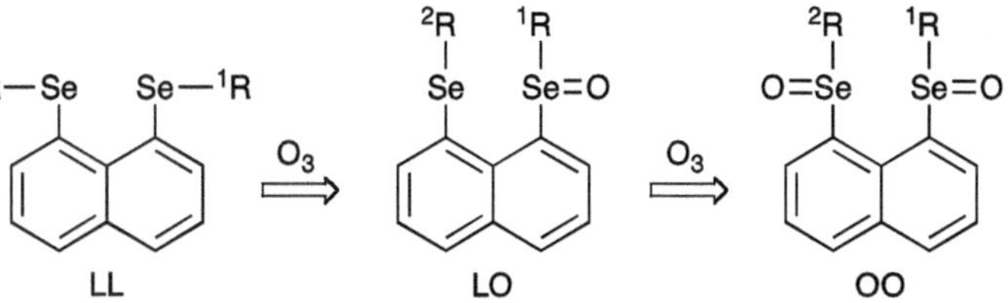

81 (^{1}R = ^{2}R = Me), **82** (^{1}R = ^{2}R = Et),
83 (^{1}R = Me, ^{2}R = Ar), **84** (^{1}R = ^{2}R = Ar)
Ar = Ph (**a**). *p*-MeOC$_6$H$_4$ (**b**), *p*-*t*-BuC$_6$H$_4$ (**c**) and *p*-O$_2$NC$_6$H$_4$ (**d**)

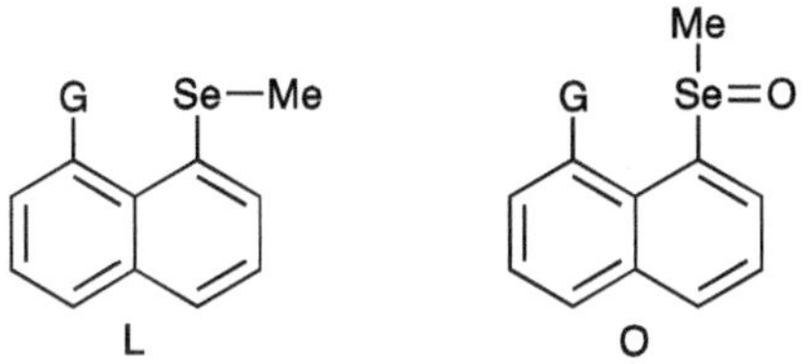

85 (G = H), **86** (G = F), **87** (G = Cl), **88** (G = Br)
(L = lone pair and O = oxide)

Scheme 12.3.7 Bis(selanyl)naphthalenes, **81**–**84**, together with **85**–**88**.

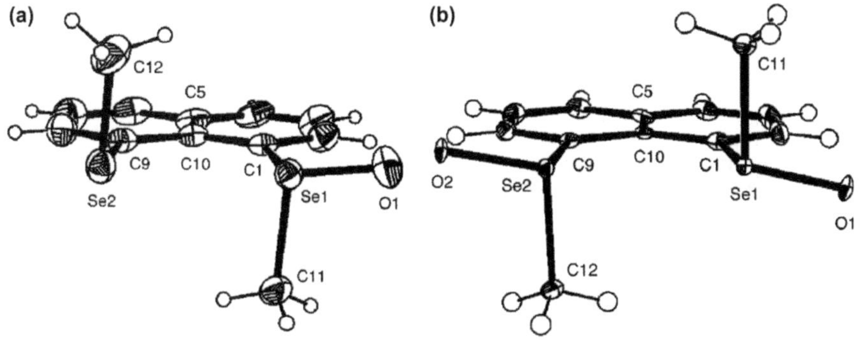

Figure 12.3.27 Molecular structures of **81** (LO) (a) and **81** (OO) (b).

[**89** (E = Se, E′ = S), **90** (E = S, E′ = Se), **91** (E = E′ = Se), and **92** (E = E′ = S)], together with 1-$C_{10}H_7EEC_{10}H_7$-1′ [**93** (E = Se) and **94** (E = S)].[57] Linear alignments of four Se_2S_2 atoms in **89** and **90** are confirmed by X-ray analysis (Figure 12.3.29). **89** was not reduced by sodium borohydride, whereas **90** was, contrary to expectation. Similarly, E–E in **91**, **93**, and **94** were cleaved, whereas that in **92** was not, when allowed to react with $NaBH_4$ in aqueous THF. The reactivity of the E–E bonds in $E_2E'_2$ in **89**–**92** is not controlled by the central E–E bonds but by the outside E′ atoms. The reason was clarified by QC calculations (see Table 12.3.6).

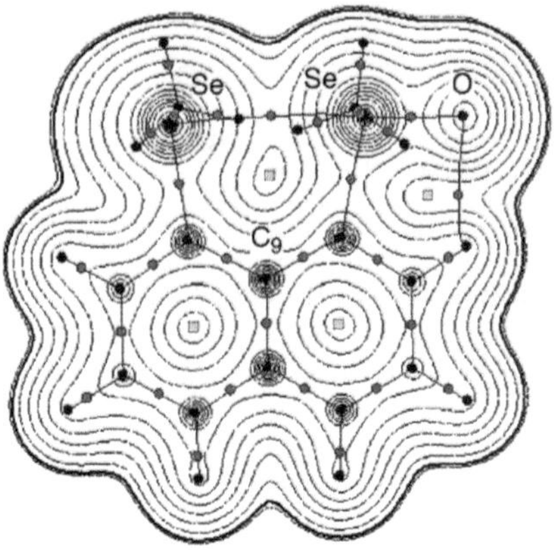

Figure 12.3.28 Contour map of $\rho_b(r_c)$ for **81** (LO) in the $SeSeC_9$ plane, together with bond-critical points (BCPs) (●), ring critical points (■), and bond paths.

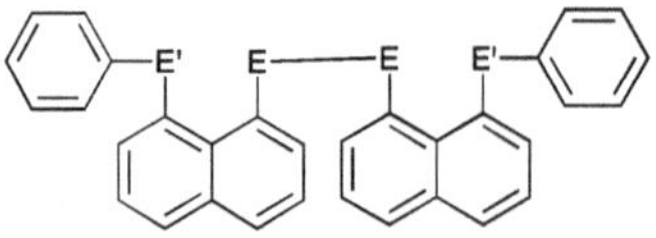

89 (E = Se, E' = S), **90** (E = S, E' = Se)
91 (E = E' = Se), **92** (E = E' = S)

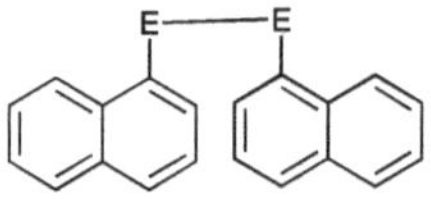

93 (E = Se), **94** (E = S)

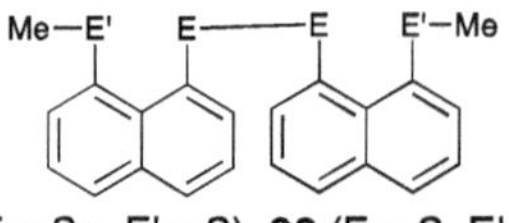

95 (E = Se, E' = S), **96** (E = S, E' = Se)
97 (E = E' = Se), **98** (E = E' = S)

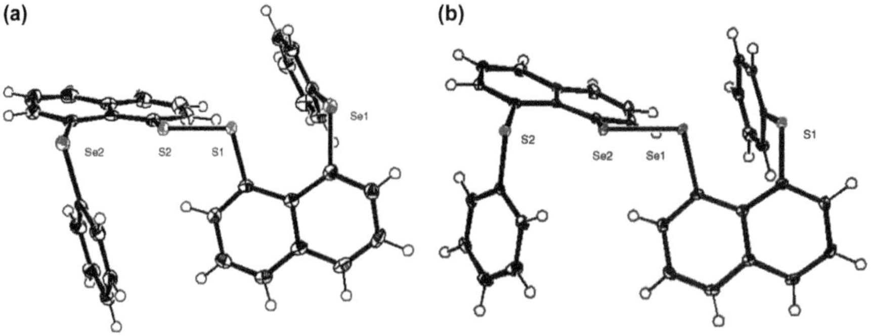

Figure 12.3.29 Molecular structures of **89** (a) and **90** (b) (50% ellipsoid).

Singh and co-workers prepared a lots of selenium compounds to examine the glutathione peroxidase-like activity, together with similar activities of the ebselen analogue. Most of the selenium compounds derived from

Table 12.3.6 Relative energies for **AA**, **AB**, and **BA** of **95**–**98**, together with those for **A** and **B** of **93** and **94** calculated at the B3LYP level.[a]

Compound	$\Delta E(AA)^{b,c}$	$\Delta E(AB)^{b,d}$	$\Delta E(BA)^{b,e}$
95 (SSeSeS)	5.8	0.0	−26.8
96 (SeSSSe)	21.0	0.0[f]	10.2[g]
97 (SeSeSeSe)	22.3	0.0	−11.0
98 (SSSS)	7.1	0.0	−4.2
93 (SeSe)[h]		0.0	11.8
94 (SS)[h]		0.0	10.2

[a]The 6-311 + G(d) basis sets being employed for Se and S and the 6-311G(d) basis sets for C and H.
[b]In kJ mol^{-1}.
[c]ΔE(**AA**) = E(**AA**) − E(**AB**).
[d]Taken as the standard for each.
[e]ΔE(**BA**) = E(**BA**) − E(**AB**).
[f]30.2 kJ mol^{-1} from **95** (**AB**).
[g]67.2 kJ mol^{-1} from **95** (**BA**).
[h]Data for **93** (**A**) and **94** (**A**) are given in the column of **AB** and those for **93** (**B**) and **94**(**B**) are in that of **BA**.

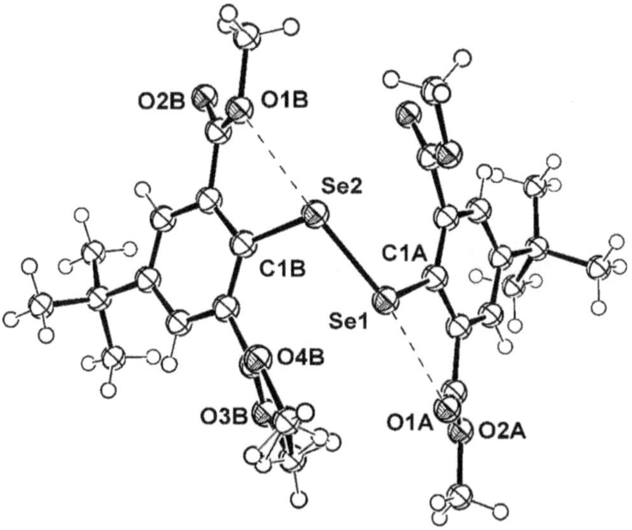

Figure 12.3.30 Molecular structure of **44** (50% ellipsoid).

2-bromo-5-*t*-butylisophthalic acid contain intramolecular Se···O interactions.[39] Among them, the intramolecular Se···O interactions are confirmed at the both sides of the Se–Se bond. Figure 12.3.30 shows the structure of **44**, determined by X-ray analysis. This must be the formation of Se_2O_2 4c–6e, where both ester- and carbonyl-oxygen take part in the construction of Se_2O_2 4c–6e.

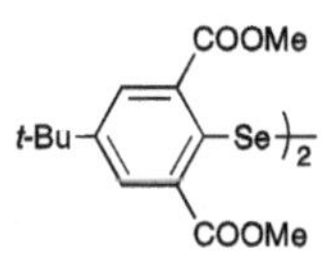

44

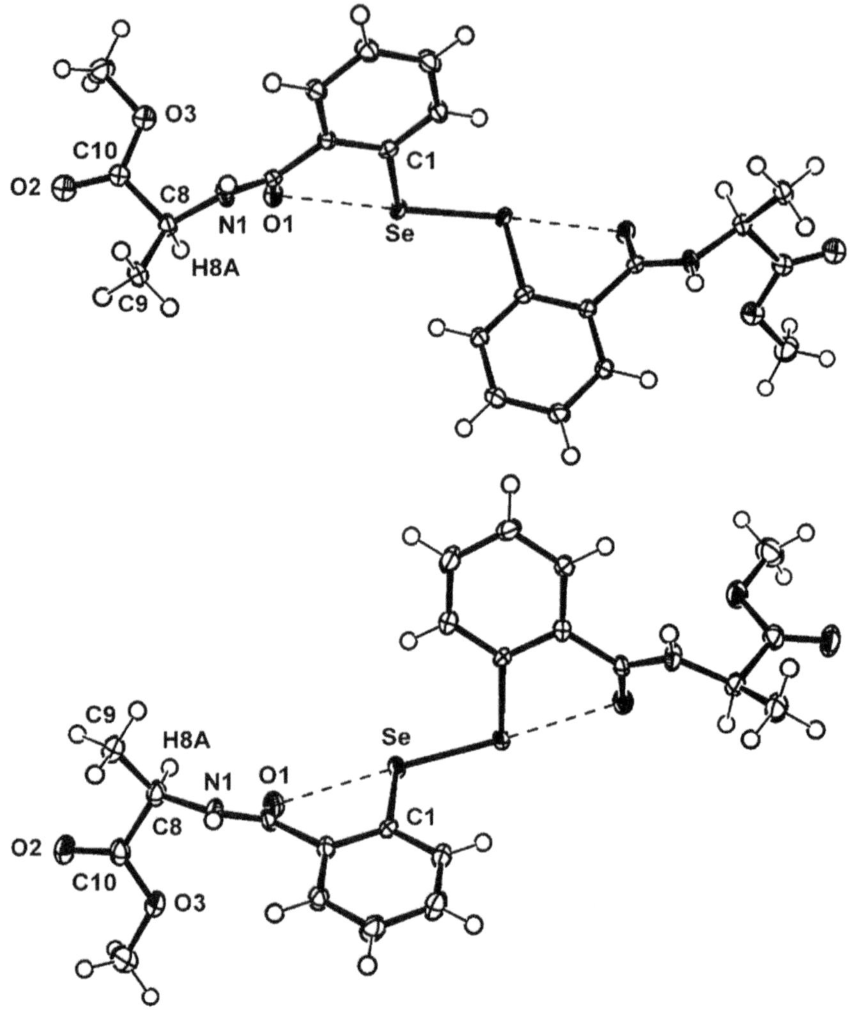

Figure 12.3.31 Molecular structures of **99a** (top) and **99b** (bottom) (50% ellipsoid).

They also encountered Se_2O_2 4c–6e with the Se–Se bonds coordinated intramolecularly by O at the both sides in **99** and **100**.[58] Figure 12.3.31 shows the structures of the enantiomers of **99** determined by X-ray analysis. The structure of **100** is close to that of **99**.

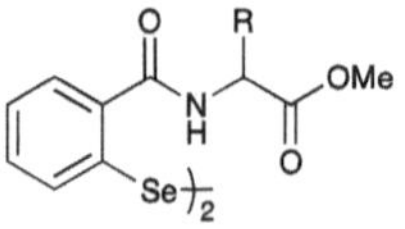

99a/99b: R = Me
100: R = H
101: R = CH_2Ph

However, they also encountered cases that did not construct E_2O_2 4c–6e (E = Se and Te) but form E_2O 3c–4e in **102** and **103**.[59] Figure 12.3.32 shows the

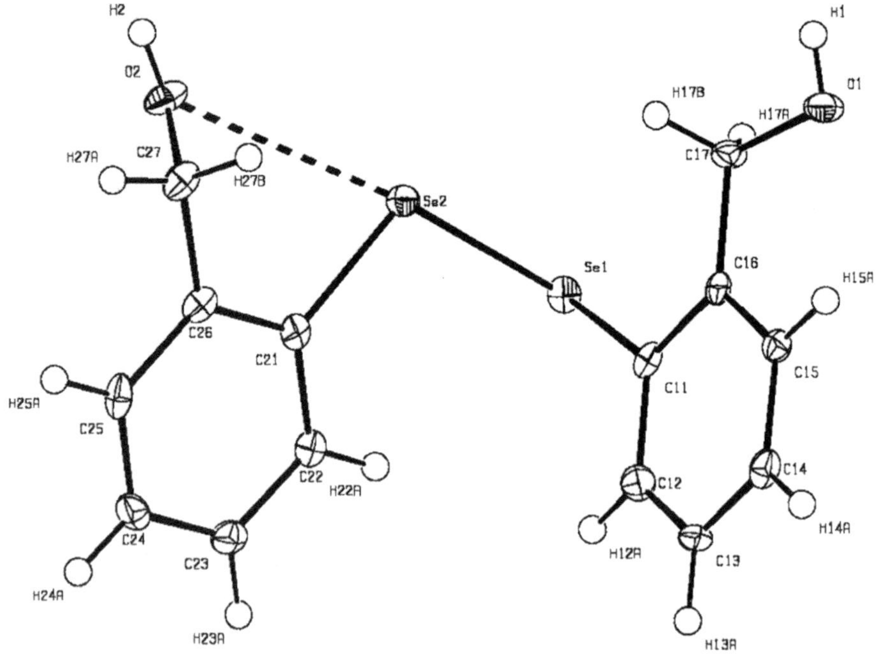

Figure 12.3.32 Molecular structure of **102**.

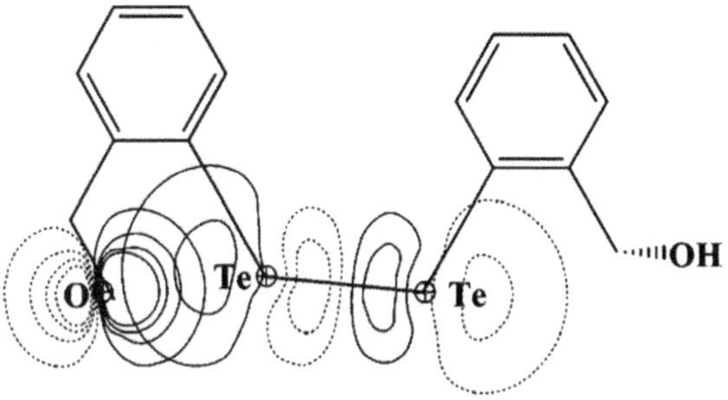

Figure 12.3.33 NBO contour diagram representing the $n_O \rightarrow \sigma^*_{Te-Te}$ orbital interaction in **103**.

structure of **102** and Figure 12.3.33 depicts the NBO contour diagram of ρ(***r***) of **103**, which clearly represents the $n_O \rightarrow \sigma^*$(Te–Te) 3c–4e interaction.

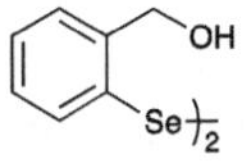

102

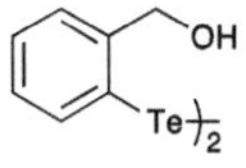

103

Oakley reported the temperature and pressure dependence of conductivity in the tetrathiophenalenyl radical, TTPLY (**104**), existing in two distinct

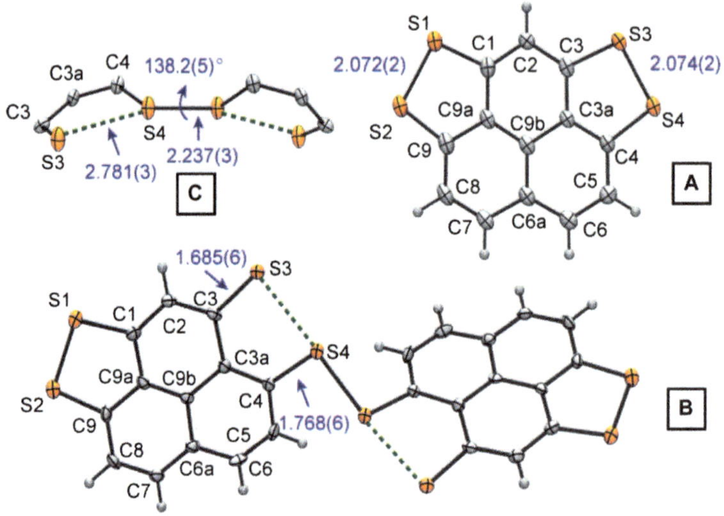

Figure 12.3.34 Ortep drawings (50% probability ellipsoids) in the **104**$^{+}$ cation in [**104**][SbF_6] (A), the dimer [**104**]$_2$ (B) and the disulfide unit in [**104**]$_2$ (C).

molecular and crystal modifications.[7a] The presence of two disulfide groups in **104** leads to a highly delocalized spin distribution and the lowest cell potential ever observed for a monofunctional phenalenyl derivative. While the heteroatom substituents successfully block C–C bond formation, **104** associates in the solid state to afford the hypervalent S–S-bonded dimer, nonetheless (Figure 12.3.34).

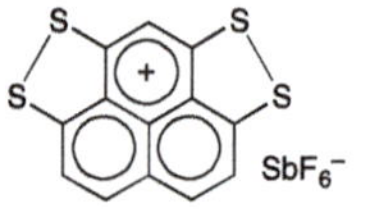

[**104**]•[SbF_6]

The α-phase crystallizes in the tetragonal space group $P4\text{-}2_1m$ and consists of π-stacked radicals. The β-phase belongs to the monoclinic space group $P2_1/c$ and is composed of π-stacked dimers at ambient temperature and pressure.[7b] The radicals in the β-phase are linked laterally by hypervalent four-centre six-electron S···S–S···S bonds. On the other hand, powder X-ray diffraction analysis of β-**105a** at 393 K has established that the phase transition corresponds to a dimer-to-radical conversion in which the hypervalent S···S–S···S bond is cleaved. Variable-temperature and -pressure conductivity measurements indicate that α-**105a** behaves as a Mott insulator, but the ambient temperature conductivity σ_{RT} increases from near 10^{-7} S cm^{-1} at 0.5 GPa to near 10^{-4} S cm^{-1} at 5 GPa (Figure 12.3.35). The response of the S···S–S···S bond in β-**105a** to heat and pressure is compared to that of related dimers possessing S···Se–Se···S bonds.

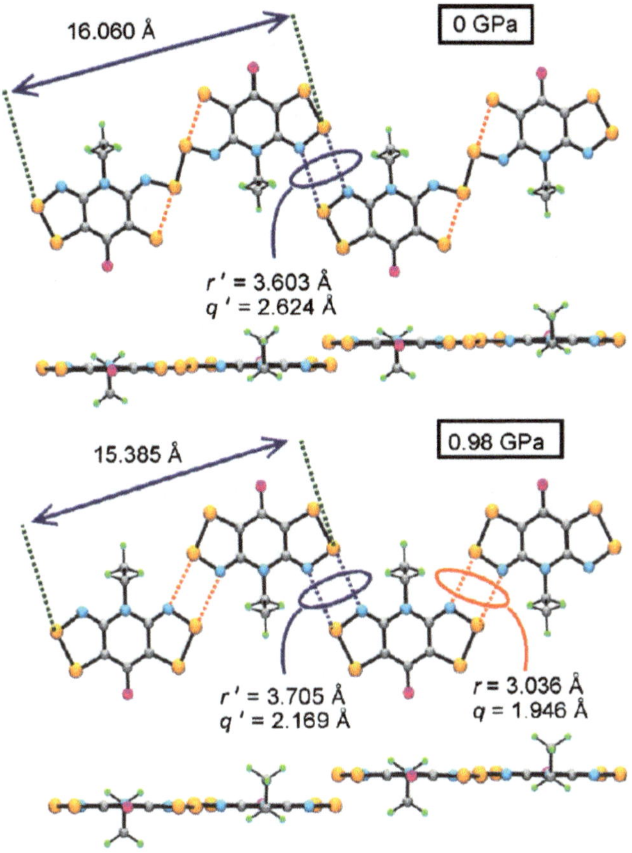

Figure 12.3.35 Supramolecular distances r' and q' along molecular ribbons of β-**105a** at 0 and 0.98 GPa, where r denotes for the separation between the E–N bonds of the two radicals and q denotes the relative slippage coordinate of the two radicals.

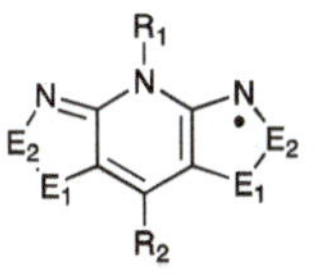

	105	**106**	**107**	**108**
E_1	S	S	Se	Se
E_2	S	Se	S	Se

105a: R_1 = Et, R_2 = F

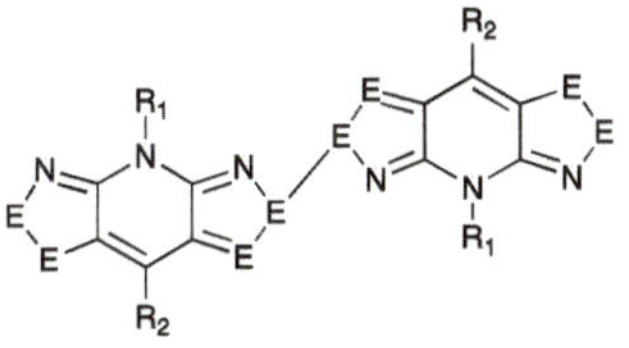

12.3.4.2 5c–6c

To clarify the nature of five-centre, six-electron (5c–6e) interactions, AIM analysis has been applied to an anthraquinone, 1,8-$(MeE)_2$ATQ (**109** (E = Se),

110 (E = S), and **111** (E = O)) and a 9-methoxyanthracene system, 9-MeO-1,8-(MeE)$_2$ATC (**112** (E = Se), **113** (E = S), and **114** (E = O)).[60] The S···O interactions in C_2E_2O 5c–6e (E = S) were predicted to be smaller than the hydrogen bond in H_2O···HOH and those for Se···O in C_2E_2O 5c–6e (E = Se) are

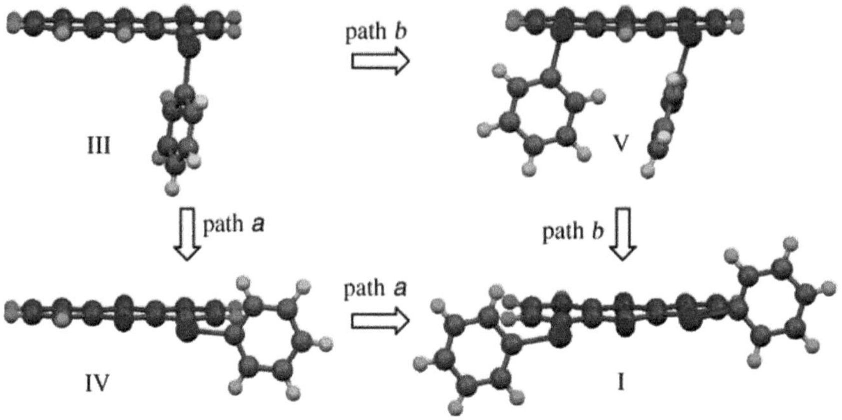

Scheme 12.3.8 Formation of 5c–6e C_2Se_2O starting from 1c–2e $n_p(Se)$, exemplified by phenylselanyl derivatives.

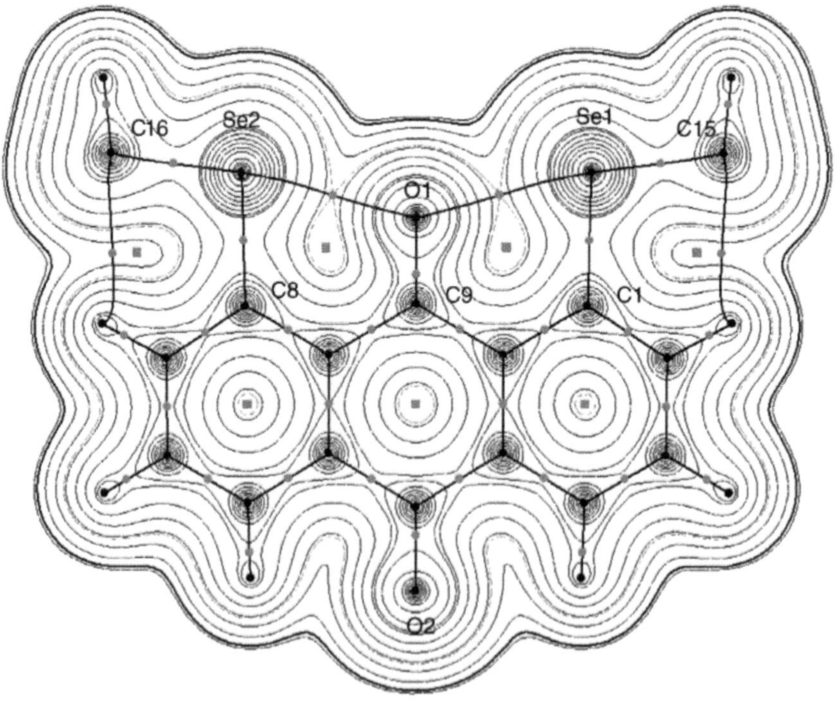

Figure 12.3.36 Contour map of $\rho_b(\boldsymbol{r}_c)$ of **109** in the anthraquinone plane, together with BCPs (●), ring critical points (■), and bond paths.

very close to or slightly smaller than that in HF···HF in both the ATQ and 9-MeOATC systems. It is well established that two n_p(O)···σ*(E–C) 3c–4e interactions effectively connect through the central n_p(O) orbital to form the extended hypervalent 5c–6e system of the σ*(C–E)···n_p(O)···σ*(E–C) type for E = Se and S in both systems (Scheme 12.3.8). Figure 12.3.36 shows the contour map of $\rho_b(\boldsymbol{r}_c)$ for **109** in the anthraquinone plane. In the course of AIM analysis, $H_b(\boldsymbol{r}_c)$ was plotted against $\rho(\boldsymbol{r})$ at the bond-critical points (BCPs) for the C_2E_2O 5c–6e (E = S and Se). The plot bent in the border area between vdW and HB interactions.

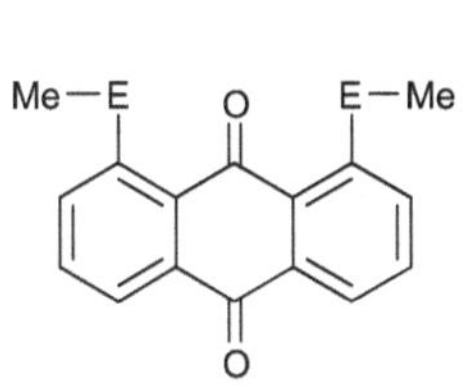

109 (E = Se), **110** (E = S), **111** (E = O)

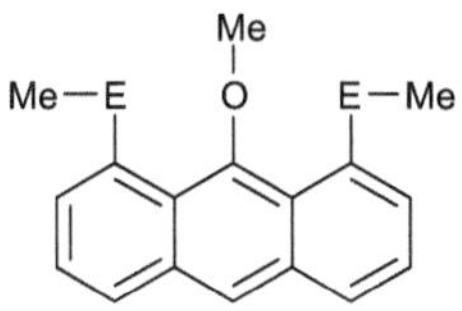

112 (E = Se), **113** (E = S), **114** (E = O)

References

1. W. Nakanishi, Hypervalent chalcogen compounds, in *Handbook of Chalcogen Chemistry: New Perspectives in Sulfur, Selenium and Tellurium*, ed. F. A. Devillanova, Royal Society of Chemistry, Cambridge, 2006, Chapter 10.3, p. 644.
2. A. J. Mukherjee, S. S. Zade, H. B. Singh and R. B. Sunoj, *Chem. Rev.*, 2010, **110**, 4357.
3. P. Kilian, F. R. Knight and J. D. Woollins, *Chem. Eur. J.*, 2011, **17**, 2302.
4. (a) F. T. Burling and B. M. Goldstein, *J. Am. Chem. Soc.*, 1992, **114**, 2313; (b) Y. Nagao, T. Hirata, S. Goto, S. Sano, A. Kakehi, K. Iizuka and M. Shiro, *J. Am. Chem. Soc.*, 1998, **120**, 3104; (c) S. Wu and A. Greer, *J. Org. Chem.*, 2000, **65**, 4883; (d) Y. Nagao, H. Iimori, S. Goto, T. Hirata, S. Sano, H. Chuman and M. Shiro, *Tetrahedron Lett.*, 2002, **43**, 1709; (e) E. Meyer, A. C. Joussef, H. Gallardo, A. J. Bortoluzzi and R. L. Longo, *Tetrahedron*, 2003, **59**, 10187; (f) Y. Nagao, T. Honjo, H. Iimori, S. Goto, S. Sano, M. Shiro, K. Yamaguchi and Y. Sei, *Tetrahedron Lett.*, 2004, **45**, 8757.
5. (a) J. C. Taylor and G. D. Markham, *J. Biol. Chem.*, 1999, **274**, 32909; (b) W. Brandt, A. Golbraikh, M. Täger and U. Lendeckel, *Eur. J. Biochem.*, 1999, **261**, 89.
6. See also K. P. Bhabak and G. Mugesh, *Acc. Chem. Res.*, 2010, **43**, 1408.
7. (a) L. Beer, R. W. Reed, C. M. Robertson, R. T. Oakley, F. S. Tham and R. C. Haddon, *Org. Lett.*, 2008, **10**, 3121; (b) K. Lekin, S. M. Winter, L. E. Downie, X. Bao, J. S. Tse, S. Desgreniers, R. A. Secco, P. A. Dube and R. T. Oakley, *J. Am. Chem. Soc.*, 2010, **132**, 16212.
8. G. M. de Oliveira, E. Faoro, E. S. Lang and G. A. Casagrande, *Z. Anorg. Allg. Chem.*, 2006, 659.

9. E. Faoro, G. M. de Oliveira and E. S. Lang, *Z. Anorg. Allg. Chem.*, 2006, **632**, 2049.
10. L. Zsolnai and H. Pritzkow, *ZORTEP, Program for Personal Computer*, University of Heidelberg, Germany, 1996.
11. E. J. Juarez-Perez, M. C. Aragoni, M. Arca, A. J. Blake, F. A. Devillanova, A. Garau, F. Isaia, V. Lippolis, R. Nunez, A. Pintus and C. Wilson, *Chem. Eur. J.*, 2011, **17**, 11497.
12. M. C. Aragoni, M. Arca, F. A. Devillanova, P. Grimaldi, F. Isaia, F. Lelj and V. Lippolis, *Eur. J. Inorg. Chem.*, 2006, 2166.
13. C. G. Hrib, F. Ruthe, E. Seppälä, M. Bätcher, C. Druckenbrodt, C. Wismach, P. G. Jones, W.-W. du Mont, V. Lippolis, F. A. Devillanova and M. Bühl, *Eur. J. Inorg. Chem.*, 2006, 88.
14. A. Kucsman and I. Kapovits, Non-bonded sulfur–oxygen interaction in organic sulfur compounds, in *Organic Sulfur Chemistry: Theoretical and Experimental Advances*, ed. F. Bernardi, I. G. Csizmadia and A. Mangini, Elsevier Scientific, Amsterdam, 1985, Chapter 4, p. 191.
15. (a) J. G. Angyan, R. A. Poirier, A. Kucsman and I. G. Csizmadia, *J. Am. Chem. Soc.*, 1987, **109**, 2237; (b) A. Kucsman, I. Kapovits, I. Kovesdi, A. Kalman and L. Parkanyi, *J. Mol. Struct.: THEOCHEM.*, 1985, **127**, 135.
16. K.-Y. Akiba, *Chemistry of Hypervalent Compounds*, Wiley-VCH, New York, 1999, Chapter 2, p. 9.
17. K. Akiba and Y. Yamamoto, *Heteroatom Chem.*, 2007, **18**, 161.
18. Y. Yamauchi, K. Akiba and H. Nakai, *Chem. Lett.*, 2007, **36**, 1120.
19. K. Ohkata, M. Ohsugi, K. Yamamoto, M. Ohsawa and K. Akiba, *J. Am. Chem. Soc.*, 1996, **118**, 6355.
20. F. Tellez, A. Cruz, H. Lopez-Sandoval, I. Ramos-Garcıa, M. Gayosso, R. N. Castillo-Sierra, B. Paz-Michel, H. Nöth, A. Flores-Parra and R. Contreras, *Eur. J. Org. Chem.*, 2004, 4203.
21. W. Nakanishi, S. Hayashi, M. B. Pitak, M. B. Hursthouse and S. J. Coles, *J. Phys. Chem. A*, 2011, **115**, 11775.
22. S. Ogawa, S. Ohwada, M. Yoshida and H. Muraoka, *Heteroatom Chemistry*, 2011, **22**, 586.
23. P. C. Srivastava, S. Bajpai, R. Lath, R. Kumar, V. Singh, S. Dwivedi, R. J. Butcher, S. Hayashi and W. Nakanishi, *Polyhedron*, 2008, **27**, 835.
24. S. Kato, K. Tani, M. Ishida, J. Nonogaki, M. Ebihara, S. Hayashi, W. Nakanishi, O. Niyomura, F. Ando and J. Koketsu, *Dalton Trans.*, 2011, **40**, 8156.
25. N. Kano, Y. Itoh, Y. Watanabe, S. Kusaka and T. Kawashima, *Angew. Chem. Int. Ed.*, 2008, **47**, 9430.
26. F. R. Knight, K. S. A. Arachchige, R. A. M. Randall, M. Bühl, A. M. Z. Slawin and J. D. Woollins, *Dalton Trans.*, 2012, **41**, 3154.
27. K. Selvakumar, H. B. Singh, N. Goel and U. P. Singh, *Organometallics*, 2011, **30**, 3892.
28. A. G. Petrovic, P. L. Polavarapu, J. Drabowicz, Y. Zhang, O. J. McConnell and H. Duddeck, *Chem. Eur. J.*, 2005, **11**, 4257.
29. S. Allenmark, *Chirality*, 2008, **20**, 544.

30. K. Srivastava, P. Shah, H. B. Singh and R. J. Butcher, *Organometallics*, 2011, **30**, 534.
31. T. G. Back, D. Kuzma and M. Parvez, *J. Org. Chem.*, 2005, **70**, 9230.
32. P. Kirsch and A. Hahn, *Eur. J. Org. Chem.*, 2005, 3095.
33. P. Kirsch and A. Hahn, *Eur. J. Org. Chem.*, 2006, 1125.
34. D. H. R. Barton, M. B. Hall, Z. Lin, S. I. Parekh and J. Reibenspies, *J. Am. Chem. Soc.*, 1993, **115**, 5056.
35. R. Kaur, S. C. Menon, S. Panda, H. B. Singh, R. P. Patel and R. J. Butcher, *Organometallics*, 2009, **28**, 2363.
36. K. Kandasamy, S. Kumar, H. B. Singh, R. J. Butcher and K. T. Holman, *Eur. J. Inorg. Chem.*, 2004, 1014.
37. K. Selvakumar, H. B. Singh, N. Goel, U. P. Singh and R. J. Butcher, *Chem. Eur. J*, 2012, **18**, 1444.
38. (a) S. S. Zade, H. B. Singh and R. J. Butcher, *Angew. Chem.*, 2004, **116**, 4613; *Angew. Chem. Int. Ed.*, 2004, **43**, 4513; (b) S. S. Zade, S. Panda, H. B. Singh, R. B. Sunoj and R. J. Butcher, *J. Org. Chem.*, 2005, **70**, 3693.
39. K. Selvakumar, H. B. Singh and R. J. Butcher, *Chem. Eur. J.*, 2010, **16**, 10576.
40. S. S. Zade, S. Panda, S. K. Tripathi, H. B. Singh and G. Wolmershäuser, *Eur. J. Org. Chem.*, 2004, 3857.
41. V. P. Singh, H. B. Singh and R. J. Butcher, *Chem. Asian J.*, 2011, **6**, 1431.
42. V. P. Singh, H. B. Singh and R. J. Butcher, *Eur. J. Inorg. Chem.*, 2010, 637.
43. S. D. Apte, S. S. Zade, H. B. Singh and R. J. Butcher, *Organometallics*, 2003, **22**, 5473.
44. M. Koketsu, T. Otsuka, D. Swensonc and H. Ishihara, *Org. Biomol. Chem.*, 2007, **5**, 613.
45. G. Morel, E. Marchand, S. Sinbandhit and L. Toupet, *Heteroatom Chemistry*, 2003, **14**, 95.
46. (a) W. Nakanishi, S. Hayashi and S. Toyota, *Chem. Commun.*, 1996, 371; (b) W. Nakanishi, S. Hayashi and S. Toyota, *J. Org. Chem.*, 1998, **63**, 8790; (c) W. Nakanishi, S. Hayashi and T. Arai, *Chem. Commun.*, 2002, 2416.
47. (a) W. Nakanishi, S. Hayashi and T. Uehara, *Eur. J. Org. Chem.*, 2001, 3933; (b) W. Nakanishi, S. Hayashi and T. Nakai, *Phosphorus, Sulfur, Silicon*, 2005, **180**, 1431.
48. (a) W. Nakanishi, S. Hayashi, A. Sakaue, G. Ono and Y. Kawada, *J. Am. Chem. Soc.*, 1998, **120**, 3635; (b) W. Nakanishi and S. Hayashi, *J. Org. Chem.*, 2002, **67**, 38; (c) W. Nakanishi and S. Hayashi, *Phosphorus, Sulfur, Silicon*, 2005, **180**, 961.
49. W. Nakanishi, S. Hayashi and T. Uehara, *J. Phys. Chem. A*, 1999, **103**, 9906.
50. See also (a) G. Gafner and F. H. Herbstein, *Acta. Cryst.*, 1962, **15**, 1081; (b) M. A. Davydova and Yu.T. Struchkow, *Zh. Strukt. Kim.*, 1962, **3**, 184; (c) M. A. Davydova and Yu.T. Struchkow, *Zh. Strukt. Kim.*, 1968, **9**, 1968; (d) R. D. Jackson, S. James, A. G. Orpen and P. G. Pringle, *J. Organomet. Chem.*, 1993, **458**, C3; (e) H. Bock, M. Sievert and Z. Havlas, *Chem. Eur. J.*, 1998, **4**, 677; (f) D. Szabo, I. Kapovits, A. Kucsman, P. Nagy, G. Argay

and A. Kalman, *J. Mol. Stru.*, 1999, **476**, 157; (g) R. S. Glass, S. W. Andruski, J. L. Broeker, H. Firouzabadi, L. K. Steffen and G. S. Wilson, *J. Am. Chem. Soc.*, 1989, **111**, 4036; (h) H. Fujihara, H. Ishitani, Y. Takaguchi and N. Furukawa, *Chem. Lett.*, 1995, 571; (i) P. Kilian, A. M. Z. Slawin and J. D. Woollins, *Chem. Eur. J.*, 2003, **9**, 215; (j) P. Kilian, A. M. Z. Slawin and J. D. Woollins, *Dalton Trans.*, 2003, 3876; (k) J. Omelanczuk, A. Karacar, M. Freytag, P. G. Jones, R. Bartsch, M. Mikolajczyk and R. Schmutzler, *Inorg. Chim. Acta*, 2003, **350**, 583.

51. S. Hayashi and W. Nakanishi, *Bull. Chem. Soc. Jpn.*, 2008, **81**, 1605.
52. R. Cosmo, T. W. Hambley and S. Sternhell, *Acta Crystallogr., Sect. B*, 1990, **46**, 557.
53. P. Nagy, D. Szabo, I. Kapovits, A. Kucsman, G. Argay and A. Kalman, *J. Mol. Struct.: THEOCHEM.*, 2002, **606**, 61.
54. S. Hayashi, W. Nakanishi, A. Furuta, J. Drabowicz, T. Sasamori and N. Tokitoh, *New J. Chem.*, 2009, **33**, 196.
55. (a) W. Nakanishi, S. Hayashi and N. Itoh, *Chem. Commun.*, 2003, 124; (b) W. Nakanishi, S. Hayashi and N. Itoh, *J. Org. Chem.*, 2004, **69**, 1676; (c) W. Nakanishi, S. Hayashi, T. Furuta, N. Itoh, Y. Nishina, M. Yamashita and Y. Yamamoto, *Phosphorus, Sulfur, Silicon*, 2005, **180**, 1351.
56. W. Nakanishi, S. Hayashi, S. Yamaguchi and K. Tamao, *Chem. Commun.*, 2004, 140.
57. W. Nakanishi, S. Hayashi, S. Morinaka, T. Sasamori and N. Tokitoh, *New J. Chem.*, 2008, **1881**, 32.
58. K. Selvakumar, P. Shah, H. B. Singh and R. J. Butcher, *Chem. Eur. J.*, 2011, **17**, 12741.
59. S. K. Tripathi, U. Patel, D. Roy, R. B. Sunoj, H. B. Singh, G. Wolmershaüser and R. J. Butcher, *J. Org. Chem.*, 2005, **70**, 9237.
60. W. Nakanishi, T. Nakamoto, S. Hayashi, T. Sasamori and N. Tokitoh, *Chem. Eur. J.*, 2007, **13**, 255.

CHAPTER 12.4

Theoretical Calculations and NMR Spectroscopy

WARO NAKANISHI* AND SATOKO HAYASHI

Department of Material Science and Chemistry, Faculty of Systems Engineering, Wakayama University, 930 Sakaedani, Wakayama 640–8510, Japan
*Email: nakanisi@sys.wakayama-u.ac.jp

12.4.1 Introduction

Experimentally determined values must be absolutely true, especially those on the horizontal axes, although they contain experimental errors. However, it is often very important to know the reason why the values are observed experimentally, not only for better understanding of the results but also for further development of the investigations. Nevertheless, the experimentally determined values do not usually tell the reason by themselves. How can we access the reason, other than by speculation? The reason will be clarified theoretically, although some possibilities could be eliminated experimentally. Quantum chemical (QC) calculations are most popular method to accomplish the purpose. It is now possible to predict the reason for usual phenomena or evaluate most properties reliably.[1–7]

This chapter explains briefly the theoretical treatment of chemical phenomena that apply to chalcogens, after a brief introduction to the calculation methods that are frequently used by the experimental chemists who need the support of calculations to understand or explain their experimental results. Theoretical treatments of nuclear magnetic resonance (NMR) parameters are also discussed from the viewpoint of experimental chemists. It is beyond our

Handbook of Chalcogen Chemistry: New Perspectives in Sulfur, Selenium and Tellurium, 2nd Edition, Volume 2
Edited by Francesco Antonio Devillanova and Wolf-Walther du Mont

Published by the Royal Society of Chemistry, www.rsc.org

ability to review the calculation methods and the results in detail, together with limitations of space. The reader should refer to more specialized books and reviews,[1–7] if necessary.

12.4.2 Background of Theoretical Calculations

12.4.2.1 Survey of Calculation Levels

The Schrödinger equation provides a method to predict the properties of atoms, molecules, and adducts, as is well known.[1–7] The Schrödinger equation can be applied in separate time-independent and time-dependent parts when steady states are discussed. In this case, the time-independent part of the Schrödinger equation is employed to solve the problem, although time-dependent phenomena are also important in the calculations. Suitable approximations are necessary when the equation is applied to real systems or models. The Born–Oppenheimer approximation is the first one that simplifies the equation for applications.[8] This approximation allows the equation to be solved separately in the two parts, which are electronic and nuclear Hamiltonians. An electronic Hamiltonian is used to solve most problems in atomic and molecular systems, neglecting the kinetic energy term for nuclear motions. In other words, the systems can be solved assuming the nuclei are at fixed positions. However, only a few systems can be solved by the Schrödinger equation,[1–7] those that involve only one electron, such as hydrogen-like atoms and the hydrogen molecular ion. Further approximations are necessary to apply the equation to real systems or models. Many electrons must be taken into account in the calculations in such systems. In this case, electrons must obey Fermi–Dirac statistics due to the electron spin of 1/2.[9] Therefore, the wave function of a many-electron system must have an antisymmetric nature, which requires that the sign of the wave function must be changed when any two electrons are changed.

The antisymmetric wave function can be represented by a determinant. The Hartree–Fock (HF) equation[10] is the basic one that satisfies nature. However, the HF equation is not adequate to describe the exchange-correlation, which is the electron correlation moving especially with opposite spins. Only one matrix presentation may not be enough to express the electronic state of the system with many electrons. Configuration interaction (CI) methods[7] improve the situation; these employ excited electronic states to express the wave function in the ground state. All exited states are employed in the full CI method, which is the most complete non-relativistic treatment within the limitations of the given basis set. However, the CI method is very time consuming. In this connection, the coupled cluster (CC) method supplies high-level calculations.[11] CCD (coupled-cluster doublets),[11a] CCSD (coupled-cluster singlets and doublets),[12] and CCSDT (coupled-cluster singlets, doublets, and triplets)[13] will be employed, together with other higher levels.

$$\hat{H} = \hat{H}_0 + \lambda \hat{H}' \tag{1}$$

$$\hat{H}' = \Sigma_{l>m}{}^{n}(1/r_{lm}) - \Sigma_{m=1}{}^{n}\Sigma_{j=1}{}^{n}(\hat{J}_j(m) - \hat{K}_j(m))$$

$$\psi = \psi^{(0)} + \lambda\psi^{(1)} + \lambda^2\psi^{(2)} + \lambda^3\psi^{(3)} + \cdots \quad (2)$$

$$E = E^{(0)} + \lambda E^{(1)} + \lambda^2 E^{(2)} + \lambda^2 E^{(3)} + \cdots (E^{\mathrm{HF}} = E^{(0)} + E^{(1)}) \quad (3)$$

$$E^{(2)} = -\Sigma_{s\neq 0}|<\psi_s{}^{(0)}|\hat{H}'|\psi_0^{(0)}>|^2/(E_s^{(0)} - E_0^{(0)}) \quad (4)$$

The Møller-Plessert (MP) perturbation theory is another approach to the electron correlation.[14] The method employs the many-body perturbation theory and adds higher excited states to the HF representation. The Hamiltonian ($\hat{H}$) consists of that of the HF level ($\hat{H}_0$) and the perturbed part ($\hat{H}'$), which is the difference between the full electron potential and the one-electron potential of the HF type ($\hat{H}_0$), as shown by Equation 1. (λ is some parameter to show that $\lambda\hat{H}'$ is small in comparison to $\hat{H}_0$.) After the treatment of the perturbation theory (Equations 2 and 3), the MP*n* energy is obtained depending on the order of the perturbation considered. The MP2 energy ($E^{(2)}$) is the first perturbation to the HF energy ($E^{\mathrm{HF}} = E^{(0)} + E^{(1)}$ in this treatment). $E^{(2)}$ is given by Equation 4, which is always negative.[5,6] Indeed, the MP2 treatment is employed very popularly in the QC calculations, but we must be careful, since the overestimation of correction terms may occur due to the non-variational treatment of the method. The higher-order corrections in MP*n* ($n \geq 3$) may improve the overestimation. The MP treatment is a very good method to obtain reliable results. However, the MP2 treatment is still time-consuming and this may prevent its application to larger systems. Density functional theory (DFT)[15–17] can apply the equation to larger systems.

$$\begin{aligned} E_0 &= -\Sigma_{A=1}^{N} Z_A \int (\rho(\mathbf{r}_1)/r_{1A})dv_1 - (1/2)\Sigma_{i=1}^{n} <\varphi_i{}^{\mathrm{KS}}(\mathbf{r}_1)|\Delta_1|\varphi_i{}^{\mathrm{KS}}(\mathbf{r}_1)> \\ &\quad + (1/2)\int (\rho(\mathbf{r}_1)\rho(\mathbf{r}_2)/r_{12})dv_1 dv_2 + E_{\mathrm{XC}}[\rho] \\ &= E_{\mathrm{V}}[\rho] + E_T[\rho] + E_J[\rho] + E_{XC}[\rho] \end{aligned} \quad (5)$$

$$E_{\mathrm{XC}}{}^{\mathrm{GGA}}[\rho^\alpha, \rho^\beta] = \int g(\rho^\alpha(\mathbf{r}), \rho^\beta(\mathbf{r}), \nabla\rho^\alpha(\mathbf{r}), \nabla\rho^\beta(\mathbf{r}))\mathrm{d}v \quad (6)$$

The DFT approach is based on the strategy of modelling electron correlation through general functionals of the electron density, ρ. The DFT energy of ground state E_0 is evaluated by Equation 5 (Δ_1: Laplacian operator for electron 1), where $E_{\mathrm{XC}}[\rho]$ means that E_{XC} is calculated through functionals of ρ, for example. The first, second, and third terms of Equation 5 show the classical Coulomb attractive interaction energies between electronic charge and external potential (nuclear charge in this case) ($E_{\mathrm{V}}[\rho]$), the kinetic energy of electrons in Kohn and Sham orbitals $\varphi_i{}^{\mathrm{KS}}$ ($E_{\mathrm{T}}[\rho]$) and the classical Coulomb repulsive energies between electronic charges ($E_{\mathrm{J}}[\rho]$), respectively. The fourth term represents an exchange-correlation term other than the first, second, and third terms, which includes the remaining parts of the electron–electron interactions. $E_{XC}[\rho]$ can be derived based on the electron densities and the gradients in the case of generalized-gradient approximation (GGA), which is approximately expressed by Equation 6, where ρ^{α}, ρ^{β}, and ρ show the α, β, and (α + β) spin densities, respectively.

$$E_{XC}[\rho] = E_X[\rho] + E_C[\rho] \tag{7}$$

$$E_{XC}{}^{B3LYP} = E_X{}^{LDA} + c_o(E_X{}^{HF} - E_X{}^{LDA}) + c_X E_X{}^{B88} + E_C{}^{VWN3} + c_C(E_C{}^{LYP} - E_C{}^{VWN3}) \tag{8}$$

$E_{XC}[\rho]$ is evaluated separately by the exchange and correlation parts, as shown by Equation 7. Both $E_X[\rho]$ and $E_C[\rho]$ are constructed by the local functionals and gradient-corrected functionals. Whereas the former is the function only of $\rho(\boldsymbol{r})$, the latter depends both on $\rho(\boldsymbol{r})$ and $\nabla^2\rho(\boldsymbol{r})$, although this is not shown.

Pure DFT methods are defined by the pairing of an exchange functional with a correlation functional. BLYP is an example, which is constructed by the gradient-corrected exchange functional proposed by Becke with the gradient-corrected correlation functional proposed by Lee, Yang, and Parr. On the other hand, hybrid DFT functionals are a mixture of the HF and DFT exchange with the DFT correlation. Equation 8 shows the typical hybrid DFT functionals of B3LYP. $E_{XC}{}^{B3LYP}$ can be evaluated at any admixture depending on the values of c_o, c_X, and c_C, although they are constants. Values of $c_o = 0.20$, $c_X = 0.72$, and $c_C = 0.81$ are usually employed in the B3LYP level. Becke gave the values through fitting to the atomization energies, ionization potentials, proton affinities, and first-low atomic energies in the G1 molecule set. The DFT method is suitable for calculating larger molecules, although it is usually difficult to evaluate weak interactions such as van der Waals interactions. The DFT method has been continually developed, as we explain next.

12.4.2.2 Development in the DFT Functionals

Most exchange-correlation functionals currently used were developed about 20–40 years ago.[18] Becke proposed the exceedingly popular Beck 3-parameter–Lee–Yang–Parr (B3LYP) functional in 1993.[19] Boese and Martin called these first-generation functionals, which exhibit a number of chemically important weakness, especially when they are applied to weakly bonded systems.[20] A number of second-generation functionals have proposed as successors, such as PBE, mPW91, VSXC, PBE0, PKZB, B97, B97-1, B97-2, HCTH, OPTX/OPTC, τ-HCTH, and its hybrid, TBSS, and B98. The BMK functional also belongs to the second-generation functionals, which show high performance in kinetics with an accuracy range of 2 kcal mol^{-1}. Two important avenues that have been pursued in this direction are (a) inclusion of new variables explicitly dependent on the Kohn–Sham orbitals, such as kinetic energy density (τ) and (b) treatment of DFT as a parameterized 'semi-*ab initio*' method. Tackling the problem of transition states, various groups attempted to develop what are called third-generation functionals. However, the early third-generation functionals do not seem accurate enough to replace the second-generation functionals.

Scuseria and Staroverov summarized the six strategies that have been widely employed for designing density functionals:[18] (1) local spin density approximation (LSDA), (2) density-gradient expansion, (3) constraint satisfaction, (4) modelling the exchange-correlation hole, (5) empirical fits, and (6) mixing HF and approximate DFT exchange. New hybrid meta-exchange-correlation functionals were developed in this line, such as M05-2X[21] and M06-2X[22] only for non-metals with M05[21] and M06[22] for both transition metals and non-metals. The M05-2X and M06-2X functionals are high non-locality ones with double the amount of non-local exchange (2X). By comparing their performance with that of other functionals, it is stressed that (1) M06-2X is recommended most highly for the study of main-group thermochemistry and kinetics, non-covalent interactions, and a combination of main-group thermochemistry, kinetics, and non-covalent interactions, as a whole. The M05-2X and BMK functionals are also recommended, although M06-HF seems the best functional for the study of long-range charge transfer via a TD-DFT approach. On the other hand, M06 is recommended for the study of organometallic and inorganometallic thermochemistry, together with M06-L and M05.

In spite of the development of the new hybrid meta-exchange-correlation functionals with very good performance in calculations, the DFT level of B3LYP seems most popularly employed to understand and explain experimental results, together with the MP2 level. DFT functionals such as B3LYP have a weakness in the performance of the long-distance interactions. This is improved by employing the r^{-1} function at long distances. Such treatments can be used in the Gaussian 09 program package[23] by use of the keyword CAM-B3LYP.[24] The LC-wPBE keyword can be applied similarly, where the prefix LC-, proposed by Hirao and co-workers,[25] may be added to any pure functional to apply the long correction.

12.4.3 Survey of Relativistic Effect

The relativistic effect is an important issue[26] when properties of atoms and molecules are calculated, especially when the system contains heavier atoms.[27,28] The relativistic effect will arise when the speed of an electron in an atom cannot be neglected relative to that of light. Such high-speed electrons exist in heavier atoms, since the energy of an electron in an atom increases proportionally to the square of nuclear charge (Z^2). The evaluation of the relativistic effect is complex and difficult; however, it has been evaluated successfully and dramatically under the zeroth-order regular approximation (ZORA) level.[29] The ZORA treatment is explained briefly, here.

$$E = ({m_0}^2c^4 + p^2c^2)^{1/2} - m_0c^2 + V \tag{9}$$

$$= (p^2/2m_0)[1 + (E - V)/2m_0c^2]^{-1} + V \quad (10)$$

$$E = V + p^2/2m_0 + p^4/8m_0{}^3c^2 + \cdots \quad (11)$$

$$E = p^2[2m_0\{1 + (E - V)/2m_0c^2\}]^{-1} + V \quad (12)$$

$$E = p^2c^2[(2m_0c^2 - V)\{1 + E(2m_0c^2 - V)\}]^{-1} + V \quad (13)$$

$$E_0 = p^2c^2(2m_0c^2 - V)^{-1} + V; \quad E_1 = -E_0p^2c^2(2m_0c^2 - V)^{-2} \quad (14)$$

The relativistic energy of a single electron in a scalar potential V is given by Equation 9. Equation 10 can be derived through dividing both numerator and denominator by c^2 after rationalizing the numerator. Equation 11, known as Pauli Hamiltonian, is obtained through the usual expansion by $(E - V)/2m_0c^2$. However, Equation 11 will not be valid near the nuclei for the Coulomb potential $V = -Z/r$, since it diverges near the nucleus. On the other hand, Equation 13 is derived from Equation 12 (derived from Equation 9 or Equation 10) through the expansion by $E(2m_0c^2 - V)$. Equation 14 shows the zeroth-order and first-order terms of Equation 13 (E_0 and E_1, respectively). One realizes that E_0 in Equation 14 is different from the non-relativistic energy ($E = p^2/2m_0 + V$). Namely E_0 contains the relativistic effect in it. The treatment of relativistic effect using E_0 in Equation 14 is called ZORA. E_0 could be successfully applied to evaluate the relativistic effect, although the perturbative treatment of the higher-order expansion of the Hamiltonian is pointed out.[30] Table 12.4.1 shows the

Table 12.4.1 Uranium orbital energies evaluated under relativistic and non-relativistic conditions.[a]

Orbital	*Dirac*	*CPD 0th order*[b]	*CPD 1st order*[b]	*Pauli 1st order*[c]	*NR*[d]
$1s_{1/2}$	–4255.55	–4872.99	–4158.88	–4114.71	–3690.78
$2s_{1/2}$	–795.00	–818.96	–793.51	–753.91	–640.21
$3s_{1/2}$	–200.69	–202.63	–200.51	–190.53	–161.26
$4s_{1/2}$	–51.09	–51.24	–51.05	–48.28	–40.57
$5s_{1/2}$	–11.32	–11.33	–11.31	–10.62	–8.81
$5p_{1/2}$	–9.073	–9.076	–9.066	–8.490	–7.006
$5p_{3/2}$	–7.057	–7.061	–7.055	–7.087	
$5d_{5/2}$	–3.465	–3.469	–3.466	–3.500	
$6s_{1/2}$	–1.720	–1.719	–1.718	–1.580	–1.298
$6p_{1/2}$	–1.069	–1.069	–1.068	–0.973	–0.795
$6p_{3/2}$	–0.7410	–0.7409	–0.7407	–0.7476	
$5f_{5/2}$	–0.1033	–0.1040	–0.1040	–0.1047	–0.3419
$5f_{7/2}$	–0.0728	–0.0735	–0.0735	–0.0647	
$6d_{3/2}$	–0.0710	–0.0711	–0.0711	–0.0728	–0.1157
$6p_{5/2}$	–0.0537	–0.0538	–0.0538	–0.0512	
$7s_{1/2}$	–0.1340	–0.1339	–0.1339	–0.1231	–0.1071

[a]In au.
[b]Based on the relativistic Chang–Pelissier–Durand Hamiltonian, which is identical to ZORA in Equation 14 (see ref. 32).
[c]Equation 11.
[d]Non-relativistic treatment.

selected orbital energies of uranium evaluated employing Dirac equation, E_0 and E_1 in Equation 14, first order in Equation 11 and non-relativistic treatment (NR). It is demonstrated that ZORA is satisfactorily accurate with more accuracy for lower energy valence electrons than deep core states.[29b,31] ZORA is applied to evaluate the relativistic effect in various phenomena, containing NMR parameters.

12.4.4 Quantum Chemical Calculations in Chalcogen Chemistry

Various types of QC calculations have been performed in chalcogen chemistry to examine the mechanisms, to analyse the reaction processes and to support experimental results containing the optimization of structures and conformer analysis. The (relative) stabilities of species containing bond dissociation energies and potential energy surfaces are also taken into account. Weak interactions became current targets of QC calculations as the reliability of the calculations increased. Both DFT and MP2 levels are frequently employed for the calculations, together with the higher levels such as the CISD[7,33] and QCISD[34] levels. The methodological verification is also contained.

The experimental results can be better understood and the reasons for the observed results are clarified if they are supported by calculations. The structures of $Se(SCH_3)_2$ and $Te(SCH_3)_2$ were investigated using gas-phase electron diffraction (GED), which were examined through *ab initio* and DFT geometry optimizations,[35] for example. The best fit to the GED structures was achieved at the MP2 level of calculations. The theoretical models agreed that each compound exists as two stable conformers, one in which the methyl groups are on different sides (g^+g^+conformer) of the S–E–S plane (E = Se, Te) and one in which they are on the same side (g^+g^-conformer) (see Scheme 12.4.1).

To clarify the nature of chemical bonds and interactions must be the first step in the study of the chalcogen chemistry. The two centre-three electron bonds (2c–3e) have been paid much attention over a long period. High-pressure equilibrium mass spectrometry experiments yield a reaction enthalpy of −104 kJ mol^{-1} for the equilibrium reaction shown by Equation 15 at 506 K, which results in the bond energy of 107 kJ mol^{-1} at 0 K.[36] The value is supported by the calculated 107.9 kJ mol^{-1} at the

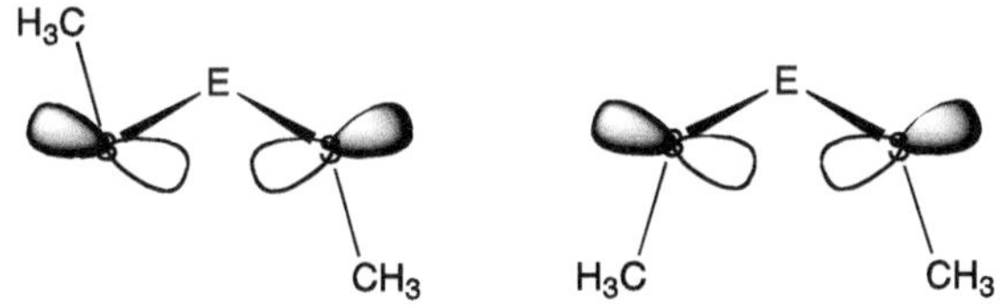

Scheme 12.4.1 The g^+g^+ (left) and g^+g^- (right) conformers of $E(SCH_3)_2$ (E = Se, Te), showing the p-type lone pairs on the S atoms.

B3LYP/6–31G(d)//B3LYP/6–31G(d)+ZPC level. Studies on a competing reaction shown by Equation 16 yield an experimental bond enthalpy of 119 kJ mol^{-1} at 506 K, which is also well supported by the calculated bond energy of 121.3 kJ mol^{-1} at 0 K.

$$Et_2S^+ + Me_2S \rightleftarrows [Et_2S \therefore SMe_2]^+ \quad (15)$$

$$Et_2S^+ + Et_2S \rightleftarrows [Et_2S \therefore SEt_2]^+ \quad (16)$$

The 2c–3e bonds between the heteronuclear atoms must be more complex than the homonuclear case. Structural, energetical, and topological investigations are also performed on the bonds in prototypical radicals, ionic and neutral $RSOH^-$, $RR'SOH_2^+$ and RR′SOH, where R, R′ = H, and CH_3.[37] Qualitative and quantitative criteria, defined from the topological analysis of the electron localization function, are proposed to characterize such a 2c–3e bond. The neutral radicals show an intermediate behaviour between the pure 2c–3e S∴O bond in anions and the electrostatic interaction in cations, despite the low energy of bond formation. As in the radical anions, the substitution favours the formation of a 2c–3e bond.

Gleiter and co-workers carried out QC calculations on homonuclear intermolecular interactions between chalcogen centres (O, S, Se, Te) employing model systems **1**–**12**.[38] The MP2/SDB-cc-pVTZ, 6–311G* level of theory was used to derive the geometrical parameters and the interaction energies of the model systems. Figure 12.4.1 shows the total energies and the contributions form the electrostatic (E_{elst}), induction (E_{ind}), dispersion (E_{disp}), and exchange correlation (E_{exch}) energies, separated by the symmetry-adapted perturbation theory (SAPT). A detailed analysis based on SAPT reveals that induction and dispersion forces contribute to the bonding in each case. The electrostatic energy also contributes to the intermolecular bonding in **1–3**, however, it does not for **4–12**.

Natural bond orbital (NBO) analysis reveals that the interaction in the dimers **1–3** is mainly due to weak hydrogen bonding between methyl groups and chalcogen centres. For the aggregates with heavier centres the chalcogen–chalcogen interaction dominates and hydrogen bonding only plays a minor role. Figure 12.4.2 shows the minimum geometries of **1** and **10**. Electron-withdrawing groups on the chalcogen centres increase the interaction energy and reduce the intermolecular distance dramatically. The one-electron picture of an interaction between the lone pair of the donor and the chalcogen–carbon σ* orbital allows a qualitatively correct reproduction of the observed trend.

They also carried out the QC calculations on the intermolecular interactions between chalcogen centres (O, S, Se, Te) of the heteronuclear cases in addition to the homonuclear cases, employing model systems.[40] NBO analysis was also employed for the calculations. Systems of O and S donors with O and S acceptors are mainly due to weak hydrogen bonding between methyl groups and chalcogen centres. These systems all contain hard chalcogen atoms as acceptors. The chalcogen–chalcogen interaction dominates for all other systems. The one-electron picture of the $n_p(E) \rightarrow \sigma^*(E–C)$ interaction serves as a model to qualitatively rationalize trends found in many of these systems.

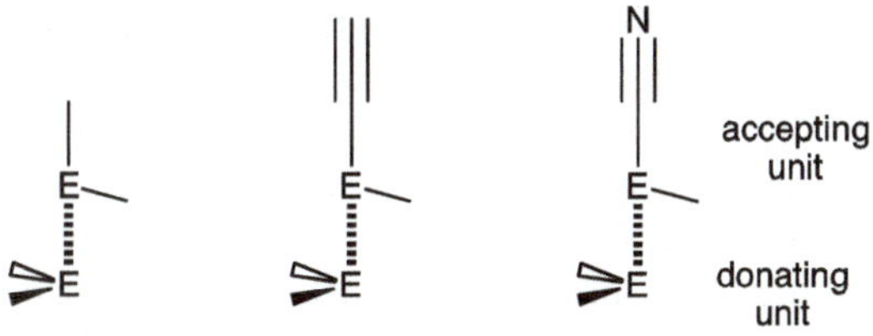

1 (E = O) **2** (E = O) **3** (E = O)
4 (E = S) **5** (E = S) **6** (E = S)
7 (E = Se) **8** (E = Se) **9** (E = Se)
10 (E = Te) **11** (E = Te) **12** (E = Te)

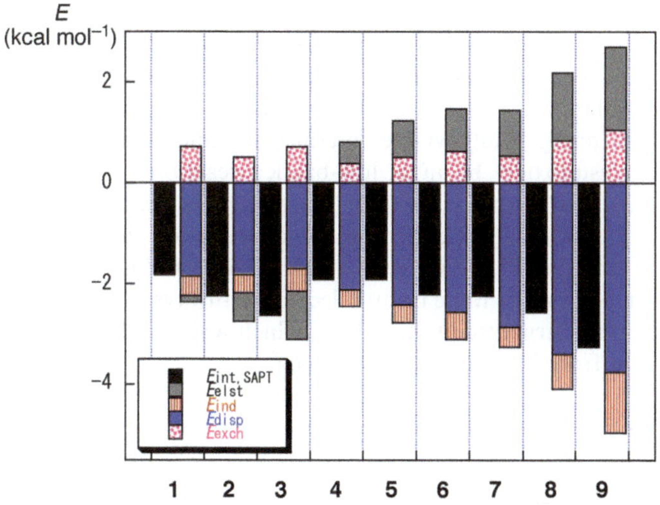

Figure 12.4.1 Contributions of the electrostatic (E_{elst}), induction (E_{ind}), dispersion (E_{disp}), and exchange correlation (E_{exch}) energies derived by the SAPT2002 program,[39] together with $E_{int,SAPT}$ containing δ_{HF}.

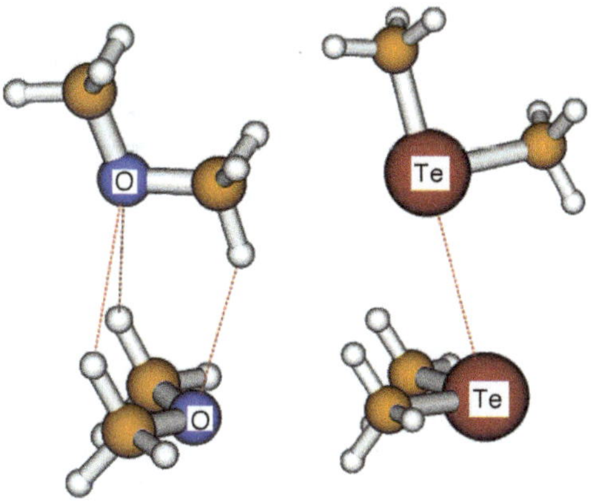

Figure 12.4.2 Minimum geometries of the dimers **1** (left) and **10** (right).

A detailed analysis based on SAPT reveals that induction and dispersion forces dominate and contribute to the bonding in each case.

It is an important process in the chalcogen chemistry to reveal the reaction mechanisms, containing the behaviours of transition states. Nucleophilic substitution occurs at the sulfur atom of diselenides and selenosulfides with various nucleophiles. The mechanism was examined theoretically with the B3LYP and MP2 methods, employing the model where the substitution occurs at Se of R_1SSeR_2 and R_1SeSeR_2 (R_1 and R_2 = H or Me) by the attack of HS^-, CH_3S^-, HSe^-, and CH_3Se^-.[41] The mechanism was demonstrated to be addition-elimination in all cases. A stable hypercoordinate Se intermediate lies in a well that is 8–14 kcal mol^{-1} deep. Nucleophilic attack at Se is both kinetically and thermodynamically more favourable than at S.

The reaction mechanism of organic sulfides with singlet oxygen (**13**) was reexamined and a revised mechanism was proposed.[42] On the basis of high-level *ab initio* calculations, it has established that an *S*-hydroperoxysulfonium ylide **15** is an energetically viable intermediate and that it can be formed from a peroxysulfoxide **14** in a low-barrier reaction (~6 kcal mol^{-1}) (see Figure 12.4.3). Furthermore, **15** can rearrange via reasonably low energy barriers (~12 kcal mol^{-1}) to either an *R*-hydroperoxide **16** or to a protonated sulfone ylide structure **17**, which leads to the sulfone product **18** via a keto–enol rearrangement. Which of these intramolecular pathways is the lowest in energy will depend on the structure of the sulfide, which will also determine the degree of competition from intermolecular trapping.

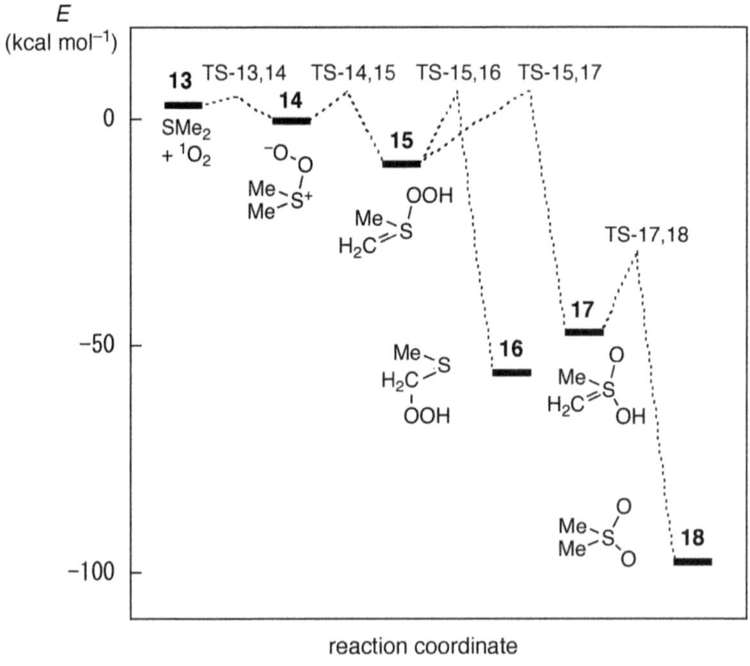

Figure 12.4.3 Energy profile for the reaction of dimethyl sulfide with singlet oxygen (**13**) in solution based on computational and experimental data.

Similarly, the mechanism of oxidation of organic sulfides in aqueous solutions by hydrogen peroxide was investigated via *ab initio* calculations.[43]

The formation of *N*-triflylsulfilimines in the reaction of triflylimino-λ^3-bromane **19** and various sulfides occurs even at 0 °C under transition-metal-free conditions. The reaction proceeds similarly with sulfoxides and the predominant retention of configuration at the stereogenic sulfur centre is confirmed using (*R*)-methyl *p*-tolyl sulfoxide. Figure 12.4.4 shows the transition state for the reaction of dimethyl sulfide with imino(phenyl)-λ^3-bromane **20**, a model of **19**.[44]

Weak interactions containing chalcogens are of current interest. The sulfur-π interactions are investigated for the H_2S–benzene adduct, as the simplest possible prototype of the interactions between aromatic rings and sulfur atoms in the side chains of amino acids in proteins.[45] High-quality potential energy curves were obtained using the aug-cc-pVQZ basis set at the CCSD(T) level. The equilibrium intermonomer distance for the hydrogens–down C_{2V} configuration is 3.8 Å with an interaction energy of −2.74 kcal mol^{-1} (see Figure 12.4.5). The binding energy to the complete basis set limit gives −2.81 kcal mol^{-1} after extrapolation. This binding energy is comparable to that of H_2O–benzene or of the benzene dimer and the equilibrium distance is in close agreement with experiment. Other orientations of the dimer were also considered at less complete levels of theory. A considerable reduction in binding for the sulfur-down configuration, together with an energy decomposition analysis, indicates that the attraction in H_2S–benzene is best thought of as arising from a favourable electrostatic interaction between partially positive hydrogen atoms in H_2S with the negatively charged π-cloud of the benzene. Other methods are also applied for further investigations.[46]

Similarly, the characteristics of the interaction between the π-cloud of naphthalene and up to two H_2O or H_2S molecules were studied.[47] Calculations show that clusters formed by naphthalene and one H_2O or H_2S molecules have similar geometric features and interaction energies. Figure 12.4.6 shows the optimized

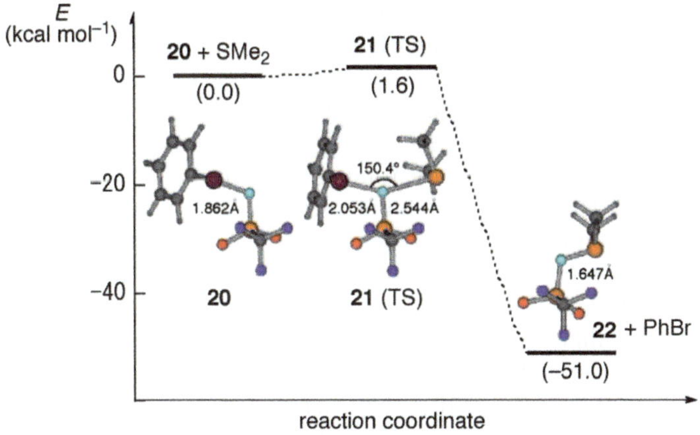

Figure 12.4.4 Energy profile for the reaction of imino-λ^3-bromane **20** with dimethyl sulfide, together with calculated structures of **20**, transition state **21**, and sulfilimine **22** at the MP2 level.

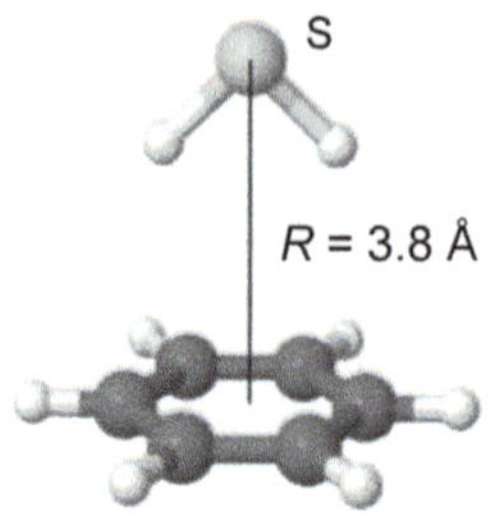

Figure 12.4.5 Orientations for H_2S–benzene adduct.

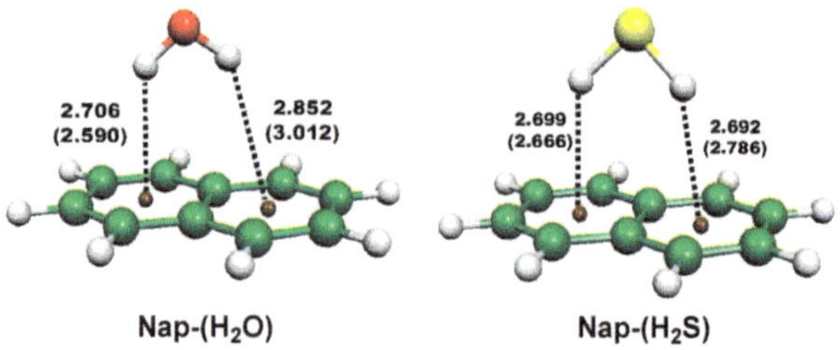

Figure 12.4.6 Optimized structures for naphthalene···H_2E clusters with E–H···π contact (E = O and S) (distances to ring centres in Å from MP2/6–31 + G* optimizations and those in parentheses correspond to MPWB1K/6–31 + G* level).

structures. Best estimates for the interaction energy amount to –2.95 and –2.92 kcal mol^{-1} for H_2O and H_2S, respectively, as obtained with the CCSD(T) method. It is pointed out that calculations at the MP2 level employing large basis sets should be avoided because they produce highly overestimated interaction energies, especially for hydrogen sulfide complexes. The MPWB1K functional, however, provides values pretty similar to those obtained with the CCSD(T) method. Although the magnitude of the interaction is similar with both H_2O and H_2S molecules, its nature is somewhat different. The H_2O complex presents electrostatic and dispersion contributions of similar magnitude, whereas for H_2S the interaction is dominated by dispersion.

In clusters containing two H_2O or H_2S molecules, several minima were characterized. In water clusters, the total interaction energy is dominated by the presence of an O–H···O hydrogen bond and structures with this contact are the most stable, as a consequence. However, clusters containing H_2S show structures with no interaction between H_2S moieties which are as stable as the hydrogen-bonded ones, because they allow an optimal H_2S···naphthalene interaction, which is stronger than the S–H···S contact. Therefore it is possible that in larger polycycles hydrogen sulfide molecules will be spread onto the surface, maximizing S–H···π interactions, rather than aggregated, forming H_2S clusters.

Lots of calculations are performed on the chalcogen compounds other than those discussed above. Experimental results are well supported through QC calculations[48] containing the structure or conformer analysis,[49] together with the potential energy surfaces.[50] Mechanisms and reaction processes[51] are widely examined, together with the (relative) stabilities and bond dissociation energies.[52] Weak interactions[53] are of high current interest. Methodological verifications are also examined.[54] QC calculations on the hypervalent chalcogen compounds are also discussed in Chapter 10; however, those on the metal complexes of chalcogen compounds[55] are not discussed.

12.4.5 Application of QTAIM to Chalcogen Chemistry

The atoms-in-molecules (AIM) method, proposed by Bader, can fractionalize the properties of molecules to atoms through fractionalizing electrons of a molecule to atoms in it.[56–58] AIM enables us to analyse, evaluate, and classify the nature of chemical bonds and interactions.[59–64] Chemical bonds and interactions are mainly analysed by examining the behaviours of AIM functions at bond critical points (BCPs; $\boldsymbol{r}_c$; *), where the electron densities $\rho_b(\boldsymbol{r}_c)$ are at the three-dimensional (3D) saddle points. Equations 17 and 18 show the relations between AIM functions of $H_b(\boldsymbol{r}_c)$ (total electron energy densities), $G_b(\boldsymbol{r}_c)$ (kinetic energy densities), $V_b(\boldsymbol{r}_c)$ (potential energy densities), and $\nabla^2\rho_b(\boldsymbol{r}_c)$ (Laplacian of charge densities) at BCPs.

$$H_b(\boldsymbol{r}_c) = G_b(\boldsymbol{r}_c) + V_b(\boldsymbol{r}_c) \qquad (17)$$

$$(\hbar^2/8m)\nabla^2\rho_b(\boldsymbol{r}_c) = H_b(\boldsymbol{r}_c) - V_b(\boldsymbol{r}_c)/2 = G_b(\boldsymbol{r}_c) + V_b(\boldsymbol{r}_c)/2 \qquad (18)$$

While $\rho_b(\boldsymbol{r}_c)$ is locally concentrated relative to the average distribution around BCPs if $\nabla^2\rho_b(\boldsymbol{r}_c) < 0$, it is depleted when $\nabla^2\rho_b(\boldsymbol{r}_c) > 0$. Electrons at BCPs are stabilized when $H_b(\boldsymbol{r}_c) < 0$, therefore, interactions exhibit the covalent nature in this region, whereas they exhibit no covalency if $H_b(\boldsymbol{r}_c) > 0$ due to the destabilization of electrons at BCPs under the conditions. Scheme 12.4.2 summarizes the classification of interactions by the signs of $\nabla^2\rho_b(\boldsymbol{r}_c)$ and $H_b(\boldsymbol{r}_c)$.[56,57] Interactions in the region of $\nabla^2\rho_b(\boldsymbol{r}_c) < 0$ with $H_b(\boldsymbol{r}_c) < 0$ are called shared-shell (SS) interactions and they are closed-shell (CS) interactions for $\nabla^2\rho_b(\boldsymbol{r}_c) > 0$. The CS interactions are especially called *pure* CS interactions for $H_b(\boldsymbol{r}_c) > 0$ and $\nabla^2\rho_b(\boldsymbol{r}_c) > 0$.[56,57] Interactions in the intermediate region between SS and pure CS are tentatively called *regular* CS, when it is necessary to distinguish from pure CS.[65–67] Electrons in the regular CS region are locally depleted but stabilized at BCPs, since $\nabla^2\rho_b(\boldsymbol{r}_c) > 0$ but $H_b(\boldsymbol{r}_c) < 0$. $\nabla^2\rho_b(\boldsymbol{r}_c)$ can be replaced by $H_b(\boldsymbol{r}_c) - V_b(\boldsymbol{r}_c)/2$ in the classification shown in Scheme 12.4.2, since $(\hbar^2/8m)\nabla^2\rho_b(\boldsymbol{r}_c) = H_b(\boldsymbol{r}_c) - V_b(\boldsymbol{r}_c)/2$ (Equation 18). The classification by the signs of $\nabla^2\rho_b(\boldsymbol{r}_c)$ (or $H_b(\boldsymbol{r}_c) - V_b(\boldsymbol{r}_c)/2$) and $H_b(\boldsymbol{r}_c)$ can be achieved only by θ if $\tan\theta = H_b(\boldsymbol{r}_c)/(H_b(\boldsymbol{r}_c) - V_b(\boldsymbol{r}_c)/2)$, as shown in Scheme 12.4.2.

Nakanishi and co-workers proposed an AIM method to analyse interactions by plotting $H_b(\boldsymbol{r}_c)$ against $H_b(\boldsymbol{r}_c) - V_b(\boldsymbol{r}_c)/2$,[66] after the proposal of $H_b(\boldsymbol{r}_c)$ *versus* $\nabla^2\rho_b(\boldsymbol{r}_c)$,[67] which incorporates the classification of interactions shown in Scheme 12.4.2. Scheme 12.4.3 shows the relation between the plots and the classification. The method is called AIM dual functional analysis. Data of perturbed structures around the fully optimized one are also employed for the

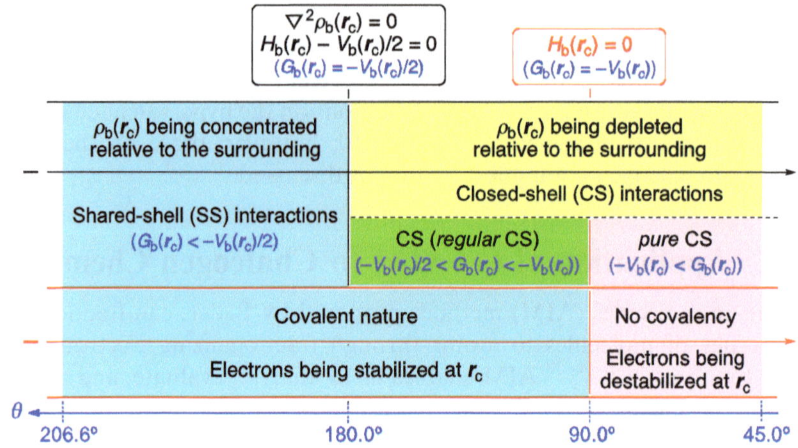

Scheme 12.4.2 Classification of interactions by AIM functions, where $H_b(\mathbf{r}_c) - V_b(\mathbf{r}_c)/2 = (\hbar^2/8m)\nabla^2\rho_b(\mathbf{r}_c)$.

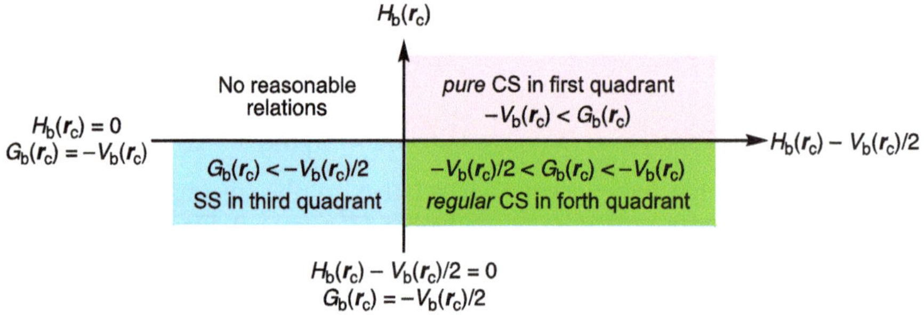

Scheme 12.4.3 Requirements for data to appear in certain quadrant in the plots of $H_b(\mathbf{r}_c)$ against $H_b(\mathbf{r}_c) - V_b(\mathbf{r}_c)/2$, where $H_b(\mathbf{r}_c) - V_b(\mathbf{r}_c)/2 = (\hbar^2/8m)\nabla^2\rho_b(\mathbf{r}_c)$.

plot in the treatment. They also proposed the concept of 'dynamic nature of interaction,' which originates from data of the perturbed structures.[68] While data at a fully optimized structure correspond to the static nature, the dynamic nature of an interaction originates from the perturbed structures at the fully optimized structure. The dynamic nature of an interaction can be recognized as the effect arising from the zero-point motion of the internal vibration corresponding to the interaction, if the perturbed structures are generated using suitable normal coordinates of the internal vibration (NIV). Table 12.4.2 shows the nature of halogen adducts of sulfides and selenides investigated by AIM dual functional analysis. Figure 12.4.7 shows the plots of $H_b(\mathbf{r}_c)$ against $H_b(\mathbf{r}_c) - V_b(\mathbf{r}_c)/2$ for the CS interactions containing data in Table 12.4.2.

The $G_b(\mathbf{r}_c)$, $V_b(\mathbf{r}_c)$, and $H_b(\mathbf{r}_c)$ values are reported to be approximately expressed using $\rho_b(\mathbf{r}_c)$ and $\nabla^2\rho_b(\mathbf{r}_c)$, as shown in Equations 19–21,[69–71] respectively, which enables us to compare the experimental results directly to

Table 12.4.2 AIM functions and parameters evaluated for weak to strong interactions calculated with NIV.[a]

No/Species[b] (E-*-X)	$r_o(E, X)$ (Å)	$c\nabla^2\rho_b(r_c)^c$ (au)	$H_b(r_c)$ (au)	R (au)	θ (°)	Freq (cm^{-1})	k_f (unit[d])	$\theta_{p:NIV}$ (°)	$\kappa_{p:NIV}$ (au^{-1})
12 Me_2S-*-Cl_2 (MC)	2.6331	0.0108	–0.0057	0.0122	117.6	104.4	0.044	162.8	51.24
13 Me_2S-*-Br_2 (MC)	2.6923	0.0093	–0.0078	0.0122	130.0	114.9	0.059	171.5	33.00
14 Me_2Se-*-Cl_2 (MC)	2.5700	0.0093	–0.0125	0.0156	143.3	123.5	0.058	182.3	14.47
15 Me_2Se-*-Br_2 (MC)	2.7286	0.0078	–0.0102	0.0128	142.7	108.8	0.078	181.0	13.82
20 Me_2ClS-*-Cl (TBP)	2.2650	0.0046	–0.0364	0.0367	172.8	334.6	0.389	192.8	4.84
21 Me_2BrS-*-Br (TBP)	2.4387	0.0048	–0.0258	0.0262	169.4	358.5	0.294	188.9	2.89
22 Me_2ClSe-*-Cl (TBP)	2.3547	0.0053	–0.0335	0.0339	171.0	307.8	0.366	186.3	0.57
23 Me_2BrSe-*-Br (TBP)	2.5196	0.0035	–0.0262	0.0264	172.3	233.4	0.946	189.0	1.95
24 Me_2S^+-*-Cl (Cov-w)	1.9791	–0.0241	–0.1197	0.1221	191.4	565.4	2.456	198.2	0.16
25 Me_2S^+-*-Br (Cov-w)	2.1433	–0.0110	–0.0798	0.0806	187.8	450.9	1.450	193.7	0.31
26 Me_2Se^+-*-Cl (Cov-w)	2.1089	–0.0070	–0.0849	0.0852	184.7	465.8	4.625	185.6	1.14
27 Me_2Se^+-*-Br (Cov-w)	2.2636	–0.0075	–0.0636	0.0640	186.7	337.8	2.952	190.0	0.37

[a]Calculated with the 6–311++G(3df,3pd) basis sets for S, Se, Cl, and Br and the 6–311+G(3d,2p) basis sets for C and H of Me at the MP2 level.
[b]Data are given for interaction at BCP, which is shown by * as in Me_2S-*-Cl_2.
[c]$c = \hbar^2/8m$.
[d]mdyn Å^{-1}.

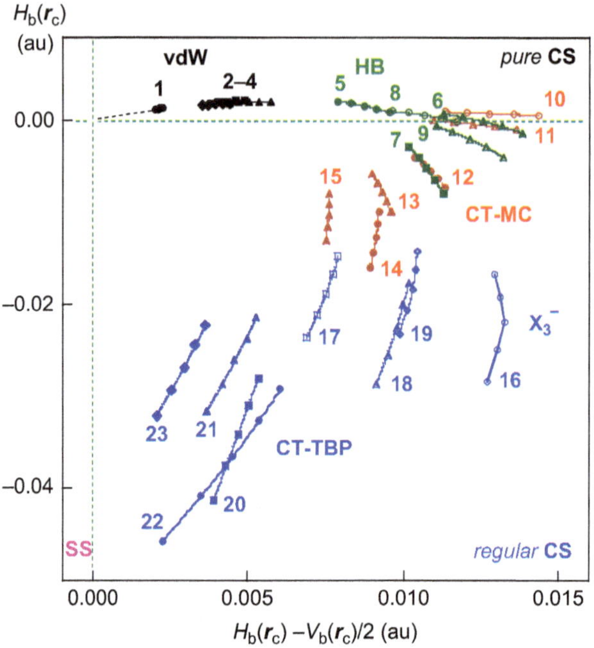

Figure 12.4.7 Plots of $H_b(\boldsymbol{r}_c)$ *versus* $H_b(\boldsymbol{r}_c) - V_b(\boldsymbol{r}_c)/2$ for chalcogen–halogen interactions of the CT-MC and CT-TBP types in Table 12.4.2, together with vdW interactions and hydrogen bondings [He--*--HF (1), Ne--*--HF (2), Ar--*--HF (3), Kr--*--HF (4), NN--*--HF (5), HF--*--HF (6), HCN--*--HF (7), H_2O---HOH (8), and Me_2O---HOH (9)]

the theoretical ones. The QTAIM (quantum theory of AIM) results are demonstrated to be very close to the experimentally determined high-resolution charge densities, employing Equations 19–21, exemplified by the N–S–N 3c–4e (three centre–four electron interactions) in 2-(2-pyridylimino)-2*H*-1,2,4-thiadiazolo[2,3-*a*]pyridine (**23**).[72] Figure 12.4.8 shows the trajectory plots obtained by QTAIM and experimental results for **23**.

$$G_b(\boldsymbol{r}_c) = (3\hbar^2/10m)(3\pi^2)^{2/3}\rho_b(\boldsymbol{r}_c)^{5/3} + (\hbar^2/6m)\nabla^2\rho_b(\boldsymbol{r}_c) \tag{19}$$

$$V_b(\boldsymbol{r}_c) = -(3\hbar^2/5m)(3\pi^2)^{2/3}\rho_b(\boldsymbol{r}_c)^{5/3} - (\hbar^2/12m)\nabla^2\rho_b(\boldsymbol{r}_c) \tag{20}$$

$$H_b(\boldsymbol{r}_c) = -(3\hbar^2/10m)(3\pi^2)^{2/3}\rho_b(\boldsymbol{r}_c)^{5/3} + (\hbar^2/12m)\nabla^2\rho_b(\boldsymbol{r}_c) \tag{21}$$

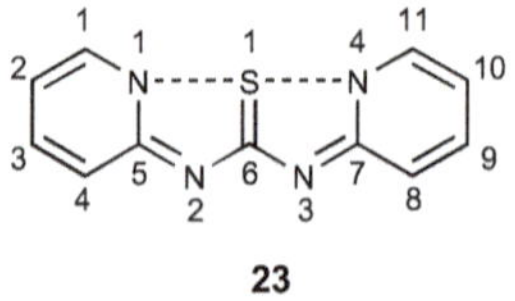

23

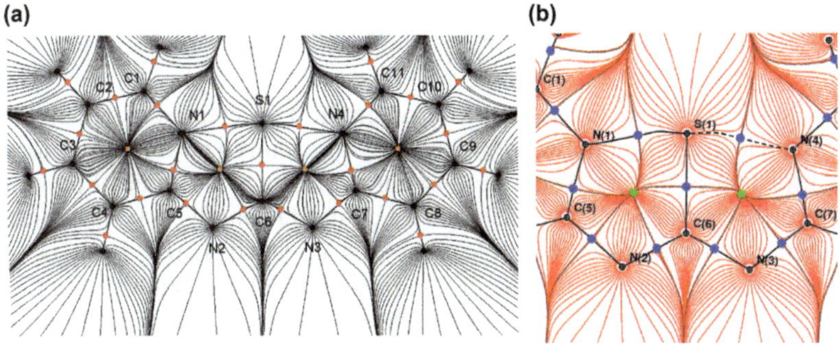

Figure 12.4.8 The negative Laplacian for **23**: (a) QTAIM calculations with BCPs (●), ring critical points (■), and bond paths. The contours (ea_o^{-3}) are at 2^l ($l = 8, 7, \ldots 0$) and (b) high-resolution X-ray diffraction determination.

Tomoda and co-workers investigated the intramolecular O---Se–X interaction in **24**–**26**.[73] AIM analysis showed that the non-bonded Se---O interactions can be characterized by the presence of a BCP and the negative $H_b(\boldsymbol{r}_c)$ values are evaluated for the interactions. The results suggested that Se---O interactions have some covalent rather than the electrostatic character.

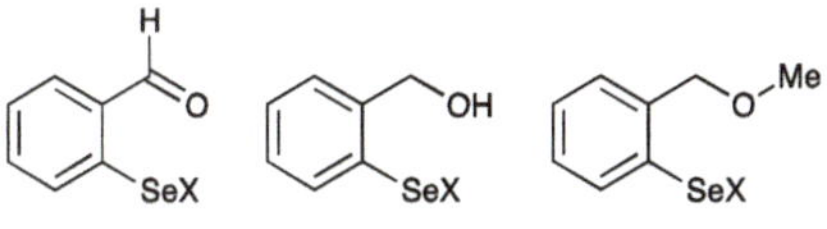

24 **25** **26**

a, X = Cl
b, X = Br
c, X = CN
d, X = SPh (**d'**, X = SMe)
e, X = SeAr (**e'**, X = SeMe)
f, X = Me

12.4.6 Role of σ-Hole in Chalcogen Chemistry: Chalcogen Bond *Versus* Halogen Bond

Before discussion of chalcogen bonds, it is instructive to explain the σ-hole in the halogen compounds first, which is the driving force for the halogen bonds.[74] Murray-Rust *et al.*[75–77] found the anomalously short intermolecular distances in the Cambridge Structural Database, less than the sum of the van der Waals radii, for example. For halogens linked to carbons, close contacts with electrophiles occurred largely at angles of 90–120° with the C–X bond (**27**), whereas the angles were primarily between 160° and 180° (**28**) with nucleophiles. The near-linear interactions with nucleophiles are now known as 'halogen bonding'; although Dumas *et al.*[78] used the term first.

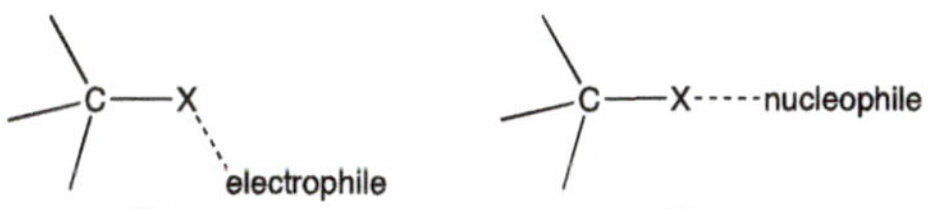

27 **28**

$$V(\boldsymbol{r}) = \Sigma_A Z_A/|\boldsymbol{R}_A - \boldsymbol{r}| - \int \rho(\boldsymbol{r}')d\boldsymbol{r}'/|\boldsymbol{r}' - \boldsymbol{r}| \quad (22)$$

Equation 22 defines the electrostatic potential $V(\boldsymbol{r})$ created at a point $\boldsymbol{r}$ by nuclei and electrons in a molecule where Z_A is the charge on nucleus A located at $\boldsymbol{R}_A$ and $\rho(\boldsymbol{r})$ is the electronic density function for a molecule. $V(\boldsymbol{r})$ can be determined experimentally[79,80] as well as computationally. $V(\boldsymbol{r})$ is normally computed on the surfaces of molecules; this is called $V_S(\boldsymbol{r})$. $V_S(\boldsymbol{r})$ is often calculated at 0.001 electrons bohr^{-3}.[81] Values of $V_S(\boldsymbol{r})$ are negative for CH_3F, CH_3Cl, and CF_4, but a positive region appears on the outermost portion of the halogen surface for CH_3Br and CH_3I. The positive halogen surface region ($V_S(\boldsymbol{r})>0$) is called 'σ-hole', which is surrounded by a negative electrostatic potential region ($V_S(\boldsymbol{r})<0$). Nucleophiles approach to the σ-hole to give the linear complexes (halogen bonds), whereas electrophiles come closer to the negative electrostatic potential region. The magnitudes of positive $V_S(\mathbf{r})$ increase in the order of Cl < Br < I for CF_3X (see Figure 12.4.9).

The three unshared pairs of each of these halogens closely approximate the configuration $s^2p_x^2p_y^2$, where p_x and p_y are perpendicular to the C–X axis. These six electrons create a belt of negative electrostatic potential around the central portion of the halogen atom. Only its outermost region, the σ-hole, retains the positive potential characteristic of the free, spherically symmetric atom.[82,83] This picture accordingly accounts for the observed preference of electrophiles to interact with the halogen laterally and nucleophiles linearly. Electronegative substituents such as in $F_3C–X$ enhance the σ-hole. Roles of electrostatics, dispersion, exchange–repulsion, charge–transfer,

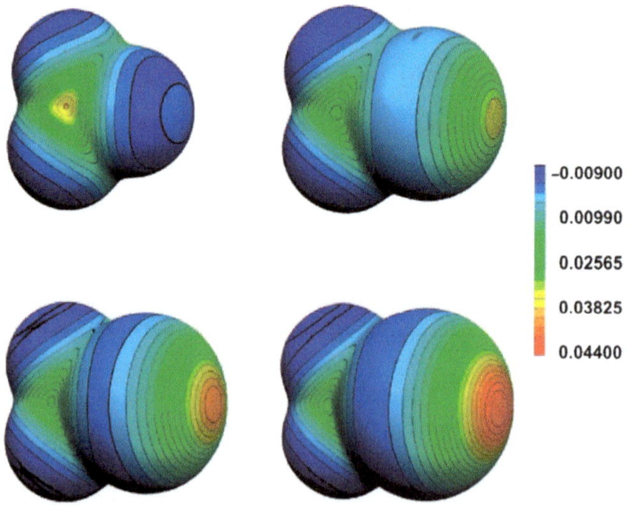

Figure 12.4.9 The molecular electrostatic potential (in Hartrees) at the isodensity surface with 0.001 electrons bohr^{-3}: CF_4 (top left), CF_3Cl (top right), CF_3Br (bottom left), and CF_3I (bottom right).

highest-occupied and lowest-unoccupied orbitals have been examined, together with others.

Similarly, the chalcogen atom (O, S, Se, Te, and Po) may also have an electropositive region at its outermost end, although it has only two lone pairs of electrons. This means that the chalcogen atom (E) can also interact with an electron donor to form a non-covalent (vdW-type) bond, which is similar to the halogen bond. Borrowing from the definitions of the halogen bond and the hydrogen bond, the E···Y contact is called the chalcogen bond.[84] The bond-length change, interaction energy, topological property of the electron charge densities, its Laplacian, and the charge transfer of the chalcogen bond were investigated theoretically in detail for the model complexes $H_2CS\cdots Cl^-$, $F_2CS\cdots Cl^-$, $OCS\cdots Cl^-$ and $SCS\cdots Cl^-$, using HF, MP2, and AIM methods.[85] Figure 12.4.10 shows the positive and negative regions of $V_S(\boldsymbol{r})$ for H_2CS, F_2CS, OCS, and SCS and Figure 12.4.11 depicts structures and molecular graphs with the critical points for the $H_2CS\cdots Cl^-$, $F_2CS\cdots Cl^-$, $OCS\cdots Cl^-$, and $SCS\cdots Cl^-$ adducts. The similarity in the properties of the chalcogen bond and the halogen bond is confirmed, although the colour in the positive region is

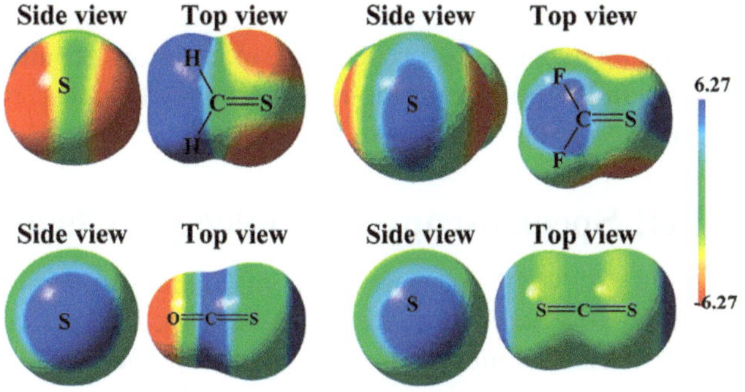

Figure 12.4.10 Electrostatic potentials of H_2CS, F_2CS, OCS, and SCS computed at the MP2/aug-cc-pVTZ theory level with a scale of –6.27 (red) to 6.27 kcal mol^{-1} (the colour at the positive region being inverse to the case of Figure 12.4.9).

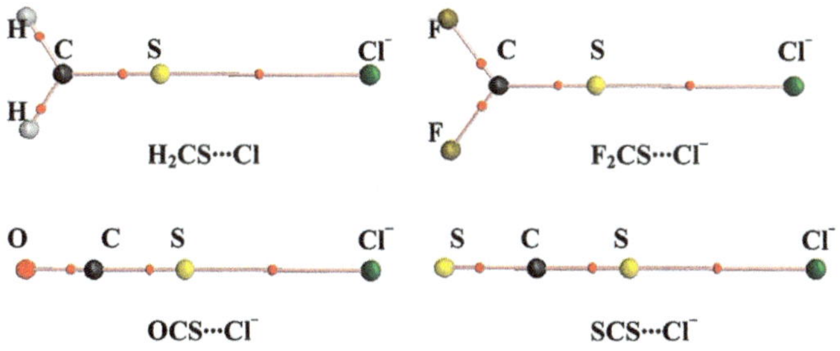

Figure 12.4.11 Structures and molecular graphs for the four chalcogen–bonded complexes. Small red dots represent the BCPs.

different. Experimental results are in good agreement with the theoretical predictions.

Clusters formed between a fluoride anion and several hydrogen sulfide molecules have been investigated *via ab initio* calculations at the MP2 level of theory, using Dunning's augmented correlation consistent basis sets.[86] Optimized geometries, vibrational frequencies, and enthalpy changes for the ligand association reactions are presented for clusters with up to five H_2S ligands interacting with an F^- anion. The minimum energy structure for the F^--H_2S 1:1 complex features proton transfer from the H_2S to the F^- anion, forming a planar C_s symmetry FH···SH^- structure. For the F^--$(H_2S)_2$ cluster, the FH···SH^- core remains and is solvated by a perturbed H_2S ligand. For the larger F^--$(H_2S)_{3-5}$ clusters, in addition to the FH···SH^--$(H_2S)_n$ cluster forms, other minima featuring a 'solvated F^-' anion are predicted. Calculated infrared spectra for the minima of each cluster size are presented to aid in assigning spectra from future experimental studies.

The characteristics and nature of the halogen bonding in a series of B···XY (B = H_2S, H_2CS, $(CH_2)_2S$; XY = ClF, Cl_2, BrF, BrCl, Br_2) complexes were analysed by means of QTAIM and NBO methodology at the MP2 level.[87] This must also be the issue of chalcogen bonding, if B of H_2S, H_2CS, and $(CH_2)_2S$ are mainly considered.

Chalcogen bonding recalls the preferential orientation around chalcogen atoms of chalcogenides with nucleophiles and electrophiles reported by Parthasarathy, a relationship which is of interest.[88,89]

12.4.7 NMR Spectroscopy in Chalcogen Chemistry

In the high-resolution NMR measurements of diamagnetic molecules, the Hamiltonian can be expressed by Equation 23, where σ_N, $\boldsymbol{B}$, $\boldsymbol{I}_N$, $\boldsymbol{I}_L$, and $\mathbf{J}_{NL}$ are magnetic shielding tensors of nucleus N, the external magnetic field, the nuclear spin momentum operators of nuclei N and L, and the indirect nuclear spin-spin coupling tensor between the nuclei N and L, respectively.[90] Equation 23 tells us that the nuclear spin I_N must be non-zero for the appearance of the NMR signal and the nuclear spins of both I_N and I_L are non-zero for the observation of nuclear spin–spin couplings. The nuclei of ^{33}S (I = 3/2), ^{77}Se (I = 1/2), ^{123}Te (I = −1/2), and ^{125}Te (I = −1/2) satisfy these conditions. We will discuss NMR chemical shifts and coupling constants employing σ_N and $\mathbf{J}_{NL}$, respectively, for ^{33}S, ^{77}Se, and ^{125}Te nuclei.

$$H = \Sigma_N (h/2\pi)\gamma_N(1 - \sigma_N)\boldsymbol{B}\cdot\boldsymbol{I}_N + \Sigma_{N>L} h\mathbf{J}_{NL}(\boldsymbol{I}_N \cdot \boldsymbol{I}_L) \tag{23}$$

McFarlane and Wood measured ^{77}Se NMR chemical shifts (δ(Se)) of various organic selenium compounds systematically using the INDOR method,[91] after the measurements of some inorganic selenium compounds. The reported values in the early stage were summarized by Lardon,[92] McFarlane,[93] and Iwamura.[94] The measurements of δ(Se) increased rapidly with the development of NMR instruments employing the pulse FT method.[95] The reported values have been summarized on a case-by-case basis.[92,96–98] Nuclear spin–spin coupling constants have also been reported and summarized.[97–99] The δ(Te) values and the nuclear couplings containing Te have also been reported, although the number seems substantially less than that for δ(Se). The δ(S) values and the nuclear couplings containing S have not been reported so much, since the measurements are much more difficult than for δ(Se) and δ(Te) due to the very low sensitivity under the natural abundance conditions with the broadening of the spectrum due to $I = 3/2$. The chemical shifts and coupling constants have been calculated variously to explain or understand the observed values.

Reported chemical shifts and coupling constants are not summarized in this chapter; the reader is referred to reviews or references.[96–99] Due to the space limitations, this chapter is restricted to showing how calculations are performed to understand and explain the observed values. Calculations are mainly explained or exemplified by δ(Se), together with ^1J(Se, Se), since relativistic corrections are necessary for δ(Te)[100] and measured values are scarce for δ(S).

12.4.8 Calculations of NMR Parameters: Magnetic Shielding Tensors σ

12.4.8.1 Theoretical Background

The absolute magnetic shielding tensors σ are calculated theoretically to explain or understand the observed chemical shifts.[101] The diamagnetic and paramagnetic shielding tensors (σ^d and σ^p, respectively) are exactly expressed by the Ramsey equation[102] and approximately calculated in the framework of the HF or DFT theory. Total absolute magnetic shielding tensors (σ^t) can be decomposed into σ^d and σ^p in the HF and DFT calculations, as shown in Equation 24.[103,104] σ^p is practically given as the averaged value of the three components, as shown in Equation 25; so are σ^t and σ^d. As shown by Equation 26, σ^d is expressed simply as the sum of contributions over the occupied orbitals. Parameters σ^p can be decomposed into the contribution of the occupied orbitals or the orbital–orbital transitions by the treatment of the coupled Hartree–Fock (CPHF) method, as shown by Equation 27. While σ^t is also evaluated at the MP2 level, σ^t cannot be decomposed into σ^d and σ^p at this level.

$$\sigma^t = \sigma^d + \sigma^p \tag{24}$$

$$\sigma^p = (\sigma^p{}_{xx} + \sigma^p{}_{yy} + \sigma^p{}_{zz})/3 \tag{25}$$

$$\sigma^d = \Sigma_i{}^{occ}\sigma^d{}_i \tag{26}$$

$$\sigma^p = \Sigma_i{}^{occ}\Sigma_a{}^{unocc}\sigma^p{}_{i\to a} = \Sigma_i{}^{occ}\sigma^p{}_i \tag{27}$$

$$\begin{aligned}\sigma^p{}_{zz}(N) = &-(\mu_o e^2/2m_e{}^2)\Sigma_i{}^{occ}\Sigma_a{}^{unocc}(\varepsilon_a - \varepsilon_i)^{-1} \\ &\times \{<\psi_i|\hat{L}_z|\psi_a><\psi_a|\hat{L}_{z,N}r_N{}^{-3}|\psi_i> \\ &+ <\psi_i|\hat{L}_{z,N}r_N{}^{-3}|\psi_a><\psi_a|\hat{L}_z|\psi_i>\}\end{aligned} \tag{28}$$

$$\sigma^{p}_{zz}(N) = -(\mu_o e^2/2m_e^2) < r_N^{-3} > \Sigma_i^{occ}\Sigma_a^{unocc}(\varepsilon_a - \varepsilon_i)^{-1} \times \{<\psi_i|\hat{L}_z|\psi_a><\psi_a|\hat{L}_{z,N}|\psi_i> + <\psi_i|\hat{L}_{z,N}|\psi_a><\psi_a|\hat{L}_z|\psi_i>\} \quad (29)$$

Based on second-order perturbation theory at the level of the HF and single-excitation CI approximation, $\sigma^{p}_{i\rightarrow a}$ on a resonance nucleus N is shown to be proportional to reciprocal orbital energy gaps $(\varepsilon_a - \varepsilon_i)^{-1}$. An approximated image can be derived from Equation 28, in which ψ_k is the kth orbital function, $\hat{L}_{z,N}$ is orbital angular momentum around the resonance nucleus, and r_N is the distance from the nucleus N. Indeed, σ^{p} is evaluated accurately by the CPHF method and can be calculated by Equation 27, but σ^{p} will be discussed with the approximated image derived from Equation 28, if suitable. Since $\sigma^{p}_{zz,N}$ contains the $\hat{L}_{z,N}$ operator, $\sigma^{p}_{zz,N}$ arises from admixtures between atomic p_x and p_y orbitals of N in various molecular orbitals.

When a magnetic field is applied to a selenium compound, mixing of unoccupied molecular orbitals (ψ_a) into occupied molecular orbitals (ψ_i) occurs. Such admixtures generate $\sigma^{p}_{zz,N}$, if ψ_i and ψ_a contain p_x and p_y of N, for example. The parameters $\sigma^{p}_{xx,N}$ and $\sigma^{p}_{yy,N}$ are similarly understood. Equation 28 is approximately rewritten as Equation 29, if $<r_N^{-3}>$ can be treated as a constant for a molecule. Equation 29 tells us that the main factors to control σ^{p} are the electron population terms around the resonance nucleus N ($<r_N^{-3}>$), reciprocal orbital energy gaps $(\varepsilon_a - \varepsilon_i)^{-1}$, and the orbital overlaps controlled by the angular momentum operators ($\{<\psi_i|\hat{L}_z|\psi_a> <\psi_a|\hat{L}_{z,N}|\psi_i> + <\psi_i|\hat{L}_{z,N}|\psi_a><\psi_a|\hat{L}_z|\psi_i>\}$.

12.4.8.2 Calculations of ^{77}Se NMR Chemical Shifts

Theoretical investigations of NMR chemical shifts were extended to ^{77}Se nuclei in the 1990s after the essential establishment of the reliability in the calculations for first- and/or second-row nuclei. Absolute magnetic shielding tensors of ^{77}Se nuclei (σ(Se)) were calculated using a variety of methods.[104–108] Nakatsuji and co-workers calculated σ(Se) of nine selenium compounds,[104] SeRR′ (R, R′ = H, Me, Et, Ph, SiH_3, GeH_3) by *ab initio* HF/finite perturbation method with the gauge origin being placed at the position of the Se atom, for example. The basis set employed was the (13sl0p4d)/[6s5p1d] plus two polarization d-functions ($\zeta = 0.144$ and 0.489) for Se, which were taken from the book of Fuzinaga *et al.*[109] The calculated values agreed well with the experimental values. The Se chemical shift was dominated by the Se valence 4p AO contribution to the paramagnetic term and shows a parallelism with the 4p-hole population and the net charge of the central Se atom. The Se chemical shift moves downfield as an increase of the hole in the 4p orbitals and therefore, as an increase in the electron-withdrawing ability of the ligands attached to the selenium atom.

The simple CPHF method[104] is not sufficient since very large basis sets would be required to overcome the 'gauge problem' and obtain reliable results.[110,111]

Ellis, Odom and co-workers proved the gauge problem through experimental and theoretical studies.[112] The theoretical results were carefully compared with $\delta_{obsd}(Se)$ for CH_3SeCH_3, $HSeCH_3$, and SeH_2 measured in the gas phase. These pioneering experiments undoubtedly represented one strong motivation for other theoretical investigations.[104–108]

Ziegler and co-workers summarized the early stage in the calculations of σ(Se).[113] The ^{77}Se chemical shifts were calculated employing the HF-based individual gauge for localized orbitals (IGLO-HF),[105] the gauge-including atomic orbitals (GIAO-HF),[105,107,112] and correlated second-order perturbation theory (GIAO-MP2) approaches,[105,107] together with the GIAO-based coupled-cluster singles and doubles (GIAO-CCSD) method.[106,114] It is concluded that correlation effects must be included for a quantitative description of ^{77}Se chemical shifts, in particular, for a proper description of the selenium lone pairs.[105–107] The inclusion of correlation effects at the MP2 level is shown to be sufficient for many selenium systems, although higher levels such as GIAO-CCSD are required to describe the electron correlation properly of cyclic dication Se_4^{2+}, for example.[105] On the other hand, DFT accounts for correlation effects implicitly[16] and should thus be able to predict the ^{77}Se chemical shifts.[108,115–117]

They tested the DFT-GIAO method[105,107,112] against other theoretical approaches and reported calculations of ^{77}Se chemical shifts and shieldings for a number of selenium-containing compounds. The calculated shifts span a range of about 2800 ppm and therefore cover almost the complete range of known ^{77}Se chemical shifts. Compared to other methods, DFT is seen to give generally results of better quality than the GIAO-HF scheme. DFT is even superior to both GIAO-HF and GIAO-MP2 for highly correlated molecules, CSe_2 and Se_4^{2-}, which are the most prominent examples. This can be explained by the overestimation of correlation at the MP2 level of theory.[105] The highly correlated and expensive GIAO-CCSD method should be able to give results of comparable or better quality for such systems. Further improvement of the method including the SOS-DFPT method (sum-over-states density-functional-perturbation-theoretical method) was also discussed.

Nakanishi and Hayashi reported σ(Se) calculated with the GIAO-DFT (B3LYP) method to examine the applicability of the basis set system to their aim of investigations,[118] employing the Gaussian 94 program package.[119] The method with the 6–311++G(3df,2pd) basis set at the B3LYP level gave excellent results[11] ($r = 0.998$), so did the 6–311++G(d,p) basis set. The results facilitated further methodological improvement, although the limited number of selenium compounds examined must also be responsible for such excellent results.

Bayse examined the basis sets and the levels suitable to predict the δ(Se) values, employing large number of selenium compounds using GIAO-MP2 and GIAO-DFT methods.[120] Reliable chemical shifts are calculated for many selenium compounds. However, the reliability of GIAO-DFT is limited in many of the same cases as MP2, but outperforms it for $[SeO_3]^{2-}$, $[SeO_4]^{2-}$, $SeCF_2$, 1,2,3-selenadiazole, $[Se_4]^{2+}$, and related ions and underperforms in several notable cases such as Me_2SeO, SeO_3, $SeCl_2$, and MeSeSeMe. It is difficult to suggest a generally reliable choice due to significant variance in performance with method and functional, but a basis set of at least triple-ζ quality would be recommended. Systems involving 3c–4e bonding should be augmented with diffuse functions, and theoretical chemical shifts of terminal selenium anions should be treated with scepticism because of the absence of solvation effects. The limited RECP basis set gives surprisingly good results—a promising development for the reliable calculation of theoretical chemical shifts for the larger molecules of interest to synthetic, biological, and medicinal chemists.

Laitinen and co-workers calculated ^{77}Se NMR chemical shifts of chalcogen cation ring systems $S_xSe_{4-x}^{2+}$ ($x = 0$–3) with CAS, BPW91, and B3PW91 methods using the GIAO formalism.[121] The hybrid functional B3PW91 shows inferior performance, but both CAS and BPW91 unquestionably confirm the experimental assignment and are able to predict the NMR chemical shifts of these computationally difficult cases with excellent accuracy. Table 12.4.3 shows the calculated ^{77}Se NMR chemical shifts and the observed values. The electronic structures and molecular properties of square-planar 6π-electron ring molecules and ions E_2N_2 and E_4^{2+} (E = S, Se, Te) were also studied using various *ab initio* methods and density functionals.

The ^{77}Se chemical shifts are also calculated for individual selenium-containing compounds with various methods to predict the structures. The conformer was successfully assigned to the anti,anti-isomer of Se[N(mes*)]$_2$ and syn- and anti-isomers of Se[NH(mes)]$_2$ through the observed ^{77}Se chemical shifts supported by the calculated values (Figure 12.4.12), where mes and mes* stand for 2,4,6-$Me_3C_6H_2$ and 2,4,6-$(Me_3C)_3C_6H_2$, respectively.[123] In these investigations, the PBE0/TZVP calculations were carried out on 28 molecules. The computed ^{77}Se shielding tensors exhibit a good linear relationship with the observed chemical shifts. (The calculated mean deviation amounts to 48 ppm.) The inclusion of solvent effects during geometry optimization improved the accuracy of the calculations, although it would be time-consuming.

The oxidation products of selenomethionine (SeMet) have been studied *via* experimental ^{77}Se NMR and theoretical ^{77}Se chemical shifts.[124] Four signals are observed: a diastereomeric pair of selenoxides at 840 ppm and two

Table 12.4.3 ^{77}Se NMR chemical shifts calculated at different levels of theory using cc-pVTZ basis set.[a]

Species	*BPW91*	*B3PW91*	*CAS*	*exp*[b]
Se_4^{2+}	1941	2120	1893	1936
SSe_3^{2+} (SeSeSe)	1924	2087	1892	1939
SSe_3^{2+} (SeSeSe)	2049	2229	1965	2032
cis-S_2Se^{2+}	2042	2198	1967	2023
trans-$S_2Se_2^{2+}$	1873	2013	1858	1890
S_3Se^{2+}	2001	2135	1941	1954

[a]Values are from MeSeMe.
[b]Ref. 122.

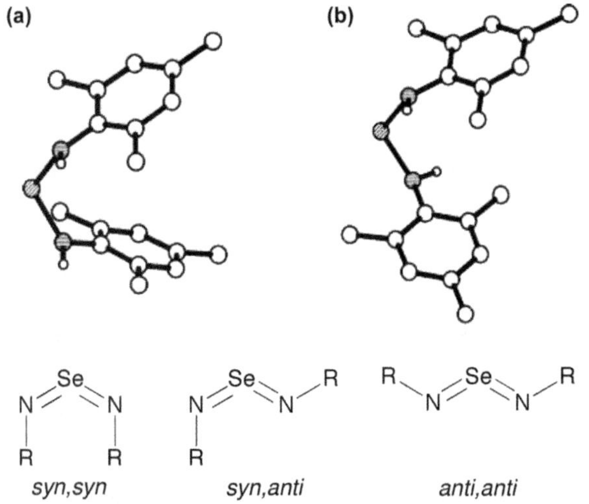

Figure 12.4.12 Crystal structures of syn- and anti-isomers of Se{NH(mes)}$_2$ (a and b, respectively) with three possible conformations of Se{N(mes*)}$_2$.

unidentified resonances at 703 and 716 ppm. Theoretical ΔG and chemical shifts suggest the 703 and 716 ppm resonances correspond to hypervalent selenuranes, formed by reaction of the selenoxide with the amine or acid group of the amino acid. The *N*-formyl SeMet formed only the selenoxide pair at 840 ppm. The oxidized SeMet methyl ester produced signals at 703 and 716 ppm, which are assigned as the Se–N selenurane.

Arylselenenyl fluorides ArSeF were obtained from Ar_2Se_2 or $ArSeSiMe_3$ with XeF_2. They are detected by low-temperature ^{19}F and ^{77}Se NMR spectroscopy.[125] Substitution in the *ortho*-position of the aromatic ring to provide electronic or steric protection is a requirement for the formation. ArSeF compounds decompose according to $3ArSeF \rightarrow [ArSeSeF_2Ar] + ArSeF \rightarrow ArSeF_3 + Ar_2Se_2$. Reaction energies for this disproportionation as well as that of the S and Te homologues have been calculated with MP2, CCSD(T), and B3LYP methods. They are increasingly exothermic in the sequence $S < Se < Te$. The ^{77}Se and ^{19}F chemical shifts were calculated by GIAO-MP2 and GIAO-B3LYP methods and are in good agreement with experimental values for selected Se–C and Se–F compounds. $XeOSeF_5^-$ is investigated similarly.[126]

Contracted basis sets of double zeta (DZ) quality were examined for the atoms from K to Kr.[127,128] This extends earlier work on a segmented contracted DZ basis set for atoms H–Ar. The segmented contracted basis set devised in this work offers the advantage of good accuracy and flexibility with few primitives. The choice of the polarization functions (d and f) was made from the correlated atomic calculations. ^{77}Se chemical shifts in H_2Se and CSe_2 were calculated using the GIAO-DFT method and compared with the experimental values. Except for chemical shift, it can be verified that the results give the best agreement with experimental and benchmark values.[129]

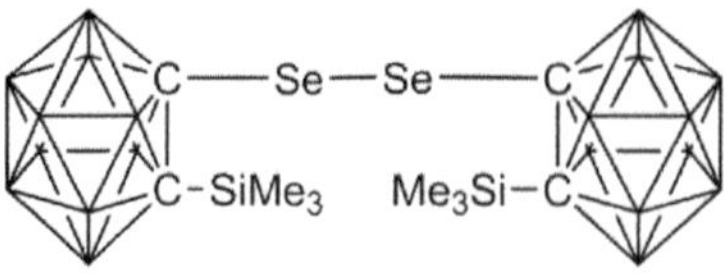

29

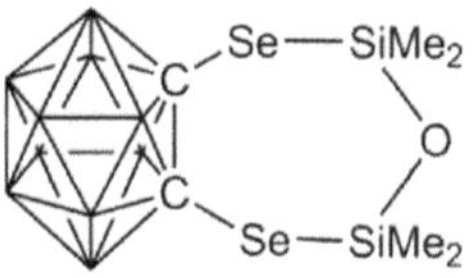

30

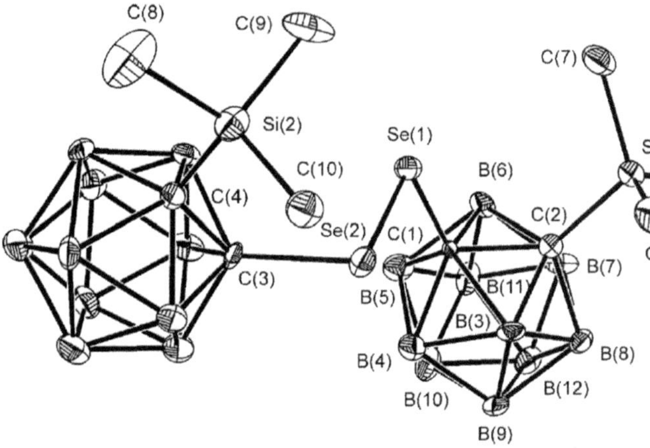

Figure 12.4.13 ORTEP plot (50% probability; hydrogen atoms being omitted for clarity) of the molecular structure of the 1,1′-diseleno-bis[2-trimethylsilyl-1,2-dicarba-closo- dodecaborane(12)] (**29**)

Molecular solid state structures were determined by X-ray diffraction for **29** and **30** and those in solution were confirmed by NMR spectroscopy (^{1}H, ^{13}C, ^{29}Si, and ^{77}Se NMR).[130] The observed results were well supported by the DFT calculations with the B3LYP/6–311+G(d,p) method (Figure 12.4.13).[130]

12.4.8.3 Information Derived from Calculated σ(Se)

12.4.8.3.1 Analysis of the α, β, γ, and δ Effects, Together with the pre-α Effect

The origin of δ(Se) is clarified to analyse δ(Se) with physical meaning. The α and β effects[94,131] are well known as the downfield shifts in the processes from H_2Se to MeSeH then to EtSeH, respectively, for example. To achieve the aim of clarifying the origin of δ(Se) in a unified form, an idea called the 'pre-α effect' is proposed.[132] The pre-α effect is defined as the downfield shift by a proton added to a lone pair orbital of Se^{2-}, as shown in Scheme 12.4.4. The δ(Se) value of H_2Se (δ(Se: H_2Se)), relative to Se^{2-}, corresponds to twice as large as the pre-α effect, which would be most practical. The γ effect of the upfield shifts is also discussed, together with the mechanism, which corresponds to the process from EtSeH to *n*-PrSeH, for example.[94,131] Scheme 12.4.4 shows the case for the vinyl effect derived from the p(Se)–π(C=C) conjugation.

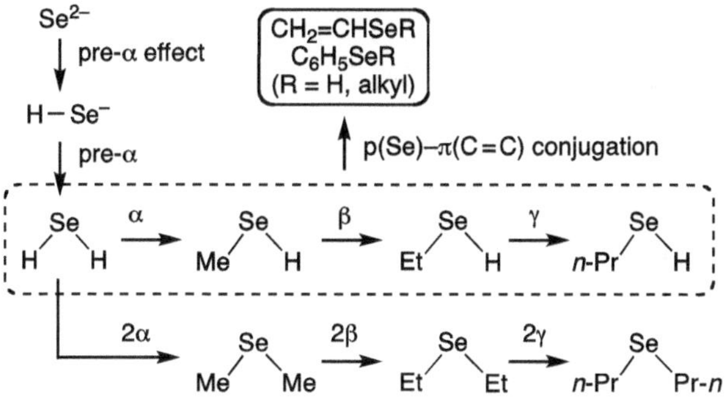

Scheme 12.4.4 Pre-α, α, β, and γ effects.

Table 12.4.4 shows σ^d(Se), σ^p(Se), and σ^t(Se) of various selenides calculated by the GIAO–DFT (B3LYP) method, together with σ^t(Se) by the GIAO–MP2 method.[133] Table 12.4.5 shows the contributions from each ψ_i of the valence orbital to σ^p(Se) and the components ($\sigma^p(Se)_{xx}$, $\sigma^p(Se)_{yy}$, and $\sigma^p(Se)_{zz}$) in H_2Se, together with the energies (ε_i) and the characters of ψ_i. The pre-α effect with two protons in H_2Se is explained by the generation of double σ(Se–H) and σ*(Se–H) through the protonation to the spherical Se^{2-}. Such orbitals lead effective transitions for the pre-α effect. It is also demonstrated that the α effect is the downfield shift caused by the replacement of Se–H by Se–Me. The extension of HOMO-2 ($4p_y$(Se)), HOMO-1 ($4p_x$(Se)) and HOMO ($4p_z$(Se)) over the whole molecule of Me_2Se is mainly responsible for the α effect. The β effect originates not from the occupied-to-unoccupied ($\psi_i \rightarrow \psi_a$) transitions but from the occupied-to-occupied ($\psi_i \rightarrow \psi_j$) transitions, although not shown in Table 12.4.5. The γ effect of the upfield shifts is also analysed, although complex. The δ effect in *n*-BuSeR relative to *n*-PrSeR is negligibly small. The effect of the p(Se)–π(C=C) conjugation,[134] caused by ethenyl and phenyl groups, is analysed, together with the mechanism, in relation to the orientational effect, although not shown in Table 12.4.4.[132]

12.4.8.3.2 Contributions from Atomic p(Se), d(Se), and f(Se) Orbitals to σ^p(Se)

Contributions from atomic p(Se), d(Se), and f(Se) orbitals to σ^p(Se) are evaluated for various selenium-containing compounds, where σ^p(Se) can be imaged according to Equation 28. The 6–311+G(3df,3pd) basis set is employed with the GIAO-DFT (B3LYP) method for the evaluation separately by the types of atomic orbitals (s(Se), p(Se), d(Se), and f(Se)) employing the orthogonality of the AOs. The contributions from s(Se) will be zero due to the spherical distribution of electrons. The contributions are evaluated for neutral and charged Se^*H_n (* = null, +, or −) and some oxides to build the image of the contributions. The effect of methyl and halogen substitutions is also examined

Table 12.4.4 Calculated pre-α, α, and β effects of various selenides, together with the observed values.[a,b]

Method		GIAO-DFT			GIAO-MP2		Observed		
Compd	$\sigma^{d}(Se)$	$\sigma^{p}(Se)$	$\sigma^{t}(Se)$	*Effect*[c,d]	$\sigma^{t}(Se)$	*Effect*[c,d]	$\delta(Se)$	*Effect*[d,e]	*Solvent*
Se^{2-} (O_h)	3005.7	0.0	3005.7		3005.1				
H_2Se (C_{2v})	2998.0	–931.3	2066.7	–469.5: pα	2252.7	–376.2: pα	–331.7[f]	222: pα	GP[g]
MeSeH (C_s)	2998.2	–1155.0	1843.2	–223.5: α	2072.7	–180.0: α	–141.6[h]	111: α	GP[g]
EtSeH (C_s)	3000.1	–1235.0	1765.1	–78.1: β	1995.5	–77.2: β	36[i]	151: β	$CDCl_3$
^{i}PrSeH (C_s)	3004.5	–1469.7	1534.8	–154.2: β	1776.2	–148.3: β	161[i]	151: β	$CDCl_3$
^{t}BuSeH (C_s)	3009.3	–1553.5	1455.8	–129.1: β	1696.2	–125.3: β	289[i]	144: β	$CDCl_3$
Me_2Se (C_{2v})	2999.1	–1349.0	1650.1	–208.3: α	1907.4	–172.7: α	0.0[j]	115: α	$CDCl_3$
Et_2Se (C_{2v})	3006.2	–1516.6	1489.6	–80.3: β	1747.6	–79.9: β	230[i]	115: β	$CDCl_3$
$^{i}Pr_2Se$ (C_2)	3015.3	–1777.1	1238.2	–103.0: β	1476.1	–107.8: β	429[i]	107: β	$CDCl_3$
$^{t}Bu_2Se$ (C_2)	3027.1	–1970.7	1056.4	–99.0: β	[k]	[k]	614[i]	102: β	$CDCl_3$

[a]The 6–311 + G(3df) basis sets being employed for Se and the 6–311+G(3d,2p) basis sets for other nuclei of the Gaussian 03 program.
[b]In ppm.
[c]In σ(Se) scale.
[d]Values and the corresponding effects are shown.
[e]In δ(Se) scale.
[f]Ref. 135 and δ(Se) = –225.5 in neat (ref. 136).
[g]In gas phase.
[h]Ref. 135 and δ(Se) = –115 in $CDCl_3$ (ref. 136).
[i]Ref. 136.
[j]δ(Se) = 13.1 in gas phase (ref. 135).
[k]Not obtained due to very large memory requirements.

Table 12.4.5 Contributions from ψ_i to $\sigma^p(Se)$ and the components in H_2Se, together with the energies and characters of ψ_i.[a–d]

i in ψ_i	ε/eV	$\sigma^p(Se)_{xx}$	$\sigma^p(Se)_{yy}$	$\sigma^p(Se)_{zz}$	$\sigma^p(Se)$	*Sym*	*Character*
$\psi_i \rightarrow \psi_j$[e]		317.9	314.8	318.7	317.1		
1–14		−15.9	−5.8	−7.0	−9.6		Inner orbitals
15	−19.58	−1.4	−17.0	−29.4	−15.9	A1	$\sigma(H_2Se$: 4s(Se))
16	−11.53	−247.4	−9.6	−482.3	−246.4	B2	$\sigma(H_2Se\ b_2$: $4p_y$(Se))
17	−9.86	−4.5	−215.4	−831.5	−350.5	A1	$\sigma(H_2Se\ a_1$: $4p_x$(Se))
18	−6.91	−1312.4	−565.0	−0.3	−625.9	B1	$4p_z$(Se)
1–18		−1581.7	−812.9	−1350.6	−1248.4		
Total		−1263.8	−498.1	−1031.9	−931.3		

[a]Optimized with the 6–311+G(3df) basis sets for Se and 6–311+G(3d,2p) basis sets for other nuclei of the Gaussian 03 program.
[b]A utility program of Gaussian 03 (NMRANAL-NH03G) is applied to separate the contributions from each molecular orbital.
[c]The contribution from each molecular orbital contains only that from the occupied-to-unoccupied ($\psi_i \rightarrow \psi_a$) transitions.
[d]σ(Se) and the components are given in ppm.
[e]Corresponding to the occupied-to-occupied transitions.

Table 12.4.6 Contributions from each molecular orbital (MO) and from atomic p(Se), d(Se), and f(Se) orbitals, together with χ(H), to $\sigma^p(Se)$ in Se^*H_n (* = null, +, or −).[a–c]

Species	$\sigma^p(Se)$	*from p(Se)*	*from d(Se)*	*from f(Se)*	*from* $\chi(H)^d$	*% p(Se)*
SeH^- ($C_{\infty v}$)	−501.5	−496.0	−11.0	0.2	5.3	98.9
SeH_2 (C_{2v})	−928.8	−906.1	−27.5	−0.4	5.2	97.6
SeH_3^+ (C_{3v})	−1079.9	−1039.0	−48.1	0.4	6.8	96.2
SeH_3^+ (D_{3h})	−1074.6	−1029.9	−48.7	−0.6	4.6	95.8
SeH_4 (T_d)	−1666.8	−1629.5	−40.4	0.5	2.6	97.8
SeH_4 (C_{2v})[e]	−966.3	−765.8	−203.7	−0.1	3.3	79.3
SeH_5^+ (D_{3h})[e]	−1207.4	−911.4	−294.8	−0.1	−1.1	75.5
SeH_5^+ (C_{4v})[f]	−1159.9	−867.9	−291.3	0.0	−0.8	74.8
SeH_5^- (C_{4v})[f]	−905.9	−674.5	−235.6	−0.2	4.3	74.5
SeH_6 (O_h)	−983.6	−514.1	−467.4	0.2	−2.3	52.3

[a]Calculated with the DFT-GIAO method employing the 6–311 + G(3df,3pd) basis sets.
[b]A utility program (NMRANAL-NH03G) being employed.
[c]In ppm.
[d]Contribution from AOs at H.
[e]Trigonal bipyramidal (TBP).
[f]Square pyramidal (SP).

employing $R_rSe^*X_xO_o$ (* = null, +, or −) where R = H or Me and X = F, Cl, or Br.

Table 12.4.6 shows the results for Se^*H_n (* = null, +, or −). The p(Se) contributions are larger than 96% for SeH^- ($C_{\infty v}$), SeH_2 (C_{2v}), SeH_3^+ (C_{3v}), SeH_3^+ (D_{3h}), and SeH_4 (T_d). Therefore, $\sigma^p(Se)$ of these compounds can be analysed based on p(Se). The p(Se) contributions are 79–75% for SeH_4 (TBP), SeH_5^+ (TBP), SeH_5^+ (SP), and SeH_5^- (SP). Methyl and halogen substitutions increase the contributions by 1–2% (per Me) and 4–7% (per X), respectively.

The contributions are 92–79% for H_2SeO (C_s), H_2SeO_2 (C_{2v}), and H_4SeO (C_{2v}). The values are similarly increased by the substitutions. Consequently, σ^p(Se) of these compounds can be analysed based on p(Se) with some corrections by d(Se). The p(Se) contribution of SeH_6 (O_h) is 52%. σ^p(Se: SeH_6 (O_h)) must be analysed based on both p(Se) and d(Se). The contributions for the Me and X derivatives of SeH_6 amount to 86–77%. Therefore, σ^p(Se) of the derivatives can also be analysed mainly based on p(Se) with some corrections by d(Se) in this case. Contributions from f(Se) are negligible.

12.4.8.3.3 Evaluation of Electron Population Terms $<r_N^{-3}>$ in Relation to Charge Effect on σ(Se)

Valence orbitals shrink or expand depending on the magnitude of the electron repulsion if those at the same kind of nuclei are compared.[2,3] The total magnitude of the electron repulsion at a nucleus (N) in a molecule (***M***) should be correlated with the electron density at N, which could be evaluated by the nuclear charge ($Q(N)$) in ***M***. On the other hand, the electron population terms $<r_N^{-3}>$ will evaluate how MOs shrink or expand and $<r_N^{-3}>$ is one of most important factors to control σ^p(Se), as expressed in Equations 28 and 29. Therefore, it must be crucial to clarify whether $<r_N^{-3}>$ or the relative values ($<r_N^{-3}>_{rel}$) correlate with $Q(N)$.

Nakanishi and co-workers proposed a method to evaluate $<r_N^{-3}>$ in ***M***, which employs pure p atomic orbitals.[137] They recognized that Equation 30 will fold for the pure p atomic orbitals for an example. Therefore, the $<r_N^{-3}>_{ij}$ values averaged by ψ_i and ψ_j, which consist of pure p atomic orbitals, can be calculated according to Equation 31, although $<r_N^{-3}>_{HOMO}$ ($\psi_i = \psi_j =$ HOMO) and $<r_N^{-3}>_{LUMO}$ ($\psi_i = \psi_j =$ LUMO) will be used in place of $<r_N^{-3}>_{ij}$, if they are suitable.

$$<4p_x|\hat{L}_{z,N} r_N^{-3}|4p_y> = <4p_x|r_N^{-3}|4p_x><4p_x|\hat{L}_z|4p_y> \quad (30)$$

$$<r_N^{-3}>_{ij} = <\psi_i|r_N^{-3}|\psi_j> \quad (31)$$

The $<r_{Se}^{-3}>_{4p}$ values are calculated for HOMO and/or LUMO constructed by the pure atomic 4p(Se) orbitals in Se^{4+}, Se^{2+}, Se^0, and Se^{2-}, together with HSe^+, H_2Se, and HSe^- of the $C_{\infty v}$ or C_{2v} symmetry, which satisfy the conditions to calculate $<r_{Se}^{-3}>_{4p}$, shown in Equations 30 and 31. Table 12.4.7 summarizes the results, which show that the $<r_{Se}^{-3}>_{4p;LUMO;rel}$ values are very close to $<r_{Se}^{-3}>_{4p;HOMO;rel}$ if those of the same species are compared. Therefore, a common $<r_{Se}^{-3}>_{4p}$ value can be substantially used for HOMO and LUMO in the NMR analysis for a selenium species. Figure 12.4.14 shows the plot of $<r_{Se}^{-3}>_{4p;HOMO;rel}$ and $<r_{Se}^{-3}>_{4p;LUMO;rel}$ *versus* $Q(Se)$ (natural charges evaluated with NPA analysis). The regression curve is given by Equation 32, which connects $<r_{Se}^{-3}>_{4p}$ to $Q(Se)$ through an exponential function.

$$y = 2.713 \exp(0.0602x) - 1.713 \ (R^2 = 0.999) \quad (32)$$

Table 12.4.7 The $<r_{Se}^{-3}>_{4p;HOMO}$, $<r_{Se}^{-3}>_{4p;HOMO;rel}$, $<r_{Se}^{-3}>_{4p;LUMO}$, and $<r_{Se}^{-3}>_{4p;LUMO;rel}$ values for Se^{4+}, Se^{2+}, Se^0, and Se^{2-}, together with HSe^+, H_2Se, and HSe^- at the singlet state and Q(N) evaluated with the B3LYP/6–311 +G(3d) method.

Species	*HOMO*[a]/a_o^{-3}	*HOMO;rel*[b]	*LUMO*[c]/a_o^{-3}	*LUMO;rel*[d]	*Q(N)*[e]
Se^{4+}			17.764	1.737	4
Se^{2+}	13.798	1.349	13.869	1.356	2
Se^0	10.228	1.000[f]	10.242	1.001	0
Se^{2-}	7.147	0.699			–2
HSe^+	11.630	1.137	11.659	1.140	0.907
H_2Se	9.760	0.954			–0.190
HSe^-	8.373	0.819			–1.030

[a] $<r_{Se}^{-3}>_{4p;HOMO}$.
[b] $<r_{Se}^{-3}>_{4p;HOMO;rel}$.
[c] $<r_{Se}^{-3}>_{4p;LUMO}$.
[d] $<r_{Se}^{-3}>_{4p;LUMO;rel}$.
[e] Calculated natural population analysis.
[f] Taken to be a standard.

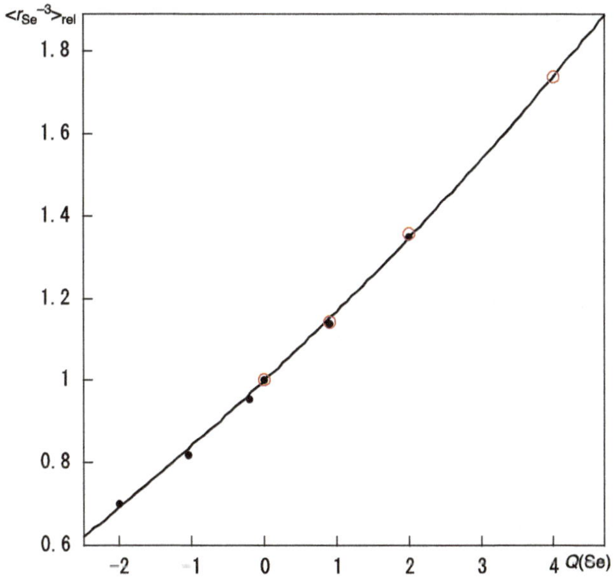

Figure 12.4.14 Plots of $<r_{Se}^{-3}>_{4p;HOMO;rel}$ (●) and $<r_{Se}^{-3}>_{4p;LUMO;rel}$ (○) *versus* Q(Se) for Se^{4+}, Se^{2+}, Se^0, and Se^{2-}, together with HSe^+, H_2Se, and HSe^- evaluated with the B3LYP/6–311 +G(3d) method.

12.4.8.4 Orientational Effect of Arylselanyl Groups on δ(Se): Theoretical Treatment

Typical conformers in relation to the orientational effect of ArSeR are planar (**pl**) and perpendicular (**pd**) conformers, where the Se–C_R bond in ArSeR is on the Ar plane in **pl** and it is perpendicular to the plane in **pd**, together with the

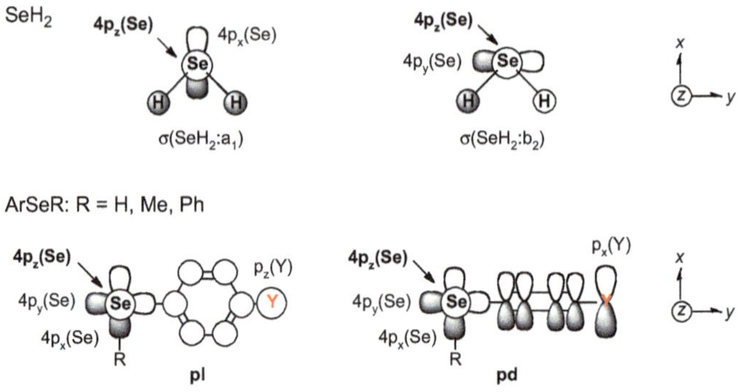

Scheme 12.4.5 Axes and some orbitals of SeH_2, ArSeH, ArSeMe, and ArSePh.

non-planar and non-perpendicular ones (**np**) (see Scheme 12.4.6). Characteristic behaviours of δ(Se) in **pl** and **pd** are explained based on the theoretical background. Scheme 12.4.5 shows the axes employed for the calculations. The main interaction in **pl** of $RSeC_6H_4Y$-*p* is the $n_p(Se)$–$\pi(C_6H_4)$–$p_z(Y)$ type, where $n_p(Se)$ is constructed by $4p_z(Se)$. Namely, $4p_z(Se)$ in $n_p(Se)$ interacts effectively with $\pi(C_6H_4Y)$ and $\pi^*(C_6H_4Y)$ in **pl**, resulting in the substantial contributions of $4p_z(Se)$ to $\pi(SeC_6H_4Y)$ and $\pi^*(SeC_6H_4Y)$. Therefore, both admixtures of $4p_z(Se)$ in $\pi(SeC_6H_4Y)$ with $4p_x(Se)$ and $4p_y(Se)$ in σ^*(CSeX: X $=C_R$ or C_{Ar}) and those of $4p_z(Se)$ in $\pi^*(SeC_6H_4Y)$ with $4p_x(Se)$ and $4p_y(Se)$ in σ(CSeX: X $=C_R$ or C_{Ar}) contribute much to $\sigma^p(Se)$ in **pl** under a magnetic field.[138] As a result, $\sigma^p(Se)$ in **pl** will be much affected by Y of both electron donors and electron acceptors, since they must play an important role in $\pi^*(SeC_6H_4Y)$ and $\pi(SeC_6H_4Y)$, respectively. On the other hand, the main interaction in **pd** is the $\sigma(SeC_R)$–$\pi(C_6H_4)$–$p_z(Y)$ type, in which Se–C_R is nearly on the *x* axis. Namely, $4p_z(Se)$ in $n_p(Se)$ remains almost pure form. Therefore, $\sigma^p(Se)$ in **pd** mainly arises from admixtures of $4p_z(Se)$ in $n_p(Se)$ with $4p_x(Se)$ in $\sigma^*(SeC_R)$ and $4p_y(Se)$ in $\sigma^*(SeC_{Ar})$. Contributions from σ(SeX: X $=C_R$ or C_{Ar}) would be very small, since it must be difficult to find $4p_z(Se)$ in some σ* orbitals. $\sigma^*(SeC_R)$ could interact effectively with $p_z(Y)$ through $\pi^*(C_6H_4)$ in **pd**. Consequently, the effect of Y on $\sigma^p(Se)$ is greater for Y of electron donors relative to the case of Y of electron acceptors in **pd**.

The characteristic orientational effect of SeAr clarified on the basis of the theoretical background will be applied to determine the structures of aryl selenides in solutions.

12.4.8.5 Relativistic Effects on σ(Se)

The contribution of relativistic terms has been pointed out for heavier atoms.[20,27] The effect is discussed under ZORA. The outline of the ZORA Hamiltonian to evaluate σ(*N*) with all electrons[141,142] is explained briefly in Equation 33,[23] where $V(\boldsymbol{r})$ is the effective Kohn-Sham potential given by the sum of the nuclear, Hartree, and exchange-correlation potentials in external magnetic field $\boldsymbol{B}$ within the DFT theory, $\boldsymbol{\sigma}$ are the Pauli matrices, and c is the speed of light. $\boldsymbol{\pi}$ and $K(\boldsymbol{r})$ are given by Equations 34 and 35, respectively. The first two terms in Equation 33 form the basis of the scalar relativistic approximation and the third term represents spin-orbit coupling. The last term is for the coupling between spin and magnetic field.

$$H_{ZORA} = V(\boldsymbol{r}) + \boldsymbol{\pi}(K(\boldsymbol{r})/2)\boldsymbol{\pi} + (K^2(\boldsymbol{r})/4c^2)\boldsymbol{\sigma} \cdot [\nabla V(\boldsymbol{r}) \times \boldsymbol{p}] - (\boldsymbol{K}(\boldsymbol{r})/\mathrm{c})\boldsymbol{\sigma} \cdot \boldsymbol{B} \quad (33)$$

$$\boldsymbol{\pi} = \boldsymbol{p} + (1/c)\boldsymbol{A}(\boldsymbol{r}), \boldsymbol{B} = \nabla \times \boldsymbol{A}(\boldsymbol{r}) \quad (34)$$

$$K(\boldsymbol{r}) = \{1 - (V(\boldsymbol{r})/2c^2)\}^{-1} \quad (35)$$

$$\sigma^{\mathrm{t}}(N) = \sigma^{\mathrm{d}}(N) + \sigma^{\mathrm{p}}(N) + \sigma^{\mathrm{so}}(N) = \sigma^{\mathrm{d+p}}(N) + \sigma^{\mathrm{so}}(N) \quad (36)$$

$\sigma^t(N)$ can be decomposed into $\sigma^d(N)$ and $\sigma^p(N)$ in the HF and DFT calculations under scalar relativistic conditions,[23,139,140] similar to the NR case. On the other hand, $\sigma^t(N)$ are expressed as the sum of $\sigma^d(N)$, $\sigma^p(N)$, and $\sigma^{so}(N)$ (the contributions from the spin-orbit interaction terms) as shown in Equation 36, if evaluated at the spin-orbit ZORA relativistic level.[23,139,140] Equation 36 can be understood based on Equation 33 by neglecting the fourth term. $\sigma^d(N) + \sigma^p(N)$ (= $\sigma^{d+p}(N)$) correspond to the first two terms in Equation 33 and are affected by the scalar relativistic effect, whereas $\sigma^{so}(N)$ originates inherently as the spin-orbit effect.

The relativistic effects are variously calculated and reported. A typical example for the relativistic effects on absolute magnetic shielding tensors (σ(Se)) is explained here, which are explicitly evaluated for 40 selenium species using the DFT(BLYP)-GIAO method.[141] Calculations are performed under relativistic and NR conditions with Slater-type basis sets in the Amsterdam Density Functional (ADF) program (version ADF 2010)[142] in the framework of ZORA, employing the optimized structures under NR conditions at B3LYP of Gaussian 03.[143] Quadruple zeta all electron with four polarization functions (QZ4Pae) are applied to evaluate σ(Se). Figure 12.4.15 shows the results. Ranges of the effect on diamagnetic (σ^d(Se)), paramagnetic shielding tensors (σ^p(Se)), and σ^{d+p}(Se) (= σ^d(Se) + σ^p(Se)) are −24 to −20 ppm, −115 to −3 ppm, and −136 to −26 ppm, respectively. The spin-orbit terms (σ^{so}(Se)) are evaluated to be 92 to 225 ppm with QZ4Pae, which clarifies the effect on total shielding tensors (σ^t(Se) = σ^{d+p}(Se) + σ^{so}(Se)) to be −8 to 152 ppm, at the spin-orbit ZORA level. The calculated σ^t(Se) values reproduced the observed values well.

12.4.8.6 Calculations of ^{33}S NMR Chemical Shifts

Jackowski and coworkers[144] presented an experimentally determined absolute shielding of 798.5 ppm for liquid OCS at 295 K, together with those for H_2S, CS_2, and SF_6. It seems necessary to employ the scaled DFT and EMPI approaches in the calculations of ^{33}S NMR chemical shifts. The scaled DFT and EMPI approaches contain factors to adjust the calculated values with the observed ones. Equations 37 and 38 explain the scaled DFT and EMPI

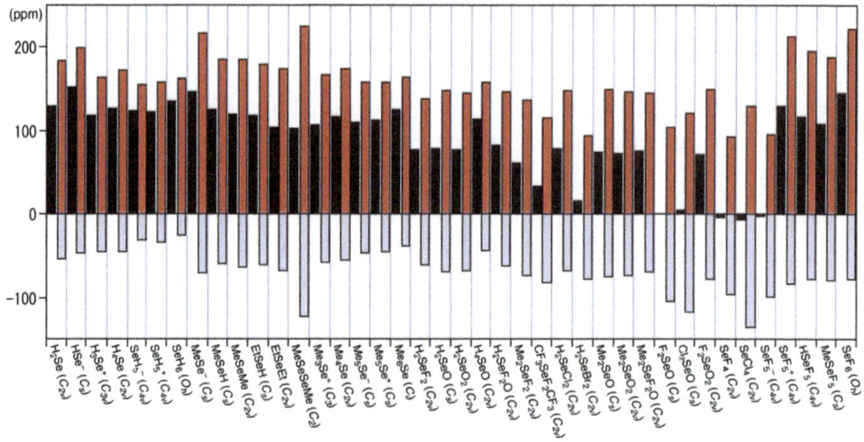

Figure 12.4.15 Relativistic effect on σ(Se) for various selenium compounds evaluated with the QZ4Pae basis sets: black, blue, and red denote the total term ($\Delta\sigma^{t}(\mathrm{Se})_{\mathrm{Rlt\text{-}so}} = \Delta\sigma^{d+p}(\mathrm{Se})_{\mathrm{Rlt\text{-}so}} + \sigma^{so}(\mathrm{Se})_{\mathrm{Rlt\text{-}so}}$), the scalar term ($\Delta\sigma^{d+p}(\mathrm{Se})_{\mathrm{Rlt\text{-}so}}$), and the spin-orbit term ($\sigma^{so}(\mathrm{Se})_{\mathrm{Rlt\text{-}so}}$), respectively.

approaches, which correct the overestimation. The k value of 0.871 (± 0.010) is proposed for Equation 37 in the B3LYP calculations.

$$\sigma_{\mathrm{S,DFT}} = \sigma_{\mathrm{dia}} + k\sigma_{\mathrm{para}} \tag{37}$$

$$\sigma_{\mathrm{TEMPI}} = \sigma_{\mathrm{RHF}} + (2/3)(\sigma_{\mathrm{MP2}} - \sigma_{\mathrm{RHF}}) \tag{38}$$

Chesnut and Quin[145–147] demonstrated that the ^{33}S NMR chemical shift of a great variety of sulfur compounds could be calculated by correlation-including scaled DFT[146] and EMPI[147] methods, generally within a standard deviation of 35 ppm (3.5% of the experimental shielding range of about 1000 ppm). Similar calculations were performed for the fluoride, chloride, and bromide of trimethylsulfonium ion (**31**) and *S*-methyltetrahydrothiophenium ion (**32**), in addition to the free cations.[150] The discrepancy between theory[145,148,149] and experiment[148,150] for the iodide of **32** was discussed.

An ultra-wideline ^{33}S NMR spectrum was recorded from an isotopically enriched sample of elemental sulfur at a high magnetic field (21.1 T) for the > 99.9% enriched sample of elemental sulfur, resulting in the largest ^{33}S quadrupolar coupling constant yet measured by solid state NMR (CQ = 43.3 MHz).[151] This investigation comes from the preliminary DFT investigations into the feasibility of using ^{33}S solid state NMR to study organic and biological molecules.

Sulfur isotropic absolute shielding constants were calculated by a DFT(B3LYP)-GIAO method to explain the observed substituent effect on ^{33}S NMR chemical shifts and line widths in $XCH_2CH_2SO_3Na$ (X = H, CH_3, OH, SH, NH_2, Cl, Br, NH_3^+).[152] Data have been interpreted with the aid of NBO[153] and NICS (nucleus-independent chemical shifts).[154,155]

12.4.8.7 Calculations of ^{125}Te NMR Chemical Shifts

Ziegler and co-workers studied ^{125}Te NMR chemical shifts systematically by first-principles electronic structure theory with the inclusion of relativity.[156] The DFT-GIAO calculations were carried out on the ^{125}Te NMR chemical shifts for a number of organic, inorganic, and organometallic tellurium-containing complexes. The calculated shifts span a range of about 3000 ppm and therefore cover almost the complete range of known ^{125}Te chemical shifts. It was found that the DFT-GIAO method is able to predict ^{125}Te NMR chemical shifts that follow the same trends as experiment.

The trimeric diphenyltin chalcogenides $(Ph_2SnE)_3$ (E = S (**33**), Se (**34**), and Te(**35**)) have been synthesized and characterized by multinuclear NMR spectroscopy (^{1}H, ^{13}C, ^{119}Sn, ^{77}Se, and ^{125}Te) as well as crystal structure analyses.[157] For **33**, DFT calculations at the B3LYP/6–31G level of theory have revealed the twisted boat conformation as the global minimum. Additionally a boat conformation (+3.4 kJ mol^{-1}) and a chair conformation (+7.9 kJ mol^{-1}) have been calculated. The mutual repulsion of the Ph rings in the chair conformation is also evident from the increased bond angles of the Sn_3S_3 ring in **33**.

A quasirelativistic two-component ZORA-DFT approach was reported to calculate PV (parity violating) resonance frequency differences between the NMR spectra of enantiomers.[158] The systematics of PV NMR shielding constants in C_2 symmetric dihydrogen dichalcogenides (H_2E_2 with E = ^{17}O, ^{33}S, ^{77}Se, ^{125}Te, and ^{209}Po) are investigated according to the theoretical treatment. The typical sin(2α)-like dependence of the PV NMR frequency splittings on the dihedral angle α is observed for the entire series. The scaling behaviour of the effect with the nuclear charge Z of E was also discussed.

The relativistic effects on absolute magnetic shielding tensors are explicitly evaluated for various tellurium species with the DFT(BLYP)-GIAO method, similarly to the case of the selenium species (see Section 12.4.8.5).[159]

12.4.9 Nucleus-Independent Chemical Shifts: Indices for Aromaticity

The concept of aromaticity has been established for the compounds with low reactivity and high stability of the planar D_{nh} structure, together with characteristic spectroscopic behaviours containing NMR. Schleyer and co-workers introduced a new concept of nucleus-independent chemical shifts (NICS) in 1996.[154,155] NICS means the magnetic shielding constants at positions other than those of the nucleus. They can be calculated but cannot usually be detected. NICS indices are the negative values of the magnetic shielding calculated at the chosen points in the vicinity of molecules. They are typically computed at ring centres, at points above and even on the grids in and around the molecules. Significantly, negative NICS values in interior positions of rings or cages indicate the presence of an induced diatropic ring current, which is interpreted as aromaticity, whereas positive values denote paratropic ring currents and anti-aromaticity.

NICS values do not depend purely on a delocalized electron system such as the π system, but also on other magnetic shielding contributions due to local circulations of electron in bonds, lone pairs, and atom cores. NICS(1) calculated at points 1 Å above the ring centre is recommended as being a better measure of the π effect than NICS(0) calculated at the ring centre.

The shape of the magnetic shielding function provides information about electron delocalization and molecular aromaticity. The direction and the size of the anisotropic effect of double, triple, or aromatic rings can be obtained by plotting the iso-chemical shielding surface (ICSS).

The aromatic ring-current shielding (ARCS) was introduced by Jusélius and Sundfolm in 1999.[160] The long-range part of the shielding function (3–20 Å from the ring, for example) is used to provide information about the ring current. The isotropic magnetic shielding function $\sigma(z)$ is given by Equation 39 where z is the perpendicular distance from the centre of the current loop, μ_0 is the vacuum permeability, $\partial I_{\text{ring}}/\partial B_{\text{ext}}$ is the current susceptibility, B_{ext} is the external magnetic field, and R is the radius of the current loop. The strength of the ring-current susceptibility can be defined as the ARCS aromaticity index, which is obtained by fitting NICS to Equation 39.

$$\sigma(z) = -(\mu_0/2)(\partial I_{\text{ring}}/\partial B_{\text{ext}})[R^2/(z^2 + R^2)^{3/2}] \quad (39)$$

The molecular aromaticity can be studied in more detail by employing orbital decomposition of the shieldings inherent in the IGLO (individual gauge for localized orbitals) method.[161,162] The total shielding can be dissected into individual contributions from each localized molecular orbitals (LMO) associated with bonds, lone pairs, and core electrons. The procedure proposed by Pipek and Mezey for LMO-NICS separates s and p systems by means of orbital symmetry[163] when applied to planar molecules. $\text{NICS}(1)_\pi$ are very similar to NICS(1) in many cases, therefore NICS(1) can also be used as the index of aromaticity. NICS can also be discussed separately by the contribution from each orbital (MO-NICS).[164] The GIAO (gauge-including atomic orbitals) method is applied in this case. NICS_π indices can also be calculated by this method.

12.4.10 Structural Determination of Arylselenides in Solutions: Applications of Orientational Effect on δ(Se)

Many series of δ(Se) have been reported for *para*-substituted phenyl selenides (p-YC_6H_4SeR; ArSeR). Odom tried to explain δ(Se) of p-YC_6H_4SeR based on those of the related compounds such as p-YC_6H_4SeMe.[165] However, the attempt seemed unsuccessful for p-$YC_6H_4SeCOPh$: the plot of δ(Se: p-$YC_6H_4SeCOPh$) *versus* δ(Se: p-YC_6H_4SeMe) did not give a good linear correlation. Nakanishi and Hayashi considered that the electronic effect on δ(Se) in p-$YC_6H_4SeCOPh$ must be very different from that of p-YC_6H_4SeMe

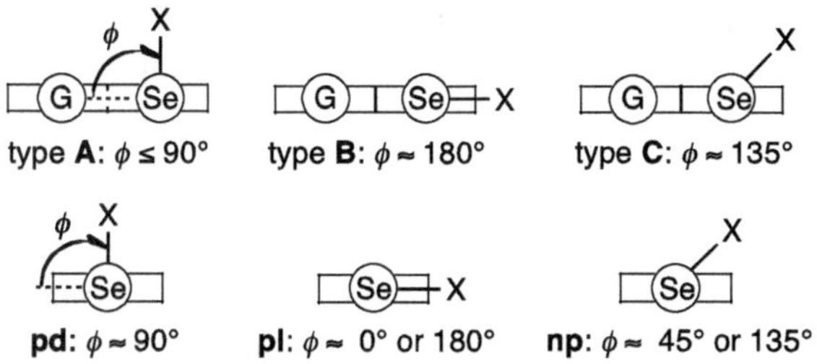

Scheme 12.4.6 The **pd**, **pl**, and **np** notation in the benzene system and types **A**–**C** in the naphthalene system.

due to the different orientation between the two. They tried to explain the structures of *p*-YC_6H_4SeR in a unified form based on the orientational effect on δ(Se) by the *p*-YC_6H_4Se group.[138,166,167] Scheme 12.4.6 shows typical conformers of ArSeR, **pl** (planar), **pd** (perpendicular), and **np** (non-planer and non-perpendicular) conformers, where the Se–C_R bond in ArSeR is on the Ar plane in **pl,** it is perpendicular to the plane in **pd**, and the Se–C_R bond in **pd** is at the intermediate between **pl** and **pd**.

They could successfully explain δ(Se) of *p*-YC_6H_4SeR (13 series) in solutions employing $\sigma^t(Se)_{rel}$ [–{σ^t(Se: *p*-YC_6H_4SeH) – σ^t(Se: MeSeMe)}] calculated for **pl** ($\angle C_oC_iSeH = 0°$) and **pd** ($\angle C_oC_iSeH = 90°$) of *p*-YC_6H_4SeH. This is a typical example of early success in determining the structure of ArSeR in solution employing calculated σ(Se) in combination with observed δ(Se).[138,166–168] This attempt at the methodology is developed further theoretically and experimentally.

It is inevitable to fix the conformation of all *p*-YC_6H_4Se groups in *p*-YC_6H_4SeR examined to clarify the relationship between δ(Se) and the structures (or conformers). 9-(Arylselanyl)anthracenes (**36**: *p*-YC_6H_4SeAtc) and 1-(arylselanyl)anthraquinones (**37**: *p*-YC_6H_4SeAtq) are chosen as the candidates for **pl** and **pd**, respectively: Y in **36** and **37** are H (**a**), NMe_2 (**b**), OMe (**c**), Me (**d**), F (**e**), Cl (**f**), Br (**g**), COOEt (**h**), CN (**i**), and NO_2 (**j**).[138] 9-(arylselanyl)triptycenes (**38**: *p*-YC_6H_4SeTpc) are also employed as another standard for $\delta(Se)_{SCS}$ of **pl.**[170] The C(sp^2)–Se–C(sp^2) bond constructs **36**, whereas the C(sp^2)–Se–C(sp^3) bond affords **38**. Little difference is detected between δ(Se: **38**)$_{SCS}$ and δ(Se: **36**)$_{SCS}$.

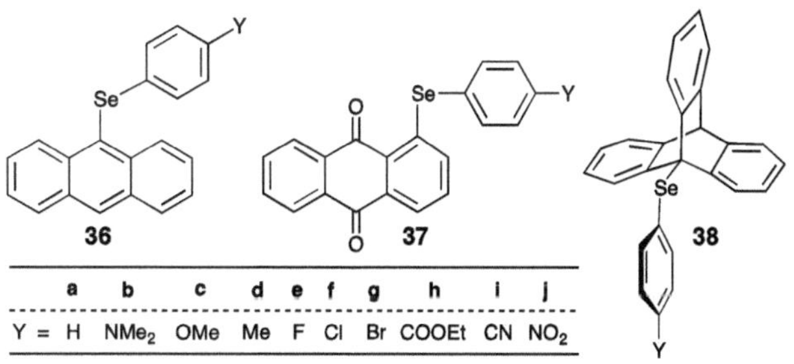

	a	b	c	d	e	f	g	h	i	j
Y =	H	NMe_2	OMe	Me	F	Cl	Br	COOEt	CN	NO_2

The notation of type **A** (**A**), type **B** (**B**), and type **C** (**C**) is applied to the conformers of the 9-anthracenyl (9-Atc) and 1-anthraquinoyl (1-Atq) groups in **36** and **37**, which is proposed for 1-(arylselanyl)naphthalenes (*p*-YC_6H_4SeNap).[118,166–169,171–174] Scheme 12.4.6 also explains the **A**, **B**, and **C** notation with **pl**, **pd**, and **np**. The structure of **36** is type **A** for 9-Atc and **pl** for Ar, which is denoted by **36** (**A**: **pl**). That of **37** is type **B** for the 1-Atq and **pd** for Ar (**37** (**B**: **pd**)). The series of δ(Se) in **36** and **37** must be typical for **pl** and **pd**, respectively, and demonstrated to serve as the standards for δ(Se) of **pl** and **pd**, respectively. The notation is also applied to the triptycyl group in **38**, although the angles between the two benzene rings are not 180° but 120°.

The driving force for (**A**: **pl**) in **36** and **38** must be the steric factor, which prevents the stabilization of (**A**: **pd**) due to the repulsive force between the C_6H_4Y group and the adjacent benzene rings in the anthryl and triptycyl groups. On the other hand, (**B**: **pd**) is stabilized by the n_p(O)---σ*(Se–C) 3c–4e-type interaction in **37**, which is further stabilized by the p–π conjugation of the p(Se)–π(Atq) type. The p–π conjugation of the p(Se)–π(Ar) type also operates in *p*-YC_6H_4SeR such as 1-(arylselanyl)naphthalenes (*p*-YC_6H_4SeNap), which controls the fine structures of *p*-YC_6H_4SeR especially in solutions. Scheme 12.4.7 illustrates the p–π conjugation in **36** and **37**, together with the steric effect in **36**.

$$\begin{aligned} &Y = NMe_2\,(\mathbf{b}) \ll OMe\,(\mathbf{c}) \ll Me\,(\mathbf{d}) < F\,(\mathbf{e}) < H\,(\mathbf{a}) \\ &< Cl\,(\mathbf{f}) \approx Br\,(\mathbf{g}) \ll COOEt\,(\mathbf{h}) \ll CN\,(\mathbf{i}) \ll NO_2\,(\mathbf{j})(\text{for } \mathbf{pl}) \end{aligned} \tag{40}$$

$$\begin{aligned} &Y = NMe_2\,(\mathbf{b}) \ll OMe\,(\mathbf{c}) \ll F\,(\mathbf{e}) < Me\,(\mathbf{d}) < Cl\,(\mathbf{f}) \approx Br\,(\mathbf{g}) \\ &\ll H\,(\mathbf{a}) \approx COOEt\,(\mathbf{h}) < NO_2\,(\mathbf{j}) \ll CN\,(\mathbf{i})(\text{for } \mathbf{pd}) \end{aligned} \tag{41}$$

Table 12.4.8 shows $\delta(Se)_{SCS}$ for **36**–**38** measured in chloroform-*d* at –60 °C. Characteristic behaviours of δ(Se) in **pl** and **pd** are summarized in Equations 40 and 41, respectively. Figure 12.4.16 shows the plot of δ(Se: **37**)$_{SCS}$ *versus* δ(Se: **36**)$_{SCS}$. One might imagine that δ(Se: **37**)$_{SCS}$ are well correlated with δ(Se: **36**)$_{SCS}$, except for the data corresponding to Y = H and NO_2. However, the characteristic differences can easily be found if δ(Se: **36**)$_{SCS}$ and δ(Se: **37**)$_{SCS}$ are

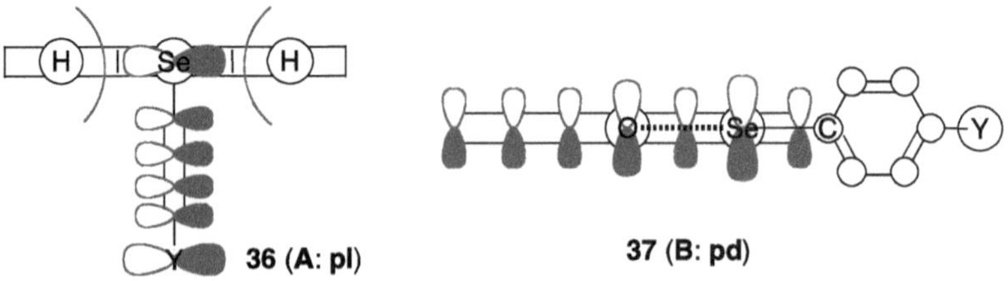

36 (A: pl) **37 (B: pd)**

Stabilized by steric repulsion Stabilized through n_p(O)---σ*(Se–C) interaction

Scheme 12.4.7 Factors to stabilize **36** (**A**: **pl**) and **37** (**B**: **pd**) conformations.

Table 12.4.8 The δ(Se)$_{SCS}$ values of **36–38**.

Compd	*NMe$_2$*	*OMe*	*Me*	*H*		*F*	*Cl*	*Br*	*CO$_2$Et*	*CN*	*NO$_2$*
36	−22.7	−12.7	−6.3	0.0	(245.3)	−3.3	1.9	2.4	17.4	27.7	32.7
37	−20.6	−15.5	−9.2	0.0	(511.4)	−10.5	−7.1	−6.4	0.1	8.5	2.7
38	−21.6	−11.6	−5.9	0.0	(257.1)	−0.7	3.7	4.2	18.7	29.9	36.6

[a]δSe)$_{SCS}$ are given for **36–38**, together with δ(Se) for **36a–38a** in parenthesis, measured in chloroform-*d* (0.050 M) at 213 K.

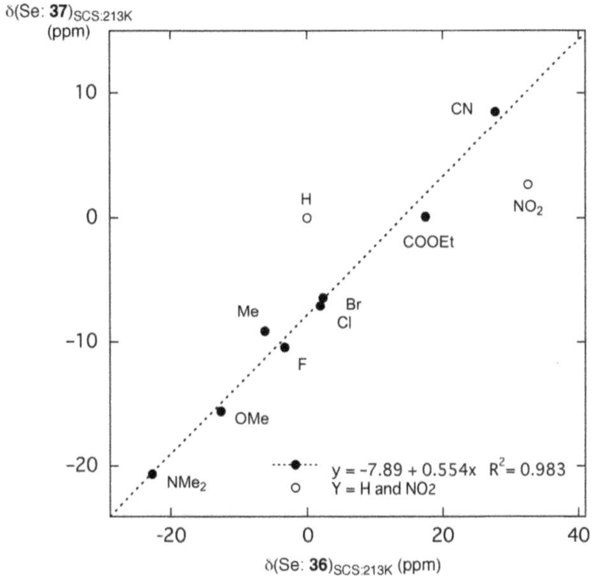

Figure 12.4.16 Plots of δ(Se:**37**)$_{SCS:213K}$ *versus* δ(Se:**36**)$_{SCS:213K}$.

compared for δ(Se)$_{SCS}$ of Y = H, NMe$_2$, and NO$_2$. The [δ(Se: Y = NMe$_2$)$_{SCS}$, δ(Se: Y = H)$_{SCS}$, δ(Se: Y = NO$_2$)$_{SCS}$] values measured at 213 K are [−22.7, 0.0, 32.7] for **36** and [−20.6, 0.0, 2.7] for **37** in ppm. The magnitudes of both δ(Se: Y = NMe$_2$)$_{SCS}$ and δ(Se: Y = NO$_2$)$_{SCS}$ are very large for **36**, whereas the magnitude of δ(Se: Y = NMe$_2$)$_{SCS}$ is also very large with a very small magnitude of δ(Se: Y = NO$_2$)$_{SCS}$ in **37**. This means that both electron-donating and electron-accepting groups greatly affect δ(Se: **36**)$_{SCS}$, whereas only electron-donating groups do much on δ(Se: **37**)$_{SCS}$, which is just as predicted theoretically for **pl** and **pd**, respectively.[140] Namely, the structures of **36** and **37** are (**A**, **pl**) and (**B**, **pd**), respectively. This observation can be confirmed through the plots of δ(Se: **36**) *versus* σ^t(Se)$_{rel}$ calculated for *p*-YC$_6$H$_4$SeH (**pl**) and δ(Se: **37**) *versus* σ^t(Se)$_{rel}$ calculated for *p*-YC$_6$H$_4$SeH (**pd**) as shown in Figure 12.4.17a and 17b, respectively. Correlations are very good for both cases, which supports above discussion and demonstrates that the structures of **36** and **37** are (**A**, **pl**) and (**B**, **pd**), respectively, for all Y examined. Figure 12.4.18 shows the plots of δ(Se: **38**)$_{SCS}$ *versus* δ(Se: **36**)$_{SCS}$. An excellent correlation is observed

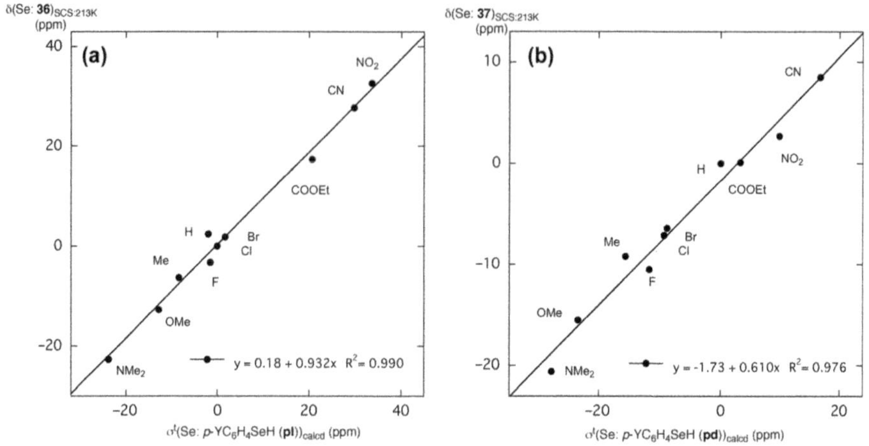

Figure 12.4.17 Plots of (a) δ(Se:**36**)$_{SCS:213K}$ *versus* σ^t(Se)$_{rel}$ calculated for *p*-YC_6H_4SeH (**pl**) and (b) δ(Se:**37**)$_{SCS:213K}$ versus σ^t(Se)$_{rel}$ calculated for *p*-YC_6H_4SeH (**pd**).

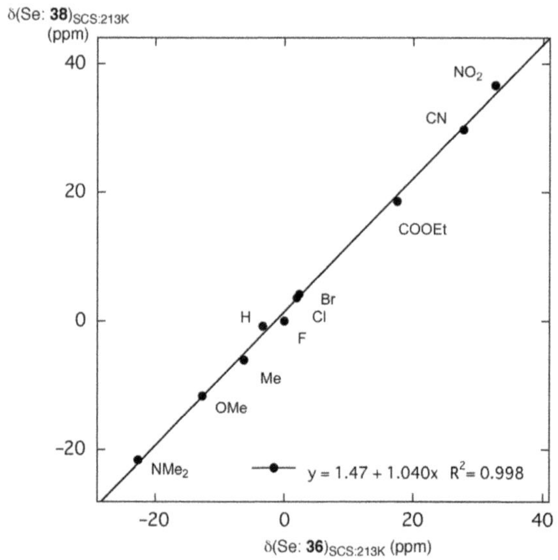

Figure 12.4.18 Plot of δ(Se)$_{SCS:213}$ of **38** versus δ(Se)$_{SCS:213}$ of **36**.

in the plot, which demonstrates that the structures of **38** are (**A**, **pl**) for all Y examined, similarly to the case of **36**.

As applications of the standard δ(Se)$_{SCS}$ for **36** (**A**, **pl**), **38**(**A**, **pl**), and **37** (**B**, **pd**), the structural determinations of arylselenides (**39**–**54**) in solutions are examined. Some of then are then discussed. An excellent correlation is also observed in the plot of δ(Se)$_{SCS}$ of 1-arylselanyl-2-methyl-naphthalenes (**39**) *versus* δ(Se: **36**)$_{SCS}$, although not shown. The results show that the structure of

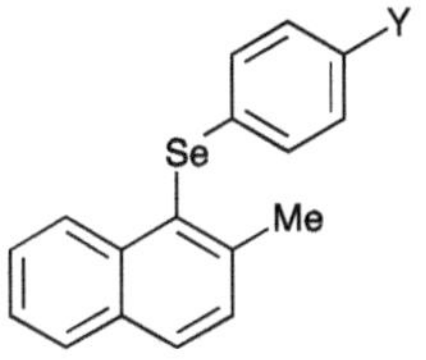

39

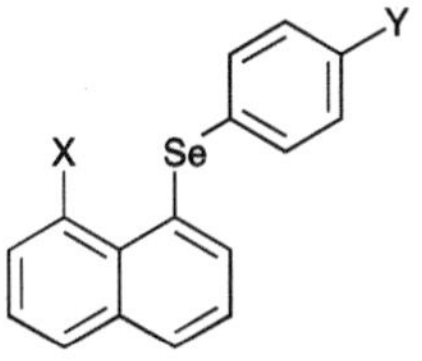

40 (X = H), **41** (X = Cl)
42 (X = Br), **43** (X = SeMe)

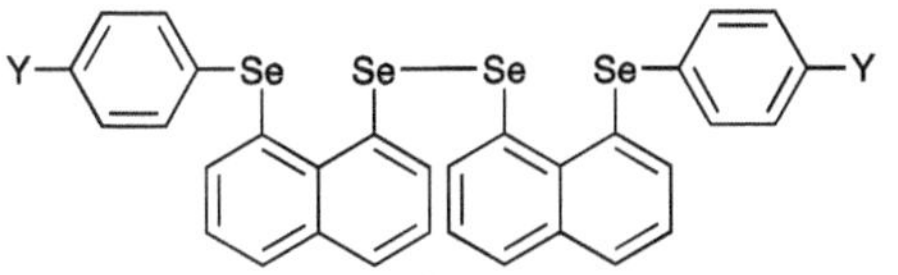

44

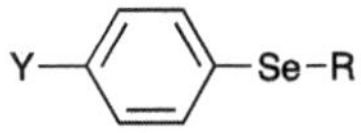

45 (R = H), **46** (R = Me), **47** (R = Ph), **48** (R = C_6H_4Y-*p*)
49 (R = CN), **50** (R = Bz), **51** (R = Br), **52** (R = $CH{=}CH_2$)
53 (R = CH=CHCl-*trans*), **54** (R = $CHCH_2CCl_2$-cyclo)

	a	**b**	**c**	**d**	**e**	**f**	**g**	**h**	**i**	**j**
Y =	H	NMe_2	OMe	Me	F	Cl	Br	COOEt	CN	NO_2

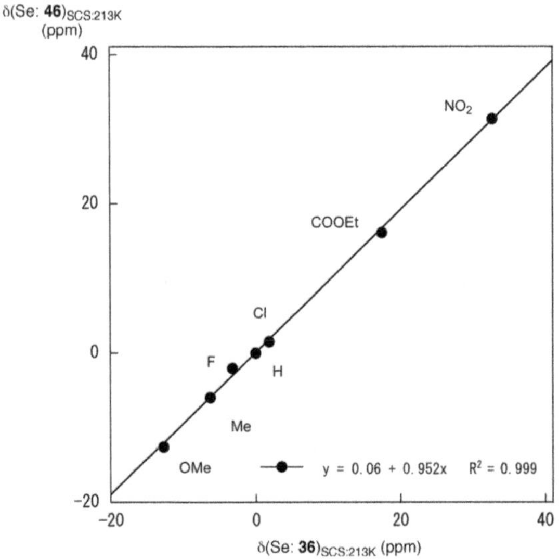

Figure 12.4.19 The plot of δ(Se: **46**)$_{SCS}$ versus δ(Se: **36**)$_{SCS}$.

39 is (**A**, **pl**) for all Y examined and δ(Se: **39**)$_{SCS}$ could also be used as the standard of **pl**, as well as δ(Se: **36**)$_{SCS}$ and δ(Se: **38**)$_{SCS}$. Figure 12.4.19 shows the plot of δ(Se: ArSeMe (**46**))$_{SCS}$ *versus* δ(Se: **36**)$_{SCS}$. The correlation is very good, which demonstrates that **46** exist in (**A**, **pl**) for all Y examined in the solutions.

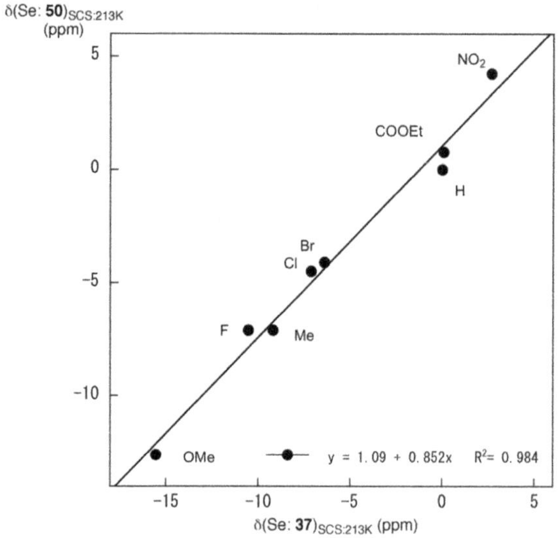

Figure 12.4.20 The plot of δ(Se: **50**)$_{SCS}$ versus δ(Se: **37**)$_{SCS}$

However, the plot of δ(Se: ArSeBz (**50**))$_{SCS}$ *versus* δ(Se: **36**)$_{SCS}$ did not give an good correlation (not shown), whereas the plot of δ(Se: **50**)$_{SCS}$ *versus* δ(Se: **37**)$_{SCS}$ gave an very good correlation (Figure 12.4.20). The results demonstrate that the structure of **50** is in (**B**, **pd**) in solutions for all Y examined. This is why the plot of δ(Se: **50**)$_{SCS}$ *versus* δ(Se: **46**)$_{SCS}$ did not show good correlation in the plot.[138]

Compounds **41** and **42** contain halogens at the 8-position of 1-arylselanylnaphthlenes (**40**). It is expected that **41** and **42** exist in (**B**, **pd**) in solutions if the electron-accepting ability of halogens is strong enough to keep the structures (**B**, **pd**) as in **37**. Figure 12.4.21 shows the plot of δ(Se: **42**)$_{SCS}$ *versus* δ(Se: **37**)$_{SCS}$. Data for Y = OMe, Me, Cl, Br, and H gave excellent correlation, showing the (**B**, **pd**) structure for the derivatives. However, data for Y = COOEt deviate slightly from the correlation, which must be a reflection of the slight equilibrium with (**A**, **pl**). The deviations in δ(Se: **42**)$_{SCS}$ seem substantial for Y = CN and NO_2. The δ(Se: **42**)$_{SCS}$ values are almost equal with each other for Y = CN and NO_2. The results must come from the substantial equilibrium with (**A**, **pl**). δ(Se: **41**)$_{SCS}$ showed a similar trend to the case of δ(Se: **42**)$_{SCS}$. The results show that the halogens seem not too effective in keeping the structure (**B**, **pd**) in solutions, especially for Y of strong electron-accepting groups, relative to the case of **37**.

Structures of 1-ArSeNap (**40**) in solutions are also examined similarly. The structures are concluded to be (**B**, **pd**) for Y = NMe_2, OMe, and F but (**A**, **pl**) for Cl, Br, COOEt, CN, and NMe_2 with in equilibrium between (**B**, **pd**) and (**A**, **pl**) for Y = Me and H, although there are some possibility of the equilibrium for Y = F. Scheme 12.4.8 shows the p–π conjugation in **40**. The relative contribution controls the predominant structure of **40** in solutions.

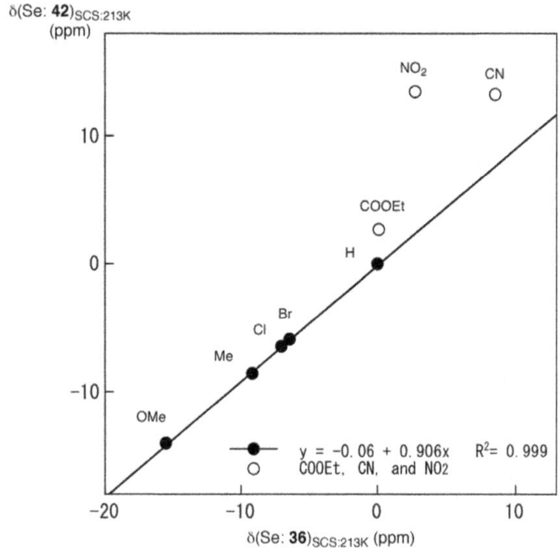

Figure 12.4.21 Plot of δ(Se: **42**)$_{SCS}$ versus δ(Se: **36**)$_{SCS:213}$.

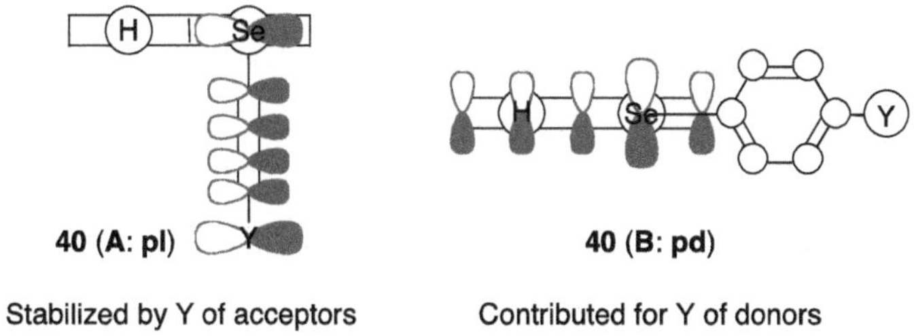

Scheme 12.4.8 Factors to stabilize (**A**: **pl**) and (**B**: **pd**) conformations.

Structures of other selenides are also examined employing the standard δ(Se)$_{SCS}$ values for **pl** and **pd**. Those for the O, S, and Te derivatives of **40** are also examined and similar conclusions are obtained.

12.4.11 Indirect Nuclear Spin–Spin Coupling Tensor *J*

12.4.11.1 Theoretical Background

According to NR theory, there are distinct several mechanisms contributing to the spin-spin coupling constants. The total value ($^nJ_{TL}$) is composed of contributions from the diamagnetic spin-orbit (DSO) term, the paramagnetic spin-orbit (PSO) term, the spin-dipolar (SD) term, and the Fermi contact (FC) term, as expressed in Equation 42. The contributions can be evaluated separately from each MO (ψ_i) and $\psi_i \rightarrow \psi_a$ transition, where ψ_i and ψ_a are occupied and unoccupied MOs, respectively.

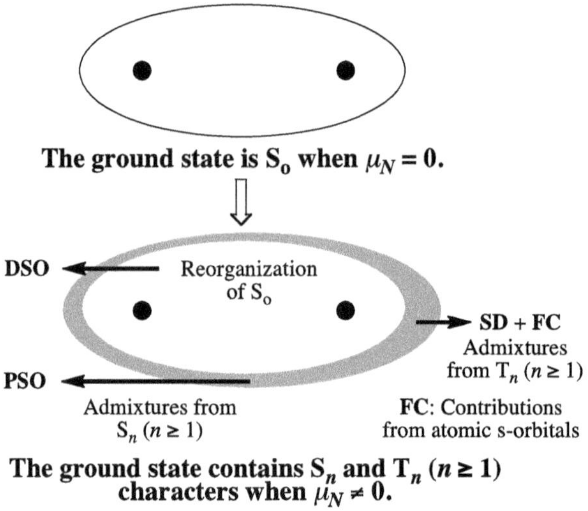

Scheme 12.4.9 Mechanisms for ${}^nJ_{\rm DSO}$, ${}^nJ_{\rm PSO}$, ${}^nJ_{\rm SD}$, and ${}^nJ_{\rm FC}$ terms, contributing to ${}^nJ_{\rm TL}$.

$$ {}^nJ_{\rm TL} = {}^nJ_{\rm DSO} + {}^nJ_{\rm PSO} + {}^nJ_{\rm SD} + {}^nJ_{\rm FC} \quad (42)$$

Scheme 12.4.9 summarizes the mechanism of the indirect nuclear spin–spin couplings and the origin of the ${}^nJ_{\rm DSO}$, ${}^nJ_{\rm PSO}$, ${}^nJ_{\rm SD}$, and ${}^nJ_{\rm FC}$ terms. The singlet state (S$_o$) of a molecule (**M**) is the ground state if the nuclei (**N**) in **M** have no magnetic moments ($\mu_N = 0$), but the ground state cannot be the absolute S$_o$ if $\mu_N \neq 0$. The DSO term arises from the reorganization of S$_o$ in the μ_N-perturbed ground state, so it is usually very small. The PSO term arises from the mixing of upper singlet states (S1, S2, S3, *etc.*) and the FC and SD terms arise from admixtures of upper triplet states (T1, T2, T3, *etc.*), in which only s-type atomic orbitals contribute to FC.

The calculations of the indirect nuclear spin–spin coupling constants containing Se are explained next.

12.4.11.2 Calculations of Nuclear Spin–Spin Coupling Constants

12.4.11.2.1 ${}^1J$(Se, X: X = C and F)

The ${}^1J$(Se, C) values were calculated by DFT methods at the B3LYP/6–311 + G(d,p) level of theory for 1-cyclohepta-2,4,6-trienyl-selanes ($Se(C_7H_7)_2$) and $RSeC_7H_7$ (R = Bu, *t*-Bu, Ph, 4-FC_6H_4), together with those for various organic selenium compounds.[175] The calculated values were confirmed to be negative due to the large negative contributions from the FC term with smaller positive magnitude contributions from the SD and PSO terms. The calculated ${}^1J$(Se, C) values reproduced the observed values well. The following correlation was obtained for approximately 20 compounds: ${}^1J(\mathrm{Se, C})_{calcd} = 0.98\,{}^1J(\mathrm{Se, C})_{obsd} + 10.3$ ($R^2 = 0.997$).

The temperature dependence of the nuclear spin coupling between Se and F ($J_{\mathrm{Se\cdots F}}$) was investigated by Tomoda and co-workers for 2-(fluoromethyl)phenylselenenyl cyanate and bis[2-(fluoromethyl)phenyl] diselenide in CD_2Cl_2 and CD_3CN, to evaluate weak intramolecular non-bonded Se···F interactions.[73,176,177] Interactions of the n(G)···σ*(Se–F) type were discussed.

12.4.11.2.2 ¹J(Se, Se)

The observed ${}^1J$(Se, Se) values (${}^1J_{obsd}$(Se, Se)) are reported to be rather small (≤64 Hz) for noncyclic RSeSeR (**55**). On the other hand, large ${}^1J_{obsd}$(Se, Se) values (330–380 Hz) are observed for 4-substituted naphto[1,8-*c,d*]-1,2-diselenoles (**56**: 379.4 Hz for **56c**, 375.9 Hz for **56d** and 330.8 Hz for **56g**), which correspond to symperiplanar diselenides. ${}^1J$(Se, Se: **56**) becomes larger as the electron density on Se increases.[178]

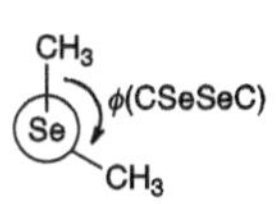

MeSeSeMe (**55a**)
RSeSeR' (**55**)

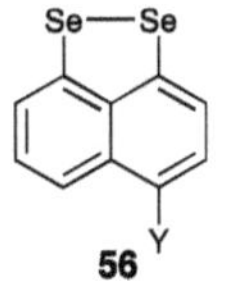

56

	a	b	c	d	e	f	g
Y =	H	OMe	Me	Cl	COOMe	CN	NO_2

The ${}^1J$(Se, Se) values are analysed on the basis of the MO theory with the triple ζ basis sets of the Slater type at the DFT level.[142] ${}^1J$(Se, Se) are calculated as modelled by MeSeSeMe (**55a**), which shows the typical torsional angular dependence on $\phi(\mathrm{C_{Me}SeSeC_{Me}})$.[178] Equation 42 is applied to analyse the total value (${}^nJ_{TL}$). The PSO terms contribute predominantly. Whereas ${}^1J_{TL}$(Se, Se: **55a**) is predicted to be less than 44 Hz for $\phi = 90° \pm 15°$, it is very large at $\phi = 0°$ (684 Hz) and 180° (628 Hz). Therefore, ${}^1J_{obsd}$(Se, Se: **55**) and ${}^1J_{obsd}$(Se, Se: **56**) are explained substantially as modelled by **55a** with $\phi \approx 90°$ and 0°, respectively, although R and R′ in **55** and **56** must also affect on the values. Figure 12.4.22 draws the plots of ${}^1J_{DSO}$, ${}^1J_{PSO}$, ${}^1J_{SD}$, ${}^1J_{FC}$, ${}^1J_{SD+FC}$, and ${}^1J_{TL}$ *versus* ϕ in **55a**. It shows well that ${}^1J_{TL}$ changes depending on ϕ, similarly to the case of ${}^3J$(H, H).[179] ${}^1J_{DSO}$ are negligible (<0.03 Hz). The results exhibit that ${}^1J_{PSO}$(Se, Se: **55a**) and ${}^1J_{SD+FC}$(Se, Se: **55a**) contribute 65% and 35% to ${}^1J_{TL}$(Se, Se: **55a**), respectively, irrespective of the ϕ(CSeSeC) values.

The contributions are evaluated separately from each MO (ψ_i) and each $\psi_i \rightarrow \psi_a$ transition, where ψ_i and ψ_i are occupied and unoccupied MOs, respectively. The separate evaluation of ${}^1J$(Se, Se) enables us to recognize and visualize the origin and the mechanism of the couplings. The mechanism of ${}^1J$(Se, Se: **55a**) is investigated employing ${}^1J_{PSO}$(Se, Se: **55a**), since J_{PSO}(Se, Se) contributes about 65% to ${}^1J_{TL}$(Se, Se: **55a**). Figure 12.4.23 shows the $\psi_{42} \rightarrow \psi_{44}$

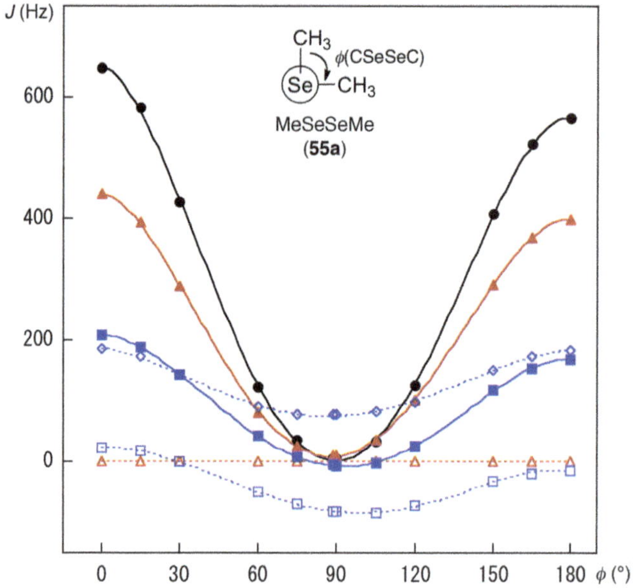

Figure 12.4.22 Plots of ${}^1J_{TL}$ (●), ${}^1J_{DSO}$ (△), ${}^1J_{PSO}$ (▲), ${}^1J_{SD+FC}$ (■), ${}^1J_{SD}$ (□), and ${}^1J_{FC}$ (◇) *versus* ϕ(CSeSeC) in ${}^1J$(Se, Se: **55a**).

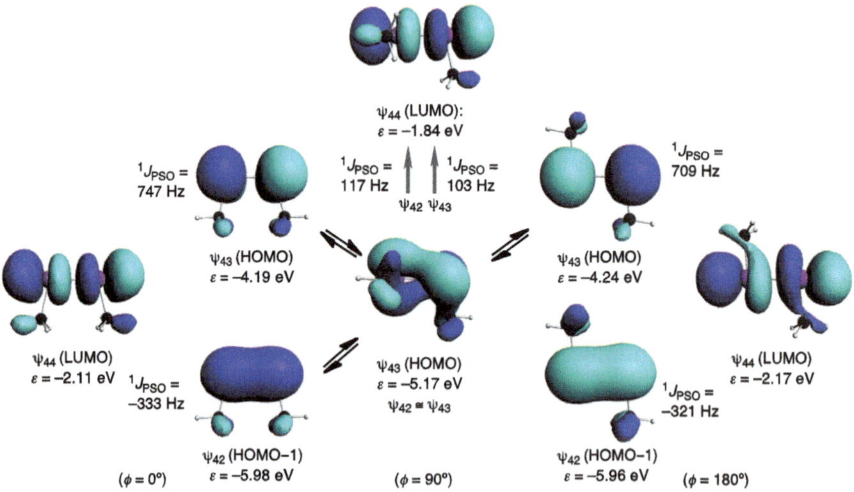

Figure 12.4.23 Contributions from the $\psi_{42}\rightarrow\psi_{44}$ and $\psi_{43}\rightarrow\psi_{44}$ transitions to ${}^1J_{PSO}$(Se, Se: **55a**) at $\phi = 0$, 90, and 180°; ψ_{42} and ψ_{43} interconvert at $\phi \approx 90°$.

and $\psi_{43}\rightarrow\psi_{44}$ transitions at $\phi = 0°$, 90°, and 180°. Characters of ψ_{42}(HOMO-1), ψ_{43}(HOMO), and ψ_{44}(LUMO) are π(Se–Se), π*(Se–Se), and σ*(Se–Se), respectively, at $\phi = 0°$ and 180°. The $\psi_{42}\rightarrow\psi_{44}$ and $\psi_{43}\rightarrow\psi_{44}$ transitions contribute almost exclusively to ${}^1J_{PSO}$(Se, Se: **55a**) at $\phi = 0°$ and 180°, whereas the transitions are almost cancelled as a whole at $\phi \approx 90°$. This is why $^1J_{PSO}$(Se, Se: **55a**) is small at $\phi \approx 90°$.

The charge dependence of ^1J (Se, Se: **56**) is also examined theoretically. $^1J_{TL}$(Se, Se: **56**) are shown to be controlled by $Qn(^1\text{Se})$ ($^1J_{TL}$(Se, Se: **56**) = –2129.7 $Qn(^1\text{Se})$ + 1013.7: $r = 0.965$). The relativistic effect should also be considered in nuclear spin-spin couplings.

12.4.11.3 Relativistic Effect on Nuclear Spin–Spin Coupling Constants for S, Se, and Te

The relativistic ZORA-DFT method was employed to investigate indirect spin–spin coupling tensors of ^1J(E, F), containing E = S, Se, and Te in SF_6, SeF_6, and TeF_6.[180] The results demonstrate that the ZORA-DFT method reproduces the magnitude of $^1J(\text{E, F})_{iso}$. $^1K(\text{E, F})_{iso}$ of EF_6 (E = S, Se, and Te) are assumed to be negative, since the negative values are observed for Et_2NTeF_5. The calculated $^1J(\text{E, F})_{iso}$ for SF_6 and SeF_6 are in good agreement with experimental values, $^1J(\text{S, F})_{iso}$ = (–)251.6 ± 0.2 Hz (liquid at 298 K) and $^1J(\text{Se, F})_{iso}$ = (–)1420.9 ± 0.5 Hz (liquid), deviating by –60 and –104 Hz, respectively. Calculations underestimate the magnitude of $^1J(^{125}\text{Te, F})_{iso}$ by 1100 Hz relative to the experimental value, where $^1J(^{125}\text{Te, F})_{iso}$ = + 3736 Hz. The FC term dominates the contribution to $^1J(\text{E, F})_{iso}$ for all three species. Contributions to ^1J(E, F) from both the PSO term and the cross-term between (FC + SD) and PSO become increasingly important. $^1J(\text{E, F})_{iso}$ and Δ^1J(E, F) ($\Delta J = J_{33} - (J_{11} + J_{22})/2$) are of similar orders of magnitude but opposite in sign. The QC calculation of ^1J(E, F) can serve as an effective tool to aid experimental NMR spectroscopists identify systems in which Δ^1J(E, F) must be considered. The ZORA-DFT method also indicates that ^1J(E, F) is of the same order of magnitude as $^1J(\text{E, F})_{iso}$.

In this connection, the isotropic average and anisotropy of the ^{77}Se–^{13}C and ^{77}Se–^{77}Se spin–spin coupling tensors were calculated for carbon diselenide (CSe_2) by applying the ZORA-DFT method.[142,181–186] Computed nuclear shielding anisotropies are also reported. The coupling tensor anisotropies were derived experimentally with the liquid crystal (LC) NMR method. Due to the fact that the number of adjustable parameters in the experimental analysis is greater than the number of anisotropic coupling constants, the ratio of the coupling tensor anisotropies, $\Delta J_{CSe}/\Delta J_{SeSe}$, had to be fixed to the value obtained from the calculations. The anisotropies of the J_{CSe} and $J_{Se,Se}$ spin–spin coupling tensors in carbon diselenide were derived from the calculations and from NMR experiments performed in liquid crystalline solutions. The computed coupling tensor anisotopies appear about 20% larger than the respective experimental ones. However, for example, the deviation between the computed and experimental results for the ratio $\Delta J_{CSe}/J_{CSe}$ is less than 4%. The confirmed non-zero ΔJ_{CSe} means that the earlier reported shielding tensor anisotropies ($\Delta\sigma_C$ and $\Delta\sigma_{Se}$) have to be revised. Both values are now more accurate than the earlier values, which decrease by about 10%. The

ZORA-DFT calculations overestimate the anisotropies, but the ratio $\Delta\sigma_{Se}/\Delta\sigma_{C}$ is well reproduced.

References

1. G. E. Scuseria and V. N. Staroverov, in *Theory and Application of Computational Chemistry: The First 40 Years*, ed. C. E. Dykstra, G. Frenking, K. S. Kim and G. E. Scuseria, Elsevier: Amsterdam, 2005, Chapter 24, p. 669.
2. W. J. Hehre, L. Radom, P. v.R. Schleyer and J. A. Pople, *Ab Initio Molecular Orbital Theory*, John Wiley & Sons, New York, 1986.
3. H. Eyring, J. Walter and G. E. Kimball, *Quantum Chemistry*, John Wiley & Sons, New York, 1944.
4. P. W. Atkins and R. S. Friedman, *Molecular Quantum Mechanics*, Oxford University Press, Oxford, 4th edn, 2005.
5. J. Simons and J. Nichols, *Quantum Mechanics in Chemistry*, Oxford University Press, New York, 1997.
6. Y. Harada, *Quantum Chemistry*, Vols. 1 and 2, Shokabo, Tokyo, 2007.
7. J. B. Foresman and Æ. Frisch, *Exploring Chemistry with Electronic Structure Methods*, Gaussian, Pittsburgh, PA, 2nd edn, 1996, p. 278.
8. M. Born and R. Oppenheimer, Zur Quantentheorie der Molekeln, Ann., *Phys. (Leipzig)*, 1927, **84**, 457.
9. (a) E. Fermi, *Rendiconti Lincei*, 1926, **3**, 145; translated as A. Zannoni, *On the Quantization of the Monoatomic Ideal Gas*, 1999, arXiv:condmat/9912229; (b) P. A. M. Dirac, *Proc. Roy. Soc. London A*, 1926, **112**, 661.
10. K. Fukui, *Modern Industrial Chemistry, Quantum Chemistry*, Asakura., Tokyo, 1982, Vol. 2, p. 106.
11. (a) J. A. Pople, R. Krishnan, H. B. Schlegel and J. S. Binkley, *Int. J. Quantum Chem.*, 1978, **14**, 545; (b) R. J. Bartlett and G. D. Purvis III, *Int. J. Quantum Chem.*, 1978, **14**, 561.
12. (a) J. Cizek, in *Advances in Chemical Physics*, P. C. Hariharan (ed), Vol. 14, Wiley Interscience, New York, 1969, 35; (b) G. D. Purvis III and R. J. Bartlett, *J. Chem. Phys.*, 1982, **76**, 1910; (c) G. E. Scuseria, C. L. Janssen and H. F. Schaefer III, *J. Chem. Phys.*, 1988, **89**, 7382; (d) G. E. Scuseria and H. F. Schaefer III, *J. Chem. Phys.*, 1989, **90**, 3700.
13. J. A. Pople, M. Head-Gordon and K. Raghavachari, *J. Chem. Phys.*, 1987, **87**, 5968.
14. (a) C. Møller and M. S. Plesset, *Phys. Rev.*, 1934, **46**, 618; (b) J. Gauss, *J. Chem. Phys.*, 1993, **99**, 3629; (c) J. Gauss, *Ber. Bunsen-Ges. Phys. Chem.*, 1995, **99**, 1001.
15. (a) P. Hohenberg and W. Kohn, *Phys. Rev.*, 1964, **136**, B864; (b) W. Kohn and L. J. Sham, *Phys. Rev.*, 1965, **140**, A1133; (c) D. R. Salahub and M. C. Zerner (eds), *The Challenge of d and f Electrons*, ACS, Washington, D. C., 1989.

16. R. G. Parr and W. Yang, *Density-Functional Theory of Atoms and Molecules*, Oxford University Press, Oxford, 1989.
17. For the various DFT functionals, see Æ. Frisch, M. J. Frisch and G. W. Trucka, *Gaussian 09 User's Reference*, Gaussian, 2009, for example.
18. See Ref. 1, for example.
19. (a) C. Lee, W. Yang and R. G. Parr, *Phys. Rev. B*, 1988, **37**, 785; (b) A. D. Becke, *J. Chem. Phys.*, 1993, **98**, 5648.
20. A. D. Boese and J. M. L. Martin, *J. Chem. Phys.*, 2004, **121**, 3405.
21. (a) Y. Zhao, N. E. Schultz and D. G. Truhlar, *J. Chem. Phys.*, 2005, **123**, 161103; (b) Y. Zhao, N. E. Schultz and D. G. Truhlar, *J. Chem. Theory Comput.*, 2006, **2**, 364.
22. Y. Zhao and D. G. Truhlar, *Theor. Chem. Acc.*, 2008, **120**, 215.
23. M. J. Frisch, G. W. Trucks, H. B. Schlegel, G. E. Scuseria, M. A. Robb, J. R. Cheeseman, G. Scalmani, V. Barone, B. Mennucci, G. A. Petersson, H. Nakatsuji, M. Caricato, X. Li, H. P. Hratchian, A. F. Izmaylov, J. Bloino, G. Zheng, J. L. Sonnenberg, M. Hada, M. Ehara, K. Toyota, R. Fukuda, J. Hasegawa, M. Ishida, T. Nakajima, Y. Honda, O. Kitao, H. Nakai, T. Vreven, J. A. Montgomery, Jr., J. E. Peralta, F. Ogliaro, M. Bearpark, J. J. Heyd, E. Brothers, K. N. Kudin, V. N. Staroverov, R. Kobayashi, J. Normand, K. Raghavachari, A. Rendell, J. C. Burant, S. S. Iyengar, J. Tomasi, M. Cossi, N. Rega, J. M. Millam, M. Klene, J. E. Knox, J. B. Cross, V. Bakken, C. Adamo, J. Jaramillo, R. Gomperts, R. E. Stratmann, O. Yazyev, A. J. Austin, R. Cammi, C. Pomelli, J. W. Ochterski, R. L. Martin, K. Morokuma, V. G. Zakrzewski, G. A. Voth, P. Salvador, J. J. Dannenberg, S. Dapprich, A. D. Daniels, Ö. Farkas, J. B. Foresman, J. V. Ortiz, J. Cioslowski and D. J. Fox, *Gaussian 09, Revision B.01*, Gaussian, Inc., Wallingford CT, 2009.
24. T. Yanai, D. Tew and N. Handy, *Chem. Phys. Lett.*, 2004, **393**, 51.
25. (a) Y. Tawada, T. Tsuneda, S. Yanagisawa, T. Yanai and K. Hirao, *J. Chem. Phys.*, 2004, **120**, 8425; (b) O. A. Vydrov and G. E. Scuseria, *J. Chem. Phys.*, 2006, **125**, 234109.
26. (a) K. G. Dyall and K. Fægli (eds), *Introduction to Relativistic Quantum Chemistry*, Oxford University Press: Oxford, NY, 2007; (b) M. Reiher and A. Wolf, *Relativistic Quantum Chemistry: The Fundamental Theory of Molecular Science*, Wiley-VCH, Weinheim, 2009.
27. (a) R. Fukuda, M. Hada and H. Nakatsuji, *J. Chem. Phys.*, 2003, **118**, 1015; (b) R. Fukuda, M. Hada and H. Nakatsuji, *J. Chem. Phys.*, 2003, **118**, 1027; (c) S. Tanaka, M. Sugimoto, H. Takashima, M. Hada and H. Nakatsuji, *Bull. Chem. Soc. Jpn.*, 1996, **69**, 953; (d) C. C. Ballard, M. Hada, H. Kaneko and H. Nakatsuji, *Chem. Phys. Lett.*, 1996, **254**, 170; (e) H. Nakatsuji, M. Hada, H. Kaneko and C. C. Ballard, *Chem. Phys. Lett.*, 1996, **255**, 195; (f) M. Hada H. Kaneko and H. Nakatsuji, *Chem. Phys. Lett.*, 1996, **261**, 7.
28. (a) M. Bühl, M. Kaupp, O. L. Malkina and V. G. Malkin, *J. Comput. Chem.*, 1999, **20**, 91; (b) J. Konu, M. Ahlgren, S. M. Aucott, T. Chivers,

S. H. Dale, M. R. J. Elsegood, K. E. Holmes, S. L. M. James, P. F. Kelly and R. S. Laitinen, *Inorg. Chem.*, 2005, **44**, 4992; (c) D. R. Urban and J. Wilcox, *J. Phys. Chem. A*, 2006, **110**, 5847; (d) C. M. Widdifield and R. W. Schurko, *Concepts Magn. Reson., Part A*, 2009, **34A**, 91.
29. (a) E. van Lenthe, A. E. Ehlers and E. J. Baerends, *J. Chem. Phys.*, 1999, **110**, 8943; (b) E. van Lenthe, E. J. Baerends and J. G. Snijders, *J. Chem. Phys.*, 1993, **99**, 4597; (c) E. van Lenthe, E. J. Baerends and J. G. Snijders, *J. Chem. Phys.*, 1994, **101**, 9783; (d) E. van Lenthe, J. G. Snijders and E. J. Baerends, *J. Chem. Phys.*, 1996, **105**, 6505; (e) E. van Lenthe, R. van Leeuwen, E. J. Baerends and J. G. Snijders, *Int. J. Quantum Chem.*, 1996, **57**, 281.
30. H. Nakano, T. Nakajima, T. Tsuneda and K. Hirao Hirao, In *Theory and Application of Computational Chemistry: The First 40 Years*, ed. C. E. Dykstra, G. Frenking, K. S. Kim and G. E. Scuseria, Elsevier: Amsterdam, 2005, Chapter 20.4.3, p. 548.
31. J. R. Yates, C. J. Pickard, M. C. Payne and F. Mauri, *J. Chem. Phys.*, 2003, **118**, 5746.
32. Ch. Chang, M. Pelissier and P. Durand, *Phy. Scr.*, 1986, **34**, 394.
33. (a) J. B. Foresman, M. Head-Gordon, J. A. Pople and M. J. Frisch, *J. Phys. Chem.*, 1992, **96**, 135; (b) M. Head-Gordon, R. J. Rico, M. Oumi and T. J. Lee, *Chem. Phys. Lett.*, 1994, **219**, 21; (c) M. Head-Gordon, D. Maurice and M. Oumi, *Chem. Phys. Lett.*, 1995, **246**, 114.
34. (a) J. Gauss and D. Cremer, *Chem. Phys. Lett.*, 1988, **150**, 280; (b) E. A. Salter, G. W. Trucks and R. J. Bartlett, *J. Chem. Phys.*, 1989, **90**, 1752.
35. H. Fleischer, D. A. Wann, S. L. Hinchley, K. B. Borisenko, J. R. Lewis, R. J. Mawhorter, H. E. Robertson and D. W. H. Rankin, *Dalton Trans.*, 2005, 3221.
36. M. A. James, M. L. McKee and A. J. Illies, *J. Am. Chem. Soc*, 1996, **118**, 7836.
37. I. Fourrê and J. Bergês, *J. Phys. Chem. A*, 2004, **108**, 898.
38. C. Bleiholder, D. B. Werz, H. Kppel and R. Gleiter, *J. Am. Chem. Soc.*, 2006, **128**, 2666.
39. R. Bukowski, W. Cencek, P. Jankowski, B. Jeziorski, M. Jeziorska, S. A. Kucharski, A. J. Misquitta, R. Moszynski, K. Patkowski, S. Rybak, K. Szalewicz, H. L. Williams, P. E. S. Wormer, *SAPT2002: An Ab Initio Program for Many-Body Symmetry-Adapted Perturbation Theory Calculations of Intermolecular Interaction Energies*. Sequential and Parallel Versions, 2003.
40. C. Bleiholder, R. Gleiter, D. B. Werz and H. Köppel, *Inorg. Chem.*, 2007, **46**, 2249.
41. S. M. Bachrach, D. W. Demoin, M. Luk and J. V. Miller, Jr., *J. Phys. Chem. A*, 2004, **108**, 4040.
42. F. Jensen, A. Greer and E. L. Clennan, *J. Am. Chem. Soc.*, 1998, **120**, 4439.
43. J.-W. Chu and B. L. Trout, *J. Am. Chem. Soc.*, 2004, **126**, 900.

44. (a) M. Ochiai, M. Naito, K. Miyamoto, S. Hayashi and W. Nakanishi, *Chem. Eur. J.*, 2010, **16**, 8713, see also; (b) M. Ochiai, K. Miyamoto, T. Kaneaki, S. Hayashi and W. Nakanishi, *Science*, 2011, **332**, 448.
45. T. P. Tauer, M. E. Derrick and C. D. Sherrill, *J. Phys. Chem. A*, 2005, **109**, 191.
46. T. Takatania and C. D. Sherrill, *Phys. Chem. Chem. Phys.*, 2007, **9**, 6106.
47. E. M. Cabaleiro-Lago, J. Rodriguez-Otero and A. Pena-Gallego, *J. Phys. Chem. A*, 2008, **112**, 6344.
48. (a) R. D. Bach, J. E. Winter and J. J. W. McDouall, *J. Am. Chem. Soc.*, 1995, **117**, 8586; (b) R. Flammang, P. Gerbaux, E. H. Morkved, M. W. Wong and C. Wentrup, *J. Phys. Chem.*, 1996, **100**, 17452; (c) P. Gerbaux, Y. Van Haverbeke, R. Flammang, M. W. Wong and C. Wentrup, *J. Phys. Chem. A*, 1997, **101**, 6970; (d) R. D. Bach, M. N. Glukhovtsev and C. Canepa, *J. Am. Chem. Soc.*, 1998, **120**, 775; (e) W. Nakanishi, S. Hayashi, Y. Kusuyama, T. Negoro, S. Masuda and H. Mutoh, *J. Org. Chem.*, 1998, **63**, 8373; (f) G. M. Li, S. Niu, M. Segi, R. A. Zingaro, H. Yamamoto, K. Watanabe, T. Nakajima and M. B. Hall, *J. Org. Chem.*, 1999, **64**, 1565; (g) A. I. Gonzalez, O. Mo and M. Yanez, *J. Phys. Chem. A*, 1999, **103**, 1662; (h) M. Arca, F. Demartin, F. A. Devillanova, F. Isaia, F. Lelj, V. Lippolis and G. Verani, Can, *J. Chem*, 2000, **78**, 1147; (i) S. Yamago, H. Miyazoe, R. Goto, M. Hashidume, T. Sawazaki and J. Yoshida, *J. Am. Chem. Soc.*, 2001, **123**, 3697; (j) T. Sasamori, Y. Arai, N. Takeda, R. Okazaki, Y. Furukawa, M. Kimura, S. Nagase and N. Tokitoh, *Bull. Chem. Soc. Jpn.*, 2002, **75**, 661; (k) M. Haranczyk, R. Bachorz, J. Rak, M. Gutowski, D. Radisic, S. T. Stokes, J. M. Nilles and K. H. Bowen, Jr., *J. Phys. Chem. B*, 2003, **107**, 7889; (l) T. Sasamori, N. Takeda and N. Tokitoh, *J. Phys. Org. Chem*, 2003, **16**, 450; (m) O. Piechaczyk, M. Doux, L. Ricard and P. Le Floch, *Organomet.*, 2005, **24**, 1204; (n) G. Fischer, S. Herler, P. Mayer, A. Schulz, A. Villinger and J. J. Weigand, *Inorg. Chem.*, 2005, **44**, 1740; (o) J. Fischer, M. Schuermann, M. Mehring, U. Zachwieja and K. Jurkschat, *Organomet.*, 2006, **25**, 2886; (p) Y. He, Z.-H. Kang, Q.-S. Li, C. H. A. Tsang, C.-H. Fam and S.-T. Lee, *Angew. Chem., Int. Ed. Engl.*, 2009, **48**, 128.
49. (a) O. Niyomura, S. Kato and S. Inagaki, *J. Am. Chem. Soc.*, 2000, **122**, 2132; (b) D. F. Plusquellic, R. D. Suenram, B. Mate, J. O. Jensen and A. C. Samuels, *J. Chem. Phys.*, 2001, **115**, 3057; (c) C. di Valentin, M. Freccero, R. Zanaletti and M. Sarzi-Amade, *J. Am. Chem. Soc.*, 2001, **123**, 8366; (d) M. W. Wong, Y. Steudel and R. Steudel, *Chem. Phys. Lett.*, 2002, **364**, 387; (e) P. Romaniello and F. Lelj, *J. Phys. Chem. A*, 2002, **106**, 9114; (f) Y.-G. Wang, C. Matta and N. H. Werstiuk, *J. Comp. Chem.*, 2003, **24**, 1720; (g) G. N. Merrill and S. P. Webb, *J. Phys. Chem. A*, 2003, **107**, 7852; (h) U. Herzog and U. Boehme, *Silicon Chem.*, 2003, **2**, 77.
50. W. Quapp and V. Melnikov, *Phys. Chem. Chem. Phys.*, 2001, **3**, 2735.
51. (a) F. Jensen, *J. Org. Chem.*, 1992, **57**, 6478; (b) K. Nahm, Y. Li, J. D. Evanseck, K. N. Houk and C. S. Foote, *J. Am. Chem. Soc.*, 1993,

115, 4879; (c) A. G. Baboul, H. B. Schlegel, M. N. Glukhovtsev and R. D. Bach, *J. Comput. Chem.*, 1998, **19**, 1353; (d) M.-D. Su and S.-Y. Chu, *J. Phys. Chem. A*, 1999, **103**, 1101; (e) E. L. Tae, Z. Zhu and M. S. Platz, *Tetrahedron Lett.*, 1999, **40**, 4921; (f) L. Wang and J. Zhang, *J. Mol. Struct.: THEOCHEM.*, 2001, **543**, 167.

52. (a) K. B. Wiberg and H. Castejon, *J. Am. Chem. Soc.*, 1994, **116**, 10489; (b) C. C. Pye, J. D. Xidos, R. A. Poirier and D. J. Burnell, *J. Phys. Chem. A*, 1997, **101**, 3371; (c) B. S. Jursic, *J. Int. Quant. Chem.*, 1997, **62**, 291; (d) M. Pettersson, J. Lundell, L. Khriachtchev, E. Isoniemi and M. Raesaenen, *J. Am. Chem. Soc.*, 1998, **120**, 7979; (e) G. Li and Z. Tang, *J. Phys. Chem. A*, 2003, **107**, 5317; (f) M. A. Silva, B. R. Bellenie and J. M. Goodman, *Organic Lett.*, 2004, **6**, 2559; (g) M. Kone, B. Illien, J. Graton and C. Laurence, *J. Phys. Chem. A*, 2005, **109**, 11907; (h) O. Mo, M. Yanez, M. Eckert-Maksic, Z. B. Maksic, I. Alkorta and J. Elguero, *J. Phys. Chem. A*, 2005, **109**, 4359.

53. (a) B. V. Cheney and M. W. Schulz, *J. Phys. Chem.*, 1990, **94**, 6268; (b) M. L. McKee, *J. Phys. Chem.*, 1993, **97**, 10971; (c) J. A. Platts, S. T. Howard and B. R. F. Bracke, *J. Am. Chem. Soc.*, 1996, **118**, 2726; (d) J. Pranata, *Bioorg. Chem.*, 1997, **25**, 213; (e) W. Nakanishi, S. Hayashi and H. Kihara, *J. Org. Chem.*, 1999, **64**, 2630; (f) R. Ludwig, *Phys. Chem. Chem. Phys.*, 2002, **4**, 5481; (g) F. Mendizabal, P. Pyykko and N. Runeberg, *Chem. Phys. Lett.*, 2003, **370**, 733; (h) O. Takahashi, Y. Kohno, K. Saito and M. Nishio, *Chem. Eur. J.*, 2003, **9**, 756; (i) M. Wierzejewska and M. Saldyka, *Chem. Phys. Lett.*, 2004, **391**, 143; (j) J. S. Wright, C. N. Rowley and L. L. Chepelev, *Mol. Phys.*, 2005, **103**, 815; (k) S. S. Zade, S. Panda, H. B. Singh, R. B. Sunoj and R. J. Butcher, *J. Org. Chem.*, 2005, **70**, 3693.

54. (a) J. A. Altmann, N. C. Handy and V. E. Ingamells, *Mol. Phys.*, 1997, **92**, 339; (b) B. Paizs and S. Suhai, *J. Comput. Chem.*, 1998, **19**, 575; (c) B. Braida, D. Lauvergnat and P. C. Hiberty, *J. Chem. Phys.*, 2001, **115**, 90; (d) F. Qi, O. Sorkhabi, A. G. Suits, S.-H. Chien and W.-K. Li, *J. Am. Chem. Soc.*, 2001, **123**, 148; (e) H. Yoshida, K. Takeda, J. Okamura, A. Ehara and H. Matsuura, *J. Phys. Chem. A*, 2002, **106**, 3580; (f) F. Marinelli and A. Lichanot, *Chem. Phys. Lett.*, 2002, **367**, 430; (g) E. P. Muniz and F. E. Jorge, *Int. J. Quant. Chem.*, 2006, **106**, 943.

55. (a) E. G. Hope and W. Levason, *Coord. Chem. Rev.*, 1993, **122**, 109; (b) J. Arnold, *Prog. Inorg. Chem.*, 1995, **43**, 353; (c) M. Uchiyama, M. Kameda, O. Mishima, N. Yokoyama, M. Koike, Y. Kondo and T. Sakamoto, *J. Am. Chem. Soc.*, 1998, **120**, 4934; (d) S. G. Raptis, M. G. Papadopoulos and A. J. Sadlej, *J. Chem. Phys.*, 1999, **111**, 7904; (e) A. K. Singh and S. Sharma, *Coord. Chem. Rev.*, 2000, **209**, 49; (f) W. Levason, S. D. Orchard and G. Reid, *Coord. Chem. Rev.*, 2002, **225**, 159; (g) L. Rulisek and Z. Havlas, *J. Phys. Chem. A*, 2002, **106**, 3855; (h) I. Papai, Y. Hannachi, S. Gwizdala and J. Mascetti, *J. Phys. Chem. A*, 2002, **106**, 4181; (i) G. Maroulis and C. Pouchan, *J. Phys. Chem. B*, 2003, **107**, 10683; (j) C. W. Liu, C.-M. Hung, B. K. Santra, Y.-H. Chu,

J.-C. Wang and Z. Lin, *Inorg. Chem.*, 2004, **43**, 4306; (k) A. Burnin, E. Sanville and J. J. BelBruno, *J. Phys. Chem. A*, 2005, **109**, 5026; (l) O. A. Wallner and K. J. Szabo, *J. Org. Chem.*, 2005, **70**, 9215; (m) E. Sanville, A. Burnin and J. J. BelBruno, *J. Phys. Chem. A*, 2006, **110**, 2378; (n) R. Jose, N. U. Zhanpeisov, H. Fukumura, Y. Baba and M. Ishikawa, *J. Am. Chem. Soc.*, 2006, **128**, 629; (o) P. Karamanis, G. Maroulis and C. Pouchan, *Chem. Phys.*, 2006, **331**, 19.

56. R. F. W. Bader (ed.), *Atoms in Molecules. A Quantum Theory*, Oxford University Press, Oxford, 1990.
57. C. F. Matta and R. J. Boyd, ed., *The Quantum Theory of Atoms in Molecules: From Solid State to DNA and Drug Design*, Wiley-VCH, Weinheim, 2007.
58. (a) R. F. W. Bader, T. S. Slee, D. Cremer and E. Kraka, *J. Am. Chem. Soc.*, 1983, **105**, 5061; (b) R. F. W. Bader, *Chem. Rev.*, 1991, **91**, 893; (c) R. F. W. Bader, *J. Phys. Chem. A*, 1998, **102**, 7314; (d) F. Biegler-König, R. F. W. Bader and T. H. Tang, *J. Comput. Chem.*, 1982, **3**, 317; (e) R. F. W. Bader, *Acc. Chem. Res.*, 1985, **18**, 9; (f) T. H. Tang, R. F. W. Bader and P. MacDougall, *Inorg. Chem.*, 1985, **24**, 2047; (g) F. Biegler-König, J. Schönbohm and D. Bayles, *J. Comput. Chem.*, 2001, **22**, 545; (h) F. Biegler-König and J. Schönbohm, *J. Comput. Chem.*, 2002, **23**, 1489.
59. J. M. Molina and J. A. Dobado, *Theor. Chem. Acc.*, 2001, **105**, 328.
60. J. A. Dobado, H. Martînez-Garcîa, J. M. Molina and M. R. Sundberg, *J. Am. Chem. Soc.*, 2000, **122**, 1144.
61. (a) M. Yamashita, Y. Yamamoto, K.-y. Akiba, D. Hashizume, F. Iwasaki, N. Takagi and S. Nagase, *J. Am. Chem. Soc.*, 2005, **127**, 4354; (b) Y. Yamamoto and K.-y. Akiba, *J. Syn. Org. Chem. Jpn.*, 2004, **62**, 1128.
62. S. K. Ignatov, N. H. Rees, B. R. Tyrrell, S. R. Dubberley, A. G. Razuvaev, P. Mountford and G. I. Nikonov, *Chem. Eur. J.*, 2004, **10**, 4991.
63. W. Nakanishi, T. Nakamoto, S. Hayashi, T. Sasamori and N. Tokitoh, *Chem. Eur. J*, 2007, **13**, 255.
64. S. K. Tripathi, U. Patel, D. Roy, R. B. Sunoj, H. B. Singh, G. Wolmershäuser and R. J. Butcher, *J. Org. Chem.*, 2005, **70**, 9237.
65. W. Nakanishi and S. Hayashi, *Current Organic Chemistry*, 2010, **14**, 181.
66. W. Nakanishi and S. Hayashi, *J. Phys. Chem. A*, 2009, **113**, 10050.
67. W. Nakanishi, S. Hayashi and K. Narahara, *J. Phys. Chem. A*, 2008, **112**, 13593.
68. W. Nakanishi and S. Hayashi, *J. Phys. Chem. A*, 2010, **114**, 7423.
69. V. G. Tsirelson, Interpretation of experimental electron densities by combination of the QTAMC and DFT, in *The Quantum Theory of Atoms in Molecules: From Solid State to DNA and Drug Design*, ed. C. F. Matta and R. J. Boyd, Wiley-VCH: Weinheim, Germany, 2007, Chapter 10.
70. J. P. Perdew, S. Kurth, Density functionals for non-relativistic Coulomb systems in the new century, in *A Primer in Density Functional Theory*, ed.

C. Fiolhais, F. Nogueira and M. Marques, Lecture Notes in Physics 620, Springer, Berlin, 2003, p. 1.
71. D. A. Kirzhnitz, *Sov. Phys. JETP*, 1957, **5**, 64.
72. W. Nakanishi, S. Hayashi, M. B. Pitak, M. B. Hursthouse and S. J. Coles, *J. Phys. Chem. A*, 2011, **115**, 11775.
73. M. Iwaoka, H. Komatsu, T. Katsuda and S. Tomoda, *J. Am. Chem. Soc.*, 2004, **126**, 5309.
74. (a) P. Politzer, P. Lane, M. C. Concha, Y. G. Ma and J. S. Murray, *J. Mol. Model.*, 2007, **13**, 305; (b) T. Clark, M. Hennemann, J. S. Murray and P. Politzer, *J. Mol. Model.*, 2007, **13**, 291; (c) P. Metrangolo, F. Meyer, T. Pilati, G. Resnati and G. Terraneo, *Angew. Chem. Int. Ed.*, 2008, **47**, 6114.
75. P. Murray-Rust and W. D. S. Motherwell, *J. Am. Chem. Soc.*, 1979, **101**, 4374.
76. P. Murray-Rust, W. C. Stallings, C. T. Monti, R. K. Preston and J. P. Glusker, *J. Am. Chem. Soc.*, 1983, **105**, 3206.
77. N. Ramasubbu, R. Parthasarathy and P. Murray-Rust, *J. Am. Chem. Soc.*, 1986, **108**, 4308.
78. J.-M. Dumas, H. Peurichard and M. J. Gomel, *Chem. Res.*, 1978, **S**, 54.
79. R. F. Stewart, *J. Chem. Phys.*, 1664, **1972**, 57.
80. P. Politzer and D. G. Truhlar (eds), *Chemical Applications of Atomic and Molecular Electrostatic Potentials*, Plenum, New York, 1981.
81. R. F. W. Bader, M. T. Carroll, J. R. Cheeseman and C. Chang, *J. Am. Chem. Soc.*, 1987, **109**, 7968.
82. H. Weinstein, P. Politzer and S. Srebrenik, *Theor. Chim. Acta.*, 1975, **38**, 159.
83. P. Politzer and J. S. Murray, *Theor. Chem. Acc.*, 2002, **108**, 134.
84. P. Metrangolo, T. Pilati and G. Resnati, *Cryst. Eng. Comm.*, 2006, **8**, 946.
85. W. Wang, B. Ji and Y. Zhang, *J. Phys. Chem. A*, 2009, **113**, 8132.
86. D. Hauchecorne, A. Moiana, B. J. van der Vekena and W. A. Herrebout, *Phys. Chem. Chem. Phys.*, 2011, **13**, 10204.
87. X. Zhang, Y. Zeng, X. Li, L. Meng and S. Zheng, *Struct. Chem.*, 2011, **22**, 567.
88. R. E. Rosenfield, Jr., R. Parthasarathy and J. D. Dunitz, *J. Am. Chem. Soc.*, 1977, **99**, 4860.
89. N. Ramasubbu and R. Parthasarathy, *Phosphorus Sulfur*, 1987, **31**, 221.
90. For more detail, see Chapter 4 of Ref. 101.
91. W. McFarlane and R. J. Wood, *J. Chem. Soc., Dalton Trans.*, 1972, 1397.
92. D. L. Klayman and W. H. H. Günther (eds), *Organic Selenium Compounds: Their Chemistry and Biology*, Wiley, New York, 1973, Chapter XV.
93. H. C. E. McFarlane and W. McFarlane, Selenium-77 and tellurium-125, in *NMR of Newly Accessible Nuclei*, ed. P. Laszlo, Academic Press, New York, 1983, Vol. 2, p . 275.
94. H. Iwamura and W. Nakanishi, *J. Synth. Org. Chem. Jpn.*, 1981, **39**, 795.

95. A. E. Derome, *Modern NMR Techniques for Chemistry Research*, Pergamon Press, Oxford, 1987.
96. S. Patai and Z. Rappoport (eds), *The Chemistry of Organic Selenium and Tellurium Compounds*, Wiley, New York, 1986; Vol. 1, Chapter 6.
97. T. M. Klapötke and M. Broschag, *Compilation of Reported ^{77}Se NMR Chemical Shifts*, Wiley, New York, 1996.
98. H. Duddeck, Chemical shifts and coupling constants for selenium-77, in *Landolt-Bornstein: Nuclear Magnetic Resonance (NMR) Data, Group III: Vol. 35, Subvolume G: Condensed Matter*, ed. R. R. Gupta, M. D. Lecher, Springer-Verlag, Berlin, 2004.
99. H. Duddeck, Sulfur, selenium, & tellurium NMR, in *Encyclopedia of Nuclear Magnetic Resonance*, ed. D. M. Grant and R. K. Harris, Wiley, New York, 1996, Vol. 7, p. 4633.
100. W. Nakansihi and S. Hayashi, unpublished result.
101. M. Kaupp, M. Bühl and V. G. Malkin (eds), *Calculations of NMR and EPR Parameters: Theory and Applications*, Wiley-VCH, Weinheim, Germany, 2004.
102. (a) N. F. Ramsey, *Phys. Rev.*, 1949, **77**, 567; (b) N. F. Ramsey, *Phys. Rev.*, 1950, **78**, 699; (c) N. F. Ramsey, *Phys. Rev.*, 1951, **83**, 540; (d) N. F. Ramsey, *Phys. Rev.*, 1952, **86**, 243; (e) N. F. Ramsey, *Phys. Rev.*, 1952, **85**, 143; (f) N. F. Ramsey, *Phys. Rev.*, 1953, **91**, 303.
103. This decomposition includes a small arbitrariness due to the coordinate origin dependence, though it does not damage usual chemical analyses and insights into ^{77}Se NMR spectroscopy.
104. H. Nakatsuji, T. Higashioji and M. Sugimoto, *Bull. Chem. Soc. Jpn.*, 1993, **66**, 3235.
105. M. Bühl, W. Thiel, U. Fleischer and W. Kutzelnigg, *J. Phys. Chem.*, 1995, **99**, 4000.
106. M. Bühl, J. Gauss and J. F. Stanton, *Chem. Phys. Lett.*, 1995, **241**, 248.
107. G. Magyarfalvi and P. Pulay, *Chem. Phys. Lett.*, 1994, **225**, 280.
108. V. G. Malkin, O. L. Malkina, M. E. Casida and D. R. Salahub, *J. Am. Chem. Soc.*, 1994, **116**, 5898.
109. S. Fuzinaga (ed.), *Gaussian Basis Sets for Molecular Calculations, Physical Sciences Data 16*, Elsevier Science, Amsterdam, 1984.
110. H. Fukui, *Magn. Res. Rev.*, 1987, **11**, 205.
111. D. B. Chesnut, in *Annual Reports on NMR Spectroscopy*, ed. G. A. Webb, Academic, New York, 1989, Vol. 21, p. 51.
112. P. D. Ellis, J. D. Odom, A. S. Lipton, Q. Chen and J. M. Gulick, in *Nuclear Magnetic Shieldings and Molecular Structure; NATO ASI Series*, ed. J. A. Tossell, Kluwer, Amsterdam, 1993, p. 539.
113. G. Schreckenbach, Y. Ruiz-Morales and T. Ziegler, *J. Chem. Phys.*, 1996, **104**, 8605.
114. J. Gauss and J. F. Stanton, *J. Chem. Phys.*, 1995, **102**, 251.
115. V. G. Malkin, O. L. Malkina and D. R. Salahub, *Chem. Phys. Lett.*, 1993, **204**, 80.

116. V. G. Malkin, O. L. Malkina and D. R. Salahub, *Chem. Phys. Lett.*, 1993, **204**, 87.
117. V. G. Malkin, O. L. Malkina, L. A. Erikson and D. R. Salahub, Modern density functional theory: a tool for chemistry, in *Theoretical and Computational Chemistry*, ed. P. Politzer and J. M. Seminario, Elsevier, Amsterdam, 1995, Vol. 2, p. 273.
118. W. Nakanishi and S. Hayashi, *J. Phys. Chem. A*, 1999, **103**, 6074.
119. M. J. Frisch, G. W. Trucks, H. B. Schlegel, P. M. W. Gill, B. G. Johnson, M. A. Robb, J. R. Cheeseman, T. Keith, G. A. Petersson, J. A. Montgomery, K. Raghavachari, M. A. Al-Laham, V. G. Zakrzewski, J. V. Ortiz, J. B. Foresman, J. Cioslowski, B. B. Stefanov, A. Nanayakkara, M. Challacombe, C. Y. Peng, P. Y. Ayala, W. Chen, M. W. Wong, J. L. Andres, E. S. Replogle, R. Gomperts, R. L. Martin, D. J. Fox, J. S. Binkley, D. J. Defrees, J. Baker, J. P. Stewart, M. Head-Gordon, C. Gonzalez and J. A. Pople, *Gaussian 94, Revision D.4*; Gaussian, Inc.: Pittsburgh, PA, 1995.
120. C. A. Bayse, *J. Chem. Theory Comput.*, 2005, **1**, 1119.
121. H. M. Tuononen, R. Suontamo, J. Valkonen and R. S. Laitinen, *J. Phys. Chem. A*, 2004, **108**, 5670.
122. M. J. Collins, R. J. Gillespie, J. F. Sawyer and G. J. Schrobilgen, *Inorg. Chem.*, 1986, **25**, 2053.
123. T. Maaninen, H. M. Tuononen, L. Kosunen, R. Oilunkaniemi, J. Hiitola, R. Laitinen and T. Chivers, *Z. Anorg. Allg. Chem.*, 2004, **630**, 1947.
124. J. A. Ritchey, B. M. Davis, P. A. Pleban and C. A. Bayse, *Org. Biomol. Chem.*, 2005, **3**, 4337.
125. H. Poleschner and K. Seppelt, *Chem. Euro. J*, 2004, **10**, 6565.
126. B. Fir, J. M. Whalen, H. P. A. Mercier, D. A. Dixon and G. J. Schrobilgen, *Inorg. Chem.*, 2006, **45**, 1978.
127. G. G. Camiletti, S. F. Machado and F. E. Jorge, *J. Comput. Chem.*, 2008, **29**, 2434.
128. The complete set of the s, p, d and f parameters of the DZP basis set for third-row atoms are available on the internet at http://www.cce.ufes.br/qcgv/pub/.
129. See also, I. D. Sadekov, V. I. Minkin, A. V. Zakharov, A. G. Starikov, G. S. Borodkin, S. M. Aldoshin, V. V. Tkachev, G. V. Shilov and F. J. Berry, *J. Organomet. Chem.*, 2005, **690**, 103.
130. B. Wrackmeyer, E. V. Klimkina and W. Milius, *Z. Anorg. Allg. Chem.*, 2011, 1895.
131. Downfield shifts of about 110–170 and 100 ppm are observed for the α and β effects, respectively, although the values depend on the conditions of measurement, such as solvents (see Table 2).[91,92,94,96,97] The γ effect is an upfield shift of about 30 ppm.
132. W. Nakanishi, S. Hayashi and M. Hada, *Chem. Eur. J.*, 2007, **13**, 5282.
133. For calculations of σ(Se: H_2Se), see (a) W. Kutzelnigg, U. Fleischer and C. van Wüllen, Shielding calculations: IGLO method, in *Encyclopedia of Nuclear Magnetic Resonance*, ed. D. M. Grant and R. K. Harris, Wiley,

New York, 1996, Vol. 7, p. 4284; (b) P. D. Ellis, J. D. Odom, A. S. Lipton, Q. Chen and J. M. Gulick, Gas phase measurement and ab initio calculations, in *Nuclear Magnetic Shieldings and Molecular Structure*, ed. J. A. Tossell, Kluwer Academic Publishers, Dordrecht, 1993, p. 539; (c) P. J. Wilson, *Mol. Phys.*, 2001, **99**, 363. See also Ref. 132.

134. The ethenyl and phenyl effects in the p(Se)–π(C=C) conjugation are reported to be downfield shifts of about 300 and 500 ppm, respectively.[91,92,94,96,97]
135. N. P. Luthra, R. B. Dunlap and J. D. Odom, *J. Magn. Reson.*, 1983, **52**, 318.
136. N. P. Luthra, A. M. Boccanfuso, R. B. Dunlap and J. D. Odom, *J. Organomet. Chem.*, 1988, **354**, 51.
137. W. Nakanishi, S. Hayashi, K. Narahara, D. Yamaki and M. Hada, *Chem. Eur. J.*, 2008, **14**, 7278.
138. (a) W. Nakanishi, S. Hayashi, D. Shimizu and M. Hada, *Chem. Eur. J.*, 2006, **12**, 3829; (b) S. Hayashi, W. Nakanishi, *Bioinorg. Chem. Appl.*, 2006, DOI: 10.1155/BCA/2006/79327.
139. (a) G. Schreckenbach and T. Ziegler, *J. Phys. Chem.*, 1995, **99**, 606; (b) G. Schreckenbach and T. Ziegler, *Int. J. Quantum Chem.*, 1996, **60**, 753; (c) G. Schreckenbach and T. Ziegler, *Int. J. Quantum Chem.*, 1997, **61**, 899.
140. (a) S. K. Wolff and T. Ziegler, *J. Chem. Phys.*, 1998, **109**, 895; (b) S. K. Wolff, T. Ziegler, E. van Lenthe and E. J. Baerends, *J. Chem. Phys.*, 1999, **110**, 7689.
141. W. Nakanishi, S. Hayashi, Y. Katsura and M. Hada, *J. Phys. Chem. A*, 2011, **115**, 8721.
142. E. J. Baerends, T. Ziegler, J. Autschbach, D. Bashford, A. Berces, F. M. Bickelhaupt, C. Bo, P. M. Boerrigter, L. Cavallo, D. P. Chong, L. Deng, R. M. Dickson, D. E. Ellis, M. van Faassen, L. Fan, T. H. Fischer, C. Fonseca Guerra, A. Ghysels, A. Giammona, S. J. A. van Gisbergen, A. W. Götz, J. A. Groeneveld, O. V. Gritsenko, M. Grüning, S. Gusarov, F. E. Harris, P. van den Hoek, C. R. Jacob, H. Jacobsen, L. Jensen, J. W. Kaminski, G. van Kessel, F. Kootstra, A. Kovalenko, M. V. Krykunov, E. van Lenthe, D. A. McCormack, A. Michalak, M. Mitoraj, J. Neugebauer, V. P. Nicu, L. Noodleman, V. P. Osinga, S. Patchkovskii, P. H. T. Philipsen, D. Post, C. C. Pye, W. Ravenek, J. I. Rodríguez, P. Ros, P. R. T. Schipper, G. Schreckenbach, J. S. Seldenthuis, M. Seth, J. G. Snijders, M. Sola, M. Swart, D. Swerhone, G. te Velde, P. Vernooijs, L. Versluis, L. Visscher, O. Visser, F. Wang, T. A. Wesolowski, E. M. van Wezenbeek, G. Wiesenekker, S. K. Wolff, T. K. Woo and A. L. Yakovlev, ADF 2010.01 SCM Theoretical Chemistry; Vrije Universiteit: Amsterdam, Netherlands, 2010.
143. M. J. Frisch, G. W. Trucks, H. B. Schlegel, G. E. Scuseria, M. A. Robb, J. R. Cheeseman, J. A. Montgomery, Jr., T. Vreven, K. N. Kudin, J. C. Burant, J. M. Millam, S. S. Iyengar, J. Tomasi, V. Barone, B. Mennucci, M. Cossi, G. Scalmani, N. Rega, G. A. Petersson, H. Nakatsuji, M. Hada, M. Ehara, K. Toyota, R. Fukuda, J. Hasegawa,

M. Ishida, T. Nakajima, Y. Honda, O. Kitao, H. Nakai, M. Klene, X. Li, J. E. Knox, H. P. Hratchian, J. B. Cross, V. Bakken, C. Adamo, J. Jaramillo, R. Gomperts, R. E. Stratmann, O. Yazyev, A. J. Austin, R. Cammi, C. Pomelli, J. W. Ochterski, P. Y. Ayala, K. Morokuma, G. A. Voth, P. Salvador, J. J. Dannenberg, V. G. Zakrzewski, S. Dapprich, A. D. Daniels, M. C. Strain, O. Farkas, D. K. Malick, A. D. Rabuck, K. Raghavachari, J. B. Foresman, J. V. Ortiz, Q. Cui, A. G. Baboul, S. Clifford, J. Cioslowski, B. B. Stefanov, G. Liu, A. Liashenko, P. Piskorz, I. Komaromi, R. L. Martin, D. J. Fox, T. Keith, M. A. Al-Laham, C. Y. Peng, A. Nanayakkara, M. Challacombe, P. M. W. Gill, B. Johnson, W. Chen, M. W. Wong, C. Gonzalez and J. A. Pople, *Gaussian 03, Revision B.05*, Gaussian, Inc., Pittsburgh, PA, 2003.

144. K. Jackowski, W. Makulski and W. Kozminski, *Magn. Reson. Chem.*, 2002, **40**, 563.
145. D. B. Chesnut and L. D. Quin, *Heteroat. Chem.*, 2004, **15**, 1042.
146. D. B. Chesnut, *Chem. Phys. Lett.*, 2003, **380**, 251.
147. D. B. Chesnut, *Chem. Phys. Lett.*, 1995, **246**, 235.
148. L. C. Dickinson, D. B. Chesnut and L. D. Quin, *Magn. Reson. Chem.*, 2004, **42**, 1037.
149. A. Bagno, *J. Mol. Struct.: THEOCHEM.*, 1997, **418**, 243.
150. R. Annunziata and G. Barbarella, *Org. Magn. Reson.*, 1984, **22**, 250.
151. L. A. O'Dell and I. L. Moudrakovski, *J. Magn. Reson.*, 2010, **207**, 345.
152. R. Musio and O. Sciacovelli, *Magn. Reson. Chem.*, 2006, **44**, 753.
153. NBO (version 3.1), E. D. Glendening, A. E. Reed, J. E. Carpenter, F. Weinhold, Theoretical Chemistry Institute, University of Wisconsin, Madison, 2001.
154. P.v.R. Schleyer, C. Maerker, A. Dransfeld, H. Jiao, N. J. Hommes and R.v.E. Hommes, *J. Am. Chem. Soc.*, 1996, **118**, 6317.
155. G. Subramanian, P.v.R. Schleyer and H. Jiao, *Angew. Chem., Int. Ed. Engl.*, 1996, **35**, 2638.
156. Y. Ruiz-Morales, G. Schreckenbach and T. Ziegler, *J. Phys. Chem. A.*, 1997, **101**, 4121.
157. H. Lange, U. Herzog, U. Bohme and G. Rheinwald, *J. Organomet. Chem.*, 2002, **660**, 43.
158. S. Nahrwold and R. Berger, *J. Chem. Phys.*, 2009, **130**, 214101.
159. W. Nakansihi, S. Hayashi and K. Matsuiwa, unpublished result.
160. (a) J. Jusélius and D. Sundholm, *Phys. Chem. Chem. Phys.*, 1999, **1**, 3429; (b) J. Jusélius and D. Sundholm, *Phys. Chem. Chem. Phys.*, 2000, **2**, 2145; (c) J. Jusélius and D. Sundholm, *J. Org. Chem.*, 2000, **65**, 5233; (d) J. Jusélius and D. Sundholm, *Phys. Chem. Chem. Phys.*, 2001, **3**, 2433; (e) J. Jusélius, M. Straka and D. Sundholm, *J. Phys. Chem. A*, 2001, **105**, 9939; (f) R. J. F. Berger, M. A. Schmidt, J. Jusélius, D. Sundholm and H. Schmidbaur, *Z. Naturforsch.*, 2001, **56b**, 979; (g) J. Jusélius, M. Patzschke and D. Sundholm, *J. Mol. Struct.: THEOCHEM.*, 2003, **633**, 123.

161. (a) W. Kutzelnigg, *Isr. J. Chem.*, 1980, **19**, 193; (b) M. Schindler and W. Kutzelnigg, *J. Chem. Phys.*, 1982, **76**, 1919; (c) M. Schindler and W. Kutzelnigg, *J. Am. Chem. Soc.*, 1983, **105**, 1360; (d) M. Schindler and W. Kutzelnigg, *Mol. Phys.*, 1983, **48**, 781.
162. For reviews, see: (a) W. Kutzelnigg, U. Fleischer and M. Schindler, in *NMR Basic Principles and Progress*, ed. P. Diehl, E. Fluck and R. Kosfeld, Springer: Berlin, 1990; Vol. 23, p. 165; (b) W. Kutzelnigg, Ch. van Wüllen, U. Fleischer, R. Franke and T.v. Mourik, in *NMR Shieldings and Molecular Structure*, ed. J. A. Tossell, Kluwer Academic Publishers, Norwell, MA, 1993.
163. J. Pipek and P. G. Mezey, *J. Chem. Phys.*, 1998, **90**, 4916.
164. (a) T. Heine, P.v.R. Schleyer, C. Corminboeuf, G. Seifert, R. Reviakine and J. Weber, *J. Phys. Chem. A*, 2003, **107**, 6470; (b) J. A. Bohmann, F. Weinhold and T. C. Farrar, *J. Chem. Phys.*, 1997, **107**, 1173; (c) C. Corminboeuf, T. Heine, G. Seifert, P.v.R. Schleyer and J. Weber, *Phys. Chem. Chem. Phys.*, 2004, **6**, 273.
165. G. P. Mullen, N,P. Luthra, R. B. Dunlap and J. D. Odom, *J. Org. Chem.*, 1985, **50**, 811.
166. W. Nakanishi and S. Hayashi, *Chem. Lett.*, 1998, 523.
167. S. Hayashi and W. Nakanishi, *J. Org. Chem.*, 1999, **64**, 6688.
168. S. Hayashi, H. Wada, T. Ueno and W. Nakanishi, *J. Org. Chem.*, 2006, **71**, 5574.
169. W. Nakanishi, S. Hayashi and T. Uehara, *Eur. J. Org. Chem.*, 2001, 3933.
170. T. Nakamoto, S. Hayashi and W. Nakanishi, *J. Org. Chem.*, 2008, **73**, 9259.
171. W. Nakanishi, S. Hayashi and H. Yamaguchi, *Chem. Lett.*, 1996, 947.
172. W. Nakanishi, S. Hayashi, A. Sakaue, G. Ono and Y. Kawada, *J. Am. Chem. Soc.*, 1998, **120**, 3635.
173. W. Nakanishi and S. Hayashi, *J. Org. Chem.*, 2002, **67**, 38.
174. W. Nakanishi, S. Hayashi and T. Uehara, *J. Phys. Chem. A*, 1999, **103**, 9906.
175. B. Wrackmeyer, Z. G. Hernandez and M. Herberhold, *Magn. Reson. Chem.*, 2007, **45**, 198.
176. (a) M. Iwaoka and S. Tomoda, Phosphorus, *Sulfur, and Silicon*, 1992, **67**, 125; (b) M. Iwaoka and S. Tomoda, *J. Am. Chem. Soc.*, 1996, **118**, 8077; (c) H. Komatsu, M. Iwaoka and S. Tomoda, *Chem. Commun.*, 1999, 205; (d) M. Iwaoka, H. Komatsu and S. Tomoda, *Chem. Lett.*, 1998, 969; (e) M. Iwaoka, T. Katsuda, H. Komatsu and S. Tomoda, *J. Org. Chem.*, 2005, **70**, 321.
177. M. Iwaoka, H. Komatsu, T. Katsuda and S. Tomoda, *J. Am. Chem. Soc.*, 2002, **124**, 1902. See also, Ref. 176d.
178. (a) W. Nakanishi and S. Hayashi, *Chem. Eur. J.*, 2008, **14**, 5645; (b) A. Tanioku, S. Hayashi and W. Nakanishi, *Bioinorg. Chem. Appl.*, 2009, DOI: 10.1155/2009/381925.
179. M. Karplus, *J. Chem. Phys.*, 1959, **30**, 11; M. Karplus, *J. Am. Chem. Soc.*, 1963, **85**, 2870.

180. K. W. Feindel and R. E. Wasylishen, *Magn. Reson. Chem.*, 2004, **42**, S158.
181. J. Jokisaari and J. Autschbach, *Phys. Chem. Chem. Phys.*, 2003, **5**, 4551.
182. J. Lounila and J. Jokisaari, *Prog. Nucl. Magn. Reson. Spectrosc.*, 1982, **15**, 249.
183. J. Vaara, J. Jokisaari, R. E. Wasylishen and D. L. Bryce, *Prog. Nucl. Magn. Res. Spectrosc.*, 2002, **41**, 233.
184. ADF 2002.03 Version SCM Theoretical Chemistry; Vrije Universiteit: Amsterdam, Netherlands, 2002.
185. G. te Velde, F. M. Bickelhaupt, E. J. Baerends, S. J. A. van Gisbergen, C. F. Guerra, J. G. Snijders and T. Ziegler, *J. Comput. Chem.*, 2001, **22**, 931.
186. C. Fonseca Guerra, J. G. Snijders, G. te Velde and E. J. Baerends, *Theor. Chem. Acc.*, 1998, **99**, 391.

Subject Index

References to tables and charts are in **bold** type.